# *Sensory Integration and Praxis Tests*

## Manual

A. Jean Ayres, Ph.D.

*Published by*

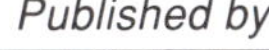

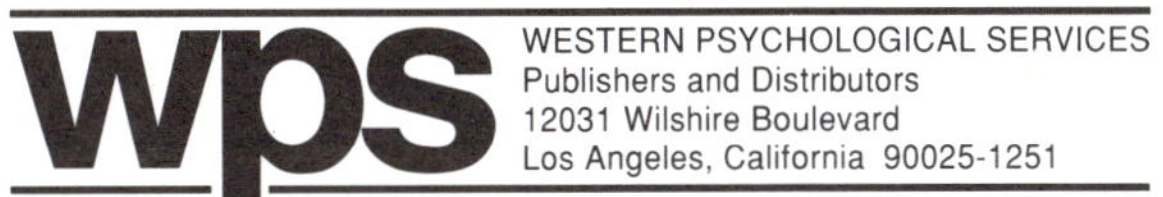


WESTERN PSYCHOLOGICAL SERVICES
Publishers and Distributors
12031 Wilshire Boulevard
Los Angeles, California 90025-1251

Third Printing .......................................................... December 1991

 2 3 4 5 6 7 8 9 Printed in U.S.A.

# TABLE OF CONTENTS

# LIST OF TABLES

# LIST OF FIGURES

# PREFACE

The Sensory Integration and Praxis Tests (SIPT) are the end product of a long period of development. An original group of perceptual-motor tests was gradually developed and refined, locally normed, and collected into the Southern California Sensory Integration Tests (SCSIT). Widespread use of the tests here and abroad attested to a need for such instruments in the fields of occupational and physical therapy.

Triggered by the offer from Rush University of Chicago to help norm the SCSIT nationally, and by the offer from Sensory Integration International to help gather validity and reliability data, the decision was made to develop and standardize the SIPT. Four new praxis tests (Praxis on Verbal Command, Sequencing Praxis, Oral Praxis, and Constructional Praxis) had been under development since 1978. These tests, plus revised versions of 12 of the more useful tests from the SCSIT as well as the Southern California Postrotary Nystagmus Test (SCPNT), constitute the new battery of 17 Sensory Integration and Praxis Tests.

The SIPT is the result of a great cooperative effort among many occupational and physical therapists. The generous time contribution of the therapists throughout the United States and Canada made norming for both countries possible. Gratitude is extended to all of the agencies and institutions that allowed testing of their students, to therapists who helped with pilot testing of early forms of the SIPT, to SIPT instructors, and to therapists gathering normative, validity, and reliability data.

Zoe Mailloux was closely associated with the SIPT throughout its development. Shay McAtee led the pilot test crew of Terri Chew, Catherine Cottermann, Katherine Inamura, Valerie Jespersen-Wheat, Moya Kinnealey, Susan Knox, Kaeko Saeki, Mary Silberzahn, Betsy Slavik, Susan Stallings, and Mary Jane Zehnpfennig.

SIPT reliability and validity data were provided by the following therapists: Sheryl Bonanno, Barbara B. Burris, Sharon Cermak, Florence Clark, Terry Crowe, Patricia Dixon, Winifred Dunn, Dorothy Ecker, Clare Guiffrida, Lois Hickman, Esther Huecker, Judith Kimball, Moya Kinnealey, Susan Knox, Shelly Lane, Violet Maas, Helen Madill, Zoe Mailloux, Shay McAtee, Elizabeth Murray, Elizabeth Newcomer, Diane Parham, Karen A. Pettit, Antje Price, Susanne Smith Roley, Gloria Scammerhorn, Virginia Scardina, Lynn Schmidt, Kay Seig, Susan Stallings, Susan Stryzewski, Patricia Wilbarger, Patricia Wood, and Susan Young.

Normative data for the SIPT were provided by Marcia Bridgeman, Susan Curtis, Robin Levy, Frieda Luftglas, Jacquelyn Ott, and Mary Margaret Windsor in the New England region; by Sheila Smith Allen, Andrea Altmayer, Kathy Arn, Kristin Bergmann, Cheryl Colangelo, Claire Daffner, Martha Frank, Gregg Hritko, Nancy Kauffman, Hermine Mishkin, Sue Seiler, and Connie Walp in the Mid-Atlantic region; by Lynn Balzer-Martin, Ricardo Carrasco, Julie Coleman, Evelyn Dales, Mike Hillis, Saburi Imara, Lynne Israel, Nina Love, Virginia Maddox, Dottie Marsh, Jane Shaw, Marcia Staples, Shelley Stowers, Susan Strzyzewski, and Sheila Tayrose in the South Atlantic region; by Lynne Barnes, Betty Bayne, Virginia Bell, Karen Bewley, Karen Cosgrove, Barbara Friedman, Dorothy Jacobsen, Nancy Wharton Jones, Robin Kirschenbaum, Gloria Levine, Jeanne Lewin, Carol Lieberman, Gail Lotak, JoAnne Maczulski, Carolyn Messina, MaryAn Nortman, Janet Puderbaugh, Marjorie Ritchey, and Susan Tenpas in the East North Central region; by Katherine Bicek, Gloria Frolek Clark, Meredith Hirsch, Esther Huecker, Susan Hietala LeVahn, Susan Merryfield, Barbara White Prudhome, and Susan Swanson in the West North Central region; by Bonnie Fellows, Pat Gailey, Geri Mailender, Sara Masur, Shirley Peganoff, Barbara Sarbaugh, and Jo Teachman Sprague in the East South Central region; by Peg Bledsoe, Jane Bowman, Debora Davidson, Kellie Dickinson, Marietta Maxey, Molly McEwen, Caryl Semmler, and Betsy Slavik in the West

South Central region; by Wendy Crandall, Debra Hines, June Holmstrom, Martha Stevens, Patricia Stutz-Tanenbaum, Diana Woods, and Eunice Zee-Chen in the Mountain region; by Marie Anzalone, Julie Bissell, Sheryl Bonanno, Janet Brown, Terri Chew, Katherine Inamura, Stephanie Levine, Wendy Mack, Rita Newman, Julie Shuer, Serena Sutherland, Patricia Wilbarger, and Mary Jane Zehnpfennig in the Pacific region; and by Linda Daniels, Jennifer Fisher, Barbara Graff, Joan Macdonald, Loree Primeau, and Phyllis Straathof in Canada.

In addition, appreciation is extended to Western Psychological Services and many of their personnel involved in SIPT production. Early assistance was given by Cathy Wendler, Gale H. Roid, and Gretchen Guiton. Later statistical analyses were provided by George J. Huba and Diana Marr. Dr. Marr also prepared portions of the Manual and coordinated the final Manual and ChromaGraphs™. Lisa Melchior also helped coordinate the final Manual and ChromaGraphs™; Barbara Schroeder designed and programmed the ChromaGraph™; Barbara Schroeder and Jill Wang programmed the SIPT computer report; Kathy Tootle designed the SIPT forms, and Diane Greene and Michael Kay were editors.

A. Jean Ayres, Ph.D.
August, 1988

# CHAPTER 1
## DESCRIPTION OF THE SIPT

### Purpose

The Sensory Integration and Praxis Tests (SIPT) contribute to the clinical understanding of children with irregularities in learning or behavior. These tests were designed to assess several different practic abilities, various aspects of the sensory processing status of the vestibular, proprioceptive, kinesthetic, tactile, and visual systems, and the major behavioral manifestations of deficits in integration of sensory inputs from these systems. The senses of olfaction, taste, and audition are not tested, nor are linguistics, although one test evaluates the ability to translate verbal directions into practic actions. Based upon the assumption that both learning and behavior are functions of the brain, the SIPT uses a neurobiological model to help define the bases for learning or behavior disorders.

The SIPT evolved over three decades, with test and item selection largely determined by their capacity to discriminate between normal children and children with sensory integrative and learning deficits. Those sensory integrative and practic deficits that most effectively distinguish between normal and dysfunctional children have been culled and collected into the SIPT. Accordingly, the SIPT is meant to serve primarily as a diagnostic and descriptive tool.

### Appropriate Use of the SIPT

The SIPT was designed for use with children from 4 years of age through 8 years, 11 months of age. It is inappropriate, in most cases, to use these tests with children falling outside that age range. Most testable children with learning, behavioral, or developmental irregularities are suitable candidates for most of the tests.

The tests included in the SIPT are of the performance type. A mild neuromotor problem (such as choreoathetosis) will not appreciably affect test performance on most of the tests, but a severe neuromotor problem (such as spasticity or athetosis) will interfere with performance on at least half of the tests. All but those tests assessing tactile and kinesthetic perception and standing balance (with eyes closed) require sight.

Only one of the tests, Praxis on Verbal Command (PrVC), is strongly and intentionally dependent upon auditory-language comprehension. The rest of the tests have been deliberately designed to minimize dependence upon linguistic competence, in order to avoid placing the child with delayed language at a severe disadvantage. Although test directions are given verbally, the nature of the task usually can be communicated through nonverbal means if necessary. Trial items assist the child with limited English-language comprehension to understand the test requirements. No verbal response is required for the SIPT. The test taker communicates to the examiner through nonverbal gestures such as nodding the head, pointing, or executing the test task. However, the test norms are based upon the performance of children with no known language disorders, whose native language is English, and who received precise test instructions in English. Therefore, if the SIPT is used to assess children whose native language is not English, or children with language-comprehension disorders such as receptive aphasia, the user should exercise caution in interpreting the results.

The SIPT scores are only one source of data in the evaluative process, and normally should be supplemented by other data and by clinical observations before any final diagnostic or treatment decisions are made. SIPT scores alone do not indicate the cause of learning or behavior problems, but may help to identify the sensory integrative or practic concomitant of those problems. SIPT scores should be accompanied by clinical observations of postural, ocular, and behavioral responses. To place the SIPT data in proper perspective, the clinician should obtain a brief history and description of the presenting problem, along with any relevant medical information about the child. Results of other tests (particularly tests of intelligence, speech and language development, and academic achievement) may also be helpful in the interpretation of SIPT scores. SIPT scores alone should not be used to determine whether or not a child with sensory integrative dysfunction or dyspraxia will profit from therapy. Such predictions require the professional judgment of the test user in addition to the inferences drawn from the SIPT scores.

The interpretation of SIPT scores requires some theoretical understanding of sensory integration and praxis. In addition, the administration of the SIPT requires considerable skill and practice on the part of the examiner. Therefore, the clinician is urged to obtain formal training in the general theory of sensory integration and praxis and in the specific SIPT administration procedures and techniques before administering the SIPT.

All tests in the SIPT are individually administered. It is recommended that the test be given in two sittings, and that the battery be used in its entirety. If it is necessary to give the SIPT in a single sitting, a break is recommended after the first testing of Postrotary Nystagmus. Detailed instructions for administration of the SIPT are provided in chapter 3 of this Manual.

The administration time for the SIPT ranges from approximately 1½ to 2 hours, with an additional 30 to 45 minutes for the test user to prepare the protocol sheets for computer scoring. The completed protocol sheets are then sent to Western Psychological Services (WPS) for final computer scoring and preparation of a full-color, plotted graph (WPS ChromaGraph™) and a WPS TEST REPORT.

## Description of the Tests

The SIPT replaces the Southern California Sensory Integration Tests (SCSIT) (Ayres, 1972a, 1980). All but four tests from the SCSIT (Position in Space, Crossing the Midline of the Body, Right-Left Discrimination, and Double Tactile Stimuli Perception) have been revised and are included in the SIPT. Several tests from the SCSIT (Design Copying, Manual Form Perception, Bilateral Motor Coordination, and Standing Balance) underwent major redesign and expansion for inclusion in the SIPT. The rest of the SCSIT were changed to a lesser extent in either design or scoring, or both. The SIPT also contains four new praxis tests: Constructional Praxis (CPr), Praxis on Verbal Command (PrVC), Sequencing Praxis (SPr), and Oral Praxis (OPr). Two measurements of hand use (Space Visualization Contralateral Use and Preferred Hand Use), derived from hand usage on the Space Visualization (SV) test, have been added. In addition, the Southern California Postrotary Nystagmus Test (SCPNT) is included in the SIPT.

The 17 tests in the SIPT are summarized in Table 1. These tests are divided into four groups, as described in the following paragraphs. It should be noted, however, that these groups are not mutually exclusive.

### The Form and Space Perception Tests

The two motor-free tests of visual perception from the SCSIT have been included in the SIPT, but with a revised and improved scoring method. These are the Space Visualization (SV) and Figure-Ground Perception (FG) tests. Two new subscores, Contralateral Hand Use and Preferred Hand Use, have been derived from the patterns of right and left hand usage on the Space Visualization test.

**Space Visualization (SV).** This test has been in use for over 20 years, and was carried over from the SCSIT to the SIPT with only minor changes in computer scoring. It uses four egg-shaped blocks, four diamond-shaped blocks, and two plastic formboards (one with an egg-shaped hollow and one with a diamond-shaped hollow). Pegs are inserted into the formboards in different locations to create different test items, each consisting of one formboard and two blocks, only one of which will fit into the formboard. The child chooses one of the two blocks and places it in the formboard or, in cases of poor coordination, the examiner places the block for the child. Variations of peg location in the formboards and use of different pairs of blocks allow the examiner to present the child with up to 30 different "puzzles."

The test is intended to require visual space perception and, in the more advanced items, mental manipulation of objects in space. Because motor performance is not required and does not enter into scoring, the test is deliberately a visual, and not a visual-motor, test. The Space Visualization test is discontinued after the fifth error, not necessarily consecutive. The Space Visualization accuracy score consists of the total number of correct choices. An additional time-adjusted accuracy score adjusts the child's accuracy score on the basis of his or her total response time. Designed to be used with children who are not test-sophisticated and who are difficult to test, it has greater appeal than many of the other tests in the SIPT and is an appropriate initial test.

Although the child is not required to put the blocks into the formboard, most children do so, providing an opportunity for gathering information about the child's hand preference and the use of each hand. The hand the child uses to pick up a block is recorded for each item. The Space Visualization Contralateral Use (SVCU) score reflects the tendency to use each hand in contralateral versus ipsilateral space. If the child uses his or her right hand to pick up a block placed to the left of the body's

**Table 1**
**Content of the SIPT Subscales**

| Test/Subscale | Description |
|---|---|
| **Space Visualization (SV)** | Test contains 30 items; testing is discontinued after 5 errors. |
| 1. *Time-adjusted accuracy* | 1. Number of correctly answered items, adjusted by the log of the child's average time per attempted item |
| 2. Accuracy | 2. Number of correctly answered items |
| 3. Time[a] | 3. Logarithm of average time per item |
| 4. Contralateral use | 4. Proportion of attempted items for which right hand was used to pick up left block, or left hand was used to pick up right block |
| 5. Preferred hand use | 5. For right-handed subjects, proportion of attempted items in which right hand was used; for left-handed subjects, proportion in which left hand was used |
| **Figure-Ground Perception (FG)** | Test contains 16 plates (48 items); testing is discontinued after 7 errors, 4 of which occur on the last 3 plates attempted. |
| 1. *Accuracy* | 1. Number of correctly identified figures |
| 2. Time[a] | 2. Logarithm of time for all items attempted |
| **Standing and Walking Balance (SWB)** | Test contains 16 items; testing is discontinued after 3 consecutive scores of 0 or 1. |
| 1. *Total score* | 1. Number of seconds for Items 1–8 & 11–16, plus number of steps for Items 9 & 10 |
| 2. Eyes open | 2. Sum of Items 1, 3, 5, 6, 9, 11, 12, 15 |
| 3. Eyes closed | 3. Sum of Items 2, 4, 7, 8, 10, 13, 14, 16 |
| 4. Right foot | 4. Sum of Items 5, 7, 11, 13 |
| 5. Left foot | 5. Sum of Items 6, 8, 12, 14 |
| **Design Copying (DC)** | Test contains 25 items; testing is discontinued after 2 consecutive accuracy scores of 0 in Part I or 3 consecutive scores of 0 on gross approximation in Part II. |
| 1. *Total accuracy* | 1. Sum of Part I accuracy and Part II "should have" parameters |
| 2. Adjusted accuracy | 2. Total accuracy, adjusted for Part II "should not have" parameters |
| 3. Part I accuracy | 3. Sum of Part I accuracy scores (Items 1–13) |
| 4. Part II accuracy | 4. Number of Part II "should have" parameters (Items 14–25) |
| 5. Atypical approach parameters: | 5. (Atypical approach parameters are not scored for 4-year-olds.) |
| a. Boundaries[a] | a. Number of items in which child's drawing touches boundary (from Items 14–25) |
| b. Additions[a] | b. Number of items containing additional lines (from Items 15–25) |
| c. Segmentations[a] | c. Number of items containing segmentation errors (from Items 17, 19–21, 23–25) |
| d. Reversals[a] | d. Number of items containing mirror-image errors (from Items 2–4, 6–9, 15–17, 19, 20, 22, 24, 25) |
| e. Right-to-left errors[a] | e. Number of items drawn from right-to-left (from Items 2, 5, 7, 14–16, 18, 19, 21, 22) |
| f. Inversions[a] | f. Number of items drawn upside-down (from Items 2, 5–9, 15, 17, 18, 22) |
| g. Jogs[a] | g. Number of items drawn with jogs or ears (from Items 17, 18, 20, 22) |
| **Postural Praxis (PPr)** | Test contains 17 items; all items are administered to all children. |
| 1. *Total accuracy* | 1. Sum of accuracy scores (0, 1, 2) for all items |
| **Bilateral Motor Coordination (BMC)** | Test contains 14 items; feet items (11–14) not administered to 4-year-olds; testing is discontinued after 4 consecutive scores of 0. |
| 1. *Total accuracy* | 1. Sum of accuracy scores for arms and feet items |
| 2. Arms accuracy | 2. Sum of accuracy scores (0, 1, 2) for Items 1–10 |
| 3. Feet accuracy | 3. Sum of accuracy scores (0, 1, 2) for Items 11–14 |

*Note.* Major score (subscale) for each test is designated by italics.

[a]For these subscales, high raw scores are represented as low *SD*-scores.

***table continued on next page . . .***

**Table 1 (Continued)**
**Content of the SIPT Subscales**

| Test/Subscale | Description |
|---|---|
| **Praxis on Verbal Command (PrVC)** | Test contains 24 items; all items are administered to all children. |
| 1. *Total accuracy* | 1. Number of correctly executed items |
| 2. Total time[a] | 2. Total time required to execute all items (includes correct and incorrect items) |
| **Constructional Praxis (CPr)** | Test uses 7 blocks for Structure I, 15 blocks for Structure II; Structure II not administered to 4-year-old children. |
| 1. *Total accuracy* | 1. Part I accuracy plus Part II accuracy |
| 2. Part I accuracy | 2. Accuracy of placement for 7 blocks in Structure I (maximum score = 1 point for Block 1, 2 points for Block 2, 4 points for Block 3, 3 points for Block 4, 2 points for Block 5, 2 points for Block 6, 2 points for Block 7) |
| 3. Part II accuracy | 3. Number of blocks placed correctly in Structure II (maximum = 15) |
| 4. Part II errors: | 4. Note that each block is scored on only the most severe error, so the sum of all errors in Part II can never exceed 15. |
| a. Displacement 1–2.5 cm[a] | a. Number of incorrectly placed blocks (scored only for Blocks 2–12 & 15) |
| b. Displacement > 2.5 cm[a] | b. Number of incorrectly placed blocks (scored only for Blocks 2–12 & 15) |
| c. Rotation > 15 degrees[a] | c. Number of incorrectly placed blocks (scored only for Blocks 2–11 & 15) |
| d. Reversals[a] | d. Number of incorrectly placed blocks (scored only for Blocks 2–12 & 15) |
| e. Incorrect but logical[a] | e. Number of incorrectly placed blocks (scored only for Blocks 2–15) |
| f. Gross mislocation[a] | f. Number of incorrectly placed blocks (scored only for Blocks 2–15) |
| g. Omissions[a] | g. Number of omitted blocks (scored for Blocks 1–15) |
| **Postrotary Nystagmus (PRN)** | Test contains 4 items; all items are administered to all children. |
| 1. *Average nystagmus* | 1. Average nystagmus, for first and second clockwise & counterclockwise rotations |
| 2. Average clockwise | 2. Average nystagmus following first and second clockwise rotations |
| 3. Average counterclockwise | 3. Average nystagmus following first and second counterclockwise rotations |
| 4. Time 1 clockwise | 4. Duration of nystagmus following first 20-second clockwise rotation |
| 5. Time 1 counterclockwise | 5. Duration of nystagmus following first 20-second counterclockwise rotation |
| 6. Time 2 clockwise | 6. Duration of nystagmus following second 20-second clockwise rotation |
| 7. Time 2 counterclockwise | 7. Duration of nystagmus following second 20-second counterclockwise rotation |
| **Motor Accuracy (MAc)** | Test contains 2 items; both items administered to all children. |
| 1. *Weighted total accuracy*[a] | 1. Average weighted accuracy, preferred and nonpreferred hand |
| 2. Unweighted total accuracy[a] | 2. Average unweighted accuracy, preferred and nonpreferred hand |
| 3. Preferred hand weighted accuracy[a] | 3. Accuracy for preferred hand, weighted by relative distance from solid line |
| 4. Preferred hand unweighted accuracy[a] | 4. Length of line segment off solid line + line segment outside short broken line + line segment outside medium broken line + line segment outside long broken line (cm), using preferred hand |
| 5. Nonpreferred hand weighted accuracy[a] | 5. Accuracy for nonpreferred hand, weighted by relative distance from solid line |
| 6. Nonpreferred hand unweighted accuracy[a] | 6. Length of line segment off solid line + line segment outside short broken line + line segment outside medium broken line + line segment outside long broken line (cm), using nonpreferred hand |

*Note.* Major score (subscale) for each test is designated by italics.

[a]For these subscales, high raw scores are represented as low *SD*-scores.

***table continued on next page . .***

**Table 1 (Continued)**
**Content of the SIPT Subscales**

| Test/Subscale | Description |
|---|---|
| **Sequencing Praxis (SPr)** | Test contains 9 items (total of 53 subitems); finger items (7–9) are not administered to 4-year-olds; testing is discontinued after the first two subitems on two consecutive items are scored as 0. |
| 1. *Total accuracy* | 1. Sum of accuracy scores (0, 1, 2) for all 9 items (total of 53 subitems) |
| 2. Hand accuracy | 2. Sum of accuracy scores (0, 1, 2) for Items 1–6 (total of 35 subitems) |
| 3. Finger accuracy | 3. Sum of accuracy scores (0, 1, 2) for Items 7–9 (total of 18 subitems) |
| **Oral Praxis (OPr)** | Test contains 19 items; all items are administered to all children. |
| 1. *Total accuracy* | 1. Sum of accuracy scores (0, 1, 2) for all 19 items |
| **Manual Form Perception (MFP)** | Test contains 20 items; Part II is not administered to 4-year-olds or to children who score less than 6 on Part I; testing is discontinued after 4 consecutive item scores of 0. |
| 1. *Total accuracy* | 1. Sum of accuracy for Part I (Items 1–10) and Part II (Items 1–10) |
| 2. Total time[a] | 2. Total time for all attempted items in Part I and Part II |
| 3. Part I accuracy | 3. Number of forms correctly identified in Part I (Items 1–10) |
| 4. Part I right accuracy | 4. Number of forms correctly identified by right hand in Part I (Items 1–5) |
| 5. Part I left accuracy | 5. Number of forms correctly identified by left hand in Part I (Items 6–10) |
| 6. Part I time[a] | 6. Time spent on all items in Part I (Items 1–10) |
| 7. Part I right time[a] | 7. Time spent on all right-hand items in Part I (Items 1–5) |
| 8. Part I left time[a] | 8. Time spent on all left-hand items in Part I (Items 6–10) |
| 9. Part II accuracy | 9. Number of forms correctly identified in Part II (Items 1–10) |
| 10. Part II right accuracy | 10. Number of forms correctly identified by right hand in Part II (Items 1–5) |
| 11. Part II left accuracy | 11. Number of forms correctly identified by left hand in Part II (Items 6–10) |
| 12. Part II time[a] | 12. Time spent on all items in Part II (Items 1–10) |
| 13. Part II right time[a] | 13. Time spent on all right-hand items in Part II (Items 1–5) |
| 14. Part II left time[a] | 14. Time spent on all left-hand items in Part II (Items 6–10) |
| **Kinesthesia (KIN)** | Test contains 10 items; all items are administered to all children. |
| 1. *Total accuracy*[a] | 1. Sum of accuracy scores (cm distances) for right and left hands |
| 2. Right hand accuracy[a] | 2. Sum of cm distances between child's right finger & target (Items 1, 3, 5, 7, 9) |
| 3. Left hand accuracy[a] | 3. Sum of cm distances between child's left finger & target (Items 2, 4, 6, 8, 10) |
| **Finger Identification (FI)** | Test contains 16 items; all items are administered to all children. |
| 1. *Total accuracy* | 1. Total number of correctly identified fingers, Items 1–16 |
| 2. Right hand accuracy | 2. Number of correctly identified fingers on right hand (odd-numbered items) |
| 3. Left hand accuracy | 3. Number of correctly identified fingers on left hand (even-numbered items) |
| **Graphesthesia (GRA)** | Test contains 14 items; testing is discontinued after 4 consecutive scores of 0. |
| 1. *Total accuracy* | 1. Sum of accuracy scores (0, 1, 2) for right & left hands (Items 1–14) |
| 2. Right hand accuracy | 2. Sum of accuracy scores for right-hand items (odd-numbered items) |
| 3. Left hand accuracy | 3. Sum of accuracy scores for left-hand items (even-numbered items) |
| **Localization of Tactile Stimuli (LTS)** | Test contains 12 items; all items are administered to all children. |
| 1. *Total accuracy*[a] | 1. Sum of accuracy scores (cm distances) for right & left items |
| 2. Right hand accuracy[a] | 2. Sum of cm distances between child's finger & targets on right hand & arm (odd-numbered items) |
| 3. Left hand accuracy[a] | 3. Sum of cm distances between child's finger & targets on left hand & arm (even-numbered items) |

*Note.* Major score (subscale) for each test is designated by italics.

[a]For these subscales, high raw scores are represented as low *SD*-scores.

midline, the response is contralateral; if the right hand picks up the block on the right, the response is ipsilateral. The SVCU score is determined by dividing the total contralateral responses by contralateral plus ipsilateral responses. The Preferred Hand Use (PHU) score is derived from the proportion of items in which the child used his or her preferred hand.

**Figure-Ground Perception (FG).** This test requires the child to separate a foreground figure from a rival background. The stimulus figures in the first few items on the test are superimposed; later test items consist of more complex embedded figures. All figures are line drawings. Each of the 16 test plates contains a stimulus figure with six possible response choices, three of which are correct and three of which are incorrect distractors. To eliminate contamination of a strictly visual perception score with motor or practic function, the child only points to the response figures. The test is discontinued after seven or more errors, four of which were made on the last three plates attempted. The time required to select the three response figures is recorded for each plate. The total test accuracy score is the sum of test items correctly chosen.

**Manual Form Perception (MFP).** The first part of this two-part test requires the child to identify, by pointing, the visual counterparts of various plastic geometric forms held and manipulated one at a time in the hand. Right and left hands alternate in the task. In the second part of the test, both hands perceive forms simultaneously. With one hand, the child feels a stationary plastic geometric form attached to a board by a stem, and with the other hand the child identifies the similar form from among five response forms, also attached to a board by a stem. On both parts of the test, a shield prevents the child from seeing the plastic forms.

The Manual Form Perception task incorporates a combination of tactile and kinesthetic perception, sometimes referred to as a "haptic" sense or as stereognosis. Part I includes a visual form perception component; there is no direct visual input in Part II. Part II requires coordinating haptic information from both sides of the body and is not administered to 4-year-old children or to those correctly identifying fewer than four forms of Part I.

Response time is recorded for both Part I and Part II. Part I is similar in concept, but not in all stimulus figures, to Manual Form Perception of the SCSIT; Part II introduces a new concept of bilateral somatosensory integration not included in the SCSIT.

Manual Form Perception differs from the other tactile tests not only by greater involvement of the sense of kinesthesia in the task and the inclusion of a visual component in Part I of the test, but also by providing active, as opposed to passive, touch stimuli. Actively touching is a different physiological process from passive touching, and requires a willfully organized and emitted action with corollary discharge. In most stereognostic functions (such as buttoning or feeling objects in the pocket), the active movement is accompanied by a sequential analysis of the stimuli which, in turn, continually modify the motor activity. Thus, as the child manipulates the Manual Form Perception forms, he or she must call upon a high order of sensory integration.

**Motor Accuracy (MAc).** A test of visuomotor coordination, Motor Accuracy requires the child to draw a red line over a heavy, curved, black line. The black line covers an area 37.5 cm wide and 23.5 cm high and is printed on a test booklet 43.1 cm by 28.0 cm (11 in. × 17 in.). This large area assures sampling of eye-hand coordination in a wide variety of positions relative to the body, including crossing the body's midline. The test is designed to make fine discriminations among young children with neurological dysfunction. Separate scores for the child's preferred and nonpreferred hands enable comparison of function of the two sides of the body. Because dysfunctional children vary widely in Motor Accuracy execution speed, three sets of normative data were gathered for each hand, with each set based on different performance times. For the preferred hand, those times were approximately 120, 60, and 30 seconds; for the nonpreferred hand, the times were approximately 90, 60, and 30 seconds.

As is appropriate in assessing children with apraxia and related disorders, more emphasis is placed on accuracy than on speed. Nevertheless, the child is encouraged to take not less than 30 nor more than 120 seconds with the preferred hand and not less than 30 nor more than 90 seconds with the nonpreferred hand. Other than those recommendations, the child is allowed to draw the line at his or her own natural pace. The test accuracy score is the amount the child's red line deviates from the black line. To determine that distance, the examiner uses a mechanical line measure. The child's time is recorded, and standard deviation scores are computed using the most appropriate normative group (i.e., 30, 60, or 120 seconds for the preferred hand; 30, 60, or 90 seconds for the nonpreferred hand). This test is essentially the same as the SCSIT Motor Accuracy test.

### The Somatic and Vestibular Sensory Processing Tests

The term "somatosensory" refers to sensory input to the brain from the body. The somatosensory tests assess: (a) muscle and joint sense (Kinesthesia), (b) the perception of the spatial and temporal qualities of a series of tactile inputs (Graphesthesia), (c) the ability to differentiate fingers from tactile stimuli only (Finger Identification), and (d) the ability to localize single tactile stimuli (Localization of Tactile Stimuli). Vestibular processing is assessed by the Postrotary Nystagmus test. Both somatic and vestibular processing are assessed by Standing and Walking Balance.

**Kinesthesia (KIN).** With the child's vision occluded by a shield, the examiner places the child's finger on a given spot, then moves that finger to a second location, advising the child to be aware of the feeling of movement and position. The examiner moves the child's finger back to the first location, then asks the child to return the finger to the second location. The distance between where the examiner placed the finger and where the child placed the finger is marked for later measurement. Accuracy on this task is dependent upon the child's ability to interpret and recall sensory input from joint and muscle receptors. To help hold the child's attention and to convey the idea of the task, each location is referred to as a house. The child goes visiting to the first house, then the second house, and so on to each of 10 different houses or locations. Two trial items help the child grasp the nature of the test. Right and left hands alternate "visiting" a house, and separate right- and left-hand scores are reported. Because any activity that occludes vision is vulnerable to attention lapse, the two test items on which the child scores most poorly are readministered, and only the more favorable score on each of these items is entered into the final score. (The repetition of the two poorest attempts is the major difference between the SCSIT and SIPT versions of Kinesthesia.)

**The Tactile Tests: Finger Identification (FI), Graphesthesia (GRA), and Localization of Tactile Stimuli (LTS).** Each of these three tests assesses a different aspect of tactile perception. All were selected from a larger battery of perceptual-motor tests (Ayres, 1965) because of their discriminating ability and their relatively low dependence upon intellectual ability and verbal communication. The Finger Identification items differ somewhat from those in the SCSIT, and Graphesthesia includes a greater number and variety of designs than the SCSIT.

In Finger Identification, the child points to or touches the finger(s) previously touched by the examiner. The total test score is the sum of the correct item responses. In the Graphesthesia test, the child is asked to duplicate a design that was traced on the back of the child's hand by the examiner. Each response receives a score of 0, 1, or 2, depending upon the accuracy of the drawing, and these scores are summed across items to yield a total accuracy score. Localization of Tactile Stimuli requires the child to place his or her finger on the spot on the child's hand or arm that previously was touched by the examiner. The distance between the stimulus spot and the child's response is measured. The total score is the sum of the distances between the stimulus locations and the child's responses. To reduce error from wandering attention, the two most poorly performed items are repeated, and the better of the two scores on each of these items is recorded. This test is the same as Localization of Tactile Stimuli on the SCSIT except for the repetition of the two most inaccurately performed items.

Localization of Tactile Stimuli and Finger Identification require only the perception and localization of passively applied stimuli. Graphesthesia taps a slightly more complex function, requiring the spatial and temporal analysis of the passively received stimuli, which must first be decoded and then motorically encoded, thereby requiring graphic praxis.

**Postrotary Nystagmus (PRN).** Observation of elicited vestibular postrotational nystagmus has long been a means of appraising the integrity of the vestibular system. The Postrotary Nystagmus version of the procedure records duration of the oculomotor reflex following 10 rotations of the body in 20 seconds. The child sits on a nystagmus board which the examiner rotates by hand. After rotation, the board is stopped abruptly, and the child looks ahead while the examiner observes and records nystagmus duration with a stopwatch. The test is administered four times, first in a counterclockwise direction (to the child's left), then in a clockwise direction (to the child's right). Scores are reported for the average duration following rotation to the left, to the right, and for all four series of rotations. Postrotary Nystagmus is believed to reflect one aspect of central nervous system processing of vestibular sensory input. This test is identical to the Southern California Postrotary Nystagmus Test.

**Standing and Walking Balance (SWB).** This test evaluates the ability to balance on one or both feet, both statically and dynamically, with eyes open and eyes closed. Standing and Walking Balance is an extensive revision of the SCSIT Standing Balance test. Standing and Walking Balance is included in the SIPT because of its strong dependence on the central nervous system processing of sensory input from proprioceptors and from the macular receptors of the saccule and utricle of the vestibular system. Test items range from the very easy to the very difficult. The child is given the opportunity to repeat each item one or two times if full credit is not reached on the first or second attempt. Only the most favorable score for each item enters into the final score. Most items are scored on the number of seconds the child's body remained balanced. Exceptions are Items 9 and 10, where the item scores are the number of heel-to-toe steps correctly made. Separate scores for items executed with eyes open and eyes closed help differentiate the contributions of the visual input from the somatosensory and vestibular input. A total test score is also determined.

### The Praxis Tests

Practic skill is appraised in six behavioral domains.

Four of the tests are completely new. The praxis tests assess: (a) the child's ability to translate verbal commands into practic acts (Praxis on Verbal Command); (b) skill in three-dimensional construction (Constructional Praxis); (c) competency in perceiving, remembering, and executing a series of hand and finger movements (Sequencing Praxis); (d) ability to imitate movements and positions of the tongue, lips, and jaws (Oral Praxis); (e) facility in assuming different and unusual body postures (Postural Praxis); and (f) accuracy and approach in copying designs (Design Copying). Both Design Copying and Constructional Praxis assess visuoconstruction skill.

**Design Copying (DC).** This test retains and increases the advantages of the Design Copying test of the SCSIT. It especially expands the capacity for differential diagnosis by increasing the number and type of visuopractic requirements and by analyzing the approach the child used when reproducing the drawing. Design Copying consists of two parts. Part I is identical to Design Copying of the SCSIT except that the orientation of one of the 13 designs (Item 8) has been reversed and the child's drawing approach on Items 2 through 9 is scored. Part II of Design Copying is an entirely new addition to Design Copying of the SCSIT. It consists of 12 designs without a dot grid, each of which is reproduced by the child within an outlined square located immediately below the stimulus figure.

While the child draws, the examiner records the approach on most items, indicating the direction in which the child draws each line, when a new line is started, and the sequence in which the lines are drawn. The examiner scores the attempted items of Parts I and II for spatial accuracy and records the child's drawing approach on all attempted items except Item 1 and Items 10 through 13 of Part I. Part I accuracy is determined from the location of the child's lines relative to the appropriate dots. Spatial accuracy of Part II designs is scored on a number of desirable criteria or Should Have (SH) parameters. The total score for Design Copying is the sum of the Part I accuracy scores and the Part II Should Have scores.

The drawings in Parts I and II are also scored for the presence of 7 different undesirable and atypical approaches, or Should Not Have (SNH) parameters. All aberrant approaches of a given type are summed to provide a score on an atypical approach scale of that type.

Performance on Part I is discontinued after two consecutive item scores of 0, and Part II is discontinued after three consecutive failures to make a gross approximation of the design. Four-year-old children are not scored on drawing approach.

**Postural Praxis (PPr).** The Imitation of Postures Test of the SCSIT was extensively revised and expanded to become the Postural Praxis of the SIPT. In this test the child is asked to assume each of 17 different postures while it is being demonstrated by the examiner, and to hold each posture for seven seconds. Hand and arm postures are most frequently used, but head, trunk, and finger positions are critical on some of the items. All items are administered to all children, and the child receives a score of 0, 1, or 2 on each item, depending upon both accuracy and speed of response. Although Postural Praxis requires visual interpretation of the demonstrated position, no memory of that position is required. The total test score is the sum of all item scores.

**Praxis on Verbal Command (PrVC).** In this test the examiner verbally requests the child to assume each of 24 different unusual positions, and each position is scored for accuracy and time. For 4- and 5-year-olds, the command is repeated at 5 seconds, and the child is allowed 15 seconds to complete the action; older children are allowed only 10 seconds. Praxis on Verbal Command introduces a new concept in testing atypical children.

**Constructional Praxis (CPr).** This test assesses practic skill in relating objects to each other in an orderly arrangement or systematic assembly through building with blocks. The older or more competent child builds two different structures requiring different levels of skill. Four-year-old children, and those failing to score sufficiently well on Structure I, do not attempt Structure II. For Structure I, the child first watches as the examiner builds a model, then tries to duplicate the examiner's model. When the child has completed Structure I, the structure is disassembled, more blocks are added, and the child is instructed to build a more complicated structure of 15 blocks, duplicating a preassembled model.

Scoring of Structure I awards one point for each of 16 relatively simple criteria of accuracy of block placement. Each block of Structure II is scored on one of eight different parameters. The first seven parameters are negative aspects of space management, and the eighth is a positive attribute of correct placement. A block showing more than one of the seven negative parameters is scored only on the more serious spatial error. A block is scored as OK if it is not scored as incorrect on any of the first seven parameters. Constructional Praxis makes available for the first time a three-dimensional constructional praxis test for children.

**Sequencing Praxis (SPr).** This test assesses the child's ability to execute a series of planned hand or finger movements demonstrated by the examiner. The test consists of nine items, each with a number of subitems. Each of the six hand items requires a specific number of motions or changes of positions of either one or both hands. Each motion consists of tapping the table, the other hand, or the head with the hand or forearm in a specified position and in a specified sequence. Although the emphasis is on sequencing ability, this task also requires visual interpretation, memory, and coordination of the two sides of the body. The three finger items require tapping the fingers in a specified sequence. Sequencing errors include perseveration, omissions of movements, repeating the same movement, or mixing up the sequence.

Each of the nine items has five or six subitems, most of which alternate between right and left hand or finger usage. The number of motions to be executed is increased from two or three on the first subitem to four or eight on the last subitem. Each subitem is scored as 0, 1, or 2 for accuracy, and the total score is the sum of these subitem values.

Each item is discontinued when two consecutive subitems have been scored as 0. When the first two subitems are scored as 0 on two consecutive hand items (Items 1 through 6), the hand items are discontinued. When the first two subitems of two consecutive finger items are scored as 0, the test is discontinued.

**Oral Praxis (OPr).** In this test, the child imitates the examiner's movements of the tongue, teeth, lips, cheeks, or jaw. Some items consist of a sequence of movements. Performance is scored on a 0-1-2 scale, depending upon accuracy of imitation. The total test score is the sum of the item scores. Although the concept of oral practic ability is not new, Oral Praxis is an important addition to the standardized tests available to therapists.

### The Bilateral Integration and Sequencing Tests

Five of the SIPT assess, as a group, the bilateral integration and sequencing (BIS) function. These tests are Bilateral Motor Coordination (BMC), described in the following paragraph, and the previously described Oral Praxis, Sequencing Praxis, Graphesthesia, and Standing and Walking Balance.

**Bilateral Motor Coordination (BMC).** This test requires the imitation of smoothly executed movements with both hands (Items 1–10) and both feet (Items 11–14). This task involves both praxis and integration of function of the two sides of the body. Reciprocal movement is emphasized. Performance is negatively influenced by incoordination of the upper motor neuron type, such as choreoathetosis. Each item is scored on a 0-1-2 scale, depending upon correctness of movement. Feet items are not administered to 4-year-old children. To avoid discouragement on the part of the child, the test is discontinued after four consecutive errors. Bilateral Motor Coordination is an elaboration and revision of the SCSIT Bilateral Motor Coordination test.

# CHAPTER 2
## THE THEORETICAL MODEL

Lacking a universally accepted conceptual framework of how the brain works as a whole, the theory presented in this chapter has been constructed to provide a unifying concept for heuristic purposes. It is a provisional theory with continued modifications anticipated as research and clinical knowledge help it to evolve.

Sensory integration is the neurological process that organizes sensation from one's own body and from the environment and makes it possible to use the body effectively within the environment. The spatial and temporal aspects of inputs from different sensory modalities are interpreted, associated, and unified. Sensory integration is information processing.

Praxis is a uniquely human aptitude that underlies conceptualization, planning, and execution of skilled adaptive interaction with the physical world. Praxis is the ability by which an individual figures out how to use his or her hands and body in skilled tasks such as playing with toys, using tools (including a pencil or fork), building a structure (whether a toy block tower or a house), straightening up a room, or engaging in many occupations. Practic ability includes knowing what to do as well as how to do it. Practic skill is fundamental to purposeful activity. It is dependent upon both sensory integrative and higher cognitive processes. It involves an internal program of ideation or concept formation and goal directed action. Praxis is more "behavior" than it is "motor coordination." Although smoothness of motor output is needed to accomplish skilled, goal-directed tasks in the environment, practic ideation and planning are required to govern these actions. Practic ability develops from experience and increases in skill as the child matures. Together, sensory integration and praxis provide one of the most critical links between the environment and human behavior. They enable the informational transactions to occur between the organism and the rest of the world which are basic to the organization of behavior.

The mere presence of sensory stimuli from the environment and the body through intact peripheral receptors and sensory channels is insufficient for purposeful action. The sensory stimuli must be processed; that is, the brain must select, enhance, inhibit, compare, and relate organism-environment interaction. The brain must select, enhance, inhibit, compare, and associate the sensory information in a flexible, constantly changing pattern; in other words, the brain must integrate it. Neuromotor skill alone also is insufficient for adaptive interaction. The brain must be able to generate transactional concepts and to plan a suitable course of action. Sensory integration and praxis are functionally interdependent. Sensations from the body, especially during purposeful activity, provide the means by which a neuronal model or percept of the body is established. The body percept, sometimes referred to as the *body scheme,* is a construct introduced by Head (1920). An accurate body scheme is necessary for practic tasks, for a sense of directionality, and for relating the body to space. At the same time, conceiving, planning, and executing adaptive action is a major means by which sensation is made meaningful and translated into a body percept.

Praxis is to the tangible, two- and three-dimensional, gravity-bound world as speech is to the social world. Both praxis and speech are learned, and both enable interactions and transactions. They also require the cognitive functions of concept formation and planning that enable expression through the neuromotor system. While speech is concerned with communicating with other people, praxis is concerned with "doing" in the world. Sensory integration and praxis make possible the organism-environment interaction that is the basis for much human development, learning, and behavior. (In this theory, "learning" is used in its widest sense, i.e., a change in behavior or in concept formation resulting from experience.)

Praxis and perception are both end products of sensory integration and usually develop so naturally they are taken for granted. It is only in the last few decades that irregularities in their development and the resultant handicapping conditions have been recognized. Comparable, but usually more severe, deficits have been observed in brain-injured adults. It is partly from studies of adult-onset agnosia, aphasia, and apraxia that this theory and child assessment methods have evolved. (The prefix "a" means "without" or "not"; the prefix "dys" means "difficult" or "disordered"; "gnosis" is Greek for "perception," and "prassein" is Greek for "to do.")

Somatosensory, vestibular, and visual input to sensory integration and praxis are essential to organism-environmental interactions. Although critical to certain types of learning, and not unrelated to somatosensory and vestibular input, auditory processing and language development are only tangentially incorporated into the theory. Also, the olfactory, gustatory, and enteroceptive senses are recognized as important, but have not been included in the present form of the theory.

It has been argued that abstract intellectual ability is, in part, an evolutionary product of the steady increase in quantity and quality of informational transaction between organism and environment. Chronologically, the gradual increase in information intake was dependent upon the increase in sensory neurons and processing centers that promoted effective perception of the environment. Enhanced perception enables more complex adaptive responses, and these responses served an important role in promoting the development of the brain and its organization. Organism-environment interaction promoted the generation of ideas which could be associated into concepts.

Some theorists also link the evolutionary development of abstract intelligence with functional adaptation of the hand. Hand-mind linkage is central to praxis; as practic ability evolved, so did cognition. Current childhood disorders of brain function that interfere with processing of sensory information or with conceiving and planning purposeful action can be expected to be associated with later reduced efficiency of normal brain information processing and ability to cope with life's physical and intellectual demands.

Recent neuropsychological theories are giving greater emphasis to the brain's tendency to function as a whole. Recognizing that tendency, sensory integrative theory holds that the function of the neocortex, while capable of more complex activity such as abstraction, reasoning, and language processing, is nevertheless, partially dependent upon the lower structures. Essential sensory reception, integration, and intersensory association still occur in the central core and in the many nuclei of the brain stem and thalamus. The cerebellum also has an important role in sensory processing.

The brain's tendency to function as a whole does not preclude the natural tendency toward functional localization or specialization of different parts of the brain. Optimal function of some complex operations is dependent upon specialization and lateralization of function. The localizing process, in turn, is dependent upon adequate intra- and interhemispheral communication.

The effectiveness of organism-environment transaction in promoting human development is partially dependent upon the inherent plasticity of the central nervous system. The brain, especially the young brain, is naturally malleable; structure and function become more firm and set with age. The formative capacity allows person-environment interaction to promote and enhance neurointegrative efficiency. A deficiency in the individual's ability to engage effectively in this transaction at critical periods interferes with optimal brain development and consequent overall ability. Identifying the deficient areas at a young age and addressing them therapeutically can enhance the individual's opportunity for normal development.

# CHAPTER 3
## ADMINISTRATION

### General Administration Guidelines

**Test Setting**

The SIPT should be administered individually in a quiet, well-lit room that is free of distractions. The room should contain a child's table and two children's chairs.

**Test Materials**

Test materials included in the SIPT Kit by Western Psychological Services are:

1. Space Visualization placement card, form board, blocks, and pegs
2. Figure-Ground Perception test plates
3. Standing and Walking Balance half-round wood dowel
4. Design Copying test booklets (25) and scoring guide
5. Constructional Praxis preassembled model, blocks, and angle guide card
6. Postrotary Nystagmus angle guide card
7. Motor Accuracy test sheets (25) and line measure
8. Manual Form Perception bases, forms of various shapes, response card
9. Kinesthesia test sheets (25)
10. Localization of Tactile Stimuli pen
11. Shield
12. Centimeter/inches ruler
13. Two red nylon-tipped pens
14. Masking tape
15. Ten sets of Answer Sheets (protocol sheets)
16. Transmittal Sheets (10)
17. Test Manual
18. Carrying case
19. Nystagmus rotation board

In addition, the examiner will need to supply the following materials that are not included in the Kit:

1. Stopwatch capable of recording 1/10 seconds (available from WPS)
2. Soft (No. 2), black-leaded pencil for completing protocol sheets
3. Two other pencils, preferably one without eraser
4. Two 8½″ × 11″ sheets of paper
5. Footstool (if child's feet cannot reach floor when he or she is sitting in the chair)

**Administration Time**

The administration time of the SIPT ranges from 1½ to 2 hours. An additional half hour is required for completion of scoring and filling in circles on the protocol sheet. It is recommended that the SIPT be administered in two sessions in the order presented in this chapter, with a break after the first testing of Postrotary Nystagmus. If the tests are administered in one session, a break after Postrotary Nystagmus is still recommended; use your judgment to determine the length of the break.

**Entering Demographic Information**

Demographic information on the child should be entered on the Transmittal Sheet. Certain identifying information—age, sex, and writing hand—must be completed in order for the tests to be scored. All other information requested on this sheet is optional, including child's name, grade, special education information, child's IQ, parents' occupations, zip code, child's ethnic background, and known neurological impairment. At the bottom of the Transmittal Sheet, tactile defensiveness exhibited during testing and clinical observations during testing (i.e., prone extension, supine flexion, and ocular pursuits) may be rated after completion of testing. A Transmittal Sheet must be sent with any number of protocol sheets submitted for the child.

In addition, *the Transmittal Number that is printed at the top of the Transmittal Sheet must be entered on each protocol sheet for the same child.* This is very important; if it is missing from any protocol sheet, that protocol sheet cannot be matched with other protocol sheets for the child.

**Administering the Tests**

Follow the specific administration procedures for each test as presented later in this chapter. Directional cues also appear on the protocol sheet for each test to remind you of administration procedures as well as the instructions to give the child.

It is highly important that the child grasp the meaning of the instructions. When necessary, directions should be repeated, possibly restated in another way, to ensure that the child performs at his or her full capacity.

On some tests, demonstrations of the task are

allowed. Demonstrate only when the directions specifically state that demonstration is acceptable on that item.

During administration, minimize verbal interaction. All the tests on the SIPT battery require close attention to sensory stimuli, usually of the visual or somatosensory type, and such tasks may be inhibited by intermittent conversation. Thus, due to the nature of the items on the SIPT, good rapport with the child requires more than just verbal reinforcement. Rapport established through empathy and a positive feeling for the child will be more conducive to reliable testing than an approach that emphasizes verbal interaction.

The discontinuance criteria vary for each test. Some tests require that you continue administering items even after the child has failed several items. Other tests allow you to discontinue testing following a specified number of failed items. The discontinuance criterion for each test is indicated on the protocol sheet.

**Marking the Computer-Scannable Protocol Sheets**

It is very important that the protocol sheets be correctly completed, since they will be machine processed. Use only a soft (No. 2), black-leaded pencil to mark responses. Sheets marked with ball-point pens, nylon- or felt-tipped pens, mechanical pencils, or pencils with harder lead cannot be machine scored. In order to ensure accurate scoring, response circles must be dark and completely filled in (see examples in Figure 1). If you wish to change a response, erase the first mark completely and fill in the correct circle.

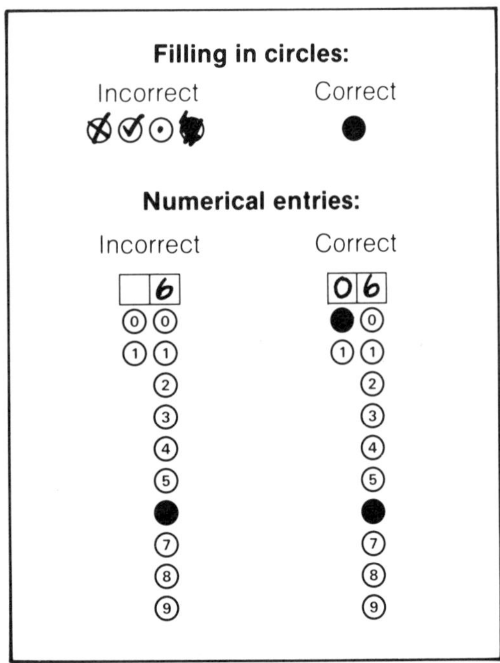


**Figure 1**
**Marking the Protocol Sheets**

When a numerical entry is required (e.g., Transmittal Number, time score), first write the number in the boxes below the heading. Then fill in the corresponding circles below the boxes (see Figure 1). Your entry must be the same number of digits as requested. If less than the requested number of digits, enter zeros in the preceding columns. For example, the Space Visualization time score is to be expressed as a two-digit whole number. If the child's time score is 6 seconds, enter it as "06."

Be careful to write *only in the specified areas.* Do not write in any margins of the protocol sheets or anywhere not specified in the directions. Mark only in the response circles or boxes clearly labeled for writing.

**Processing the Protocol Sheets**

**Checking for problems.** When you have finished administering the SIPT to a child, make sure that all the protocol sheets are properly completed and prepared for scoring; improperly completed sheets may cause errors in scoring. Check for the following problems:

1. *Incomplete information.* Age, sex, and writing hand must be entered on the Transmittal Sheet. In addition, the same Transmittal Number must be entered on each protocol sheet for the child so that it will be possible to match the forms for each child. Several tests require you to write the score for each item and to fill in the corresponding circles after the test is completed. Be sure that all such scores have been filled in.
2. *Stray marks.* Stray marks may cause errors in scoring when the sheets are "read" by the optical scanning machine. The machine sends light rays through the paper and these rays are blocked by pencil marks. Thus, stray marks may be read as responses. Erase any stray marks or writing in inappropriate spaces. Correction fluid should never be used on the protocol sheets.
3. *Light marks.* Marks that are too light will not be able to block the optical scanning light, again resulting in errors. Darken the marks on such sheets. A good rule of thumb is to darken the circles so that the printed number or letter inside the circle can no longer be seen.
4. *Improper writing materials.* Sheets must be filled in with a soft (No. 2) pencil in order to be read properly by the optical scanning machine. If you complete a sheet with other writing materials, you will need to redo it in pencil. It is generally easiest to go over all marks with a pencil, making sure the pencil marks completely cover the original marks.
5. *Damaged or torn sheets.* If a protocol sheet has been torn, do not use tape to repair it. If the sheet is still basically intact, it can be sent in for process-

ing. However, if there is extensive damage to the sheet, all demographic information and responses must be transferred to a new protocol sheet prior to sending it.

**Preparation for shipment.** Protocol sheets should never be folded. Leave them flat and in good order for shipment without damage to edges or surfaces. *Do not staple, tape, or otherwise fasten the sheets together.*

Enclose a Transmittal Sheet with each set of protocol sheets sent to WPS TEST REPORT at one time. Even if only one protocol sheet is sent in, the Transmittal Sheet must be completed. On the front of the Transmittal Sheet, enter the address to which the reports are to be sent. Print clearly in ink. (Note that this is the *only* place on the Transmittal Sheet and protocol sheets where ink should be used; the rest of the forms must be completed in pencil.) This will be used as the shipping label to mail your reports.

Also, indicate which tests are being submitted *for this child* in the box next to the shipping label. Fill in the circle for each test enclosed.

Processing of the protocol sheets usually takes less than 8 hours. However, the complete turnaround time for the individual user will vary depending on location and the postal service. For the average user in the United States, the process usually takes from 2 to 6 working days. Users outside the United States should allow more time.

## 1. Space Visualization (SV)

### Materials

- 2 plastic form boards (1 with an egg-shaped hole, 1 with a diamond-shaped hole)
- 8 blocks (4 egg-shaped and 4 diamond-shaped)
- 2 pegs
- placement card
- protocol sheet
- stopwatch

In this test, the child is asked to choose which of two blocks will fit the hole in the form board. If you are right-handed, seat the child to your left at the table and place the protocol sheet to your far right. If you are left-handed, you may wish to seat the child to your right and place the protocol sheet to your far left.

### Placement and Handling of Materials

If you are right-handed, arranging the materials and manipulating the blocks as detailed in the following paragraphs below will result in optimal savings of time and motion. Arrange the blocks appropriately on the placement card in the orientations illustrated on the card; the circles inside the illustrations represent the holes in the blocks. Then place the card over the lower section of the protocol sheet. You will move it down during testing as you need to see items.

The correct orientations of the form board, blocks, and pegs *as they should be placed in front of the child* are illustrated on the protocol sheet (see example in Figure 2). The upper figure represents the form board, and the circle in the figure represents the location of the peg in the form board. The two lower figures represent the blocks and the circles represent the locations of the holes in the blocks. The letters used in the pictures of the blocks correspond to the letters used to label the blocks on the placement card and indicate which blocks are to be used. It may be easier to identify and pick up a block by recognizing its form and noting its designated letter than by noting the location of the hole. The correct block on each item is indicated by "L" (left block) or "R" (right block) below the item number.

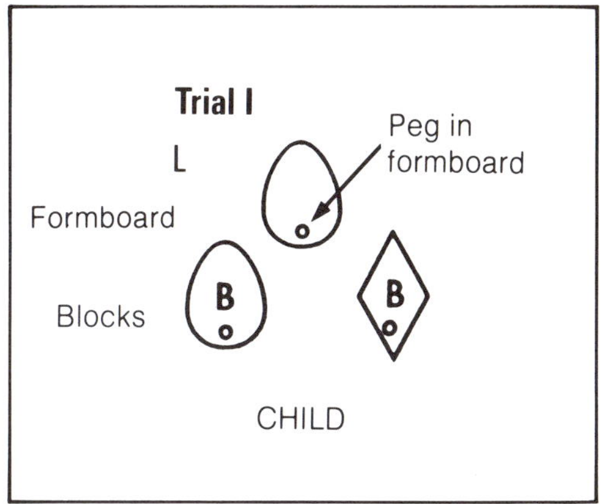


**Figure 2**
**Space Visualization: Trial I**

During administration, one form board is always in your lap while the other is on the table in front of the child. The peg is inserted in the form board, according to the illustration on the protocol sheet, before the form board is placed in front of the child. As much as possible, the child should look directly down on the form board. Place the form board with enough space between it and the edge of the table so that the blocks can be placed between the form board and the table edge.

From the placement card, pick up the left block with your left hand and the right block with your right hand, positioning them appropriately before placing both blocks simultaneously in front of the child. Place the blocks with a 5 cm (2 inch) space between them. They should be placed with the white side up. The child is not allowed to turn over the blocks. If the child wants to turn over a block to make it fit, tell the child that he or she cannot do so because the block is the wrong color on the other side.

If you can manipulate the form board and blocks rapidly and the child does not try to look at the placement card or protocol sheet, there is no need to use a shield to prevent the child from seeing the items. However, if the child appears to be trying to solve each item by looking at the protocol sheet, a shield should be used to obstruct the child's view.

After the responses to each item are recorded (see "Recording Responses," pp. 17–18), the blocks and form board are removed with the left hand while the form board for the next item is placed with the right hand. If the same form board is to be used, leave the form board on the table and change the peg in front of the child.

Detailed directions for administering the two trial items and the first three items from the test proper are given below to assist a beginning examiner. The same procedure is followed for the remaining test items. Although some variation in handling the blocks is allowed, it is best to follow the prescribed procedure as closely as possible. If you are left-handed, you may wish to reverse hand usage.

**Trial Items**

The egg form board and blocks for Trial I are in place when the child sits down at the examining table (see Figure 2). The diamond form board is in your lap. Say:

***"Which of these blocks*** (point in general direction of the two blocks) ***fits this big black hole?"*** (Point to hole in form board.)

If the child does not automatically pick up the block and put it in the form board after indicating a choice, say:

***"Put the block in the hole."***

If the child does not appear to understand the nature of the task, demonstrate by picking the correct block. If the child picks the wrong block, say:

***"Try the other one."***

After the child correctly places the egg block in the egg form board, pick up the diamond block, place it on top of the egg form board, and carry the board and the two blocks to your lap with your left hand. Meanwhile, use your right hand to simultaneously place the diamond form board, with the peg placed according to the upper figure of Trial II (see Figure 3), in front of the child.

Using your left hand for the diamond block and your right hand for the egg block, return the blocks from Trial I to their appropriate places on the placement card.

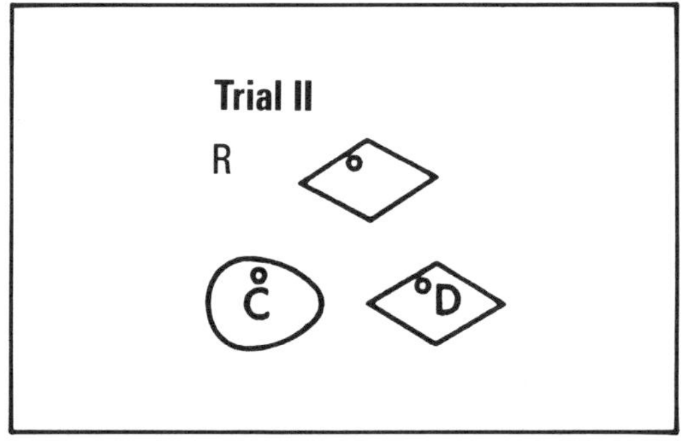


**Figure 3**
**Space Visualization: Trial II**

Then pick up the C egg block with your left hand and the D diamond block with your right hand and orient them correctly in space as you carry them to the table, positioning the blocks in front of the child. Say:

***"Here's another. Look first; then show me which one fits."***

After the child indicates which one fits, say:

***"Put it in the hole."***

After the child puts the block in the hole, say:

***"That's right; think about it first; then pick up the block that fits the hole."***

**Test Items**

While the child is completing Trial II, hold the egg form board with your left hand and reinsert the peg according to the top figure of Item 1 (see Figure 4).

After the child correctly inserts the diamond block on Trial II, pick up the egg block, place it on top of the form board, and carry it to your lap. Simultaneously

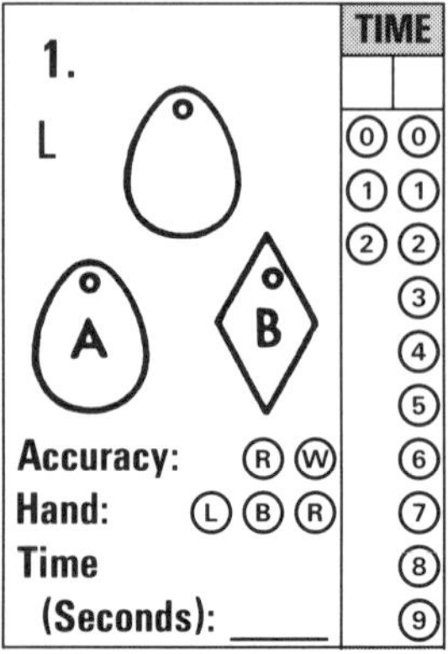


**Figure 4**
**Space Visualization: Item 1**

place the egg form board in front of the child. This placement begins the administration of the test items. Return the blocks from Trial II to the placement card.

Pick up the A egg block with your left hand and the B diamond block with your right hand, place them before the child, and *immediately* start the stopwatch or push the switch-back button. Note that the child's merely placing hands on a block or even unintentionally moving a block through involuntary motion does *not* constitute making a decision. The movement of the block must be deliberate to indicate a choice by the child.

After the child has selected a block, record the accuracy of that response, the hand used to move the block, and the elapsed time between the time the stopwatch was started and the time the child made the block selection (see "Recording Responses," this page).

As on the trial items, if the child makes an incorrect choice, ask him or her to try the other block. The correct block is always in the form board before the next item is presented.

Then hold the egg form board in place with one hand while changing the peg location and orienting the egg form board to conform with the upper figure in Item 2 (see Figure 5) with the other hand. Return the blocks used on Item 1 to the placement card. Pick up the B diamond block with your left hand and the B egg block with your right hand, orienting them in the correct positions, and place them below the form board. Immediately start the stopwatch. After the child makes a decision, record accuracy, hand used, and time. Do not make other marks on the protocol sheet that will interfere with optical scanning. This includes crossing through the box to keep your place.

When the correct block has been placed in the egg form board, pick up the diamond block with your left hand, place it on the egg form board, and carry it to your lap. Meanwhile, the diamond form board is placed in front of the child as on Item 3 (see Figure 6) using your right hand. Return the blocks from Item 2 to the placement card. Pick up the D diamond block with your left hand and the A egg block with your right hand and turn the blocks to correctly orient them while placing them in front of the child. Start the stopwatch. After the child has made his or her choice, record accuracy, hand used, and time.

After Item 3 or 4 is given, reinforce the necessity of *choosing* first, then *moving* a block by explaining in words appropriate to the child's level. For example, you might say:

***"Look at* both *blocks. Choose carefully. The first one you move counts as your choice. You want to be right the first time."***

Similar directions should be used any time the child appears not to be giving careful consideration to his or her choice. In some cases it is appropriate to hold the blocks down with your fingers until the child looks carefully.

No reference should be made to the child about his or her response being timed. However, if the child asks about the watch, the child should be told that he or she is being timed, but that it is more important to be accurate than fast. On the other hand, if the child is not trying to solve the puzzle with due diligence, tell the child that time is also important.

The rest of the test items are handled in the same manner as for Items 1, 2, and 3. Items 1 through 12 are on the front of the protocol sheet and Items 13 through 30 are on the back.

**Recording Responses**

After the child chooses a block on each item, record accuracy, hand used, and time. If the child changes his or her mind after moving a block, the *original* response, not the changed response, is recorded, regardless of whether it was correct or incorrect.

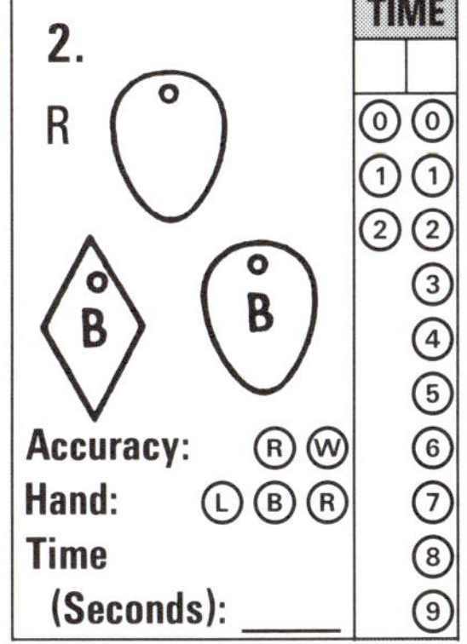


**Figure 5**
**Space Visualization: Item 2**

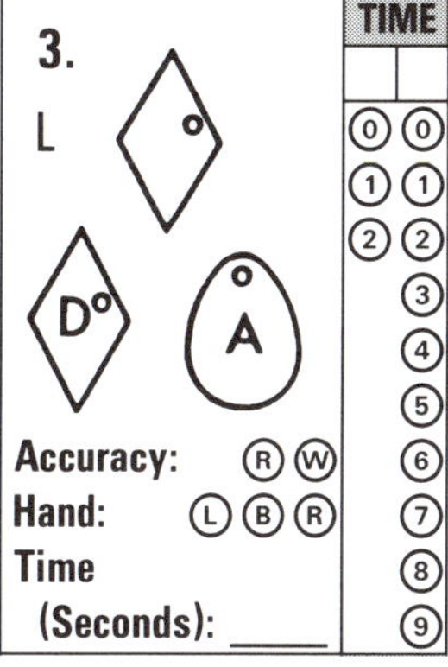


**Figure 6**
**Space Visualization: Item 3**

**Accuracy.** On the protocol sheet, record the accuracy of the child's response to each item. If the child chooses the correct block, as indicated by L or R on the protocol sheet, fill in the R (right) circle. If the child chooses the incorrect block, fill in the W (wrong) circle. If no response is made within 25 seconds, score accuracy as incorrect.

**Hand.** Record the hand used by the child to purposely move the block on each item. Fill in the L circle if the child first moved the block with the left hand or the R circle if the child first moved the block with the right hand. If the child *simultaneously* used both hands to move the block, fill in the B circle. However, if the child picked up the block with one hand and then quickly used the other hand or both hands, only the hand that *first* moved the block is marked.

**Time.** The time the child requires to make a decision on each item is counted from the time the stopwatch is started to the time the child indicates a choice by pointing to, moving, or picking up a block. Record the time required to make a decision, *not* the time required for the child to place the block in the form board. Time recorded for each test item is rounded to the nearest whole second. For example, 0.4 seconds is considered to be 0 seconds and 1.6 seconds is considered to be 2 seconds. Since time elapses between the time that blocks are placed on the table and the starting of the stopwatch, it is possible for the child to respond in 0 seconds.

During the test, it is easiest to write the number of seconds in the indicated space below the drawings for each item. After the test is completed, fill in the circles corresponding to the number of seconds recorded for each item. Be sure to fill in two digits for each item—for a time under 10 seconds, enter zero for the first digit (e.g., enter 6 seconds as "06").

It must be remembered that the time indicates how long it takes the child to reach a decision when attention is focused on the problem. If possible, the child should not be penalized by including time when the child is not paying attention, even though the blocks have been set in front of him or her. Watch closely and record the actual time elapsed when the child has indicated a choice, usually by picking up the block. Remember that the movement of the block must be deliberate to indicate a choice by the child.

**Discontinuing the Test**

As soon as five errors are made, not necessarily consecutively, the test is discontinued. Thus, it is important to keep track of the number of errors made on the test items. Whenever the child makes an incorrect choice (i.e., one scored W), indicate this in the box labeled "Number of Errors" on the front of the protocol sheet, recording 1 for the first error, 2 for the second error, 3 for the third error, and so on. When you turn over the protocol sheet after Item 12, enter the number of errors obtained so far in the box labeled "Number of Errors" to remind yourself of the number to record after the next error. Errors made on the trial items are not counted as testing errors.

**Modifications for Special Circumstances**

Although the specific directions given above are important, testing difficult children sometimes requires modification and extra directions. To stay within standard administration procedures as much as possible, remind such a child that: (a) accuracy is more important than speed of response; (b) the item is being timed (if child takes unduly long to respond); and (c) the child should determine mentally the correct block *before* moving it.

An impulsive child can be restrained by holding the blocks down inconspicuously with your fingers. A child who is incapable of the mental visualization required will naturally reach impulsively for a block. Rather than overly frustrate such a child, it is often better to let him or her choose items at random until the test is terminated. The child should also know that as soon as he or she purposefully moves a block, the block is considered to be his or her choice. Touching a block or slight movements from involuntary motion are not considered moving the block. Occasionally, with an impulsive child, it is appropriate to ask the child to point to his or her choice before moving the block. In this case, the hand used for pointing is recorded as the hand used. Such impulsivity should be noted as a clinical observation.

If the child is unable for motility reasons to place the block in the form board for any of the items, place the block for the child so that he or she will have a learning experience comparable to other children. Because this is not a test of motor performance, it is permissible to assist the child in placing the blocks in the form board or to do it for the child.

Occasionally, a child will select blocks from only one side. In this case, encourage the child to look at both blocks before deciding which one will fit. This tendency should also be noted as a clinical observation.

## Key Points to Remember

1. Seat the child to your left at the table if you are right-handed.
2. Place the protocol sheet and placement card to your right. It may be necessary to shield these materials from older or very sophisticated children.
3. One form board is always on the table in front of the child; the other is in your lap.
4. From the placement card, pick up the left block with your left hand and the right block with your right hand, positioning them appropriately before placing both blocks simultaneously in front of the child.
5. Place the blocks at the midline of the child with a 5 cm (2 inch) space between them.
6. Place the blocks with white side up; do not allow the child to turn over the blocks.
7. Start the stopwatch immediately following placement of the blocks.
8. Note the time as soon as the child indicates his or her choice.
9. Restrict impulsive action; if necessary, remind the child to look at both blocks before deciding.
10. The child must understand that moving a block counts as a choice. (After Item 3 or 4, reinforce the importance of looking carefully.)
11. If the child does not try the chosen block in the hole, ask him or her to do so.
12. If the child is unable to handle block placement, assist.
13. Record accuracy, hand used, and time.
14. It is possible for the child to respond in 0 seconds.
15. Selection of the wrong block, or no selection in 25 seconds, constitutes an error. Indicate number of errors on the protocol sheet as they occur.
16. If the child selects the wrong block, ask the child to try the other block.
17. The correct block is always in the form board before the next item is presented.
18. If you are right-handed, remove the blocks and form board from the last item with your left hand and place the new form board at the same time with your right hand.
19. If the same form board is to be used, leave it on the table and change the peg in front of the child.
20. Discontinue the test after five errors, not necessarily consecutively.
21. After the test is completed, fill in the circles corresponding to the time on each item. Time must be recorded as two digits (e.g., "06" for 6 seconds).

## 2. Figure-Ground Perception (FG)

**Materials**

- book of test plates (2 pairs of trial plates and 16 pairs of test plates, each pair containing a stimulus plate and a response plate)
- protocol sheet
- stopwatch

In this test, the child is asked to look at the upper stimulus picture of superimposed or embedded figures and then find three figures from the six figures in the lower response plate that are part of the upper design. The other three pictures in the response plate do not appear in the upper stimulus figure; they serve as distractors to the child. The numbers corresponding to the response choices in the test booklet are placed in the same spatial arrangement on the protocol sheet, with the correct choices shaded.

Seat the child to your left at the table if you are right-handed. Follow the directions below for administering the items. The child responds by pointing to the figures on the response plates that correspond to and are like the stimulus figures. No verbal response is required. The child's responses and response time for each item are recorded.

**Series I Trial Item**

The plates in Series I contain pictures of common objects. Begin by opening the book of test plates to the first trial item, Plates IA and IB (see Figure 7), and place the book on the table in front of the child. The main concept the child is to grasp during this trial item, if possible, is that three pictures or designs on one plate are correct and three are not. Pointing appropriately, say:

***"Three of these pictures*** (point to Response Plate IB) ***are up here.*** (Point to Stimulus Plate IA.) ***Which three are they?"***

After the child points to the three correct pictures in Response Plate IB, point out the three pictures *not* in Stimulus Plate IA and say:

***"These three are not up here, are they?"***

Then say:

***"That is the way it will be each time I turn the page. Find three pictures down here which are up here. Look carefully because it can be tricky."***

If an error is made on the demonstration Plates IA and IB, correct the child and give further explanation. Responses on the trial item are not recorded.

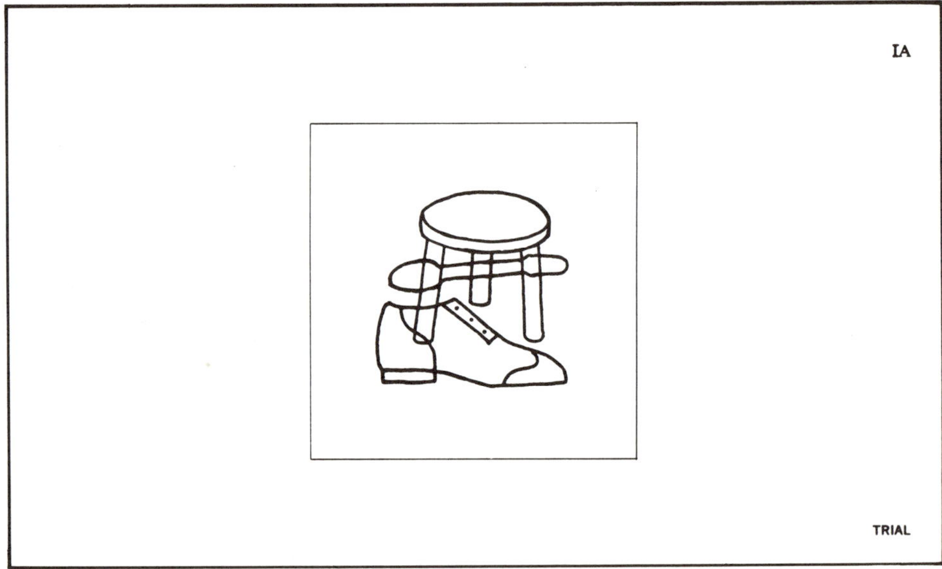


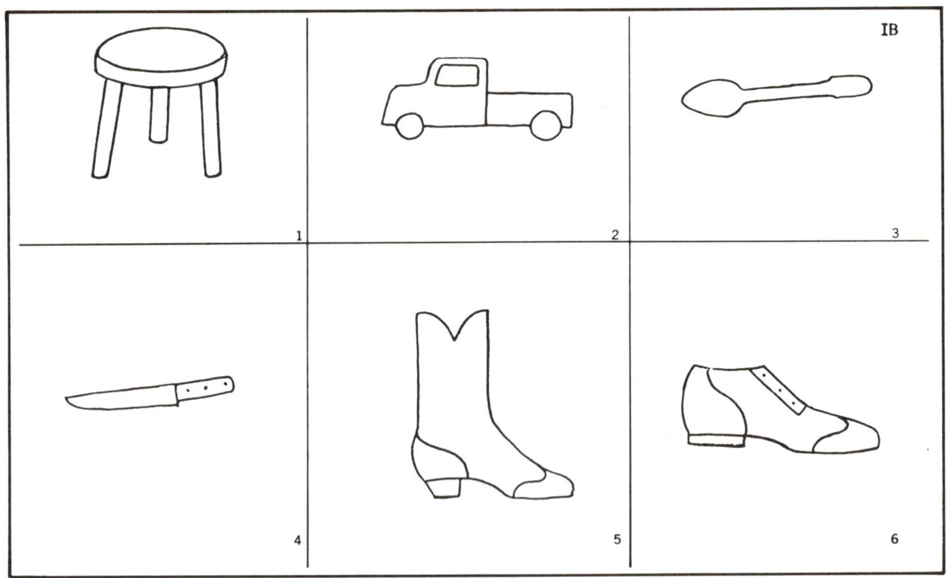


**Figure 7**
**Figure-Ground Perception: Trial I**

**Series I Test Items**

Then turn the page to Plates 1A and 1B and immediately start the stopwatch. Allow 60 seconds for the child's responses on each plate.

Remind the child frequently to be sure to find the picture in the upper plate before he or she points to it in the lower plate. The child is allowed to trace the stimulus figures with his or her finger if the child wants to do so. If you suspect that the child is guessing when the child is actually capable of detecting the correct response, ask the child to point to (or even outline) the response figure within the stimulus figure. However, when the items become too difficult for the child, encourage him or her to choose rather than simply omit an item. The child should also be encouraged to make a choice quickly if 50 seconds have elapsed and he or she has made only two choices.

The child is free to change his or her mind during the 60-second period for each plate. However, if the child does not clarify which three figures constitute the final choice, remind him or her to do so. *If the child gives more than three choices, ask him or her to choose only three; if the child does not, the first three choices are recorded as the selection.* If the child insists on knowing whether or not he or she gave the correct responses, tell the child that the responses will be explained at the end of the test. Only in the demonstration period is the child told which pictures are correct.

On the protocol sheet, fill in the circles corresponding to the child's choices (see sample in Figure 8). *Be sure to fill in only three circles for each item.* Erase any marks indicating choices made but then changed to another choice. Only response circles representing final choices should contain pencil marks.

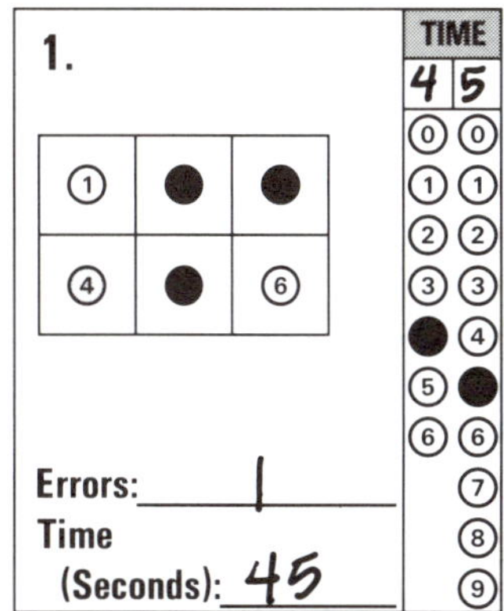


**Figure 8**
**Figure-Ground Perception:**
**Marking the Child's Responses**

Because each plate gives an opportunity for three choices, it is possible to obtain up to three errors on each item. An omission is considered to be an error. Thus, a child making two correct choices and no incorrect choices within the time limit on any plate is still considered to have made one error. If the child makes only one choice (and that one correct) on a plate within 60 seconds, two errors are recorded for the child.

Because the discontinuance criterion is based on both the number of cumulative errors throughout the test and the number of errors per plate (see "Discontinuing the Test," p. 22), it is important to keep track of the number of errors. Each time the child makes an *incorrect* response or an omission, write the number of cumulative errors in the box labeled "Number of Errors" at the bottom of the protocol sheet, recording 1 after the first error, 2 after the second error, and so on. After each item, enter the number of errors made on that plate in the space at the bottom of the item box.

If two errors are made on Plates 1A and 1B, or fewer than three choices are made within the 60-second limit, do not record the child's response, but return to the trial plates and reinstruct the child. After further instruction, turn to Plates 1A and 1B again and readminister them. Record time and responses for the readministration.

After the child makes three selections of the pictures or after the time limit has expired, write the elapsed time in seconds at the bottom of the box for each item on the protocol sheet. Time should be rounded to the nearest whole second. A maximum of 60 seconds is allowed on each item. Any response made after 60 seconds is not recorded. If only two choices are made within 60 seconds and the child fails to make a third choice, the time scored for that plate is 60 seconds. The circles corresponding to the number of seconds for each item are filled in after the test is completed. Be sure to fill in two digits for each item—for a time under 10 seconds, enter zero for the first digit (e.g., enter 6 seconds as "06").

After recording the child's responses and response time for Item 1, turn to Plates 2A and 2B and wait for or ask for a response from the child. The rest of the items are given in sequence. If all plates through 8A and 8B have been attempted and the child has not made seven errors, four of which were on the last three plates attempted, Series II is given. Turn over the protocol sheet and write the number of errors obtained so far in the box labeled "Number of Errors" to remind yourself of the number to record after the next error.

**Series II Trial Item**

The plates in Series II contain embedded geometric forms. Series II begins with Trial Plates IIA and IIB (see Figure 9). During this trial, the main concept the child needs to understand is that not all the lines in the stimulus plate are in the response plate. Say:

***"Now you will look at designs instead of pictures of things. Three of these designs*** (point to Response Plate IIB) ***are part of this one.*** (Point to Stimulus Plate IIA.) ***They are hidden in this upper design, just as some of these pictures*** (turn back to Plate 8B) ***were part of this upper figure."*** (Point to Plate 8A.)

Point to Design 1 of Plate IIB and say:

***"This design is a cross, but not like the crossed lines up here, so it is not part of the upper design. This one*** (point to Design 2) ***is part of the design up here. Can you see it?*** (Trace Design 2 with finger.) ***Can you see this one up here?*** (Point to Design 3.) ***Some of the lines in this upper picture*** (point to stimulus plate) ***are not always in the choices down here."*** (Point to response plate.)

After the child studies each design, point it out in the stimulus plate and make sure the child sees it. Indicate that not all the lines in the embedded or stimulus plate are in the designs in the response plate. Then say:

***"Which one of these*** (point to Designs 4, 5, and 6) ***is hidden in this design?"*** (Point to stimulus plate.)

Make sure the child sees that Design 5 is the correct choice and Designs 4 and 6 are incorrect choices. Give any additional instructions necessary for the child to grasp

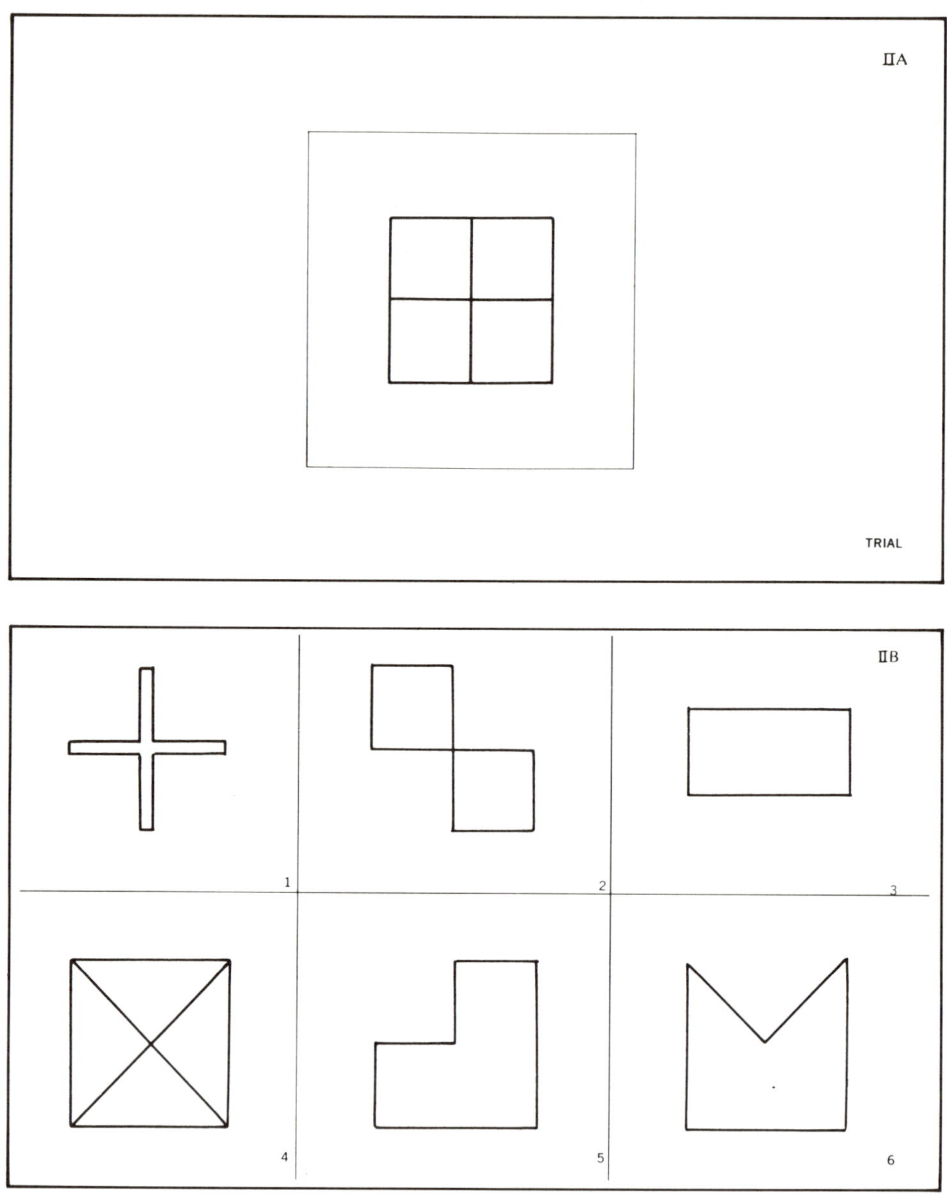


**Figure 9**
**Figure-Ground Perception: Trial II**

this idea, using the trial plates as the only visual aids. Then say:

> ***"The rest of the designs will be something like this one. Find the three designs here*** (point to the response plate) ***that are up here."*** (Point to the stimulus plate.)

Responses on the trial item are not recorded.

### Series II Test Items

Begin Series II with Plates 9A and 9B. The time required for response and the cumulative errors are recorded for Series II just as for Series I. A time limit of 60 seconds per plate still exists for Series II. In recording the cumulative errors, do not start over with Series II. For example, if a child made five errors on Series I, the first error made on Series II is counted as the sixth error.

### Discontinuing the Test

The test is discontinued after the child has made his or her seventh error, if the last four errors were made on the last three plates attempted. If the four errors were not made on the last three plates attempted, the test is continued until four errors are made on three, or parts of three, consecutive plates. An incorrectly identified design is an error. Failure to attempt a design and a design correctly identified after 60 seconds have elapsed are also errors. Time recorded for the final plate is the time at which the child made the final error. Do not record responses following the final error, even if the child changes a response, but allow the child to finish the item.

**Modifications for Special Circumstances**

Special administration instructions may be required for impulsive, very young, or mentally retarded children. Impulsive children tend to err because they insufficiently study the embedded figures. In such cases, children should be encouraged to use more care, with comments such as, "They are getting harder. Find the picture here before you point to it down here"; "Look at all the pictures before you choose"; or "Take more time; look very carefully. You have more time to find the right picture." Directions may be repeated, but only as needed, as the child proceeds with the task. If the child inquires about being timed, tell the child that he or she is being timed, but that it is more important to be correct than to be fast. Very young or mentally retarded children may also need special instruction. The following directions are suggested to help such children grasp the concepts.

The book of test plates is opened to Plates IA and IB. Say:

***"Here is a picture of a stool. Is there a picture of the same stool up here?"*** (Point to the stimulus picture.)

If the child does not see the stool in the stimulus plate, point it out. Then, pointing to each picture separately, say:

***"Here is a truck. There is no truck up here. . . . Is this one up here? . . . This one?"***

After the child grasps the idea, present Plates 1A and 1B. Say, pointing to the appropriate picture:

***"Here is a picture of a mouse. Is there a picture of this mouse up here? . . . Here is a cat. Is there a picture of this cat up here? . . . Is there a duck like this one up here?"***

If in doubt, ask the child to point to the stimulus plate for verification. The subsequent plates are presented in a similar manner until the child selects pictures without additional guidance.

## Key Points to Remember

1. Seat the child to your left at the table if you are right-handed.
2. There are two trial items: The first is before Item 1 and the second is before Item 9.
3. The child should understand that three figures are correct.
4. Start the stopwatch immediately after turning the page.
5. A maximum of 60 seconds is allowed for each test plate.
6. The child is allowed to change his or her mind.
7. If the items become too difficult, encourage the child to choose rather than omit an item. Also, encourage the child to make a choice quickly if 50 seconds have elapsed before three choices are made.
8. An error is scored for an incorrect choice or a choice which was not made within 60 seconds.
9. Do not tell the child which items are correct (except on trial items).
10. Record the child's three selections. Do not mark any selection made after 60 seconds. Record the cumulative number of errors and the number of errors per item.
11. If the child chooses four or more figures on one plate, ask the child to choose just three.
12. Note the time at which the child has made his or her choices for that item. Record time to the nearest whole second.
13. Discontinue after seven errors, if four errors were made in the last three plates attempted. If the four errors were not made in the last three plates attempted, continue until four errors are made on three consecutive plates.
14. Do not record responses following the final error, but let the child finish the item. The time recorded for the final plate is the time at which the final error was made.
15. After the test is completed, fill in the circles corresponding to the time on each item. Time must be recorded as two digits (e.g., for a time of 6 seconds, enter "06").
16. Be sure to erase any marks indicating choices made but then changed to another choice. Only circles representing final choices should contain pencil marks.

## 3. Standing and Walking Balance (SWB)

**Materials**

- half-round wood dowel (2 inches wide by 10 inches long)
- shield
- protocol sheet
- stopwatch

This test assesses the child's ability to balance while performing a series of standing and walking tasks, with eyes open and with eyes closed. The child is allowed two or three attempts to reach the maximum score for each task.

The test should be administered on a noncarpeted surface or on indoor-outdoor carpeting. A floor covered with thick carpeting should be especially avoided. The area in which the child is asked to stand or walk should not be close to walls or furniture.

The child is in his or her stocking feet. If the child did not wear socks, he or she may be tested barefoot. You should be wearing socks or nylon hose.

**Item Sequence**

Items are arranged in difficulty level. To reduce the length of the test, not all items are given to all children. It is intended that not more than two consecutive items that are too easy and three consecutive items that are too hard be administered. To meet this objective, there are *too-easy* and *too-hard* criteria. The too-easy criterion is a maximal score on either attempt of each of two consecutive items. The too-hard discontinuance criterion is a score of 0 or 1 on each of three consecutive test items. Test items are given until both criteria are reached or assumed. A child with poor balance may not reach the too-easy criterion.

If you think the child may have a severe balance problem, begin on Item 1 and continue in sequence until the discontinuance criterion of a score of 0 or 1 (indicating that the item is too hard) is given on each of three consecutive items. However, for *most* children, begin on Item 4. If the child receives the maximum score listed on the protocol sheet for Item 4 (10 seconds) on either the first or second attempt, proceed to Item 5 and continue in sequence until the discontinuance criterion of scores of 0 or 1 on each of three consecutive items is reached. It is assumed that the child who scores at 10 seconds on Item 4 also would receive the maximal score on Item 3, thus meeting the criterion of two too-easy items. If the child does not receive the maximum score on Item 4, administer the preceding items backwards (i.e., Items 3, 2, and 1, in that order) until the child receives the maximum score on two consecutive items, thus meeting the criterion of two too-easy items. Then continue with Item 5 and the remaining items in order until the discontinuance criterion of scores of 0 or 1 on three consecutive too-hard items is reached. If the child is scored as better than 0 or 1 on either two or three of the first three items but does not receive a maximum score on two consecutive items, go on to Item 5 and subsequent items. The too-easy criterion may not be reached, but the too-hard discontinuance criterion must be reached. If the child scores 0 or 1 on Items 4, 3, and 2, do not administer Item 5 or other more difficult items; the too-hard discontinuance criterion has been reached. However, you should still administer Item 1, as the child may be able to perform that item.

**Administration Procedures.**

Demonstrate Items 1, 3, 4, 5, 9, 10, 11, and 15, and any other items required by the child and give appropriate directions for each task (see "Specific Item Directions," p. 25). The protocol sheet indicates whether the child is to do each task with eyes open or eyes closed and provides cues for instructions to give to the child. If necessary, you may help position the child prior to each item. If the child cannot close his or her eyes, hold the shield close to the child's eyes to occlude as many visual stimuli as possible.

While the child is performing the tasks, his or her arms are free to move in any manner except to touch an object to help balance. You should be ready to catch the child if necessary.

All items except Items 9 and 10 are timed. For the timed items, start the stopwatch as soon as the correct position is assumed, and stop timing as soon as the child moves out of position or, for the eyes-closed items, opens his or her eyes.

On all items, the child is allowed two or three attempts (two attempts on Items 1–4 and three on Items 5–16) to reach the maximum score listed on the protocol sheet. For example, on Item 5 the child is asked to balance on the right foot for 10 seconds. If the child does this on the first attempt, the child does not make a second attempt. However, if the child is able to hold the posture for only 8 seconds, he or she makes a second attempt. If the time for the second attempt is 10 seconds, the child does not make a third attempt. Before each repeated attempt, say:

***"Good. Try again. Try to stand longer (walk further)."***

You may also repeat the demonstration for each attempt. Each attempt should be followed by a brief pause. If the child does need more than one attempt, the same foot does the same task until that item is completed. In other words, each item is completed before going on to the next item.

Each item is discontinued either when the child meets

the maximum score listed on the protocol sheet or when the required number of attempts has been made, whichever comes first. To avoid fatigue, the child should not stand longer or walk further than what is indicated as the maximum score for each item.

**Specific Item Directions**

Demonstrate Items 1, 3, 4, 5, 9, 10, 11, and 15, and any other items required by the child.

*Item 1:* With eyes open, stand with feet together, toes even. Discontinue at 5 seconds. Allow two attempts.

Begin by demonstrating while saying:

***"Stand like this. How long can you stand there?"***

You may need to position the child's feet and advise him or her to try not to move. As soon as the child's feet are in the correct position, start the stopwatch and record the length of time the child stands. The child may move at the ankles to maintain balance, but the location of the soles of the feet on the floor cannot change. As soon as the sole of either foot changes its location on the floor, stop timing. If the child stands for 5 seconds, that time is recorded and you go on to the next item. If the child stands longer than 5 seconds, tell the child that he or she has stood long enough and go on to the next item. If the child stands with feet together less than 5 seconds, say:

***"Good. Try again. Try to stand there longer."***

If the child does not assume the correct position, help him or her to do so and repeat the item. The better of the two time scores, but never more than 5 seconds, is marked in the corresponding circle after the test is completed.

*Item 2:* With eyes closed, stand with feet together, toes even. Discontinue at 5 seconds. Allow two attempts.

Begin by saying:

***"Now try it with your eyes closed. Feet together; now close your eyes."***

If the child cannot close his or her eyes, hold the shield close to the child's eyes to occlude as many visual stimuli as possible. The length of time the child maintains his or her balance without changing the location of his or her feet or opening his or her eyes is entered as on Item 1. If balance is maintained for 5 seconds or more, enter "5" in the first box on the protocol sheet, say "good," and go on to the next item. If the child maintains his or her balance with eyes closed for less than 5 seconds, say:

***"Good. Try again. Try to stand longer."***

The item is repeated, the time recorded in the second box on the protocol sheet, and the better of the two scores is marked in the corresponding circle after the test is completed.

*Item 3:* With eyes open, stand heel to toe. Discontinue at 10 seconds. Allow two attempts.

Begin by saying:

***"Now stand like this.*** (Demonstrate.) ***How long can you stand there?"***

Demonstrate by placing your right heel against your left toe so that the feet make a straight line, and help the child to assume a similar position. It does not matter which foot the child puts in front. Record the time the child stands in that position without moving his or her feet. If the child stands for 10 seconds, record that number and go on to the next item. Do not waste time by letting the child stand longer than necessary to get full credit. If the child stands less than 10 seconds, a second attempt is made. The better of the two scores is marked in the corresponding circle after the test is completed.

*Item 4:* With eyes closed, stand heel to toe. Either foot may be in front. Discontinue at 10 seconds. Allow two attempts.

Begin by saying:

***"Now try it with your eyes closed."***

The shield may be used if the child cannot close his or her eyes. Follow the same procedure as on Item 3.

If Item 4 is the first item administered, give these directions:

***"Let's see how well you can balance. Stand like this. Close your eyes. Try not to move."***

If the child stands less than 10 seconds, say:

***"Good. Try again. Try to stand longer."***

The better of the two scores (but never more than 10 seconds) is marked in the corresponding circle after the test is completed. If this score is less than 10, administer Items 3, 2, and 1 until the maximum score is reached on two items.

*Item 5:* With eyes open, stand balancing on the right foot. Discontinue at 10 seconds. Allow three attempts.

Begin by saying:

***"How long can you stand on one foot? Lift this foot;*** (touch child's left leg near the foot) ***don't hop or move around."***

Start the stopwatch as soon as the left foot is lifted and note the time when that foot is placed on the floor, even momentarily, or the right foot moves from its position. The left leg may be held at any place except on the floor or other stable place, and any movements of the body may be made as long as the sole of the right foot does not move from its position on the floor.

On this item, the child is given three attempts unless he or she stands for 10 seconds on the first or second attempt, at which time the item is discontinued. If the child stands for less than 10 seconds on either the first or second trial, say:

***"Good. Try again. Try to stand longer."***

The best of the three scores is marked in the corresponding circle after the test is completed.

*Item 6:* With eyes open, stand balancing on the left foot. Discontinue at 10 seconds. Allow three attempts.

This item is administered and scored in the same manner as on Item 5, but with the child balancing on his or her left foot. Begin by saying:

***"Now stand on the other foot. Lift this foot."*** (Touch child's right leg near foot.)

*Item 7:* With eyes closed, stand balancing on the right foot. Discontinue at 10 seconds. Allow three attempts.

This item is administered and scored in the same manner as on Item 5, except that the child's eyes are closed. Begin by saying:

***"Try it with your eyes closed. Close your eyes; lift this foot."*** (Touch left leg.)

*Item 8:* With eyes closed, stand balancing on the left foot. Discontinue at 10 seconds. Allow three attempts.

This item is administered and scored in the same manner as on Item 7, but with the child balancing on his or her left foot. Say:

***"Now stand on the other foot. Close your eyes; lift this foot."*** (Touch right leg.)

*Item 9:* With eyes open, walk with a heel-to-toe gait. Discontinue at 15 correct steps. Allow three attempts.

While demonstrating, say:

***"Walk like this. Each time, touch your heel to your toe, but don't step on your toe."***

Record the number of correct steps. A step is correct if one foot is in front of the other foot and the heel is within 2½ cm (one inch) in front of or to the side of the toe. In other words, the child is given credit for steps until he or she does not place the forward foot completely ahead of the other foot or does not place the heel close to the toe. The heel need not be exactly lined up with the toe. It is within the required distance if an imaginary line drawn even with the heel is within 2½ cm of another imaginary line drawn even with the toe and the edge of the heel of the forward foot is within 2½ cm of being in line with the edge of the stance foot. As soon as the child makes any of these errors, the score at that point is recorded. Steps made after the first error, even if correct, do not enter into scoring.

A common response is for the child to take a step and then slide the front foot back to the stance foot. If, after the child slides the front foot back, the criterion for a correct step is met, the step is counted as correct. If, however, the child slides the back foot forward, the last correct step is taken from the position of the stance (back) foot before it was brought forward.

*Item 10:* With eyes closed, walk with a heel-to-toe gait. Discontinue after 15 correct steps. Allow three attempts.

This item is administered in the same manner as on Item 9, except that the child's eyes are closed. While demonstrating, say:

***"Walk the same way with your eyes closed. Touch your heel to your toe, but don't step on your toe. Start here."***

Count and record the number of correct steps before the first error is made. A step is correct if one foot is in front of the other foot and not further than 2½ cm (one inch) in front of it or to the side of it. The child is allowed three attempts, unless he or she takes 15 steps correctly, at which time the item is discontinued. See Item 9 for scoring a step in which the front foot slides back to meet the stance foot.

*Item 11:* With eyes open, stand balancing with right foot on wood dowel. Discontinue at 15 seconds. Allow three attempts.

A 25-cm (10-inch) piece of half-round wood molding is placed on the floor, the flat side down. The child is expected to place one foot lengthwise on the dowel and balance on that foot. While demonstrating, say:

***"Stand on this stick with this foot;*** (point to right foot) ***get your balance; lift this foot.*** (Point to left foot.) ***How long can you balance?"***

You may need to assist younger children in assuming the desired position. Record, in whole seconds, the length of time that the left foot is off the floor and the right foot has not moved on the dowel. Discontinue at 15 seconds. Three attempts are allowed.

*Item 12:* With eyes open, stand balancing with left foot on wood dowel. Discontinue at 15 seconds. Allow three attempts.

This item is administered and scored in the same manner as on Item 11, except the child balances on his or her left foot instead of the right foot. Begin by saying:

***"Now try the other foot. Get your balance; lift this foot."*** (Point to right foot.)

*Item 13:* With eyes closed, stand balancing with right foot on wood dowel. Discontinue at 10 seconds. Allow three attempts.

This item is administered and scored in the same manner as on Item 11, but the child's eyes are closed and 10 seconds is the maximum score. Begin by saying:

***"Do the same thing with your eyes closed. Get your balance; close your eyes; lift this foot."*** (Touch left leg.)

*Item 14:* With eyes closed, stand balancing with left foot on wood dowel. Discontinue at 10 seconds. Allow three attempts.

This item is administered and scored in the same manner as on Item 13, but the child balances on the left foot. Begin by saying:

***"Now try the other foot. Get your balance; close your eyes; lift this foot."*** (Touch child's right leg near foot.)

*Item 15:* With eyes open, stand balancing with both feet on wood dowel. Discontinue at 15 seconds. Allow three attempts.

While demonstrating, say:

***"Now stand on the stick with your toes across it like this."***

Demonstrate by balancing with the first three metatarsal-phalangeal joints of both feet resting crosswise on the dowel and no other part of the foot touching either the dowel or the floor. The child is allowed to find the most comfortable position for balancing his or her toes on the dowel. Allow three attempts to reach 15 seconds.

*Item 16:* With eyes closed, stand balancing with both feet on wood dowel. Discontinue at 15 seconds. Allow three attempts.

This item is administered in the same manner as on Item 15, but the eyes are closed and the item is discontinued when the child has balanced with both feet for 15 seconds. Say:

***"Now do the same thing with your eyes closed."***

#### Recording Responses

Scoring is either by whole seconds (Items 1–8 and 11–16) or by the number of steps taken (Items 9–10). Position and/or movements of the head, trunk, and arms do not enter into scoring. Write the number of seconds or number of steps in the box indicated on the protocol sheet for each attempt made. After the test, fill in the circle corresponding to the best response (but never more than the maximum score listed).

Do not fill in any circle for an item that was not administered. For example, if you started on Item 4 and did not need to administer Items 3, 2, and 1, leave those circles blank. Similarly, if you started on Item 4 but needed to administer Items 3 and 2, do not fill in a circle for Item 1.

#### Discontinuing the Test

Discontinue the test when the best response (highest score) on three consecutive items is 0 or 1 (0–1 second or 0–1 step taken). Note that, on this test, a box for recording the cumulative number of errors is not provided; the discontinuance criterion is based not on *cumulative* low scores but on *consecutive* low scores.

### Key Points to Remember

1. Test on a noncarpeted surface or indoor-outdoor carpeting. Do not use a thickly carpeted floor.
2. The child should not be near walls or furniture.
3. The child should be tested wearing socks (not in shoes), or else barefoot.
4. Wear socks or nylons.
5. Begin most children on Item 4. If score on Item 4 is not maximal, give Items 3, 2, and 1, in that sequence, until maximal score is reached on two items. Then go to Items 5, 6, and so on. Also, if the child didn't receive maximal scores on Items 3, 2,

and 1, but did not yet meet the discontinuation criterion of 0 or 1 on three consecutive items, go on to Items 5, 6, and so on.

6. Demonstrate Items 1, 3, 4, 5, 9, 10, 11, and 15, or any other item when needed by the child, and give appropriate verbal directions.
7. You may help position the child prior to the item.
8. On items requiring closed eyes, ask the child to close eyes before assuming the posture or walking. On these items, a shield can be used if the child is unable to keep his or her eyes closed.
9. The child's arms are free to move in any manner. Position and/or movement of the head, trunk, and arms does not enter into scoring.
10. For timed items, start the stopwatch as soon as the correct position is assumed (e.g., foot leaves ground) and stop timing as soon as the child moves out of position (e.g., foot touches ground) or opens eyes during eyes-closed items.
11. On standing balance items, the child may move at ankles to maintain balance, but the location of soles of feet on floor cannot change.
12. The child is allowed two attempts on Items 1–4, and three attempts on Items 5–16.
13. You may repeat the demonstration and verbal directions for each attempt.
14. Each attempt is followed by a brief pause.
15. Items are discontinued at the number of seconds or number of steps indicated on the protocol sheet. No further attempts are necessary once the child fulfills the criterion.
16. On Items 1-8 and 11–16, record time to the nearest whole second.
17. On Items 9–10, record the number of the last correct step taken. Toe must be 2½ cm (1 inch) or closer to heel.
18. The child's best attempt is marked as the final score.
19. A score of 0 is possible on all items.
20. Stop test when three consecutive items are scored as 0 or 1.
21. Do not fill in circles of nonadministered items.

## 4. Design Copying (DC)

**Materials**

- test booklet
- 2 sharp pencils (preferably one without an eraser)
- clear plastic scoring guide (includes centimeter ruler)
- protocol booklet

Sit next to the child at the testing table. Because you will be replicating the designs drawn by the child, sitting on the side of the child's nondominant hand may assist observation and scoring. Place the test booklet in front of the child and the protocol booklet in front of you. Say:

***"Here are some pictures (lines, designs) for you to copy. The pictures are up here.*** (Gesture using three fingers and sweep across the designs on the first page.) ***You draw them down here."*** (Gesture to the response areas in the same general way.)

**Trial Item**

Point to the trial item (see Figure 10), saying:

***"Here is a line. It begins on this dot*** (point to top left dot and, using finger, draw line between dots from left to right) ***and stops on this dot.*** (Point.) ***You draw it down here.*** (Point to response area.) ***Begin on this dot*** (point to top left dot) ***and draw the line to this dot."*** (Draw line with finger from top left dot to top right dot.)

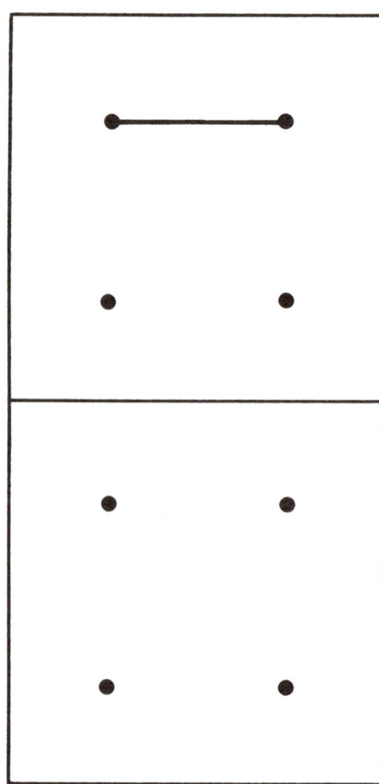

**Figure 10**
**Design Copying: Trial Item**

If the child starts to draw the line in any but the correct place, intervene (while pointing appropriately and, if necessary, guiding the child's hand). Say:

***"Begin on this dot. Draw the line to this dot."***

If the child does not begin exactly on a dot and stop exactly on a dot, guide the child's hand to redraw the line, beginning and stopping deliberately on dots. This act is particularly important in testing unsophisticated children. After the child draws the line correctly, say, pointing appropriately:

***"That's right. This line*** (point to stimulus figure) ***is between this dot*** (point to top left dot) ***and this dot*** (point to top right dot). ***There is no line here or here.*** (Point to space between lower dots in stimulus and response areas, running finger between lower dots from left to right.) ***Your picture looks like this one."***

If there is any doubt about the child's having grasped the idea, erase the child's line and ask him or her to draw one again. Add any explanation needed to help the child get the idea that he or she is copying a line to make the lower section look like the upper section. If the child already knows how to do the task, directions may be shortened accordingly.

**Part I Test Items**

Then gesture to the rest of the designs with a sweeping motion and say:

***"Now draw these pictures (lines). Draw them down here. Begin on a dot and stop on a dot. Draw carefully. We can't erase. Start with this one.*** (Using index finger, trace the stimulus line of Item 1, starting on the top dot and stopping on the bottom dot; see Figure 11.) ***Draw a line down here."*** (Move the same finger from the top dot to the bottom dot where the response line should be made.)

After the child finishes Item 1, you may say or may omit:

***"I am going to watch you and I will draw the same thing you draw. You watch your pictures."***

No additional help is given. If the child stops, point to the next item and say:

***"Now draw this one. Draw it here."*** (Gesture to response area.)

If the child is careless, remind him or her to be careful

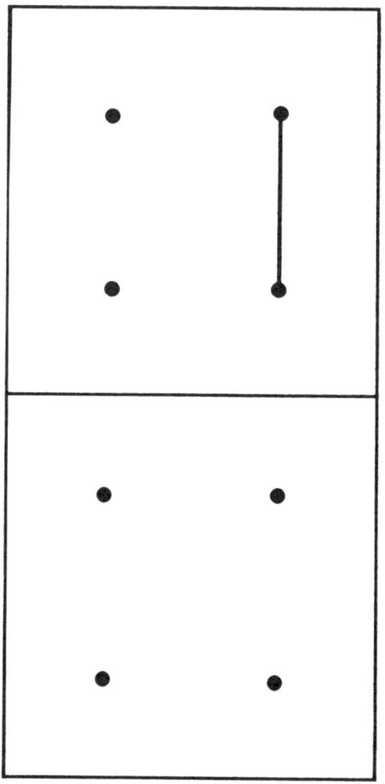

**Figure 11**
**Design Copying: Item 1**

and that erasures are not permitted. The child is not allowed to rotate the test booklet more than 45 degrees. Slight rotation of the paper is normal and accepted. When proceeding from Item 5 to Item 6, open the test booklet, then fold the second page under so that the child has before him or her only the page which is to be drawn on.

Watch the child's response to each item. On the last page of the protocol booklet, record the hand used by the child. Draw the child's approach to the drawing of Items 2 through 9, indicating the direction in which the lines of the design are drawn and the order in which the lines are drawn (see "Recording Approach to Drawing," p. 30). If the child draws too rapidly for you to record accurately, stop the child between items while you catch up. Approach is not recorded on Items 1, 10, 11, 12, and 13 of Part I. Do not record approach for 4-year-olds on Part I.

When the child makes a mistake on an item on which approach is scored, encourage him or her to complete the item, since it may be scorable on other parameters, such as right to left approach. If the child becomes discouraged, tell the child that the test is for older boys and girls, too, and that he or she is not expected to get them all just right.

Part I is discontinued when two consecutive items are scored as 0 for accuracy. Thus, when the child finishes each drawing, check it for accuracy (see "Scoring Accuracy," p. 31). If you are unsure of the accuracy of a drawing, continue to administer items until two consecutive items are clearly incorrect. When the test is completed, you will have more time to score accuracy and find the proper discontinuance point; the computer program will ignore any scores marked past this point. Even if Part I is discontinued before all items are administered, go on to Part II.

**Part II Test Items**

Open the child's test booklet and the protocol booklet to Part II. Without indicating where to start, say:

***"Now draw these designs (pictures). Draw them in the empty spaces."***

The child may choose to start with any item, but if he or she starts to duplicate a design at any place other than in the space directly under the design, he or she is told where to draw the figure. Note the first item of Part II attempted by the child in the box labeled "First Item Drawn" on page 3 of the protocol booklet for later scoring on the parameter of first item drawn. Once the child starts, encourage him or her to draw items in sequence. No erasing is allowed and no instruction is given as to staying within borders. The child is encouraged to do the best he or she can.

Unless the child is 4 years old, record the child's approach in the protocol booklet in the spaces below the stimulus figure, using the same method as for Part I.

Each item in Part II includes an accuracy score for gross approximation ("Should Have" Parameter 1 on all items; see definition on pp. 31 and 37). Part II is discontinued when each of three consecutive items are scored "No" (N) for gross approximation. Thus, when the child finishes each drawing, check for gross approximation. Other parameters are scored after the test is completed.

**Recording Approach to Drawing**

On Items 2 through 9 of Part I and all items of Part II, record the child's approach to drawing by indicating the direction in which the lines of the design are drawn and the order in which the lines are drawn. Do not record approach of 4-year-olds.

It is important that your lines be clear for scoring later. The direction and sequence in which the lines were drawn should be clearly indicated. Also, the lines should not touch each other, especially where the child picked up the pencil and put it down again. Furthermore, the crossing of one line over another should be clearly shown. A specific method for drawing approach is described below.

The direction is indicated by ending each line with a half arrow (⇀). A half arrow is used to save time and should appear only at the end point of each continuous line, no matter how complex the configuration (¬⌙⇀). When the child has completed the test and before going on to the next test, go back to complete all half arrows (⟶). Also, check all the drawings to make sure that they appropriately illustrate the child's approach.

The order in which the child's lines are drawn is indicated by numbering the lines next to each half arrow:

Each time the child picks up the pencil to begin a new line, you do the same. If the child draws the entire picture using one continuous line, you do the same, ending with the number 1 next to the half arrow. As much as possible, you should not look away from the child's paper while the child is drawing, as it may be easy to miss the beginning of a new line.

Leave slight gaps between each individual line to assist in scoring, even if the child's separate lines connect:

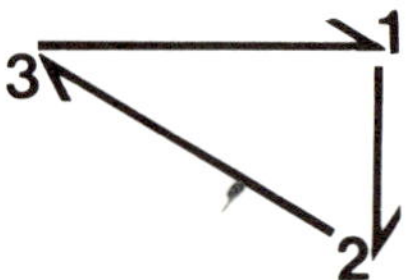


Your lines should not be drawn through the dots but close to them (this also aids in scoring):

If one line crosses another, indicate with a "bump" in the line which does the crossing:

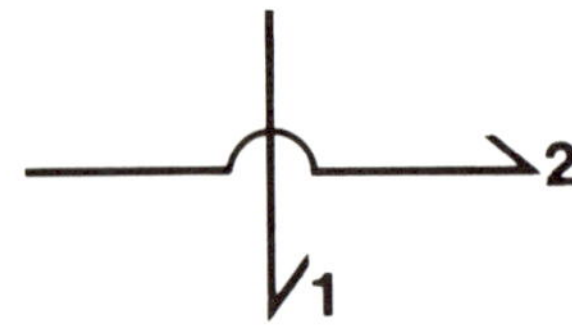


This method is preferred; however, if you find this difficult, the bump can be added after the item is completed:

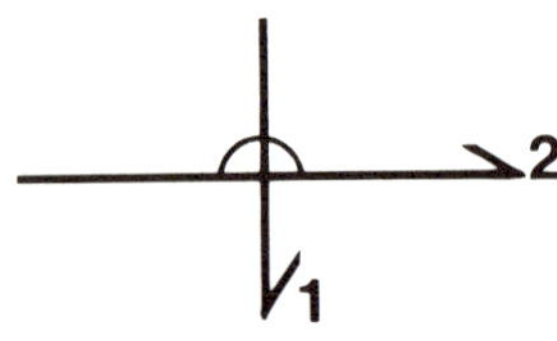

However, do not go on to the next item until this indication of crossing is added. Do not go to the next test until all arrows, numbers, and bumps are completed.

The approach is recorded in a manner that makes it easy for you to interpret later, and is not intended to replicate the child's figure exactly. If the child draws too rapidly for you to follow with your drawing, the child may be asked to wait between items while you catch up. If you miss the child's drawing and are unable to record approach, ask the child to redraw that item on a new test booklet at the end of the test. Use the second attempt for scoring only those parameters on which approach is measured. Careful observation of the child should decrease the need for using this procedure.

**Scoring Accuracy on Part I**

The child's ability to perceive and execute aspects of visuo-spatial composition is measured by the spatial accuracy of his or her drawing. In Part I, accuracy is determined by how close the child's lines are to the dots of the grid. Accuracy scores are 0 for incorrect, 1 for partially correct, and 2 for correct.

To score each Part I item for accuracy, first determine if the child's figure is approximately correct in configuration and no lines connect with incorrect dots. When there is a question as to whether the child's lines come close enough to an imaginary line between dots or to the correct dots to warrant a score of 2, the parallel lines of the scoring guide are used. The parallel lines on the scoring guide are 2, 3, and 4 mm apart. Children ages 4-0 to 4-11 are allowed an accuracy deviation of 4 mm from a line or dot; children ages 5-0 to 5-11 are allowed a deviation of 3 mm; and children age 6-0 or older are allowed a 2 mm deviation from an accurate line or dot.

To measure the amount of acceptable deviation of a 4-year-old's line from an imaginary completely accurate line, use the two lines on the scoring guide that are 4 mm apart. Place one line in the middle of the two dots between which the line should have been drawn. If all parts of the child's analogous line fall within the 4 mm leeway, that line is considered correct and meets the criterion for a score of 2. Similarly, if a line of a 4-year-old child's drawing extends past a dot or fails to reach a dot, measure the distance with the scoring guide. If the distance is less than 4 mm, that aspect of the drawing meets the requirements for a score of 2. Measure from the *edge* of the dot, not from the middle of the dot. To determine the accuracy requirements for a score of 2 by 5-year-olds, use the two horizontal lines on the scoring guide that are 3 mm apart and follow the above directions. Similarly, use the lines that are 2 mm apart when scoring drawings of children ages 6 years and older.

A response figure of generally correct form with lines approaching correct dots more closely than incorrect dots, but not lying within the space leeway described above, is scored as 1.

A response figure with lines that connect incorrect dots or are grossly misplaced is scored as 0.

When a child redraws a line because he or she was not satisfied with the initial line, score the least accurate attempt. Erasing is not allowed, but should it occur, an erased line is still an attempt and is so scored. When the child makes a mistake on an item on which approach is scored, he or she is encouraged to complete the item, for it may be scorable on approach parameters such as drawing right to left.

The illustrations in Figure 12 differentiate scores of 1 and 0. The principle underlying the difference is the degree to which the child appeared to be able to reproduce the design and to relate it to the dot grid. In some cases, the difference between a score of 1 and 0 is determined by the child's line being closer to an imaginary accurate line or correct dot than to an incorrect dot, suggesting that the child reoriented his or her erroneous line sooner for a score of 1 than he or she did for a score of 0. In one of the illustrations of Item 1, the child's line is closer to the incorrect than correct printed dots, but the performance was scored as 1 because the child, who drew in his own dots, indicated the ability to draw a line between specified dots. The child who draws his or her own dots should be reinstructed to use the printed dots in subsequent items. The illustrations typify imperfect attempts at copying designs, but not all kinds of partially correct or incorrect designs are illustrated. In case of doubt about how to score, find an illustration that is similar in quality and score similarly. A design of better quality than illustrations of 0 is scored as 1.

**Scoring Accuracy on Part II**

Spatial accuracy of Part II is scored according to a number of parameters indicating desirable characteristics of space management and graphic praxis (Should Have parameters). The parameters scored on each item are listed in the left column of Table 2. For each Should Have parameter on Part II, score as "Yes" if the parameter is present or "No" if it is not present.

Generally (but only generally), the Should Have parameters listed first are less strict than the ones further down the list. However, credit on one parameter is not dependent upon having received credit on another parameter unless the parameter so specifies.

The first Should Have parameter on all Part II items calls for a *gross approximation* of the stimulus figure. The objective of giving credit for a gross approximation is to discriminate between the child who gives a random response and one who appears to perceive the stimulus

**Table 2**
**Item Scoring on the Design Copying Parameters**

| Does Have and Should Have | Item | Does Have and Should Not Have |
|---|---|---|
| None scored | **Item 1** | None scored |
| None scored | **Item 2** | R. *Reversal:* Vertical line drawn to the left of a horizontal line. (If there is no horizontal line or no vertical line, score as Unscorable.)<br>L. *Right to Left:* A drawing begun on the right side of the figure.<br>I. *Inversion:* A horizontal line drawn to approach lower half of vertical line. (If there is no horizontal line or no vertical line, score as Unscorable.) |
| None scored | **Item 3** | R. *Reversal:* Distal end of an oblique line to the left of the proximal end of the line. An oblique line is greater than 20 degrees from the vertical axis and greater than 20 degrees from the horizontal axis. (To measure, place apex of scoring guide at proximal end of line and outer 10-degree line parallel to printed vertical or horizontal line.) |
| None scored | **Item 4** | R. *Reversal:* A figure reversed right for left where the lines intended to replicate the lower and upper lines of the stimulus figure are reversed in the child's figure. |
| None scored | **Item 5** | I. *Inversion:* A figure with an approximate angle lower than the beginning and end of the line(s).<br>L. *Right to Left:* A drawing begun in the middle with the right side drawn before the left side. |
| None scored | **Item 6** | R. *Reversal:* A figure roughly approximating the stimulus figure in which the vertical line is to the right of the other lines.<br>I. *Inversion:* A figure roughly approximating the stimulus figure in which the oblique line is below the horizontal line. |
| None scored | **Item 7** | R. *Reversal:* A figure roughly approximating the stimulus figure in which a vertical line is on the right.<br>L. *Right to Left:* A figure started on the right side (not middle of the figure.<br>I. *Inversion:* A figure roughly approximating the stimulus figure in which an approximate angle points upward. |

*table continued on next page . .*

**Table 2 (Continued)**
**Item Scoring on the Design Copying Parameters**

| Does Have and Should Have | Does Have and Should Not Have |
|---|---|
| **Item 8** | |
| None scored | R. *Reversal:* A figure roughly approximating the stimulus figure in which the oblique line is on the right side.<br>I. *Inversion:* A figure roughly approximating the stimulus figure in which the opening faces upward rather than downward. |
| **Item 9** | |
| None scored | R. *Reversal:* A rough approximation of the stimulus figure with portion meant to replicate square drawn on right side.<br>I. *Inversion:* A rough approximation of the stimulus figure with portion meant to replicate square drawn above short line(s). |
| **Items 10–13** | |
| None scored | None scored |
| **Item 14** | |
| 1. Gross approximation of figure.<br>2. A line parallel plus or minus 20 degrees to lower boundary.<br>3. Entire line drawn within horizontal middle half of area provided for drawing and not touching side boundaries.<br>4. A horizontal line between 2.5 and 3.5 cm long. | B. *Boundary:* Figure touches boundary.<br>L. *Item Right to Left:* Neither Item 14 nor 18 is first item drawn on this page. |
| **Item 15** | |
| 1. Gross approximation of figure.<br>2. Exactly two straight (within 4 mm) lines joined to form an approximate angle.<br>3. Exactly two straight (within 4 mm) lines same length (within 5 mm) joined to form a sharp angle.<br>4. Exactly two straight (within 2 mm) lines converging within 2 mm and not overlapping by more than 2 mm.<br>5. A vertical line parallel plus or minus 10 degrees to side boundary. | B. *Boundary:* Figure touches boundary.<br>A. *Additions:* Grossly inappropriate extra lines or additions to the figure.<br>R. *Reversal:* Vertical line is to left of horizontal line.<br>L. *Right to Left:* Horizontal line begun (not continued) on right and drawn to left or figure started on right side of the figure.<br>I. *Inversion:* Horizontal line drawn below vertical line. |

***table continued on next page . . .***

**Table 2 (Continued)**
**Item Scoring on the Design Copying Parameters**

| Does Have and Should Have | Does Have and Should Not Have |
|---|---|
| **Item 16**  | |
| 1. Gross approximation of figure. | B. *Boundary:* Figure touches boundary. |
| 2. A straight (within 4 mm) line meant to replicate a shaft with exactly two short lines forming an angle less than 180 degrees (arrowhead) that points away from the long line (shaft), and<br><br>the ends of the two short lines meet (within 2 mm) and both are within 2 mm of touching the long line (shaft). | A. *Additions:* Grossly inappropriate extra lines or additions to the figure.<br><br>R. *Reversal:* Short lines are at right end of long line or short lines form reverse angle at left end of long line.<br><br>L. *Right to Left:* Long horizontal line (shaft) drawn right to left. |
| 3. A line longer than arrowhead lines and representing arrow shaft is straight within 2 mm. | D. *Distortion:* Apex of "angle" is formed by a single short line that is rounded or flat or apex is more than 4 mm from end of long line; or jogs at arrow point. |
| **Item 17** | |
| 1. Gross approximation of figure. | B. *Boundary:* Figure touches boundary. |
| 2. Exactly four straight (within 4 mm) lines comprising what is meant to be the cross portion of the figure measuring within 5 mm of the same length, and<br><br>two of those lines protruding from an enclosed area, such as a quadrangle. | A. *Additions:* Grossly inappropriate extra lines or additions to the figure.<br><br>S. *Segmentation:* One or more short lines formed by means other than by intersection of two long unbroken lines or square redrawn as separate intact figure. |
| 3. All vertical lines that should be present are present and parallel to side boundary plus or minus 10 degrees. | R. *Reversal:* Portion of figure meant to replicate square drawn to left of short line(s).<br><br>I. *Inversion:* Portion of figure meant to replicate square drawn in lower part of figure.<br><br>J. *Jogs or Ears:* Jogs in lines or at corners. |
| **Item 18** | |
| 1. Gross approximation of figure. | B. *Boundary:* Figure touches boundary. |
| 2. Exactly four appropriate, straight (within 4 mm) lines forming three appropriate approximate corners, and<br><br>two of those lines of the same length (within 2 mm) forming an upward pointing middle angle. | A. *Additions:* Grossly inappropriate extra lines or additions to the figure.<br><br>I. *Inversion:* Portion of drawing meant to represent the right angle points downward. (Entire figure need not be inverted.) |
| 3. Exactly two straight (within 2 mm) oblique lines forming an upward pointing angle measuring 90 degrees plus or minus 20 degrees. | L. *Right to Left:* Drawing begun on the right side (not middle) of the figure. |
| 4. Exactly four lines meeting appropriately within 2 mm and not overlapping by more than 2 mm. | J. *Jogs or Ears:* Jogs or ears at corners. |

*table continued on next page . . .*

**Table 2 (Continued)**
**Item Scoring on the Design Copying Parameters**

| Does Have and Should Have | Does Have and Should Not Have |
|---|---|
| **Item 19** | |
| 1. Gross approximation of figure.<br>2. A quadrangle formed by four long lines meant to represent the sides of the quadrangle and four approximate corners and divided into four smaller quadrangles by two or more lines, and upper and lower small quadrangles the same height (within 2 mm) at vertical intersecting line. (To measure height, draw an imaginary line through intersection and perpendicular to lower boundary.)<br>3. A quadrangle formed by four approximate corners and four sides the same length (within 5 mm) and divided into four smaller quadrangles by two or more lines. | B. *Boundary:* Figure touches boundary.<br>A. *Additions:* Grossly inappropriate extra lines or additions to the figure.<br>S. *Segmentation:* Either: (a) two lines within a large square which are meant to intersect do not; (b) pencil is lifted at point of intersection of these lines; (c) two or more small quadrangles are drawn or redrawn as separate units; (d) the four sides of the large quadrangle are not drawn in succession; or (e) there is avoidance of drawing intersecting lines.<br>R. *Reversal:* Both rectangles on the right are larger than those on the left at horizontal intersecting line.<br>L. *Right to Left:* A horizontal line within any quadrangle started at right and drawn to the left or large quadrangle started on the right half of the figure. |
| **Item 20** | |
| 1. Gross approximation of figure.<br>2. Lines of a figure intended to replicate a triangle meeting at corners by 2 mm and not overshooting by more than 2 mm; corners not rounded.<br>3. Both long oblique lines (or combination of lines meant to replicate those lines) which are 45 degrees plus or minus 10 degrees relative to lower border and straight (within 4 mm). (To measure, place apex of scoring guide on one end of child's oblique line and 45-degree line of scoring guide parallel to printed vertical or horizontal line.) | B. *Boundary:* Figure touches boundary.<br>A. *Additions:* Grossly inappropriate extra lines or additions to the figure.<br>S. *Segmentation:* One or more lines meant to be replication of short oblique line(s) formed by a method other than intersection and continuation (without lifting pencil) of two long oblique lines; or a figure meant to represent triangle or an "X" drawn as a separate figure.<br>R. *Reversal:* Figure meant to represent a triangle situated to right of short oblique arms.<br>J. *Jogs or Ears:* Jogs or ears at corners or line intersections. |
| **Item 21** | |
| 1. Gross approximation of figure.<br>*If printed boundaries are used in place of drawn lines for quadrangle, score Parameters 2–4 as No.*<br>2. At least one quadrangle with four straight (within 4 mm) outer lines, and four approximate corners, and either two, three, or four lines drawn between sides of the quadrangle.<br>3. Exactly three horizontal lines, within a single quadrangle, which are within 2 mm of touching sides of figure and do not extend past sides by more than 2 mm.<br>4. Parameter 2 present and area within a quadrangle divided horizontally into four increasing larger areas distal to proximal. | B. *Boundary:* Figure touches boundary.<br>A. *Additions:* Grossly inappropriate extra lines or additions; more than four horizontal lines inside figure.<br>S. *Segmentation:* More than one quadrangle drawn as an intact figure (may or may not share a boundary); or rectangles formed by lines within a square are redrawn (drawn over) as separate figures.<br>L. *Right to Left:* Horizontal line(s) inside square drawn right to left.<br>Q. *Quadrangle Right to Left:* Quadrangle started on the right side of figure. |

***table continued on next page . . .***

**Table 2 (Continued)**
**Item Scoring on the Design Copying Parameters**

| Does Have and Should Have | Does Have and Should Not Have |
|---|---|
| **Item 22**  | |
| 1. Gross approximation of figure.<br>2. Four straight (within 4 mm) lines forming a figure pointing in generally correct direction and exactly three approximate corners.<br>3. Two straight (within 2 mm) oblique lines differentiated from what are meant to be horizontal and vertical lines and measuring the same length within 5 mm. | B. *Boundary:* Figure touches boundary.<br>A. *Additions:* Grossly inappropriate extra lines or additions to the figure.<br>R. *Reversal:* Opening of the right angle is more toward the left than the right or reversal of figure in which entire figure points more toward the right than the left.<br>L. *Right to Left:* Figure started with either the right oblique line or right end of horizontal line.<br>I. *Inversion:* Right angle is at upper instead of lower end of figure or figure opens downward instead of upward.<br>J. *Jogs or Ears:* Jogs or ears at corners. |
| **Item 23** | |
| 1. Gross approximation of figure.<br>2. Four oblique lines or eight shorter oblique lines (not necessarily intersecting) forming two appropriate and approximate angles, and<br><br>an imaginary line drawn between the two upper ends of two of the oblique lines parallel plus or minus 10 degrees to the upper boundary and a similar line between the two lower ends of two oblique lines parallel plus or minus 10 degrees to the lower boundary.<br>3. A quadrangle, no side of which is more than 1.5 times the length of any other side. | B. *Boundary:* Figure touches boundary.<br>A. *Additions:* Grossly inappropriate extra lines or additions to the figure.<br>S. *Segmentation:* Figure is drawn in a segmented or disjointed fashion such that: (a) one or two halves of the figure are drawn independently; (b) a number of short lines are used rather than four long intersecting lines; (c) two separate V-shaped figures do not intersect; or (d) figure meant to replicate a diamond is drawn or overdrawn as a separate unit from rest of lines. |
| **Item 24**  | |
| 1. Gross approximation of figure.<br>2. Four straight (within 4 mm) long lines the same length (within 5 mm), each clearly intersecting another line in a way that produces a quadrangle with four approximate corners and four lines.<br>3. Two recognizable arrowheads, one at each end of what are meant to be two different horizontal lines, each pointing in correct direction. (Lines need not completely touch.)<br>4. Parameter 3 present and lines of each arrowhead are straight (within 2 mm), touch each other, and shaft of arrow touches at least one line of each arrowhead within 1 mm of apex of arrowhead.<br>5. Four straight (within 4 mm) long lines each clearly intersecting another line in a way that produces a quadrangle with four lines and four sharp corners. | B. *Boundary:* Figure touches boundary.<br>A. *Additions:* Grossly inappropriate extra lines or additions to the figure.<br>S. *Segmentation:* Either: (a) two separate figures drawn, (b) quadrangle portion drawn as an intact unit, (c) part of vertical line of one large angle and part of horizontal line of the other large angle drawn in succession, disrupting the continuity of one or both angles, or (d) one or more short lines meant to replicate or be part of one of the long lines drawn as an independent line (i.e., not continued as part of the rest of the figure without lifting pencil).<br>R. *Reversal:* One or more arrowheads reversed.<br>D. *Distortion:* Arrowhead distorted. Short lines forming arrowhead not straight (within 2 mm); or arrowhead points rounded, flat, rotated, or there is a distance (gap) of 2 mm or more between shaft of arrow and either one of short lines of arrowhead. If arrowhead is reversed, score as Reversal, not Distortion. |

*table continued on next page . .*

**Table 2 (Continued)**
**Item Scoring on the Design Copying Parameters**

| Does Have and Should Have | Does Have and Should Not Have |
|---|---|
| **Item 25**  | |
| 1. Gross approximation of figure. | B. *Boundary:* Figure touches boundary. |
| 2. Two touching quadrangles, with approximate corners, each quadrangle drawn obliquely in opposite directions. (Oblique is defined as at least 1° off perpendicular.) | A. *Additions:* Grossly inappropriate extra lines or additions to the figure. |
| 3. Parameter 2 present and both quadrangles have four sharp corners. | S. *Segmentation:* Two figures of any shape drawn independently of the other, or connected by adding line(s), or by extending one of the quadrangles. |
| 4. Parameter 2 present and drawn as though it is one undivided figure with two parts sharing at least one common side at least 5 mm in length. (A figure drawn as a single figure divided by a line does not qualify.) | R. *Reversal:* The shorter of two figures (or parts of one figure) is to the right of the other part. |
| 5. Parameters 2 and 4 present and the lower short edge of one quadrangle appears to be the lower short edge of other, giving an illusion of a piece of folded paper. (The ends of the lines forming the right sides of each quadrangle must meet within 2 mm.) | |

figure and can and does attempt a replication (see examples in Figure 13). To determine whether a response figure is a gross approximation, ask yourself the following question: Having laid all the stimulus and response figures of the child on that page before you, can you correctly match the child's response figure to the stimulus figure? Such matching requires that the child make different types of response figures for most of the stimulus items so that they can be differentiated from each other. You must guard against accepting an accidental replication as a consciously intended one. Often you can tell while the child is drawing the figure whether the child is making a considered attempt at replication, although a deliberate attempt alone is not a sufficient basis for giving credit for gross approximation. One indication of an adequate gross approximation is the ability to score with confidence another Should Have parameter (or another Should Not Have parameter; see the following section, "Scoring Approach on Parts I and II").

If the response figure is not a gross approximation of the stimulus figure, all Should Have parameters are scored "No," because the figure should have the characteristics but does not have them.

Most other parameters are scored on the basis of measurement, logic, and the definitions provided below.

**Measurement of lines.** Scoring of many parameters is based on the measurement of lines and angles for length, straightness, and degrees of verticality and horizontality. The scoring guide is used for measurement purposes. It contains a centimeter ruler for measuring the length of lines. Also on the guide are a 90-degree angle with lines 10 degrees on either side of the lines forming the 90-degree angle and one line at 45 degrees from those lines. In addition, the guide has parallel lines 2 mm, 3 mm, and 4 mm apart for the purpose of determining straightness of lines.

To measure the length of a line, use the centimeter ruler on the scoring guide to measure from tip to tip in a straight line, regardless of the curvature of the line. If a parameter calls for measuring the side of a figure, measure from the point the line forming the side intersects one side to the corresponding point where it intersects the other side. If the line being measured extends beyond the intended intersection, do not measure the line beyond that intersection. Similarly, if the line forming the side of a figure fails to meet the intended intersection point, include an imaginary extension of that line to its intended destination. When measuring the *side* of a figure, it is the distance between those two points rather than the length of the line between those two points that counts. This method of measuring is particularly applicable to judging squares where sameness of length of sides is important. In Figure 14, the length of the line forming the left border of the square is measured between points A and B. The length of the line forming the upper border of the square is measured between points A and C.

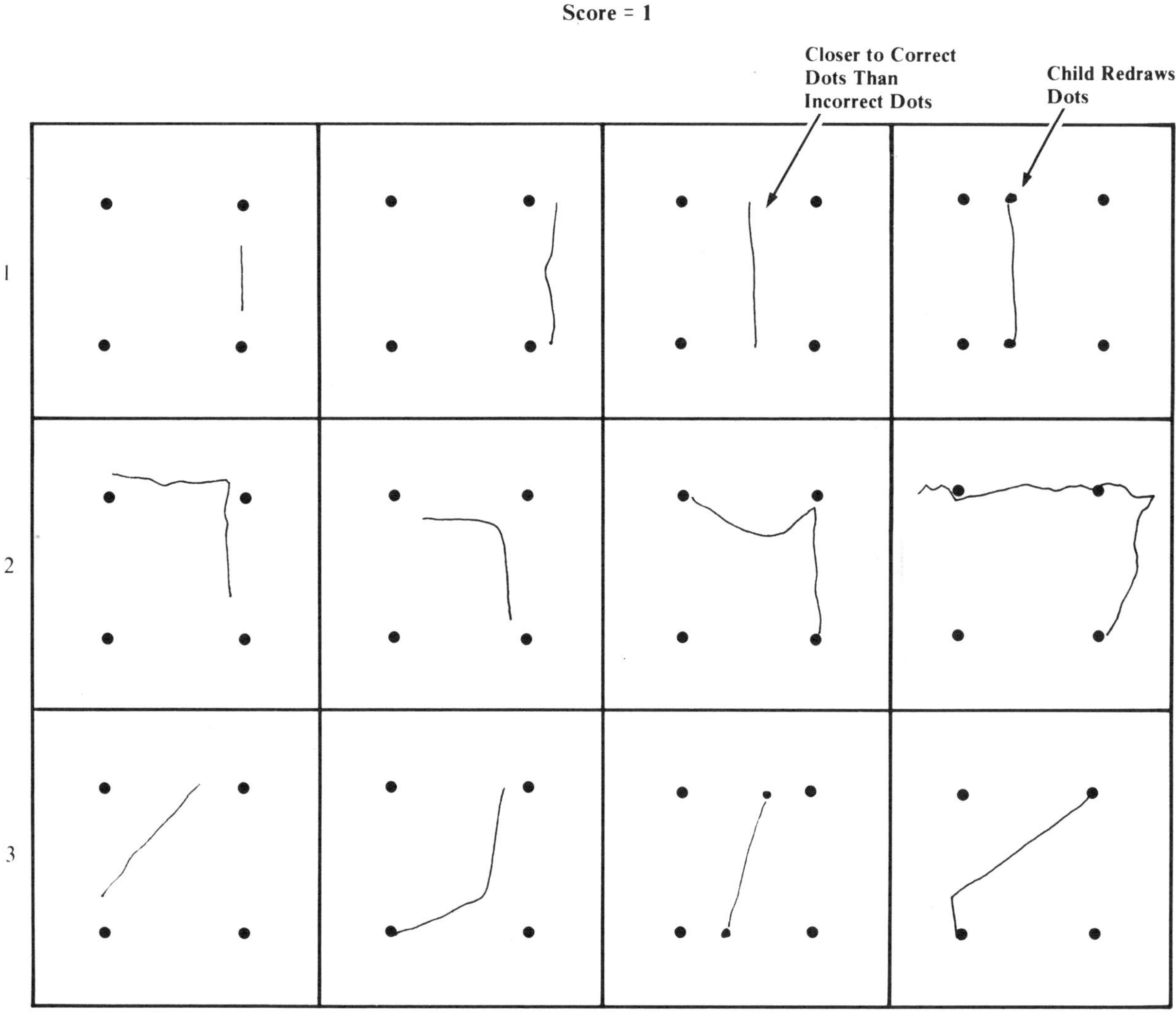


*figure continued on next page . . .*

**Figure 12**
**Design Copying: Differentiation Between Accuracy Scores of 1 and 0 on Part I**

The straightness of a line is determined by placing the scoring guide over the line in the child's drawing, using the parallel lines either 2 mm or 4 mm apart. *All* parts of the response line must fall within the boundaries specified. If any portion of the child's line is visible beyond the guidelines, the child's line fails to meet the criterion of straightness.

The degree of verticality and horizontality of a child's line is determined by the angle formed by the child's line and a line of the scoring guide placed parallel to the printed horizontal or vertical boundary lines of the child's test booklet. Figure 15 represents the scoring guide placement.

To determine whether a line is vertical or horizontal to the printed boundary line, plus or minus 10 degrees, place the apex of the angle on the scoring guide at the end of the child's line or a projection of that line (at the same point as in Figure 15) and one of the arms of the 90-degree angle of the guide parallel to a printed boundary. If the *beginning* and *end* of the child's line fall completely within the 10-degree lines on the guide, the child's line meets the specified criterion even though some portion of his or her line between the beginning and the end of the line falls outside the 10-degree boundary line. In Figure 15, the child's horizontal line qualifies as horizontal, but the vertical line does not qualify as vertical.

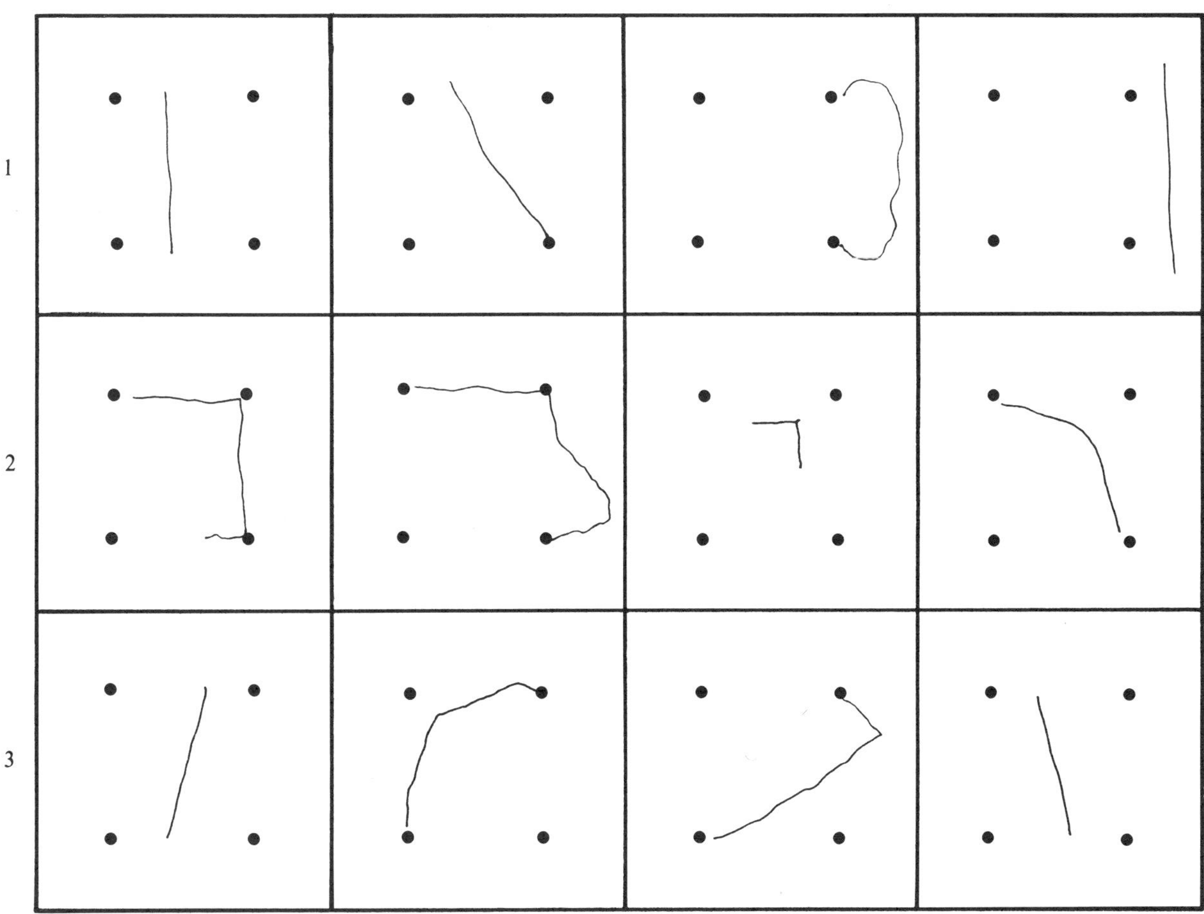


*figure continued on next page . . .*

**Figure 12 (Continued)**
**Design Copying: Differentiation Between Accuracy Scores of 1 and 0 on Part I**

To determine whether an angle or corner of a figure is 90 degrees plus or minus 10 degrees, measure from the apex of the angle or corner or projection of lines to form an imaginary apex or sharp corner. If the beginning and end of both lines forming the child's corner or angle fall within the guide's 90 degrees plus or minus 10 degrees, the child's angle or corner meets the criterion. The drawing in Figure 15 does not qualify as a 90-degree angle plus or minus 10 degrees.

Note that on the scoring guide there are actually three 90-degree angles. One is formed by the center, bolder vertical and horizontal lines. Another is formed by the outer vertical 10-degree line and the inner horizontal 10-degree line. The last is formed by the inner vertical 10-degree line and the outer horizontal 10-degree line. The center 90-degree angle is used to determine whether a line is vertical or horizontal plus or minus 10 degrees, or whether an angle is 90 degrees plus or minus 10 degrees. The other two 90-degree angles are used to determine whether a line is vertical or horizontal plus or minus 20 degrees or whether an angle is 90 degrees plus or minus 20 degrees. The same basic methods of measuring are used whether there is a leeway of 10 degrees or 20 degrees.

To measure an angle 45 degrees relative to a printed boundary, place the line that divides the 90-degree angle into two 45-degree angles parallel to one of the printed

**Score = 1**

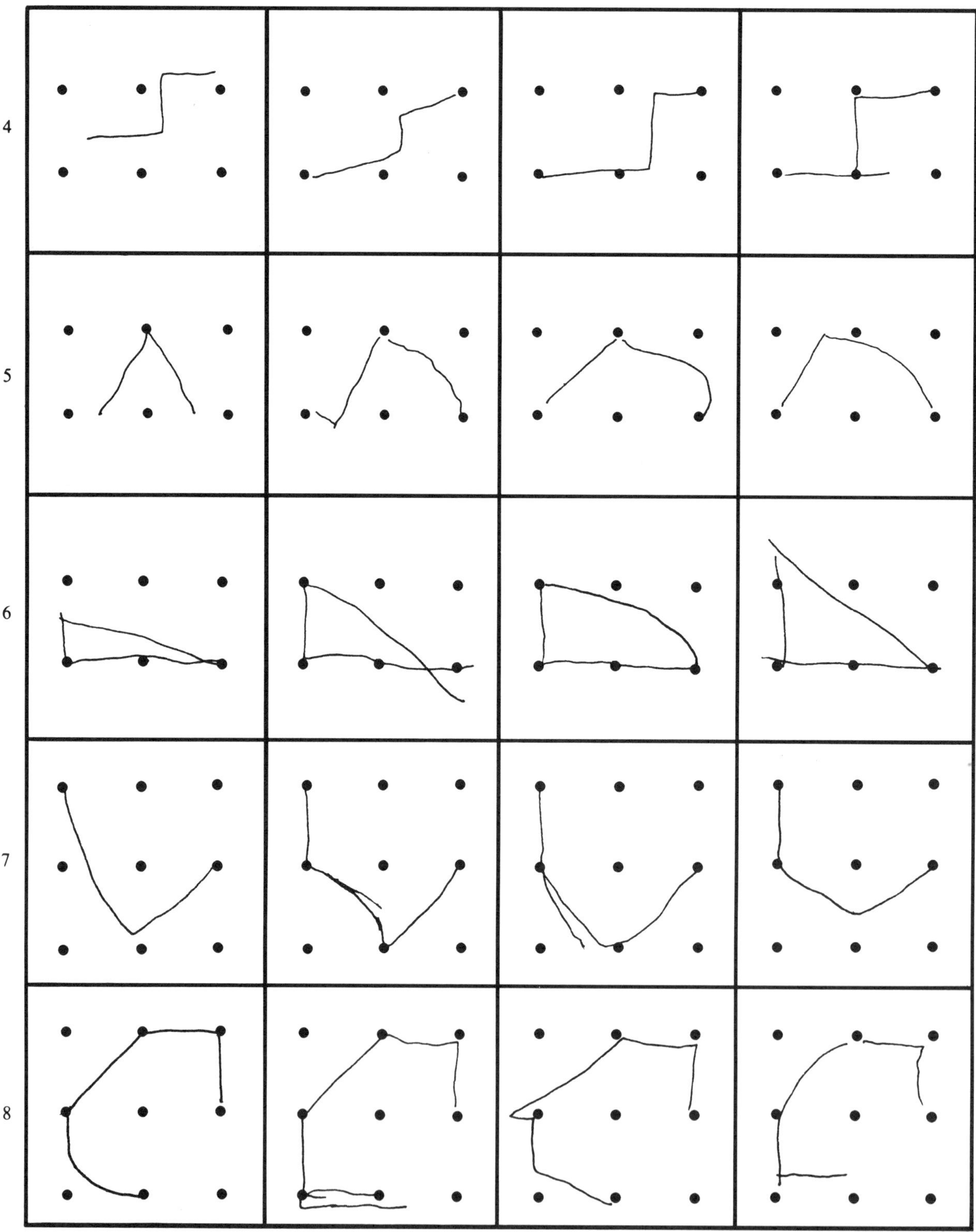


*figure continued on next page . . .*

**Figure 12 (Continued)**
**Design Copying: Differentiation Between Accuracy Scores of 1 and 0 on Part I**

**Score = 0**

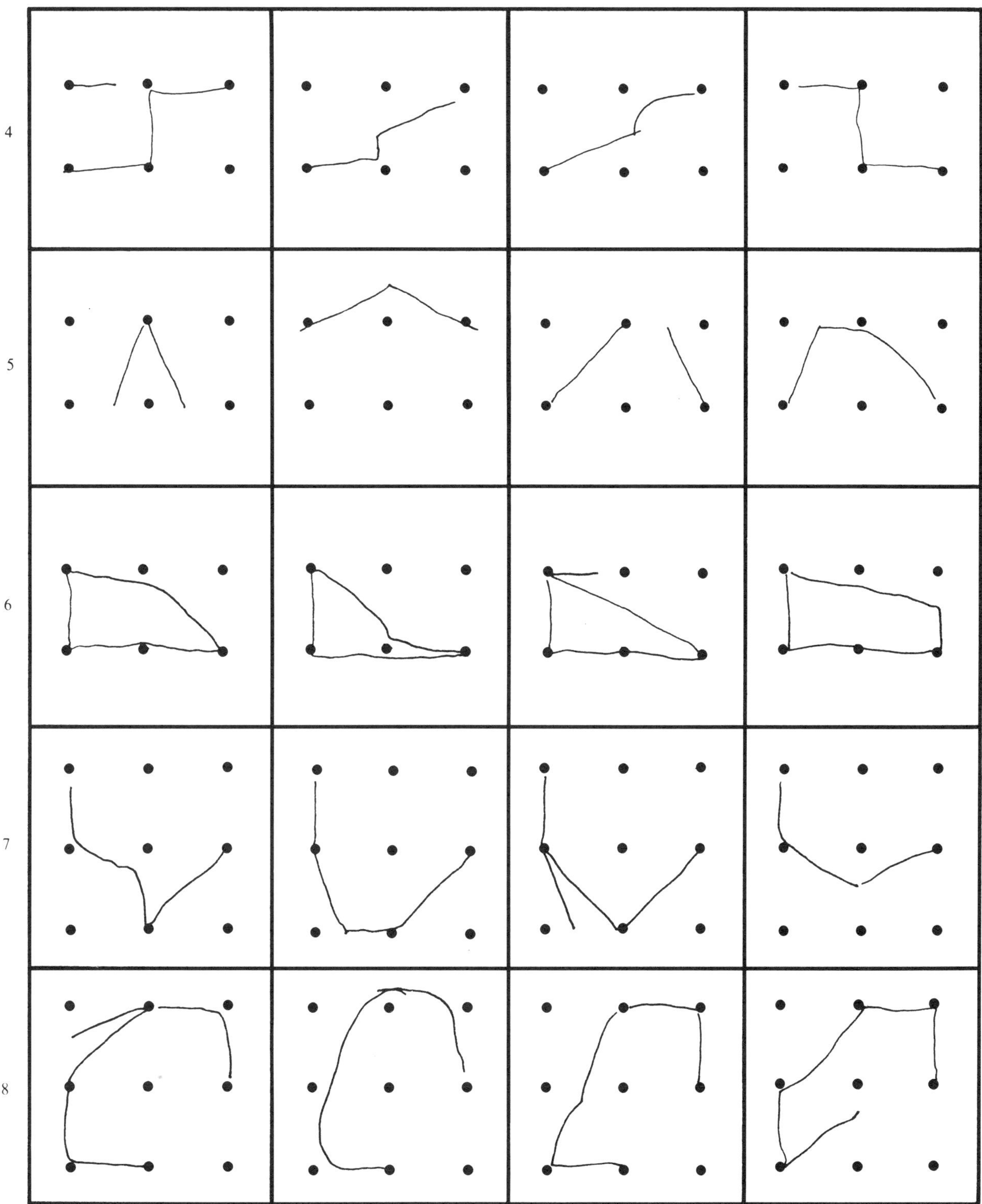


*figure continued on next page . . .*

**Figure 12 (Continued)**
**Design Copying: Differentiation Between Accuracy Scores of 1 and 0 on Part I**

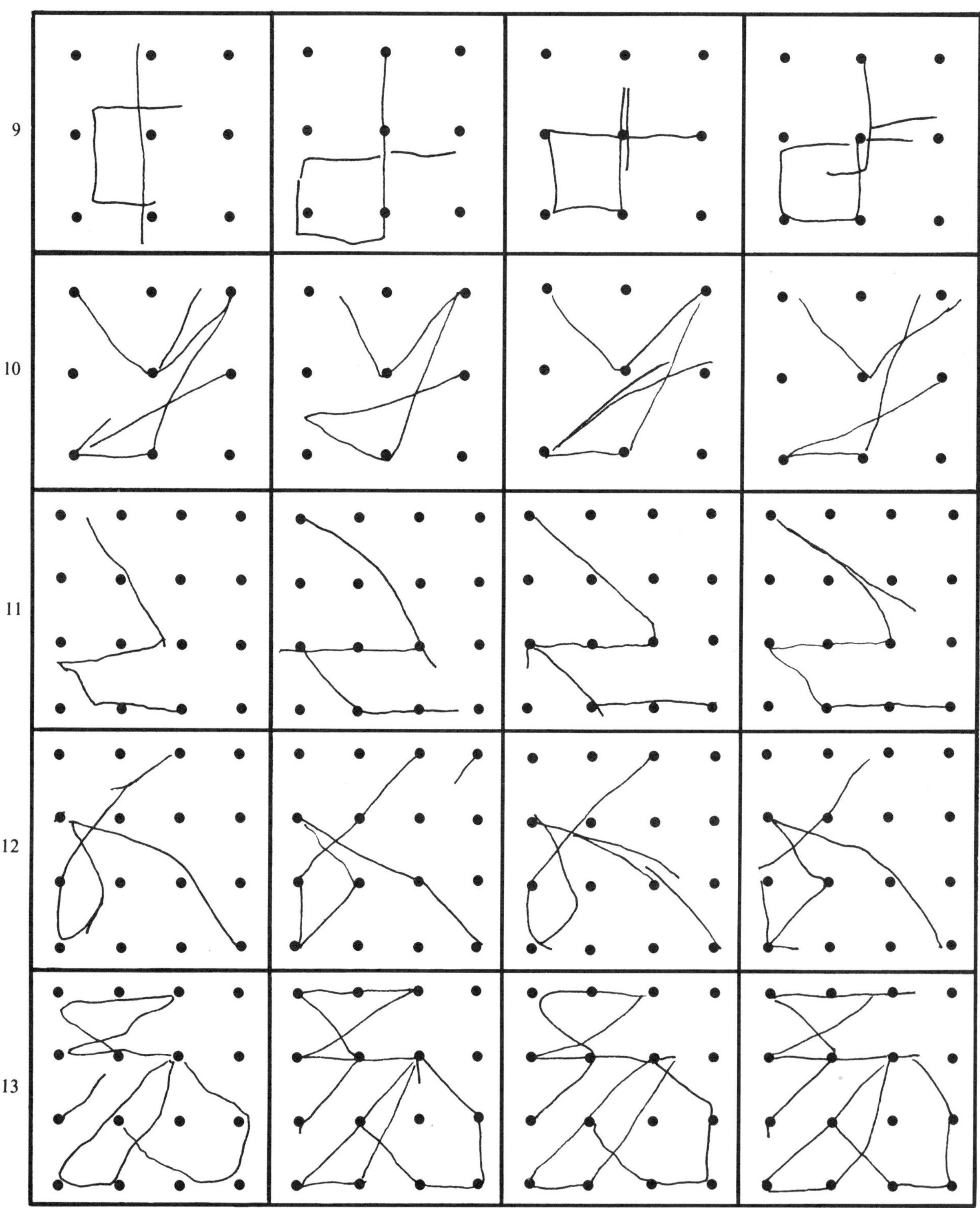


figure continued on next page . . .

**Figure 12 (Continued)**
**Design Copying: Differentiation Between Accuracy Scores of 1 and 0 on Part I**

Score = 0

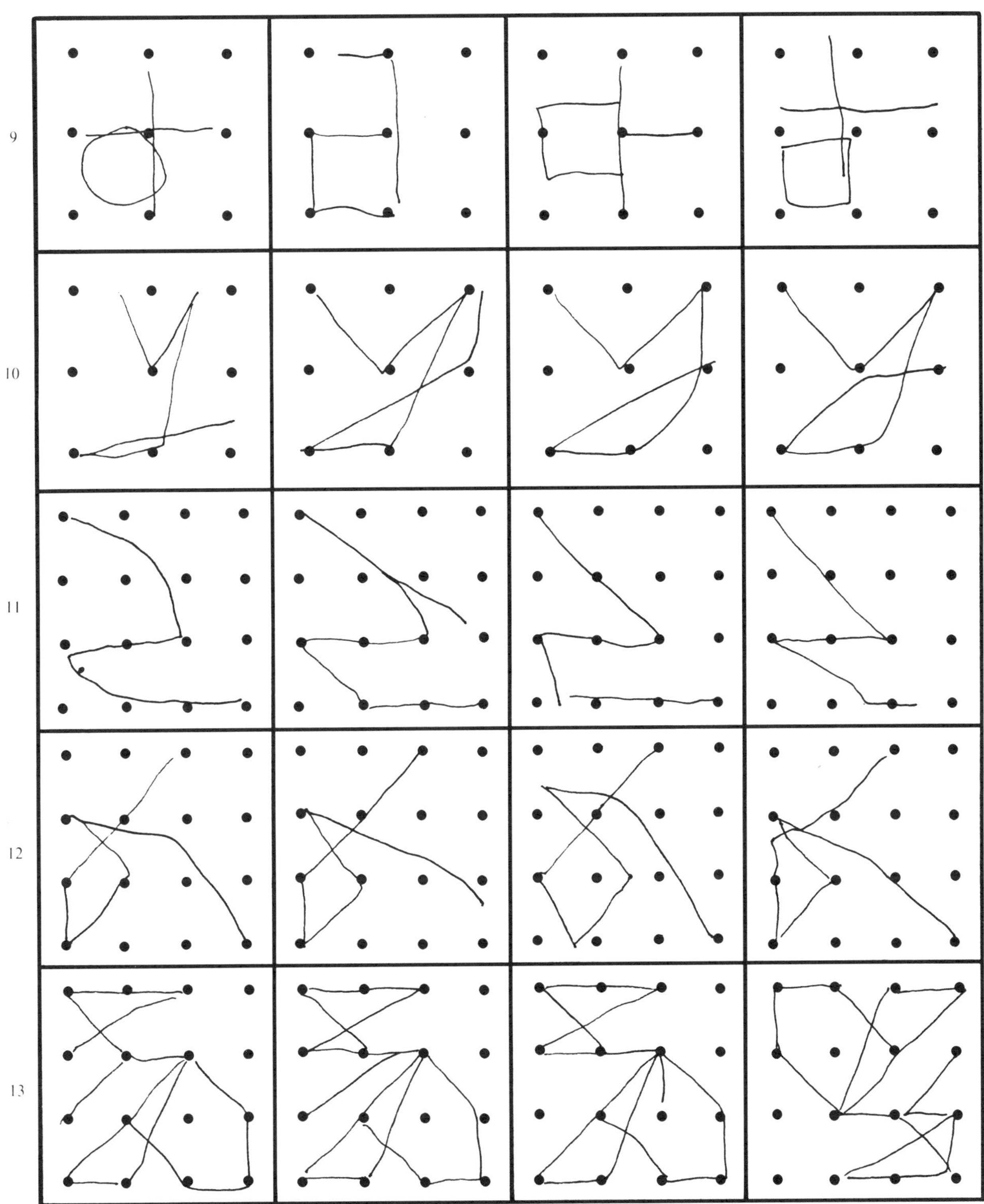


**Figure 12 (Continued)**
**Design Copying: Differentiation Between Accuracy Scores of 1 and 0 on Part I**

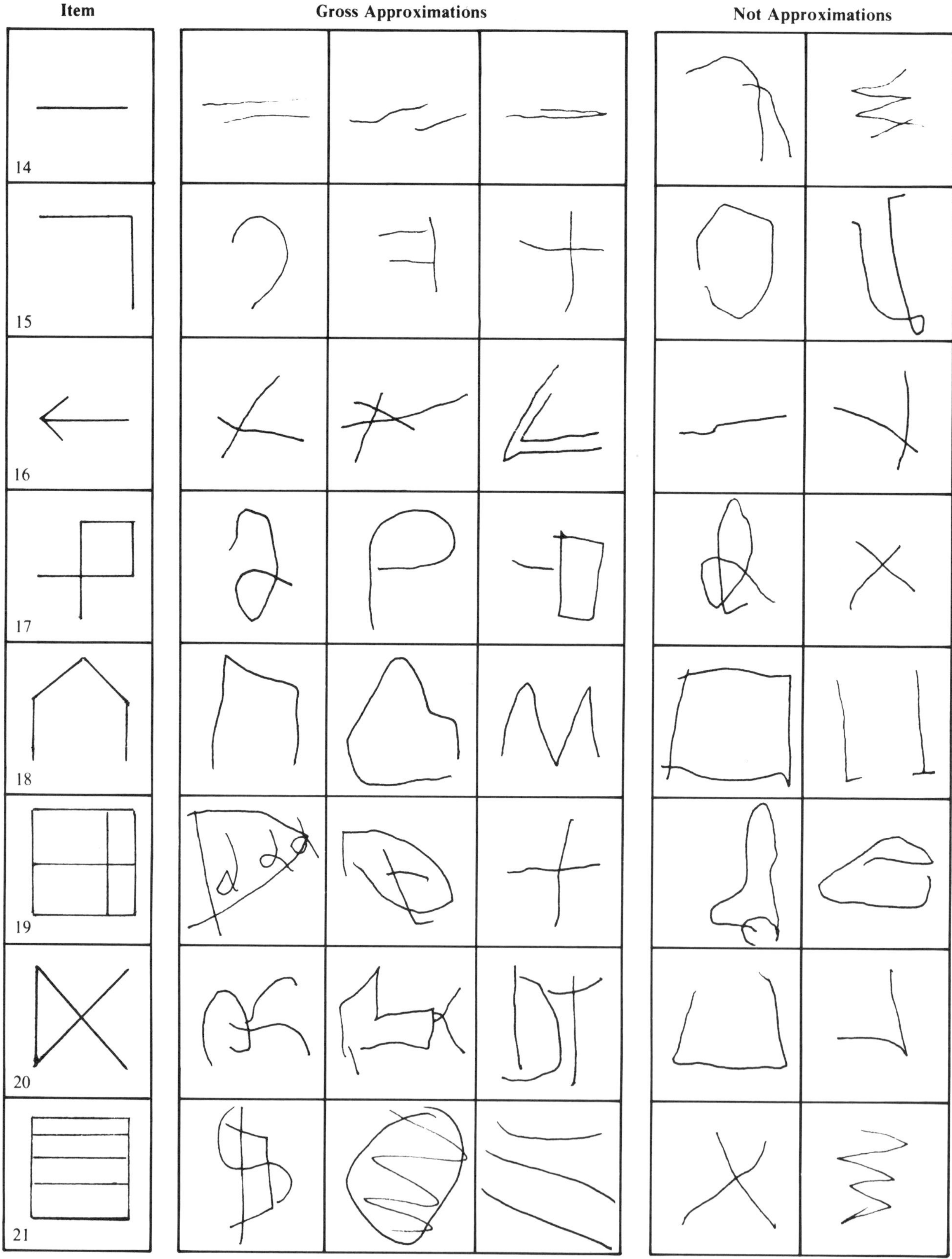


*figure continued on next page . . .*

**Figure 13**
**Design Copying: Examples of Gross Approximations**

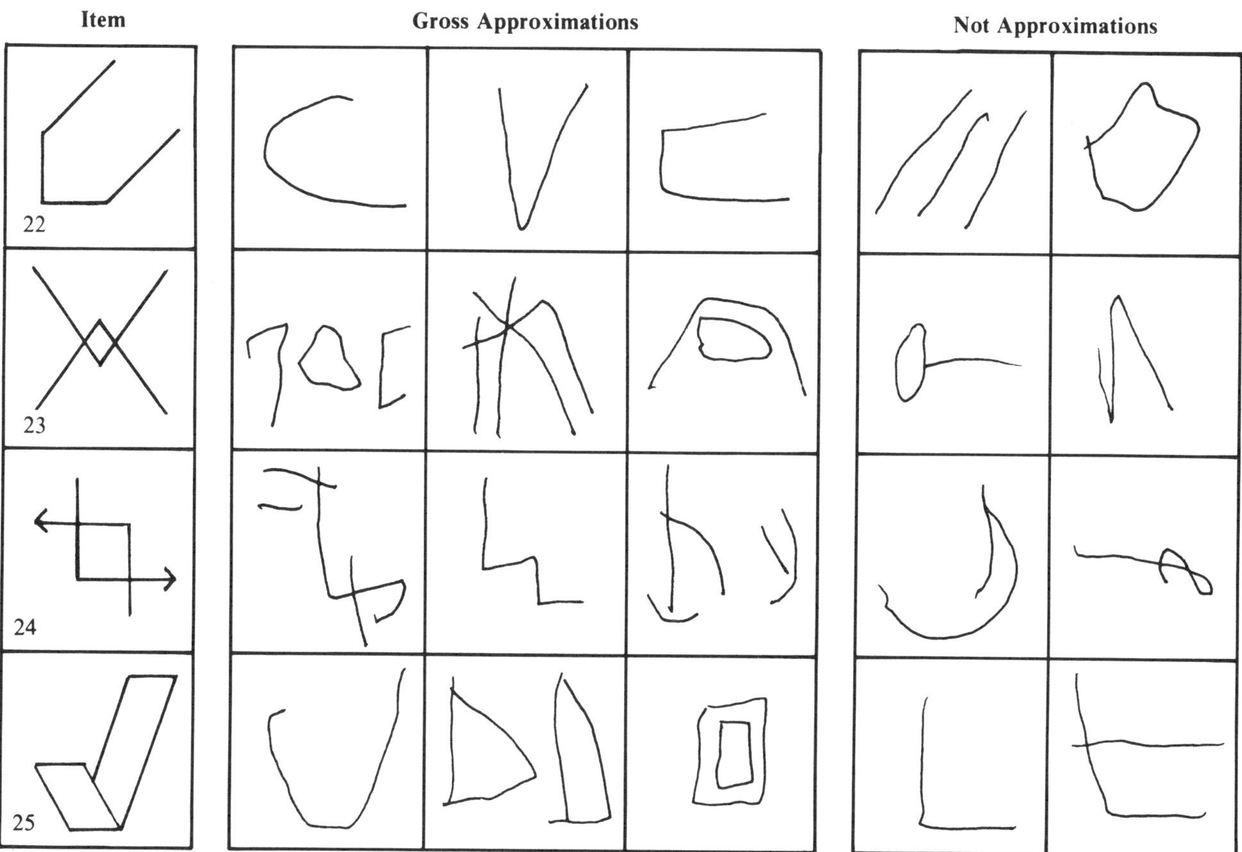


**Figure 13 (Continued)**
**Design Copying: Examples of Gross Approximations**

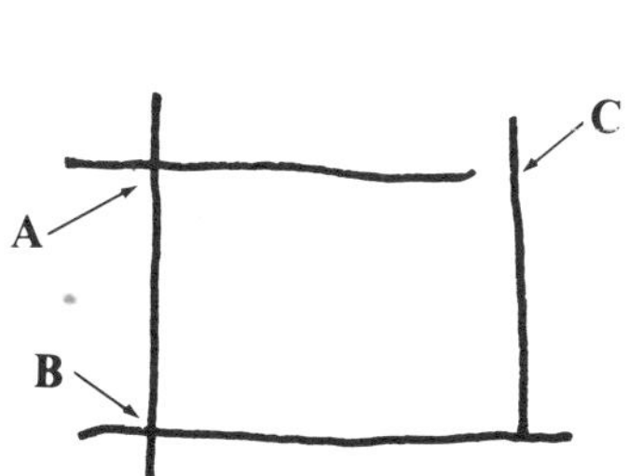


**Figure 14**
**Design Copying: Measuring Line Length**

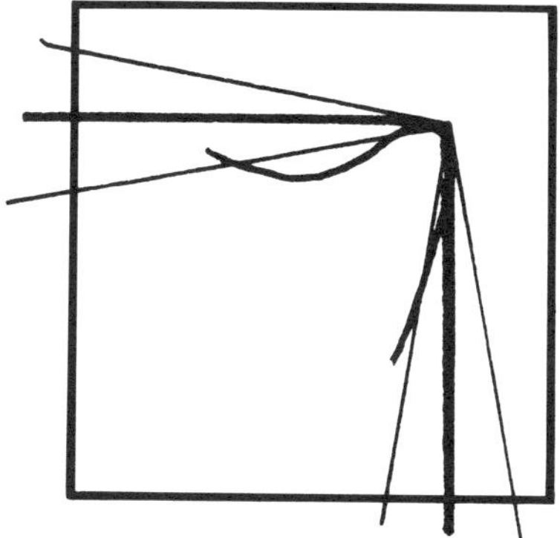

**Figure 15**
**Design Copying:**
**Measuring Horizontality and Verticality of Lines**

boundaries and the apex of the angle at the beginning of the child's line. The child's line should then fall within the area bounded by the two 10-degree lines which lie on either side of the line marking 90 degrees.

**Definition of terms.** Several terms used in the parameters are defined below. (For definitions of the Should Not Have parameters, see p. 48.)

*Angles and corners:* An angle and a corner are the same thing as far as scoring is concerned. To qualify as an angle or a corner, that portion of the figure must be either sharp or approximate, as specified by the parameter. The lines forming a *sharp* angle meet within 0.5 mm and do not overlap more than 0.5 mm. A definite angle is formed. Sharpness is not a function of straightness of the line before reaching the corner. Two curved lines may form a sharp corner. A sharp corner is not necessarily an acute angle. Sharp is precise or exact—not rough or sloppy. The two lines forming an *approximate* angle change direction at the apex of the angle, but the specific point of change in direction is not clearly identifiable or lines do not meet within 0.5 mm or they overlap by more than 0.5 mm. When an approximate corner or angle is specified, one better than an approximate angle is also acceptable. A corner with a small jog or an ear may qualify as an approximate corner but not as a sharp corner. Illustrations in Figure 16 help define qualities of angles. When a scoring parameter designates a required number of angles or corners, the number designated must be exactly that number, not more or less, unless otherwise specified.

*Appropriate:* This refers to specific aspects of the child's drawing and is used so that a figure will not receive credit on a parameter because of a technicality. For example, Should Have Parameter 4 of Item 18 reads: "Four lines meeting appropriately within 2 mm and not overlapping by more than 2 mm." Without the word *appropriate,* a quadrangle could meet the criterion. Professional reason should not be sorely taxed.

*Lines:* The presence and number of lines are determined by an identifiable point in change of direction of a line. When a line changes direction at an identifiable point, another line is drawn. Thus, Item 18 may be drawn without the pencil lifted from the paper, but if there are three identifiable points of change in direction of the line, then, by definition, there are four lines. Should one of those three angles or corners be rounded so much that the point of change in direction could not be specifically identified, the figure has only three lines. If Item 15 is drawn without lifting the pencil and with a corner slightly rounded, but you can identify the point of change in direction of the child's line, for scoring purposes, the figure has two lines.

A line that is *continued* is one that started at the same point where the line drawn immediately before left off. A line that is *started,* as opposed to a line that is continued, is one that begins independently of or not very near where the immediately preceding line stopped. Thus, in drawing a square where the first line is a vertical one on the left, ending at the lower left corner, and the next line begins at the lower left corner and forms a horizontal line drawn to the right, the horizontal line is continued, not started on the left and drawn to the right. Should that horizontal line be started at the lower right corner and drawn to the left to meet the already drawn vertical line, the horizontal line was started, not continued, on the right and drawn to the left. When a scoring parameter designates a required number of lines, the number designated must be exactly that number, not more or less, unless otherwise specified.

## Scoring Approach on Parts I and II

The child's approach to drawing is scored according to a number of parameters indicating nonfavorable, atypical, or immature qualities or graphic dyspraxia (Should Not Have parameters). Should Not Have parameters are scored on Part I, Items 2 through 9, and all Part II items, except for 4-year-olds.

The Should Not Have parameters scored on each item are listed in the right column of Table 2. These parameters, defined below, include the nonfavorable characteristics of reversal, drawing right to left, inversion, boundary, additions, distortion, segmentation, and jogs and ears.

Score each Should Not Have parameter as "Yes" if the parameter is present, "No" if it is not present, or "Unscorable" if you cannot determine whether or not it is present on an attempted item. In recording scores in the protocol booklet, it may be easiest to cut the child's test booklet (not protocol booklet) so that his or her drawings can be placed directly above your drawings of approach in the protocol booklet. Then read each parameter, asking yourself, "Does this drawing have (parameter)?" If the answer is "Yes," fill in the Y circle; if the answer is "No," fill in the N circle; and if you are unable to make a judgment, fill in the U circle.

A hypothetical example of a Should Not Have parameter on which the examiner could not score "Yes" or "No" (score is "Unscorable") is on Item 2 of Part I, where only a horizontal line and no vertical line is drawn. Since a horizontal line cannot be reversed, the examiner is unable to score on the reversal parameter. Similarly, even if the horizontal line were drawn between the lower dots, there is insufficient evidence without a vertical line to assume the figure is an inversion. Thus, the score for the inversion parameter would be "Unscorable."

Furthermore, on Part II the Should Not Have parameters are scored "Yes" or "No" only if the drawing is a gross approximation (Should Have Parameter 1; see "Scoring Accuracy on Part II," p. 31). If there is no gross approximation, score all Should Not Have parameters as

**Sharp Angles or Corners**

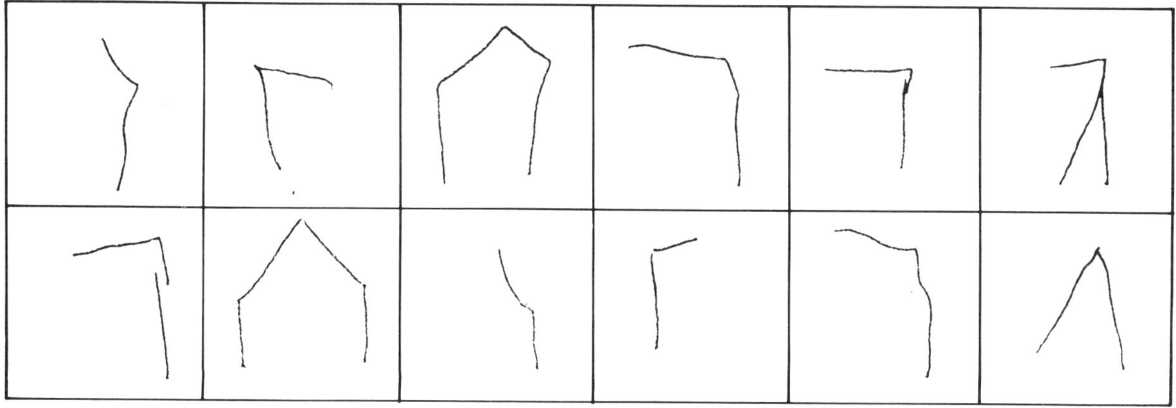

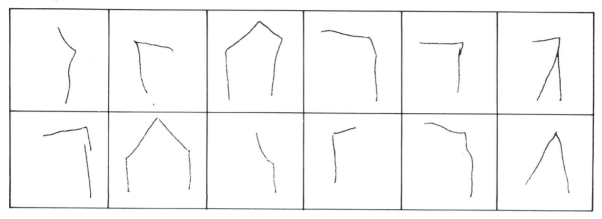

Approximate Angles or Corners

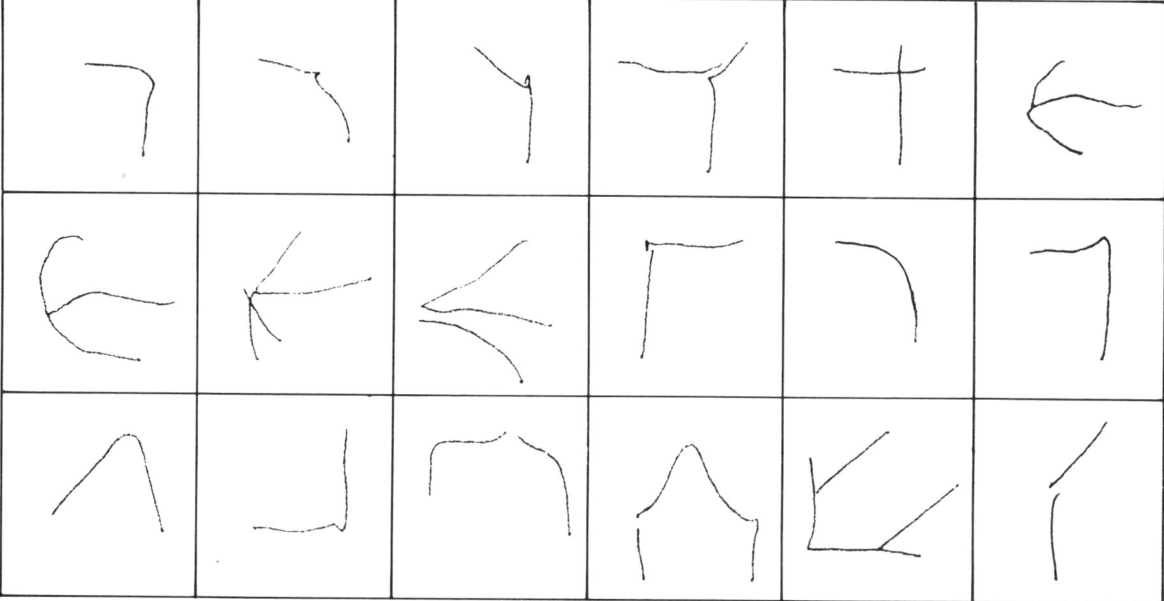

Do Not Qualify for Approximate Angles or Corners

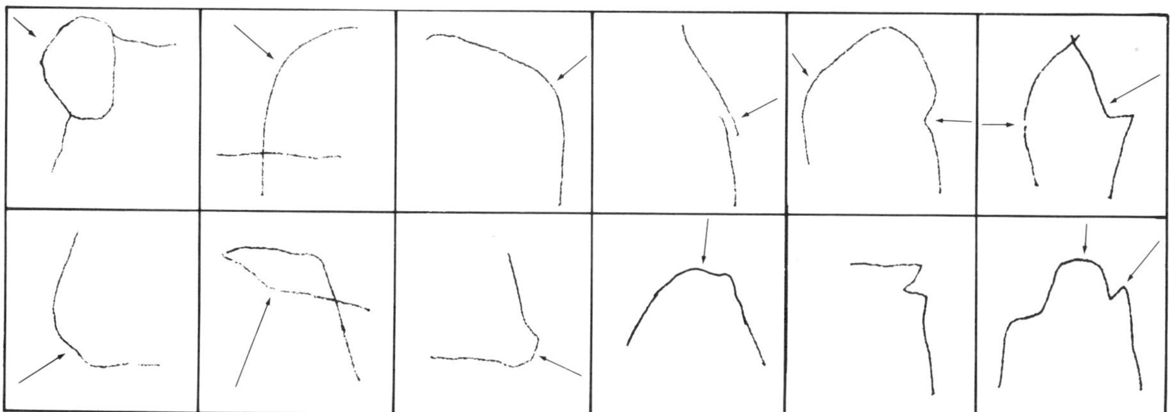

**Figure 16**
**Design Copying: Examples of Angles and Corners**

Unscorable. Should Not Have parameters on Part I are scored as "Yes" or "No" if the examiner can make a judgment about the approach, even though the item's accuracy was scored as 0. If the examiner is unable to make a judgment about the approach on a given Should Not Have parameter, regardless of the accuracy score, the parameter is scored as "Unscorable."

Some situations will arise in which you must make a scoring decision on the basis of your own judgment. In these cases, you should consider the child's intent. For example, suppose the child draws, in a left to right direction, a horizontal line between two vertical lines but leaves a short gap between the horizontal and left vertical lines. The child later adds a line to fill the gap, drawing right to left. The line closing the gap is not an inappropriate addition to the figure nor is it a correction of an error, because it was not making a new line in place of a more erroneous line. It is considered part of the line of which it is an extension. Neither is that short line scored as a line drawn right to left, for it is not a major line of the figure and it is normal to close that small gap in a right to left manner. Scoring must at all times be reasonable.

The Should Not Have parameters are defined as follows:

*Additions (A):* Additions, or extra lines, are grossly inappropriate, non-normal, deliberate, and illogical additions to a figure (see Figure 17). They are definitely inconsistent with figure accuracy. Extra lines or additions may be enclosed areas. Some appear to be a function of perseveration, some are embellishments, and some look like random attempts at some kind of response to the stimulus figure.

Corrections are not scored as additions or extra lines. If a child makes an attempt to correct or improve the drawing by redrawing one or more lines, that attempt is not scored as extra lines or additions, for such an attempt is, in a way, normal. The child is not allowed to erase, and the more accurate of the corrected lines is scored. The less accurate line is disregarded. Lines drawn back on themselves or redrawn are usually corrections, not additions, unless the redrawing is excessive, in which case they are additions. Arrows are sometimes normally drawn without lifting the pencil and by back-tracing over one of the arms of the arrow point; such a drawing back over the line is not an addition. Examples of corrections are shown in Figure 17 to help distinguish corrections from additions.

Lines made accidentally as a function of involuntary motion are not considered extra lines or additions. Such lines often appear as light lines at the end or beginning of a line and are disregarded both in duplicating the child's drawing and in scoring it. Thus, you must make a judgment about whether a child's line is deliberate or involuntary at the time of its execution. Sometimes a child will respond to the frustration of the task by avoiding it and by drawing his or her own picture. Such an event should be avoided, but should it occur and the child's drawing is a meaningful symbol, it is not scored as extra lines or additions. Frustration can also be expressed by forsaking the task and scribbling, usually in repeated circular motions. Those lines are not scored as extra lines or additions. Drawing for fun is discouraged. Should it occur, it is not considered an addition.

A segmented drawing may or may not also have extra lines or additions. For example, a quadrangle divided by two lines (Item 19) may be represented as four small quadrangles, in which case there are no additions, but if six small quadrangles are drawn, the two extra quadrangles are additions. If a section of the drawing is redrawn emphasizing segmentation, as in overdrawing the smaller quadrangle as part of a larger one, the resultant additional lines are not considered extra lines—the drawing is segmented. A jog or an ear is scored as such and not as an extra line, segmentation, or addition. A line which overshoots its destination should not be considered an extra line.

*Boundary (B):* All Should Not Have parameters of Part II begin with "Figure touches boundary." *Boundary* refers to the printed lines which surround each response area in the test booklet. If the child uses the printed boundary as part of the drawing, he or she may receive a "Yes" score for gross approximation if there are other indications of recognizable replication. The printed boundary is not acceptable, however, as a substitute for his or her own line for meeting other Should Have criteria such as horizontality, verticality, or straightness of line. If part of a figure is drawn outside the printed boundary but not touching the boundary, it is still scored as "Yes" on the boundary parameter.

*Jogs (J) or Ears:* A jog or an ear is a definite identifiable quality of irregularity at a corner or in a line, apparently due to confusion on the part of the child (see Figure 18). It is not due to involuntary motion or over- or undershooting. When jogs do occur, they are most apt to be where the child is approaching a corner or attempting to make a corner with one continuous line. They seldom occur where a child is attempting to join two lines to make an angle when approaching the apex from two separate directions. Careful observation and indication when recording the child's approach help to determine whether an irregularity is a jog. Also, help in such differentiation can be obtained by inspecting the rest of the drawing for indications of involuntary motion. A corner with a jog or an ear does not qualify as a sharp corner nor is it scored as extra lines, additions, or corrections. If the jog at a corner is small and does not distort the corner appreciably, the

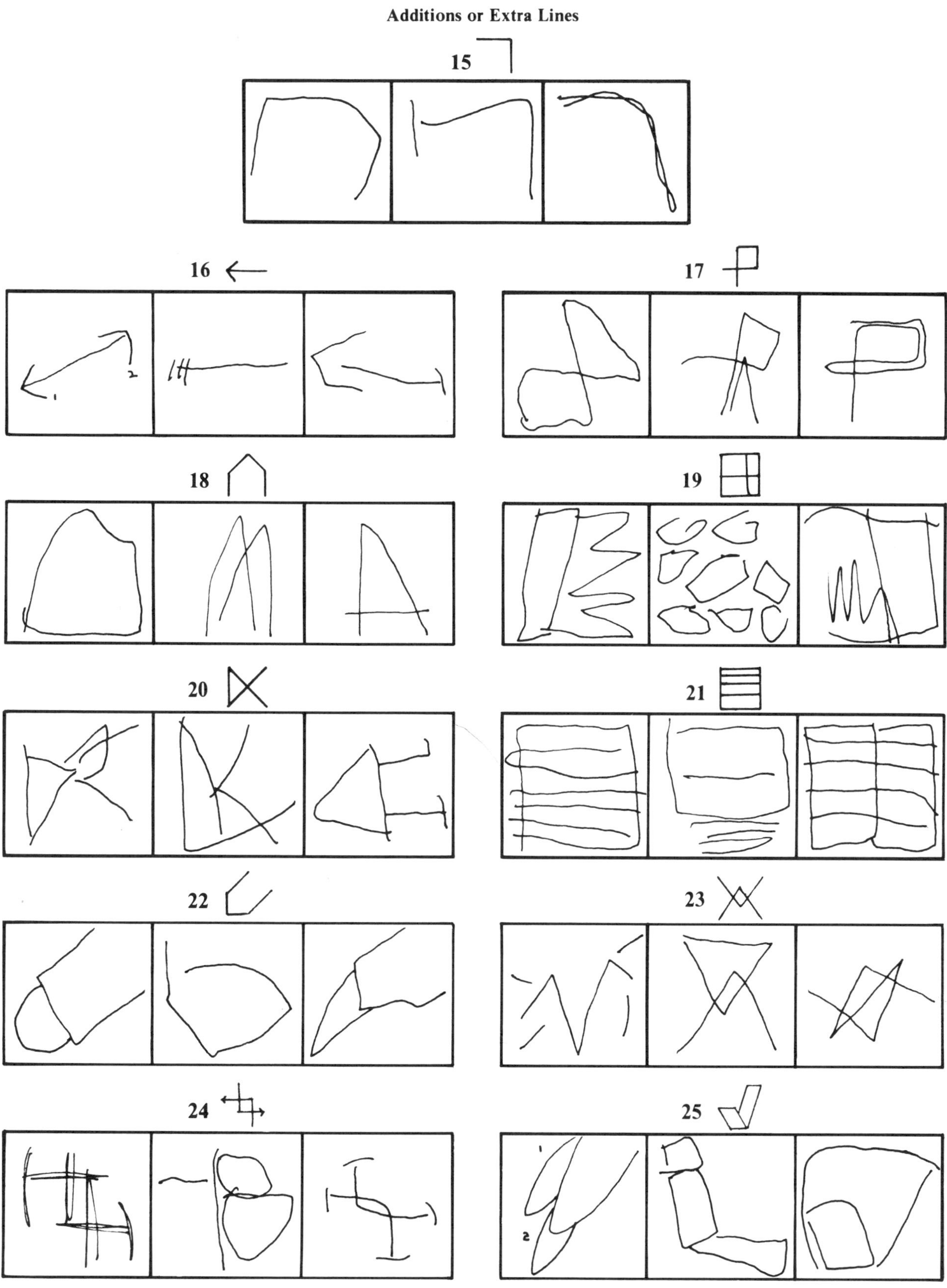


figure continued on next page . . .

**Figure 17**
**Design Copying: Examples of Additions or Extra Lines**

**Corrections: Not Additions**

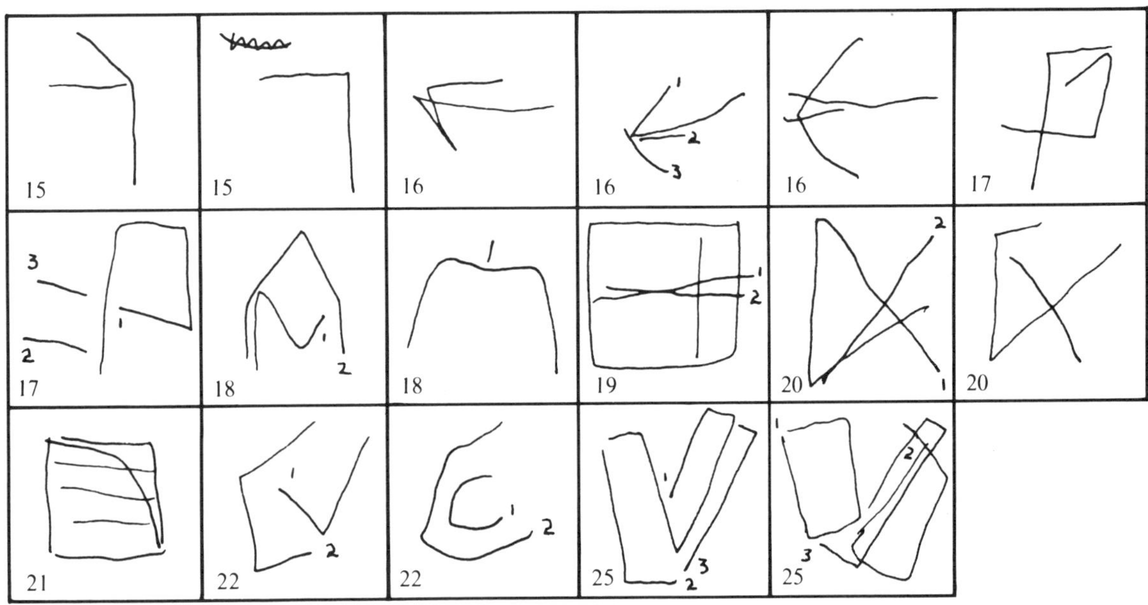

**Figure 17 (Continued)**
**Design Copying: Examples of Additions or Extra Lines**

**Jogs or Ears**

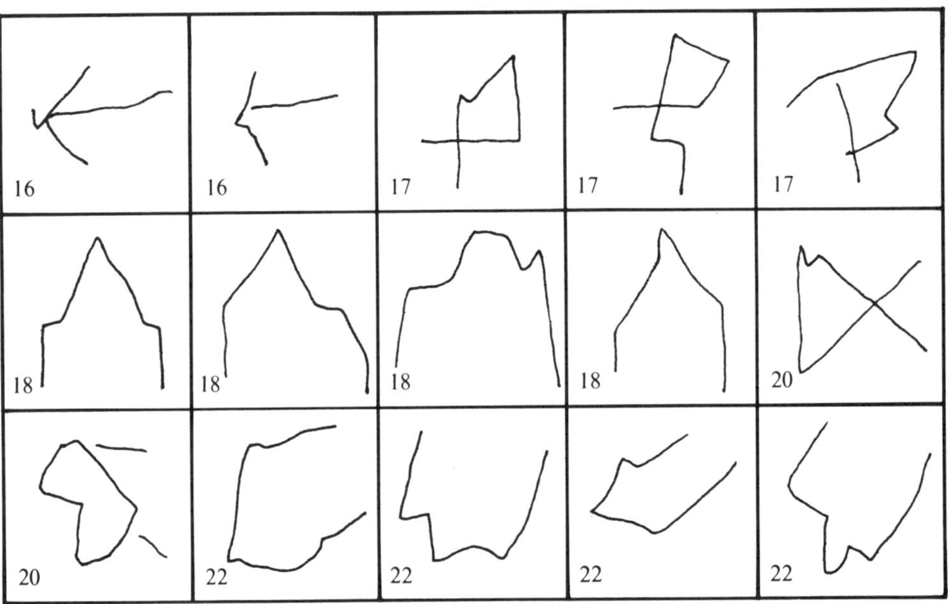

**Figure 18**
**Design Copying: Examples of Jogs or Ears**

**Segmentations**

**Figure 19**
**Design Copying: Examples of Segmentation**

corner or angle may be considered an approximate corner or angle.

*Reversal (R), Right to Left (L), and Inversion (I):* A figure that is *reversed* is one in which details on one side of the drawing are drawn in the opposite side, creating a mirror image. Only asymmetrical figures can be reversed. A figure or a line drawn *right to left* is one in which the child starts the figure or line on its right side and then proceeds to its left side. If a child should draw a line to which he or she adds a short extension, even though drawn right to left, it is not scored as a right to left line. A figure or line drawn right to left may or may not also be reversed. An *inverted* figure is upside down. A figure can be both inverted and reversed.

*Segmentation (S):* Many of the Part II test items have been designed especially to tap the tendency of some dysfunctional children to draw a figure in a segmented manner. Segmentation as a Should Not Have parameter is defined for each item for which it is appropriate. A drawing that is segmented is not drawn in a logical sequence, suggesting the child does not perceive the figure in the same manner as do most children of the same age. It may be that the child perceives a whole drawing as separated parts joined together and, hence, depicts them as such. When a scoring criterion defines segmentation as a portion of the figure drawn independently, the criterion is referring to spatial independence or intactness. For example, if Item 25 is drawn as two separate figures, the drawing is segmented, but if one quadrangle is drawn first, and then lines are added to make a second quadrangle, the approach is not segmented. Examples of segmentation are presented in Figure 19.

## Key Points to Remember

1. Sit on the side of the child's nondominant hand.
2. Be sure the child successfully executes the trial item of Part I.
3. Erasures are not permitted. If child does erase, score the more erroneous line in Part I and the more accurate line in Part II.
4. The child may not rotate the test booklet more than 45 degrees.
5. If the child draws too rapidly for you to record approach, stop the child between items.
6. The child completes items even if a mistake is made so that approach can be scored.
7. When recording approach, do not draw your lines through dots.
8. Do not look away from the child while drawing approach. Remember to number lines.
9. Do not record approach for 4-year-olds.
10. Discontinue Part I after two consecutive accuracy scores of 0 and go to Part II.
11. Do not indicate where the child is to start on Part II. Record where the child starts. After starting, encourage the child to draw figures in sequence.
12. Discontinue Part II after three consecutive items show no gross approximation.
13. Measure accuracy of lines by placing the scoring guide line in the middle of the dot for divergent lines and at the far edge of the dot for a line drawn too far.
14. Four-year-olds have 4 mm leeway for scoring line straightness on Part I; 5-year-olds have 3 mm leeway; and 6-year-olds and older have 2 mm leeway.
15. Credit for one Should Have parameter is not dependent upon having received credit (scored "Yes") on another parameter unless the parameter so specifies.
16. If the response figure in Part II is not a gross approximation of the stimulus figure, score all Should Have parameters as "No."
17. Only when the child draws a gross approximation do you score Should Not Have parameters as "Yes" or "No." If there is no gross approximation, score all Should Not Have parameters as "Unscorable."
18. Score a Should Not Have parameter as "Unscorable" if response is such that you cannot determine whether or not a parameter is present on an attempted item.
19. Do not fill in circles on items not attempted.
20. Do not score Should Not Have parameters for 4-year-olds.

## 5. Postural Praxis (PPr)

**Materials**

- protocol sheet
- stopwatch (optional)

In this test, the child is asked to imitate postures that you demonstrate. In chairs without arms, sit opposite the child. Place the protocol sheet (and the stopwatch, if you use it) on a table to the side of you. Arms and knees of both the child and yourself must be visible and identifiable (e.g., not hidden by loose sleeves or a skirt).

**Trial Item**

Begin by saying:

***"You make your arms and hands do the same thing that mine do. See how fast you can do it."***

Assume the *mirror image* of the trial posture illustrated on the protocol sheet (e.g., your hands are over your left ear and not over your right ear as on the protocol sheet; see Figure 20).

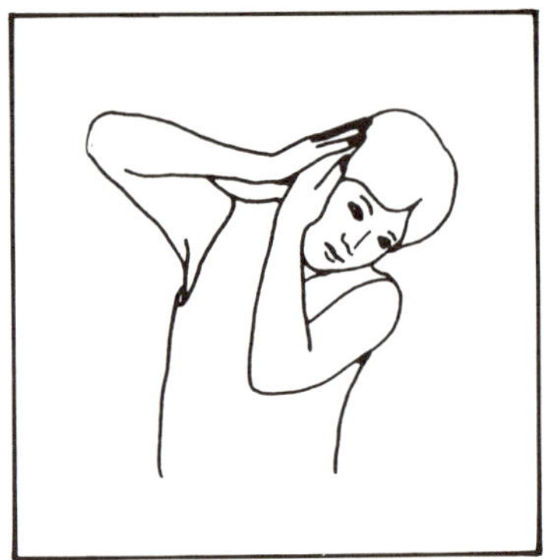

**Figure 20**
**Postural Praxis: Trial Item**

If the child does not imitate you, again ask him or her to do what you did. If the child correctly imitates the trial posture, reinforce it verbally. Regardless of whether the child imitates the posture correctly or incorrectly, gesture appropriately and/or touch the child's hands, saying:

***"This hand is over my ear, this hand is on top of it, and I'm leaning to the side. It's as though you were looking in a mirror."***

Help the child if the posture is not correct, placing him or her in the correct position. Emphasis is placed on an exact mirror imitation by the child. Since you have assumed the mirror image of the illustration on the protocol sheet, the child's position should look exactly like the illustration.

After the child has successfully demonstrated the trial posture, return to a neutral position with hands resting in your lap. The child will usually imitate you, bringing his or her hands back to the lap. If the child fails to do this, ask the child do so.

**Test Items**

Begin Item 1 by saying:

***"Now do this one. Do it quickly."***

Assume the mirror image of the posture illustrated on the protocol sheet. It is essential that you assume each posture so quickly that the child cannot follow your movements as they are made but must plan his or her movements based on the observed final posture. It is also important that the posture is accurately assumed so that repositioning is not necessary. Details must be correct (e.g., on Item 17, the hands are tilted downward slightly). Do not give the child clues with your eyes or hands.

Start timing the child's response time the moment you have assumed the position. Since you must watch the child's response to determine the moment the child has matched the position, timing is done by counting to yourself. However, check with a stopwatch periodically to assure that your speed of counting is correct. Checking may be done at a nontesting time.

Hold the posture for 7 seconds or until the child has assumed the correct posture, whichever occurs first. No corrections of the child's position are allowed after the trial item. However, if necessary, the child may be admonished during testing to move quickly and to be sure that he or she looks exactly like you. For example, you may say:

***"Watch my hands carefully and make yours do exactly the same thing. Move quickly."***

After the child correctly assumes the position or after 7 seconds, fill in the circle corresponding to the child's score (0, 1, or 2) as detailed in the section, "Recording Item Scores." If unsure about the child's score, you may draw any deviations of the posture over the illustration on the protocol sheet to assist in scoring later. In this case, note the time of the child's response, as it will be necessary in scoring. *Before sending in for processing, be sure to erase the recorded time as well as any marks made on the illustrations;* the protocol sheet cannot be processed with marks in these areas.

The rest of the test items are given in sequence. Administer all items to all children. Remember that each posture is begun from a neutral position with hands resting in your lap.

Be sure to have the child's attention before assuming each position. If necessary, the child's attention can be obtained by saying, for example:

***"Now watch this one."***

The first time the child gives a nonmirrored response (excluding the trial), remind the child to give a mirrored response. If the child appears to be deliberately trying to assume a nonmirrored position, you may explain that, since it takes more time to figure out that position, it should be avoided.

**Recording Item Scores**

**General scoring criteria.** Praxis is evaluated by comparing specific elements of the child's posture with those of the stimulus illustration on the protocol sheet. In some cases, the child may be scored as assuming the position adequately even though his or her position is less precise than that shown in the illustration. The child is not penalized in scoring for assuming a nonmirrored posture, even though he or she is encouraged once to assume a mirrored posture. Aspects of postures not mentioned in the specific item scoring criteria below do not enter into scoring.

To assist the experienced examiner in scoring efficiently, cues briefly describing scoring criteria are listed on the protocol sheet. When administering this test, fill in the circle for 0, 1, or 2 to the left of these scoring cues.

A score of 0 is given if the child fails to assume some aspects of the posture within 7 seconds, or assumes the position after 7 seconds.

A score of 1 is given if all aspects of the posture are assumed and maintained within 4 to 7 seconds. A score of 1 is also given if a child assumes some but not all the aspects of the posture correctly in 7 seconds. Unless a variation for an aspect of a posture is specifically listed in the following section, "Specific Item Scoring," the criterion listed under a score of 2 for that aspect must also be met for a score of 1.

A score of 2 is given if the child imitates all aspects of the position, as described in the scoring criteria, *within 3 seconds* after you assume the position being demonstrated. If the child changes position frequently, the appropriate position must be maintained for at least 2 of the first 3 seconds to earn a score of 2. Otherwise, a score of 1 or 0 is given depending upon the position the child is in at 7 seconds.

**Specific item scoring.** Detailed scoring criteria for scores of 0, 1, and 2 are presented for each item in the following section. Refer to Figure 21 for illustrations of the correct postures.

*Item 1:* Arm is abducted more than 45 degrees; one or more fingers touch shoulder on or medial to acromion process (but not on neck). When you assume the posture, be sure to touch all five fingers to shoulder.

0 = Posture is incorrect or not completed within 7 seconds.

1 = Posture is assumed correctly in 4 to 7 seconds, or one or more fingers touch shoulder on or medial to acromion process, but one or both of the following irregularities exist: (a) arm is abducted 45 degrees or less; (b) elbow is brought forward by 45 degrees or more.

2 = Posture is assumed correctly within 3 seconds.

*Item 2:* Fingers of one hand are curved over opposite shoulder joint so that no fingernails (except thumb) are visible when your eyes are even with the top of the child's head (other hand does not enter into scoring).

0 = Posture is incorrect or not completed within 7 seconds.

1 = Posture is assumed correctly in 4 to 7 seconds, or one or more fingernails (besides thumb) are visible but all five fingernails are above a point even with the sternal end of the clavicle.

2 = Posture is assumed correctly within 3 seconds.

*Item 3:* Arms are crossed and ventral surface of fingers and some portion of palms are on opposite sides of any part of the head. No portion of fingers and palms which touch the face should extend beyond midline of the face.

0 = Posture is incorrect or not completed within 7 seconds.

1 = Posture is assumed correctly in 4 to 7 seconds, or arms are crossed, but only fingers (any number of either hand) are placed on opposite sides (and completely beyond midline) of head.

2 = Posture is assumed correctly within 3 seconds.

*Item 4:* Ventral surface of distal phalanxes of thumbs and ventral surface of distal phalanxes of one or more fingers of each hand touch to form opening (of any shape) held in front of eyes so that you can see pupils of child's eyes without changing position. However, it is all right if the child moves head or hands so that both pupils are seen one at a time. Hold your hands 2½ cm (1 inch) from forehead and look at child.

0 = Posture is incorrect or not completed within 7 seconds.

1 = Posture is assumed correctly in 4 to 7 seconds, or thumb and one or more fingers of each hand touch tip to tip to form two openings (of any shape) held in front of eyes; this position does not permit you to see one or two pupils of the child's eyes through openings without moving the position of your head.

2 = Posture is assumed correctly within 3 seconds.

*Item 5:* Arms are crossed and hands cover opposite knees. Some portion of each hand must cover the center of the patella of the opposite leg.

0 = Posture is incorrect or not completed within 7 seconds.

1 = Posture is assumed correctly in 4 to 7 seconds, or arms are crossed and both entire hands are placed on distal half of opposite thighs.

2 = Posture is assumed correctly within 3 seconds.

*Item 6:* Each elbow is cupped by the hand from opposite arm. Cupping requires that at least one finger be to the right and one finger to the left of the tip of the ulna.

0 = Posture is incorrect or not completed within 7 seconds.

1 = Posture is assumed correctly in 4 to 7 seconds, or only one elbow is cupped by opposite hand.

2 = Posture is assumed correctly within 3 seconds.

*Item 7:* One hand is on ipsilateral side of head, trunk and head are leaning any amount to that side, and other hand is deliberately placed on ipsilateral hip or waist. (You can assess trunk leaning quickly by looking at tilt of shoulders.)

0 = Posture is incorrect or not completed within 7 seconds.

1 = Posture is assumed correctly in 4 to 7 seconds, or one hand is placed on ipsilateral side, but one or both of the following variations are present: (a) head or trunk alone is leaning; (b) other hand is placed above the waist, on thigh, or on buttock.

2 = Posture is assumed correctly within 3 seconds.

*Item 8:* Thumb and index finger of one hand touch each other inside an opening formed by touching the other thumb and index finger.

0 = Posture is incorrect or not completed within 7 seconds.

1 = Posture is assumed correctly in 4 to 7 seconds, or any finger other than, or in addition to, index finger is used with thumb to form linked shapes.

2 = Posture is assumed correctly within 3 seconds.

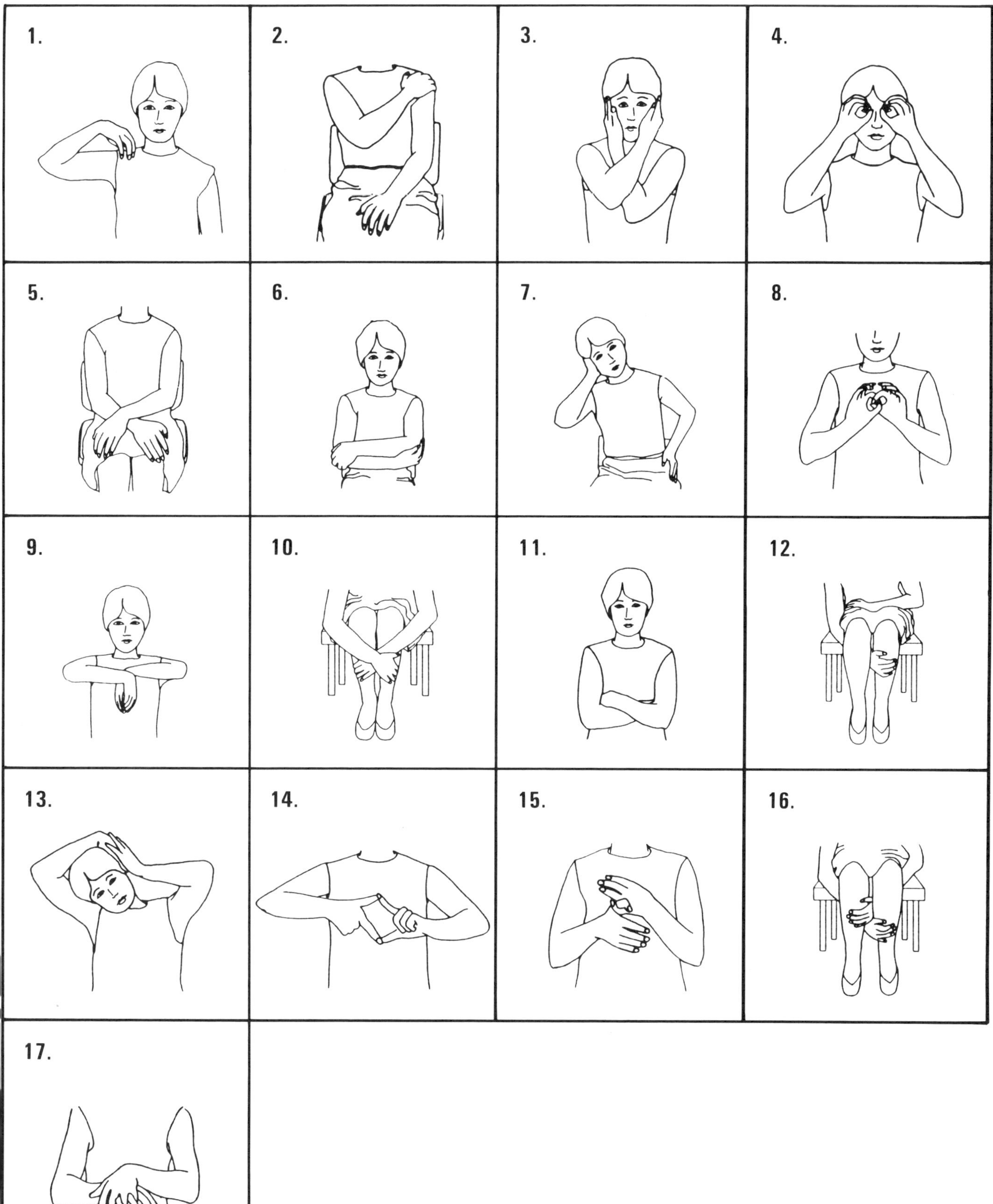


**Figure 21**
**Postural Praxis: Test Items**

*Item 9:* Wrists are crossed and at least two fingers (exclusive of thumb) of each hand touch on palmar surfaces.

0 = Posture is incorrect or not completed within 7 seconds.
1 = Posture is assumed correctly in 4 to 7 seconds, or wrists or hands are crossed and only one finger (exclusive of thumb) of either hand touches finger(s) of other hand.
2 = Posture is assumed correctly within 3 seconds.

*Item 10:* Arms, wrists, or hands are crossed and all five fingers of each hand touch anterior or lateral surface of contralateral leg anywhere between (but exclusive of) knee and ankle. Some portion of all five fingers must be over tibia or lateral to it. Not all fingers need be visible from your location.

0 = Posture is incorrect or not completed within 7 seconds.
1 = Posture is assumed correctly in 4 to 7 seconds, or wrists or arms are crossed and some but not all of five fingers of each hand touch contralateral leg.
2 = Posture is assumed correctly within 3 seconds.

*Item 11:* Forearms are folded; hands are against body and underneath opposite arms, so no part of knuckles or any other part of fingers or thumbs is visible medial to arms. Fingers may show below or lateral to the arms.

0 = Posture is incorrect or not completed within 7 seconds.
1 = Posture is assumed correctly in 4 to 7 seconds, or forearms are folded and hands are against body and some part of hand is underneath opposite arm, but knuckles or more of either hand are visible medial to arms.
2 = Posture is assumed correctly within 3 seconds.

*Item 12:* One arm is behind ipsilateral leg and hand grasps (in any manner) the anterior surface of the other leg between (but exclusive of) ankle and knee. (Other arm does not enter into scoring.) *Grasping* is defined as some degree of flexion in the two interphalangeal joints of at least two fingers of the grasping hand.

0 = Posture is incorrect or not completed within 7 seconds.
1 = Posture is assumed correctly in 4 to 7 seconds, or arm position is correct, but hand merely touches instead of grasps leg.
2 = Posture is assumed correctly within 3 seconds.

*Item 13:* One arm encircles top of head with palm and fingers flat on opposite side of head; some portion of palm of other hand is on top of hand of arm encircling head, overlapping in any amount; head tilts any degree past vertical toward encircling arm. Top of head is defined as that area above an imaginary plane drawn through the head halfway between eyes and top of head (excluding hair) and horizontal to eyes.

0 = Posture is incorrect or not completed within 7 seconds.
1 = Posture is assumed correctly in 4 to 7 seconds, or one or both of the following conditions are present: (a) position of hands is assumed, but head is not tilted in appropriate direction; (b) fingers only (rather than a portion of palm) are on top of hand of encircling arm.
2 = Posture is assumed correctly within 3 seconds.

*Item 14:* Tip of thumb of one hand touches tip of index finger of other hand and other thumb touches tip of other index finger.

0 = Posture is incorrect or not completed within 7 seconds.
1 = Posture is assumed correctly in 4 to 7 seconds, or one or two middle fingers are substituted for index finger(s).
2 = Posture is assumed correctly within 3 seconds.

*Item 15:* Thumb of one hand and fifth finger of other hand are hooked and the approach is palm to palm. Hands do not have to remain with palms parallel. There must be some flexion in both fifth finger and thumb, but not all joints of hooked fingers need be flexed.

0 = Posture is incorrect or not completed within 7 seconds.
1 = Posture is assumed correctly in 4 to 7 seconds, or fifth finger and thumb are hooked, but the approach is palm to dorsum.
2 = Posture is assumed correctly within 3 seconds.

*Item 16:* Arms or wrists are crossed behind legs; each hand comes forward between the legs to grasp the front of the opposite lower leg anywhere between (but exclusive of) the ankle and the knee; all 10 fingers are visible. Grasping is defined as some degree of flexion in the two interphalangeal joints of at least two fingers of each hand. Be careful that all your fingers are visible to the child and not covered by clothing.

0 = Posture is incorrect or not completed within 7 seconds.
1 = Posture is assumed correctly in 4 to 7 seconds, or arms or wrists are crossed behind legs, but one or more of the following variations of the

correct posture are present: (a) some but not all fingers of each hand are visible; (b) hands touch but do not grasp opposite lower legs.

2 = Posture is assumed correctly in 3 seconds. Child's response should not be discounted if fingers are in required location but not visible because clothing has covered them.

*Item 17:* Palm of one hand is on dorsum of other hand; both palms are down; ring and little finger of one hand are interposed with ring and little finger of other hand.

0 = Posture is incorrect or not completed within 7 seconds.

1 = Posture is assumed correctly in 4 to 7 seconds, or one or three fingers from the ulnar side of hands are intertwined. When three fingers are intertwined, both index fingers are touched on one side only by fingers of the opposite hand. (Anytime the index fingers are bordered on both sides by fingers from the opposite hand, a score of 0 is warranted.)

2 = Position is assumed correctly in 3 seconds. When position is correct, the index fingers are not touching any fingers from the opposite hand.

## Key Points to Remember

1. Sit opposite the child.
2. Assume mirror image of the illustration on the protocol sheet; the child's mirror image should look exactly like the illustration.
3. Place the child in the correct position during the trial item if he or she has difficulty. If the child is correct on the trial item, reinforce it verbally. Do not correct or reinforce after the trial item.
4. Make sure the child is paying attention before you assume the posture.
5. Assume postures quickly so that the child imitates the postures, not the movements.
6. Do not give cues with your eyes or hands.
7. Because you must watch child's position, time the child's response by counting to yourself rather than using a stopwatch.
8. Maintain the posture for 7 seconds or until the child assumes the correct posture.
9. During testing, you may ask the child to move quickly and make his or her hands do the same thing that yours do.
10. If unsure of the child's score, draw any deviation of the posture to assist in scoring later. Also write down the response time in such cases. Erase these marks later.
11. Return hands to your lap after the child's response. Make sure that the child also returns hands to lap.
12. On first nonmirrored response, encourage the child to assume a mirrored position. However, nonmirrored responses are not penalized in scoring.
13. Score the item as 0 if the posture is incorrect or not completed within 7 seconds.
14. Score the item as 1 if the correct posture is assumed in 4 to 7 seconds, or if the posture contains the deviations listed in the specific scoring criteria for a score of 1.
15. Score the item as 2 if the correct posture is assumed within 3 seconds.
16. On Items 2 and 12, the alternate hand is in lap and does not enter into scoring.
17. Give all items to all children.

## 6. Bilateral Motor Coordination (BMC)

**Materials**

- two 8½" × 11" sheets of paper
- tape
- footstool (in case the child's feet do not rest comfortably on the floor when the child is seated)
- protocol sheet

In this test, the child is expected to imitate a series of movements after you have demonstrated them. Each item consists of a motor pattern that is executed two or three times. Movements consist of touching the palms of the hands to the thighs or the feet to the floor, making an audible, but not loud, sound. Scoring is based on the degree of smoothness, reciprocity of movement, and sequencing.

Because this test follows Postural Praxis in which the child imitates postures, children usually have no difficulty in grasping the essential nature of the task. In contrast to administration of the Postural Praxis test, however, the child imitates your motions *after you have completed them.*

The main behavioral dimension under evaluation is the ability of the two arms or feet to move together in a smoothly integrated pattern. Reciprocal action of the two arms or two feet is the primary skill under evaluation. In reciprocal action, one arm or foot raises as the other lowers. The movement is largely at the elbows or ankles; on the arms items, the wrists should not flex and extend. Reciprocal action is especially important, since it particularly reflects interaction of the two sides of the body. Timing and sequencing of movements are essential elements of the movement pattern. Movements are expected to be rhythmical, with the arms or the feet in motion from the time the item is started to the time it is finished, as in dancing.

**Demonstrating the Items**

Sit facing the child. It is particularly important that the child's attention be obtained before demonstrating each item, since only one demonstration is allowed.

In executing these items, it is essential that you be rhythmical, natural, and well coordinated. Practice the patterns before attempting to administer this test, because if your movements are stiff or uncoordinated, the child may imitate the incoordination. A disconnected demonstration will yield a segmented response, and a segmented response is scored as incorrect. A segmented response occurs when one hand is raised then lowered to the thigh before the other hand is raised, or when there is jerkiness of motion.

Some examiners find it helpful in promoting rhythm to sing to themselves as they execute the movement. (In singing, the words are vocalized with a continuous sound.) Although the sounds made by contacting the thigh or floor are not the objective of the movements, they do help to establish the rhythm of the test items. The movements must not be a series of thigh slapping or floor hitting. The rhythm is partly established by the distance the arms or feet travel and partly by the speed of the movement.

On the arms items, the arm moves over a distance of either 15 cm (6 inches) or 2 to 4 cm (1 to 1½ inches). Since moving only 2 to 4 cm takes less time than moving 15 cm, some sounds are closer together than others. On the feet items, the toes raise from the floor 1 to 4 cm.

On the protocol sheet, the letter L indicates that the left arm or foot is used and the letter R indicates that the right arm or foot is used. Note that the letters on the protocol sheet indicate the hands or feet *you* are to use. The child's response will be the opposite. For example, on the trial item, the protocol sheet indicates that the pattern you demonstrate is L, R, L, R; the child's response should be R, L, R, L. Each item's pattern, determined by which hand is used and the distance and speed it travels, is indicated on the protocol sheet. On the arms items, when a comma follows L or R or a word (e.g., "both, both"), that hand returns to the position 15 cm above the thigh. When there is no space between the letters or words (e.g., "RR"), then two taps of the thigh are given in quick succession, with the hand raising only 2 to 4 cm above the thigh.

It is more important to be well coordinated than it is to be perfect in total timing of the movements. Use a stopwatch only for practice and not for administration. The approximate time required to execute each item is listed on the protocol sheet. Execution time includes the time required to return both hands or feet to the starting position on completion of all contacts with thighs or floor.

If the child starts to respond before you have completed the pattern, stop moving and say:

***"Wait until I finish. Then you do it."***

An item interrupted in this way is demonstrated again and the response that the child started is ignored. Otherwise, only one demonstration is given and only one attempt at response is allowed.

**Arms Items**

To administer the arms items (Items 1–10), start with forearms pronated and held 15 cm (6 inches) above the thighs. The fingers are slightly flexed in a natural position. The wrists are held in a neutral position and do not change from that neutral position as the movements are executed. Except when making quick movements, the hands return to a position 15 cm above the thigh after each arm movement and remain in that position after your demonstration while the child executes the item. After the child completes the item, put one hand in your lap while marking the item score with the other hand (see "Recording Item Scores," p. 60).

**Trial item.** Begin by demonstrating Trial I (L, R, L, R), saying:

***"Watch my hands move. When they are through moving, you do the same thing."***

Then move your left hand to gently touch your left thigh and return it to the original position as your right hand lowers to your right thigh. As your right hand returns to its original position, lower your left hand to your left thigh. As your left hand returns to its original position, lower your right hand to your right thigh, and then return it to its original position.

Movements must be reciprocal and smooth. Four seconds are required to execute the trial item. The child is then given time to replicate the pattern, after which you say:

***"(That's correct.) Be sure to move smoothly like this.*** (Move the child's hands smoothly through the pattern.) ***When I begin with this hand,*** (hold up your left hand) ***you begin with this hand.*** (Touch child's right hand.) ***When I begin with this hand,*** (hold up your right hand) ***you begin with this hand."*** (Touch child's left hand.)

If the child performs the trial item incorrectly, the trial item is demonstrated again and the child tries it one more time. If the child performs an insufficient number of movements, say:

***"Do it as many times as I did it."***

**Test items.** Continue with the test items by saying:

***"Watch me do another one."***

Demonstrate Item 1 in a manner similar to the trial item. Then, if necessary, say:

***"Now you do it."***

Give the rest of the items in sequence as described below, repeating the directions only as necessary. After the first item that the child starts with the wrong hand (excluding the trial item), remind the child to start with the mirror-image hand. After that, no further emphasis is placed on beginning with a specific hand. Any time the child performs an insufficient number of movements, after scoring the item remind the child to do it as many times as you did. The child is not penalized in scoring for starting with the wrong hand.

*Item 1:* R, L, R, L (4 seconds). This item is administered in the same manner as the trial item, but reversing hands.

*Item 2:* L, R, L, R (4 seconds). This item is administered in the same manner as the trial item.

*Item 3:* Both, clap clap, both, clap clap (4 seconds). Both hands are used symmetrically. The palms touch the ipsilateral thighs simultaneously, then rise to the 15 cm position and clap (in typical palms-together position) twice in quick succession. The pattern is repeated once.

*Item 4:* LR, LR, LR (3 seconds). The left palm touches the left thigh, immediately followed by the right palm touching the right thigh. Each hand is raised to the 15 cm position after each contact with the thigh. The motions are repeated twice, for a total of three executions of the LR pattern.

*Item 5:* Both, both both, both, both both (4 seconds). Both hands are used symmetrically. The palms touch the ipsilateral thighs simultaneously, then return to the 15 cm position, then return to the thigh and slap the same thigh twice in quick succession. The pattern is repeated once.

*Item 6:* Crossed RL, RL, RL (3 seconds). This item is administered in the same manner as Item 4, except the left arm is crossed over the right arm and the right palm touches the left thigh, followed immediately by the left palm touching the right thigh. Both hands are raised to the 15 cm position after the thigh contact. The motions are repeated twice, for a total of three executions.

*Item 7:* LL, RR, LL, RR (4 seconds). The left hand touches the left thigh twice in quick succession, then returns to the 15 cm position as the right hand lowers to touch the right thigh twice in quick succession. The pattern is repeated once, with the left hand descending as the right rises.

*Item 8:* RR, LL, RR, LL (4 seconds). This item is administered in the same manner as Item 7, except the right hand initiates the item.

*Item 9:* L, RR, L, RR (4 seconds). The left hand touches the left thigh, and as it is raised to the 15 cm position, the right hand lowers and touches the right thigh quickly twice. The motion is repeated once, with the left hand lowering again as the right rises.

*Item 10:* R, LL, R, LL (4 seconds). This item is administered in the same manner as Item 9, but reversing hands.

**Discontinuing the arms items.** Discontinue the arms items when four *consecutive* items are incorrect (i.e., scored as 0) and go on to the feet items.

**Feet Items**

Do not administer the feet items (Items 11–14) to 4-year-olds. Administer the feet items wearing socks or nylon hose, without shoes. The child is wearing socks or is barefoot. For sanitary reasons, two 8½″ × 11″ pieces of paper are taped to the floor, one on which you place your feet and one on which the child places his or her feet during these items. Place feet with a 10 cm (4 inch) space between them. The angle of the knee is about 135 degrees, making it easier to raise the toes from the floor. Your toes are raised to a point 4 cm (1½ inches) above the floor to begin each item and remain in this position while the child attempts each item. Your feet may be placed flat on the floor while the item is scored.

As in the administration of the arms items, it is essential that you be rhythmical and well coordinated, with special attention to reciprocal action. In demonstrating the items, a barely audible sound is made by the feet touching the paper on the floor. Items 11, 13, and 14 are performed in 2 seconds (twice as rapidly as comparable arms Items 1, 10, and 9). Item 12 is performed in approximately 3 seconds.

**Trial item.** One trial item precedes the actual test items for the feet. Begin by saying:

***"Now we'll do the same thing with our feet. Watch me."***

Assume the appropriate position and demonstrate Trial II (L, R, L, R). With the heels resting comfortably on the floor 10 cm (4 inches) apart, raise the toes of both feet to a point about 4 cm (1½ inches) above the floor. Then lower the toes of your left foot to touch the floor, and return it to the original position as the toes of the right foot are lowered to the floor. As the right toes are raised to position, lower the left toes to the floor. As the left toes are raised to position, lower the right. Then raise the right foot so that the toes of both feet resume the starting position. In other words, reciprocally tap the toes of your feet on the floor, beginning with the left foot. Then say:

***"You do it."***

If the child has trouble executing the movements, move the child's feet through the motions, being sure the action is reciprocal. Then demonstrate the trial item again and allow a response one more time. Then administer the test items.

**Test items.** Give the test items in sequence as described in the following section.

*Item 11:* R, L, R, L (2 seconds). This item is administered in the same manner as the trial item, but reversing feet.

*Item 12:* LL, RR, LL, RR (3 seconds). The initial position of the feet is the same as on the trial item. The left foot taps the floor twice quickly, raising only 1 to 2 cm between taps. As the toes of that foot are raised, the right toes are lowered to tap the floor twice quickly. These motions are repeated once.

*Item 13:* R, LL, R, LL (2 seconds). The right foot taps the floor once and, as it is raised, the left foot is lowered and taps the floor twice quickly. The pattern is repeated once.

*Item 14:* L, RR, L, RR (2 seconds). This item is administered in the same manner as on Item 13, but reversing feet.

**Discontinuing the feet items.** Discontinue the feet items when two *consecutive* items are incorrect (i.e., scored as 0).

**Recording Item Scores**

The child's performance on each item is scored as 0, 1, or 2, depending on the quality of the performance. On the protocol sheet, fill in the circle corresponding to the child's score on each item.

A score of 0 is given if the movements are definitely nonreciprocal (if reciprocity is required), dysrhythmic, segmented, or incorrect, or if the sequence is incomplete.

A score of 1 indicates that the movements are approximately correct or slightly irregular, such as the hands hitting the thighs a fraction of a second apart on Item 3, detected only by a difference in sound.

A score of 2 indicates that the movements are unquestionably well coordinated and without error. Sometimes a child will continue a movement pattern for a longer period of time than that demonstrated. If the child imitates the entire pattern correctly (including the number of consecutive sequences), the score is 2, as long as the additional movements correctly replicate the item or part of an item. If the added movements do not replicate the item correctly, the item is scored as 0 or 1, depending on the quality of the additional movements. The child is not penalized for correctly making more than the required number of movements. The child is also not penalized for repeating the sequence slower or faster than you did it.

Once the child begins his or her imitation, scoring is begun. Thus, if the child starts a motion and then corrects it, he or she is nevertheless scored from the beginning of the first movement. A corrected movement, therefore, cannot receive a score of 2, and may or may not receive a score of 1, depending on the quality of those first movements. The only exception is if the child starts a motion with the wrong hand or foot, obviously recognizes that he or she has started with the wrong hand or foot, and then begins again to complete a pattern which may meet the critera for a score of 2 or 1, depending on the quality. In other words, the child is not penalized for beginning with the wrong hand or foot.

## Key Points to Remember

1. It is important to practice the patterns, because if you are stiff or uncoordinated, the child may imitate the stiffness.
2. Practice with a stopwatch or by counting to yourself until the timing is correct. However, coordination is more important, so do not use the stopwatch during actual administration.
3. Sit opposite the child.
4. Be sure the child is paying attention before you demonstrate the pattern.
5. R and L on the protocol sheet refer to *your* right and left. The child's response should be the opposite.
6. Use a light, quick tap which makes an audible, but not loud, sound.
7. The child does not imitate the movements until after you have completed the demonstration. The child may be told: "Wait until I finish. Then you do it." An item interrupted in this manner should be redemonstrated.
8. On Trial I, take the child's hands through the movements following the child's attempt.
9. Start each arms pattern with hands 15 cm (6 inches) above thighs and return to this position at the end of the demonstration.
10. Remind the child to begin with the correct hand the first time he or she produces a nonmirrored response. From then on, a nonmirrored response is accepted.
11. Do not administer feet items to 4-year-olds.
12. On the feet items, the child wears socks; you wear socks, nylons, or something similar.
13. For sanitary reasons, two 8½″ × 11″ pieces of paper are taped to the floor.
14. Feet are placed so that the angle of the knee is about 135 degrees.
15. If needed, take the child's feet through Trial II following the child's attempt.
16. To score, listen carefully for rhythm and timing. Smoothness of execution and interaction of the hands or feet are under consideration. Be strict in scoring.
17. If slightly slower or faster, do not penalize.
18. The child must execute the pattern at least as many times as demonstrated. If he or she does more, it is acceptable.
19. If incorrect, score as 0.
20. If there is a slight irregularity in execution, score as 1.
21. If correct and smooth, score as 2.
22. Scoring is begun from the first movement. Therefore, if the child begins incorrectly and then starts over, the item cannot receive a score of 2. However, if the child begins with the incorrect hand or foot, then obviously corrects and executes the item correctly, score as 2. In other words, do not penalize for using the wrong hand or foot.
23. Discontinue arms items after four consecutive items are scored as 0 and go to feet items; discontinue feet items after two consecutive items are scored as 0.

## 7. Praxis on Verbal Command (PrVC)

**Materials**

- protocol sheet
- stopwatch

In this test, you will read aloud, one at a time, a series of motor commands which the child must first understand, then plan and execute. The commands all require the child to place part of his or her body in a specific position. Both accuracy of execution and the time required to complete each command are recorded.

Sit facing the child, on the same side of the table as the child, with the table on your dominant side. Hold the stopwatch in your nondominant hand so that you can record the scores with your dominant hand. Both the stopwatch and the child should be in your line of vision.

**Trial Item**

To begin, face the child, get his or her attention, and read the trial command aloud:

***"Put one hand on your nose."***

If the child responds correctly, say:

***"That's correct. Now we'll do another."***

If the child does not complete the trial item successfully, place the child's hand on his or her nose, saying:

***"Put your hand on your nose like this. Now we'll do another."***

Then continue with the test items.

**Test Items**

Administer all items to all children. For each item, read the command exactly as printed on the protocol sheet. *Do not demonstrate or assist the child during the administration of the test items.* Do not give cues with your voice, eyes, or hands. As the last word of the command is spoken, start the stopwatch.

It is important that the child's attention be obtained before administration of each item so that the response time reflects auditory language processing and praxis rather than failure to attend. If the child is distracted by extraneous stimuli while the command is being read, do not score that item. Readminister it later in the test.

When testing children ages 4-0 through 5-11, repeat the command after about 5 seconds if the action has not yet been successfully completed. If the child is close to assuming the correct posture at 5 seconds, wait to determine if it will be necessary to repeat the command. However, if the child has not attempted to assume the posture, you do not need to wait the full 5 seconds before repeating the command. The stopwatch should be left running during any repetition of the command. For children from age 6-0 through 8-11, give the command only once unless an environmental distraction makes repetition of the command necessary.

The time limit for completing the action is 15 seconds for children ages 4 and 5, and 10 seconds for ages 6 through 8. After each item, record the accuracy and execution time as described in the following section, "Recording Item Scores." Provide no feedback regarding the correctness or incorrectness of the child's response.

Before proceeding to the next item, the child should be in a neutral position. If the child does not automatically resume a neutral position after executing each item, advise him or her to do so.

Following Item 10, turn over the protocol sheet. At this point, ask the child to stand. The remaining items are administered with the child standing. The child should stand near the chair, as the chair is needed to complete two standing items. If the child needs to maintain his or her balance by holding onto the table or chair, he or she may do so, but is not advised to do so.

**Recording Item Scores**

During administration, record accuracy and execution time for each item.

**Accuracy.** If the child's response is incorrect, or correct but not within the time limit (15 seconds for ages 4–5, 10 seconds for ages 6–8), fill in the "0" circle. If the child completes the action correctly *within the time limit,* fill in the "1" circle.

There are many variations of the correct response. Accuracy is scored as correct if the child's position incorporates all parts of the command; parts of the body that are not included in the command are not scored. Also, for most items, the child does not have to touch a specific area of the body part or physical location mentioned. For example, the child may touch any part of his or her head and knee when those locations are mentioned. Furthermore, it is not necessary to hold the position for any length of time; a correct position held only momentarily (especially on items requiring balance) is still scored as accurate.

Some items may be more difficult to score for accuracy. The specific responses needed for an accuracy score of 1 (correct) on these items are listed below:

*Item 3:* Arms may be out to each side or both arms may be out to the same side.

*Item 9:* One elbow must touch dorsum of other hand at any point distal to wrist.

*Item 12:* Some part of the hands must touch and no part of the feet may touch.

*Item 13:* Some part of the toes must touch, but no part of the heels may touch.

*Item 14:* Some part of the feet must touch, but no part of the hands may touch.

*Item 15:* Some part of the knees must touch, but no part of the feet may touch.

*Item 16:* Some part of the heels must touch, but no part of the toes may touch.

*Item 19:* Both knees must bend.

*Item 23:* Knees must touch, feet must touch, and both knees must bend.

If you are not sure of the accuracy of the position, note the child's response in the Observations box on the back of the protocol sheet and score later.

**Time.** The execution time is measured from the time the last word of the first command is spoken until the time the child correctly completes the action. Because the child may begin to assume the posture before you have completed the command, it is possible to correctly complete the action in 0 seconds.

If the action is *incorrect* (accuracy score of 0), do not record the actual number of seconds taken to complete the incorrect action. Rather, *record the maximum time score* (15 seconds for ages 4 and 5, 10 seconds for ages 6 through 8).

During administration, write the time in the space labeled "Seconds." After the test, fill in the circle corresponding to each time score.

## Key Points to Remember

1. Sit on the same side of the table as the child, with the table near your dominant side.
2. Position yourself so that both the stopwatch and the child are within your line of vision.
3. Hold the stopwatch in your nondominant hand and record the scores with your dominant hand.
4. All items are given to all children.
5. Give the commands exactly as printed on the protocol sheet.
6. Do not give cues with your eyes, hands, or voice.
7. For children ages 4 and 5, repeat the command at 5 seconds (unless the action has already been completed correctly). Restate the command immediately to 4- and 5-year-olds if the child asks for the item to be repeated; you do not need to wait 5 seconds.
8. For children ages 6 through 8, directions are given only once (unless repetition of directions is required by environmental distraction).
9. For children ages 4 and 5, a total of 15 seconds is allowed per item.
10. For children ages 6 through 8, a total of 10 seconds is allowed per item.
11. Remind the child to return to the neutral position after an item.
12. If accuracy is not immediately clear, note the response and refer to the scoring criteria later.
13. Accuracy may be correct even if the position is held only momentarily (especially for items requiring balance).
14. If the action is correctly performed in the allotted time or less, record an accuracy score of 1. Record the number of seconds at which the response is completed.
15. It is possible to assume the position correctly in 0 seconds.
16. If the child is unable to complete the action correctly in the allotted time, record an accuracy score of 0. Record the time as the maximum time score (15 seconds for ages 4 and 5, 10 seconds for ages 6 through 8).

## 8. Constructional Praxis (CPr)

**Materials**

- box of blocks of various shapes and sizes, used for both Structures I and II (layout sheets inside box show the correct orientations of the blocks in the box)
- preassembled model for Structure II
- angle scoring guide (15-degree angle for scoring Structure II)
- centimeter ruler
- protocol sheet
- stopwatch

Prepare the blocks by making inconspicuous marks (e.g., penciled dots or small nicks with a knife) at specified points, as shown in Figure 22. Mark Blocks 1 and 2 of Structure I on long edges at the centers and at 2½ cm from the ends of the blocks, and Blocks 3 and 4 at the midpoint of each of the edges of the two longest sides of the blocks, as well as in the middle of each of the two shorter of the two long sides of the blocks. Marks on the shorter of the two long sides of Blocks 3 and 4 will not be opposite the marks of the longer sides. These marks enable rapid judgment of accuracy when scoring.

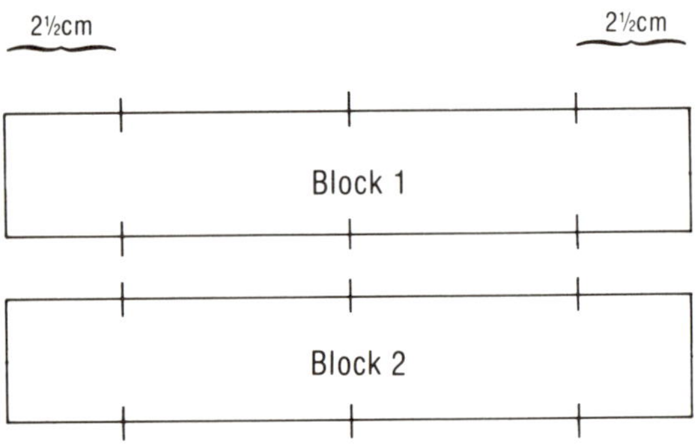


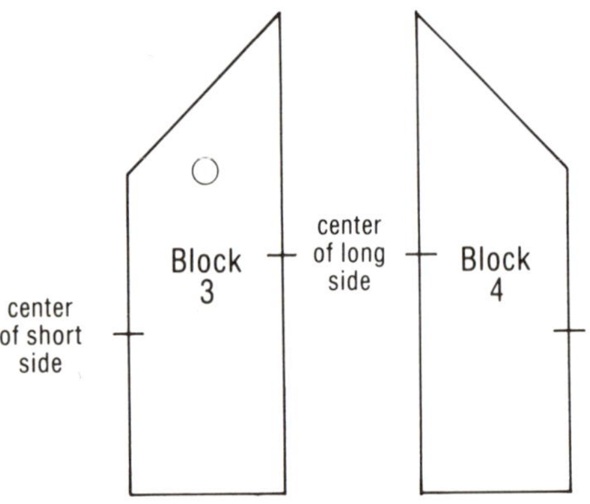


**Figure 22**
**Constructional Praxis: Marking Structure I Blocks**

This test consists of two parts. In the first part, the child builds Structure I with seven blocks in imitation of a structure you build with seven identical blocks. Structure I is designed so that you can easily build the structure accurately without having to estimate block location. This structure is scored on 16 yes-or-no criteria. On completion of that part of the test, the child (excepting 4-year-olds) builds Structure II with 15 blocks in imitation of a preassembled model. Each of the 15 blocks of Structure II is scored on one of eight different parameters. The diagrams on the protocol sheet serve as a guide for constructing Structure I and for scoring both Structures I and II. The time the child takes to build the structures is observed, but *not* recorded.

**Building Structure I**

Seat the child at the table with at least 60 cm (2 feet) of space on the tabletop to the left of his or her midline. Sit to the right of the child and, if necessary, move your chair to the side to give the child more tabletop room. Structure I is built in two phases. Blocks 1, 2, 3, and 4 of Structure I are put into place first by you and then by the child, and then the rest of the blocks are added. The structure is illustrated, with the blocks numbered, on the protocol sheet (see also Figure 23).

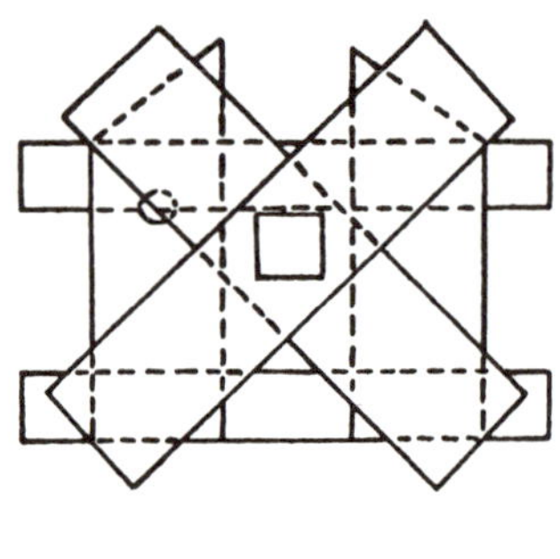


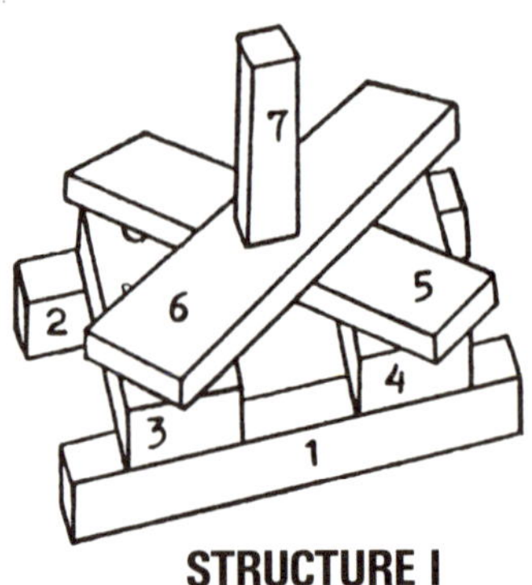


**Figure 23**
**Constructional Praxis: Diagram of Structure I**

Take the necessary blocks as numbered on the protocol sheet, one at a time, from the box, saying:

***"I'm going to build a house."***

In order for Structure I to be visible but out of reach of the child, place Block 1 25 cm (10 inches) from and parallel to the front edge of the table and 8 cm (about 3 inches) to the right of the child's right shoulder (see Figure 24). Then place Blocks 2, 3, and 4 in position according to the diagram of the structure on the protocol sheet, saying:

***"Here is the floor.*** (Place Block 2.) ***Here are the walls.*** (Place Blocks 3 and 4.) ***We line up the blocks VERY carefully.*** (Finger corners of blocks to emphasize lining up blocks.) ***Now you build a house like mine here.*** (Point to the appropriate location on the table.) ***I'll put the first block here."***

Complete the verbal directions for the first four blocks before the child begins building. Your house is left standing as is. Then give the child his or her blocks (see Figure 24). Place Block 1 in front of the child, parallel to and 10 cm (4 inches) from the table edge and at the child's midline. Next, place Blocks 3, 4, and 2, in that order, 20 cm (8 inches) distal to the child's Block 1. Place Blocks 3 and 4 next to each other with Block 2 behind them. Make sure that each individual block is placed on the table in the same right-left and distal-proximal orientation as it was in the box, with space between each block. The blocks need not necessarily be placed on the table in the same grouping as they were in the box. The important thing is that each individual block is placed in the same orientation as it was in the box, and not turned upside down or rotated.

Start the stopwatch as soon as the child starts building (usually indicated by picking up a block). If the child begins to build his or her house aligned with your structure, as opposed to diagonally to the side of it, or places Block 1 in a position nonparallel to the table edge, replace the child's block in the original location and say:

***"Keep this one here. Now you build the rest."***

If the child again moves Block 1, return it to its original position and hold it in position, repeating:

***"Keep this one here. Now you build the rest."***

If the child starts to add to your structure, tell the child to build his or her own house. If the child begins to build proximal to Block 1 instead of distal to it, return Block 2 to its position above Blocks 3 and 4, and say:

***"Build your building here."*** (Indicate space above Block 1.)

While the child arranges the four blocks in any manner, score the child's construction as described in the following section, "Scoring Structure I." Discourage the

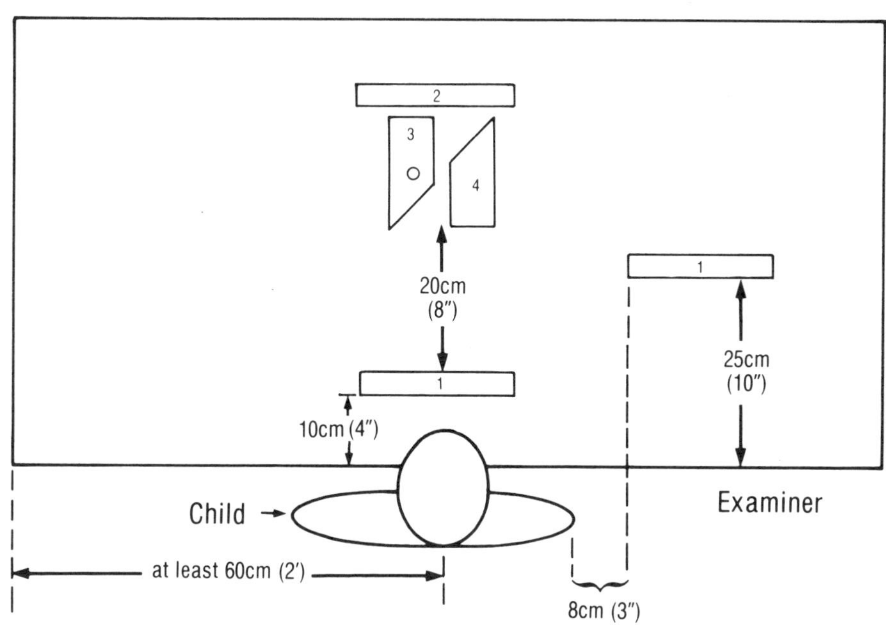


**Figure 24**
**Constructional Praxis: Placement of Structure I Blocks**

child from disassembling his or her house, but permit the child to improve upon the arrangement of the blocks. If, at this point, the child disassembles his or her structure, ask the child to rebuild it, saying:

***"We are going to build some more."***

If, at this or a later time, you need more time for scoring than that for which the child has patience, ask the child to tell you about the house while you score.

The child is allowed a maximum of 3 minutes of construction time to complete this portion of Structure I. If the child has not finished within 3 minutes, go on with the next portion of Structure I. After the following directions, the child may return to building with Blocks 2, 3, and 4, if so inclined.

Place Blocks 5 and 6 on your original structure, saying:

***"Here is the roof."***

Adding Block 7, say:

***"And here is the chimney."***

Place the child's Blocks 5, 6, and 7, in that order, behind the child's structure, saying:

***"Now build the rest of your house with these."***

Score as the child builds, changing the score if the child changes block position. For the second part of Structure I, a maximum of 3 minutes is allowed. Discontinue Structure I at the time limit or when it is evident that the child is finished building. However, if the child is nonproductive for 1 minute at any time in building Structure I, assume the child is finished and score the structure. Nonproductivity is defined as an activity that does not contribute to the further accurate construction of the structure. If appropriate, go on to Structure II.

When scoring is completed on Structure I, place all your blocks in the empty box lid or elsewhere and say:

***"Now, take your building down. Put the blocks there*** (gesture to the upper left corner of the table) ***and get ready to build a different building."***

Encourage the child to place all the blocks on the table and not stacked on top of each other.

**Scoring Structure I**

Scoring for Structure I is designed to be simple and objective enough to do quickly as the child builds the structure. This enables you to move on to the administration of Structure II without having to wait to score Structure I. Each block is scored on several criteria listed on the protocol sheet in approximately increasing complexity. If the criterion is not met, fill in the "0" circle. If the criterion is met, fill in the "1" circle.

Sometimes, failure to meet an early criterion prevents meeting subsequent ones; however, it is possible to perform incorrectly on an early criterion and correctly on a later criterion. For example, lower blocks must be fairly well constructed before Block 7 can be placed in the appropriate location. On the other hand, the angled ends of Blocks 3 and 4 may be placed distally and in correct orientation relative to the child even if Blocks 1 and 2 were not well aligned.

Note that Block 1 will almost always receive a score of 1 since it is placed by you. Block 1 is scored on only one criterion, but other blocks are scored on two or more criteria. The specific scoring criteria for each block are listed in the following section. Scoring examples for Structure I are presented in Figure 25.

*Block 1:*

Criterion 1: Block 1 is left in the position in which you placed it (i.e., parallel to edge of table plus or minus 15 degrees).

*Block 2:*

Criterion 1: Block 2 is placed so that Blocks 1 and 2 are not touching but are capable of supporting both Blocks 3 and 4 up off the table.

Criterion 2: Ends of Block 2 line up within 2½ cm (note the marks on the block) of the ends of Block 1. The distance *between* Blocks 1 and 2 is not measured.

*Block 3:*

Criterion 1: Block 3 (the block with the hole) is placed any distance to the left of Block 4. It may not be on top of or under Block 4. If both Blocks 3 and 4 are haphazardly placed, even nearly horizontally placed, a score of 1 is still given if any part of Block 3 is farther to the left than all of Block 4. If Blocks 3 and 4 are reversed in placement so that Block 4 is farther to the left than all of Block 3, subsequent items are scored as though Block 3 were Block 4 and Block 4 were Block 3. Subsequent items must not be scored as 0 on the basis of reversal of placement of Blocks 3 and 4.

Criterion 2: Block 3 rests on both Blocks 1 and 2.

Criterion 3: Entire block is placed on left half of both Blocks 1 and 2 (including midpoints), with some portion of both Blocks 1 and 2 extending beyond the left side of Block 3.

Criterion 4: Angled end is distal to the child and correctly oriented (i.e., not reversed).

*Example A*

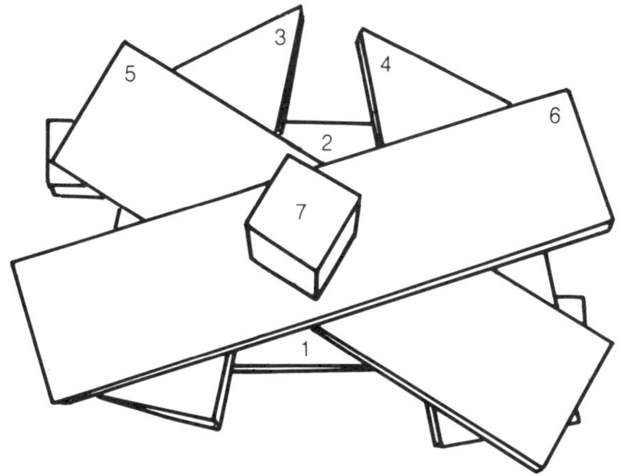


| BLOCK | CRITERION | SCORE No | SCORE Yes |
|---|---|---|---|
| **1** | Parallel to edge of table + or - 15° | ⓪ | ● |
| **2** | 1. **1** and **2** not touching; can support **3** and **4** | ⓪ | ● |
| | 2. Ends line up within 2½ cm of ends of **1** | ⓪ | ● |
| **3** | 1. **3** to the L of **4** (If to the R, score **3** as **4** and **4** as **3** hereafter) | ⓪ | ● |
| | 2. Rests on both **1** and **2** | ⓪ | ● |
| | 3. On L half of **1** and **2**; **1** and **2** extend to L of **3** | ⓪ | ● |
| | 4. Angled end distal; correct side up | ⓪ | ● |
| **4** | 1. Rests on both **1** and **2** | ⓪ | ● |
| | 2. On R half of **1** and **2**; **1** and **2** extend to R of **4** | ⓪ | ● |
| | 3. Angled end distal; correct side up | ⓪ | ● |
| **5** | 1. One end on **3** only, other on **4** only | ⓪ | ● |
| | 2. L intersection upper half of **3**, R intersection lower half of **4** | ⓪ | ● |
| **6** | 1. One end over **3** only, other over **4** only | ⓪ | ● |
| | 2. R intersection upper half of **4**, L intersection lower half of **3** (If **5** and **6** rotated 90°, read "upper" as "lower" and vice versa) | ⓪ | ● |
| **7** | 1. Entirely on **6**; any orientation | ⓪ | ● |
| | 2. Upright, entirely on **6** and entirely over **5** | ⓪ | ● |

*Example B*

| BLOCK | CRITERION | SCORE No | SCORE Yes |
|---|---|---|---|
| **1** | Parallel to edge of table + or - 15° | ⓪ | ● |
| **2** | 1. **1** and **2** not touching; can support **3** and **4** | ⓪ | ● |
| | 2. Ends line up within 2½ cm of ends of **1** | ⓪ | ● |
| **3** | 1. **3** to the L of **4** (If to the R, score **3** as **4** and **4** as **3** hereafter) | ⓪ | ● |
| | 2. Rests on both **1** and **2** | ⓪ | ● |
| | 3. On L half of **1** and **2**; **1** and **2** extend to L of **3** | ⓪ | ● |
| | 4. Angled end distal; correct side up | ⓪ | ● |
| **4** | 1. Rests on both **1** and **2** | ⓪ | ● |
| | 2. On R half of **1** and **2**; **1** and **2** extend to R of **4** | ⓪ | ● |
| | 3. Angled end distal; correct side up | ● | ① |
| **5** | 1. One end on **3** only, other on **4** only | ⓪ | ● |
| | 2. L intersection upper half of **3**, R intersection lower half of **4** | ● | ① |
| **6** | 1. One end over **3** only, other over **4** only | ⓪ | ● |
| | 2. R intersection upper half of **4**, L intersection lower half of **3** (If **5** and **6** rotated 90°, read "upper" as "lower" and vice versa) | ● | ① |
| **7** | 1. Entirely on **6**; any orientation | ⓪ | ● |
| | 2. Upright, entirely on **6** and entirely over **5** | ⓪ | ● |

*Example C*

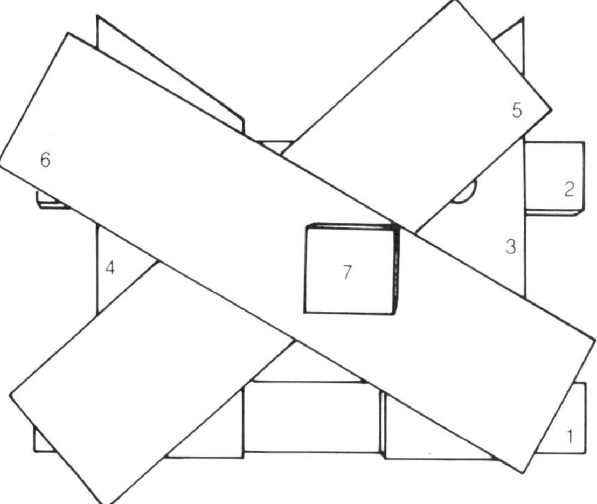


| BLOCK | CRITERION | SCORE No | SCORE Yes |
|---|---|---|---|
| **1** | Parallel to edge of table + or - 15° | ⓪ | ● |
| **2** | 1. **1** and **2** not touching; can support **3** and **4** | ⓪ | ● |
| | 2. Ends line up within 2½ cm of ends of **1** | ⓪ | ● |
| **3** | 1. **3** to the L of **4** (If to the R, score **3** as **4** and **4** as **3** hereafter) | ● | ① |
| | 2. Rests on both **1** and **2** | ⓪ | ● |
| | 3. On L half of **1** and **2**; **1** and **2** extend to L of **3** | ⓪ | ● |
| | 4. Angled end distal; correct side up | ● | ① |
| **4** | 1. Rests on both **1** and **2** | ⓪ | ● |
| | 2. On R half of **1** and **2**; **1** and **2** extend to R of **4** | ⓪ | ● |
| | 3. Angled end distal; correct side up | ● | ① |
| **5** | 1. One end on **3** only, other on **4** only | ● | ① |
| | 2. L intersection upper half of **3**, R intersection lower half of **4** | ● | ① |
| **6** | 1. One end over **3** only, other over **4** only | ⓪ | ● |
| | 2. R intersection upper half of **4**, L intersection lower half of **3** (If **5** and **6** rotated 90°, read "upper" as "lower" and vice versa) | ⓪ | ● |
| **7** | 1. Entirely on **6**; any orientation | ⓪ | ● |
| | 2. Upright, entirely on **6** and entirely over **5** | ● | ① |

**Figure 25**
**Constructional Praxis: Scoring Examples for Structure I**

***figure continued on next page . . .***

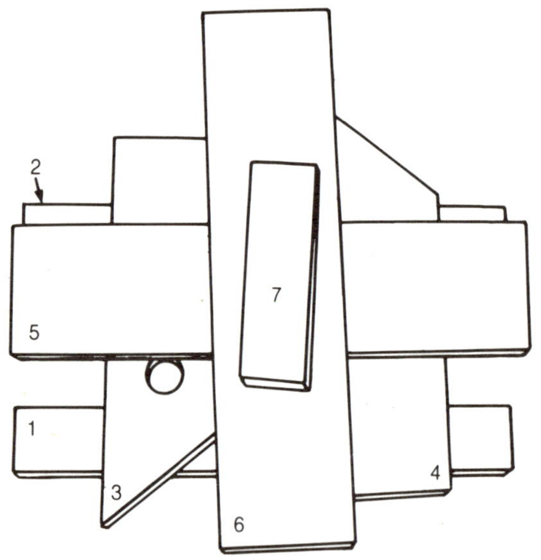


*Example D*

| BLOCK | CRITERION | SCORE No | SCORE Yes |
|---|---|---|---|
| 1 | Parallel to edge of table + or - 15° | ⓪ | ● |
| 2 | 1. **1** and **2** not touching; can support **3** and **4** | ⓪ | ● |
| | 2. Ends line up within 2½ cm of ends of **1** | ⓪ | ● |
| 3 | 1. **3** to the L of **4** (If to the R, score **3** as **4** and **4** as **3** hereafter) | ⓪ | ● |
| | 2. Rests on both **1** and **2** | ⓪ | ● |
| | 3. On L half of **1** and **2**; **1** and **2** extend to L of **3** | ⓪ | ● |
| | 4. Angled end distal; correct side up | ● | ① |
| 4 | 1. Rests on both **1** and **2** | ⓪ | ● |
| | 2. On R half of **1** and **2**; **1** and **2** extend to R of **4** | ⓪ | ● |
| | 3. Angled end distal; correct side up | ⓪ | ● |
| 5 | 1. One end on **3** only, other on **4** only | ⓪ | ● |
| | 2. L intersection upper half of **3**, R intersection lower half of **4** | ● | ① |
| 6 | 1. One end over **3** only, other over **4** only | ● | ① |
| | 2. R intersection upper half of **4**, L intersection lower half of **3** (If **5** and **6** rotated 90°, read "upper" as "lower" and vice versa) | ● | ① |
| 7 | 1. Entirely on **6**; any orientation | ⓪ | ● |
| | 2. Upright, entirely on **6** and entirely over **5** | ● | ① |

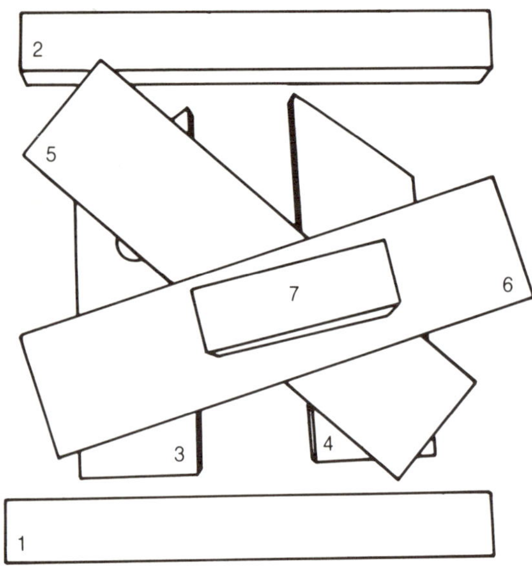


*Example E*

| BLOCK | CRITERION | SCORE No | SCORE Yes |
|---|---|---|---|
| 1 | Parallel to edge of table + or - 15° | ⓪ | ● |
| 2 | 1. **1** and **2** not touching; can support **3** and **4** | ● | ① |
| | 2. Ends line up within 2½ cm of ends of **1** | ⓪ | ● |
| 3 | 1. **3** to the L of **4** (If to the R, score **3** as **4** and **4** as **3** hereafter) | ⓪ | ● |
| | 2. Rests on both **1** and **2** | ● | ① |
| | 3. On L half of **1** and **2**; **1** and **2** extend to L of **3** | ● | ① |
| | 4. Angled end distal; correct side up | ⓪ | ● |
| 4 | 1. Rests on both **1** and **2** | ● | ① |
| | 2. On R half of **1** and **2**; **1** and **2** extend to R of **4** | ● | ① |
| | 3. Angled end distal; correct side up | ⓪ | ● |
| 5 | 1. One end on **3** only, other on **4** only | ⓪ | ● |
| | 2. L intersection upper half of **3**, R intersection lower half of **4** | ⓪ | ● |
| 6 | 1. One end over **3** only, other over **4** only | ⓪ | ● |
| | 2. R intersection upper half of **4**, L intersection lower half of **3** (If **5** and **6** rotated 90°, read "upper" as "lower" and vice versa) | ● | ① |
| 7 | 1. Entirely on **6**; any orientation | ⓪ | ● |
| | 2. Upright, entirely on **6** and entirely over **5** | ● | ① |

***figure continued on next page . .***

**Figure 25 (Continued)**
**Constructional Praxis: Scoring Examples for Structure I**

*Block 4:*

Criterion 1: Block rests on both Blocks 1 and 2.

Criterion 2: Entire block is placed on right half of both Blocks 1 and 2 (including midpoints), with some portion of Blocks 1 and 2 extending beyond the right side of Block 4.

Criterion 3: Angled end is distal relative to the child and correctly oriented (i.e., not reversed).

*Block 5:*

Block 5 must be scored by viewing the structure directly from above.

Criterion 1: Any part of one end is on Block 3 and not on Block 4 and any part of the other end is on Block 4 and not on Block 3.

Criterion 2: The intersection of the left edge of Block 5 and the left edge of Block 3 (or Block 4 if it is in the Block 3 position) is between and including the middle (note mark) of the left edge of Block 3 and the distal (upper) left corner of Block 3. Also, the intersection of the right edge of Block 5 and the right edge of Block 4 (or Block 3 if it is in the Block 4 position) is between and including the middle of the right edge of Block 4 and the proximal right corner of Block 4, including the right corner of Block 4.

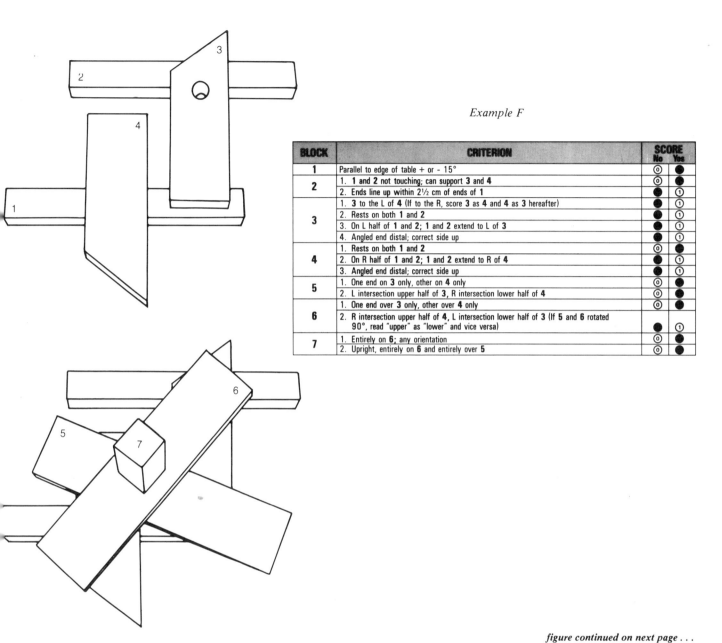

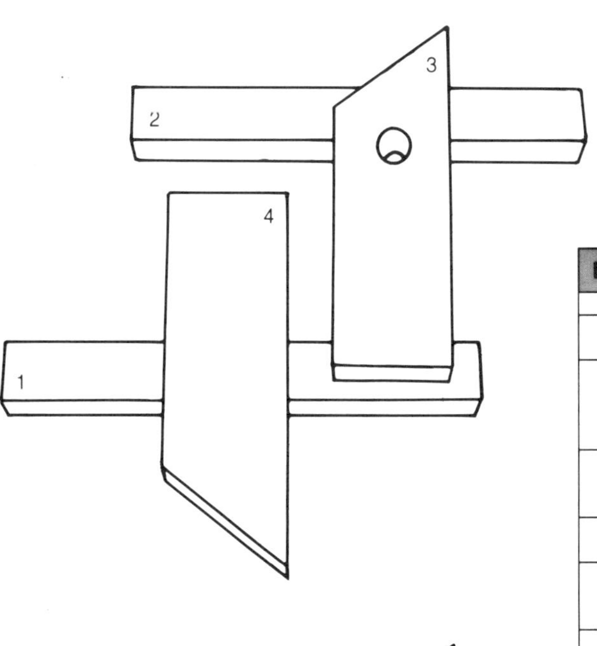


*Example F*

| BLOCK | CRITERION | SCORE No | SCORE Yes |
|---|---|---|---|
| 1 | Parallel to edge of table + or - 15° | ⓪ | ● |
| 2 | 1. **1** and **2** not touching; can support **3** and **4** | ⓪ | ● |
| | 2. Ends line up within 2½ cm of ends of **1** | ● | ① |
| 3 | 1. **3** to the L of **4** (If to the R, score **3** as **4** and **4** as **3** hereafter) | ● | ① |
| | 2. Rests on both **1** and **2** | ● | ① |
| | 3. On L half of **1** and **2**; **1** and **2** extend to L of **3** | ● | ① |
| | 4. Angled end distal; correct side up | ● | ① |
| 4 | 1. Rests on both **1** and **2** | ⓪ | ● |
| | 2. On R half of **1** and **2**; **1** and **2** extend to R of **4** | ● | ① |
| | 3. Angled end distal; correct side up | ● | ① |
| 5 | 1. One end on **3** only, other on **4** only | ⓪ | ● |
| | 2. L intersection upper half of **3**, R intersection lower half of **4** | ⓪ | ● |
| 6 | 1. One end over **3** only, other over **4** only | ⓪ | ● |
| | 2. R intersection upper half of **4**, L intersection lower half of **3** (If **5** and **6** rotated 90°, read "upper" as "lower" and vice versa) | ● | ① |
| 7 | 1. Entirely on **6**; any orientation | ⓪ | ● |
| | 2. Upright, entirely on **6** and entirely over **5** | ⓪ | ● |

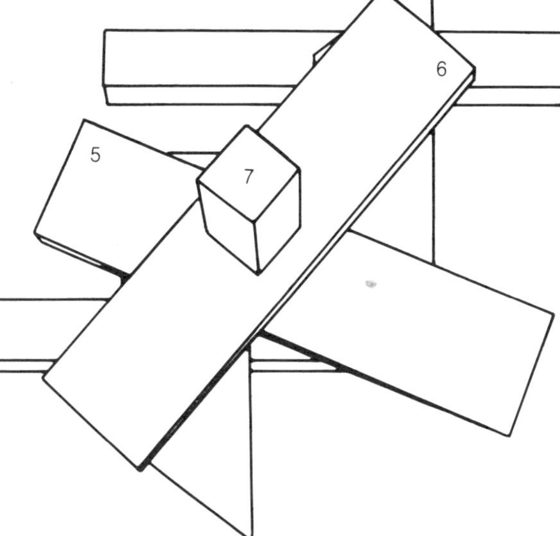


***figure continued on next page . . .***

**Figure 25 (Continued)**
**Constructional Praxis: Scoring Examples for Structure I**

Both conditions must be met.

Note that the scoring procedure for Block 5 is the same even if Blocks 3 and 4 are upside down or the angled end is proximal rather than distal to the child. Block 5 is still scored in relation to the midpoints of the outer edges of Blocks 3 and 4, even if the midpoint of the long edge of Blocks 3 and 4 does not correspond with the midpoint of the shorter side.

*Block 6:*

Block 6 must be scored by viewing the structure directly from above.

Criterion 1: Any part of one end extends over Block 3 (and not Block 4) and any part of the other end extends over Block 4 (and not Block 3).

Criterion 2: The intersection of the right long edge of Block 6 and the right edge of Block 4 is between and including the middle of the right edge of Block 4 (or Block 3 if it is in the Block 4 position) and the distal (upper) right corner of Block 4, including the corner (regardless of Block 4's orientation). Also, the intersection of the left long edge of Block 6 and the left edge of Block 3 (or Block 4 if it is in the Block 3 position) is between and including the middle of the left edge of Block 3 and the left proximal

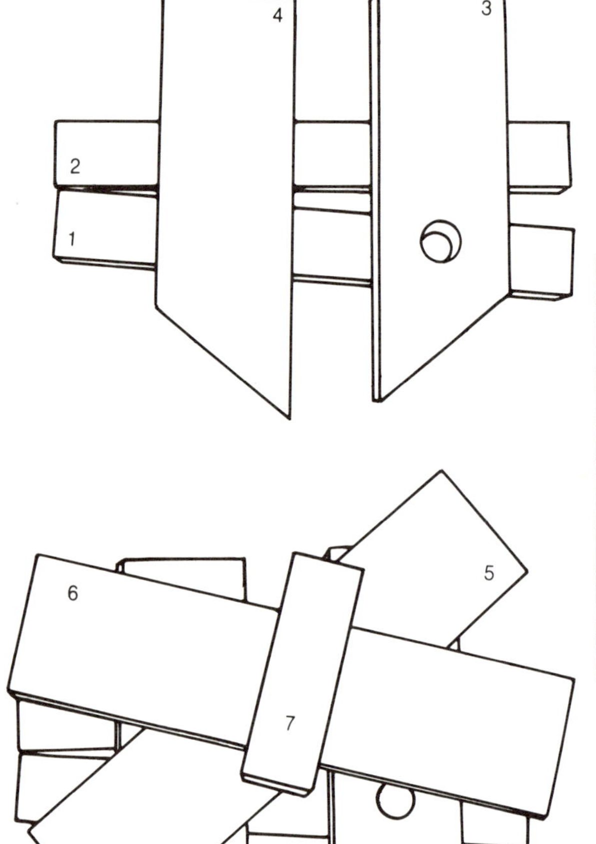


*Example G*

| BLOCK | CRITERION | SCORE No | SCORE Yes |
|---|---|---|---|
| 1 | Parallel to edge of table + or - 15° | ◎ | ● |
| 2 | 1. **1** and **2** not touching; can support **3** and **4** | ● | ① |
| | 2. Ends line up within 2½ cm of ends of **1** | ◎ | ● |
| 3 | 1. **3** to the L of **4** (If to the R, score **3** as **4** and **4** as **3** hereafter) | ● | ① |
| | 2. Rests on both **1** and **2** | ◎ | ● |
| | 3. On L half of **1** and **2**; **1** and **2** extend to L of **3** | ◎ | ● |
| | 4. Angled end distal; correct side up | ● | ① |
| 4 | 1. Rests on both **1** and **2** | ◎ | ● |
| | 2. On R half of **1** and **2**; **1** and **2** extend to R of **4** | ◎ | ● |
| | 3. Angled end distal; correct side up | ● | ① |
| 5 | 1. One end on **3** only, other on **4** only | ◎ | ● |
| | 2. L intersection upper half of **3**, R intersection lower half of **4** | ● | ① |
| 6 | 1. One end over **3** only, other over **4** only | ◎ | ● |
| | 2. R intersection upper half of **4**, L intersection lower half of **3** (If **5** and **6** rotated 90°, read "upper" as "lower" and vice versa) | ● | ① |
| 7 | 1. Entirely on **6**; any orientation | ● | ① |
| | 2. Upright, entirely on **6** and entirely over **5** | ● | ① |

**Figure 25 (Continued)**
**Constructional Praxis: Scoring Examples for Structure I**

corner of Block 3 (regardless of the orientation of Block 3). Both conditions must be met.

If Blocks 5 and 6 are placed perpendicular to each other, forming the appropriate *X*, but are rotated 90 degrees in their relationship to Blocks 3 and 4, then in the above description read "distal" (upper) as "proximal" (lower) and vice versa. In other words, scoring Block 6's location is not penalized because Block 5 was placed in a rotated manner.

*Block 7:*

Block 7 must be scored by viewing the structure directly from above.

Criterion 1: Block rests entirely on Block 6 in any orientation.

Criterion 2: Block is upright and rests entirely on Block 6 and entirely over Block 5.

**Building Structure II**

Four-year-olds do not build Structure II. You may elect not to ask certain other children to build Structure II. Structure II is illustrated on the protocol sheet (see also Figure 26). To allow adequate room for both sets of blocks, the child's structure, and the preassembled model for Structure II, you may need to move out of the way and observe from the side of the table.

Add to the child's Structure I blocks the additional blocks the child needs for making Structure II. Make sure that each individual block is placed on the table in the same left-right and distal-proximal orientation as it was in the box, with space between each block. The blocks need not necessarily be placed on the table in the same grouping as they were in the box. The important thing is that each individual block is placed in the same orientation as it was in the box, and not turned upside down or

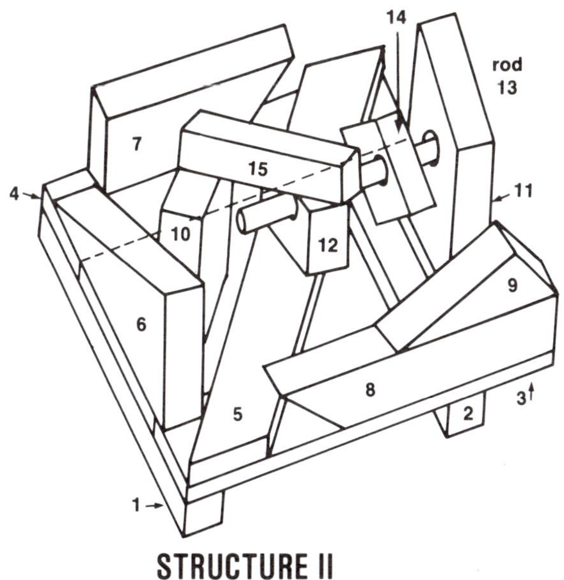


**Figure 26**
**Constructional Praxis: Diagram of Structure II**

rotated. Then place the preassembled model of Structure II on the table so that Block 3 is parallel to the edge of the table and its left-front corner is 7½ cm (about 3 inches) to the right of the child's right shoulder and about 25 cm (10 inches) from the front of the table (see Figure 27). Say:

*"**Now build one like this.*** (Place the model on the table.) ***Here are the blocks.*** (Point to the blocks.) ***Make them look just like this.*** (Point to model.) ***I'll start the building. I'll put this one here."***

Place Block 1 perpendicular to table front edge and parallel to table side edge (assuming a rectangular or square table). The proximal end of the block (i.e., nearest to the child) is positioned 10 cm (4 inches) from table front edge and 10 cm to the left of the child's midline (see Figure 27). Then say:

***"Now you build the rest here."*** (Gesture toward the space to the right of Block 1.)

If the child begins to move Block 1, say:

***"Keep this one here.*** (Return Block 1 to its original position.) ***Now you build the rest."***

If the child again moves Block 1, return it to position and hold it in position with your hand, repeating:

***"Keep this one here. Now you build the rest."***

Do not let the child begin to build until after all the directions have been given. If the child begins to build on the wrong side (i.e., left side) of Block 1, return the block used to the block area and say:

***"Build your building HERE."*** (Gesture toward the right side of Block 1.)

You may tell the child that it is all right to stand up to look at the Structure II model, but he or she may not move the model. This should be mentioned early in the test rather than when it appears that the child is having trouble.

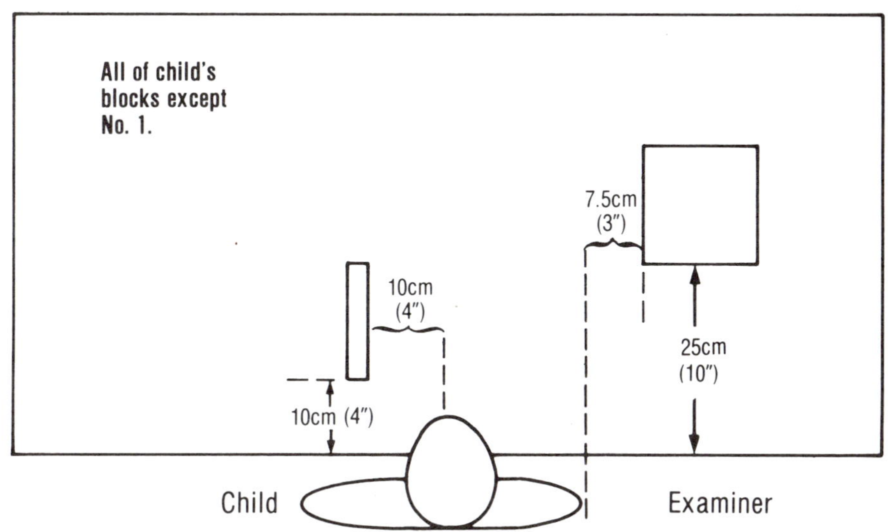


**Figure 27**
**Constructional Praxis: Placement of Structure II Model and Blocks**

Start the stopwatch as soon as the child starts building (usually indicated by picking up a block). You may need to move the child's pile of unused blocks so that the child can reach them without knocking over his or her structure. Begin scoring unobtrusively as the child builds; however, you will often have to complete scoring after the child has finished the structure. If necessary, encourage the child as he or she builds, saying, for example:

***"Make yours look like this one."***

Discourage the child from knocking down the structure either accidentally or on purpose. If the child starts to take the structure apart, say:

***"Don't take it down unless you can build a better one."***

If the child deliberately knocks down the structure and you have not finished scoring, score from memory as best you can and score blocks not used as omitted. If the child accidentally knocks down one or more blocks, he or she is asked to rebuild. Timing is continued without stopping. If you are aware that the child is making a conscious error (usually indicated by verbal expression of a preferred alteration), say:

***"No. Make yours like this one."***

The time limit for Structure II is 15 minutes. Discontinue Structure II at the time limit or if the child is nonproductive for 2 minutes (verified by asking the child, "Are you finished?"). The child is moved away from the table and you complete the scoring. On completion of scoring Structure II, place the loose blocks back in the box according to the layout sheets.

**Scoring Structure II**

The child may be dismissed while you score at leisure. Score each block on one of the eight parameters described below and fill in the circle for the parameter on the protocol sheet. Seven of the parameters describe different aspects of incorrect usage of a block; the eighth parameter indicates correct usage. *Although a block may fulfill the criteria for more than one parameter, it is scored on only one parameter.* If a block is used incorrectly in more than one way, it is scored for the parameter that will carry the heaviest penalty in scoring. The higher the number of the parameter, the greater the penalty (except for Parameter 8). For example, if a block is displaced more than 2½ cm (Parameter 2) and also rotated more than 15 degrees (Parameter 3), score for rotation. Solid boxes indicate that the block cannot be scored on a given parameter.

The blocks are scored in numerical order as designated on the diagram of the structure on the protocol sheet. Block 1 is scored as correct unless it is not used as part of the structure, in which case it is scored as omitted. Other blocks are scored in relation to the blocks upon which they are supposed to rest or to which they are supposed to be adjacent, or in relation to the general area in which they should be placed on the structure. Occasionally a child may attempt to build only a portion of the structure, omitting several lower-numbered blocks. In that case, the lowest numbered block utilized is scored first and is then used as the reference point for the scoring of the other blocks that were used. Any blocks that were not utilized or were added to the structure after the time limit are scored as omitted. If referent block(s) are displaced, score in reference to the lowest numbered block, back to Block 1 if necessary.

In the course of constructing the structure, the child may improve upon or worsen the placement of one or more blocks. The final placement of a block is the basis upon which it is scored.

Scoring examples for Structure II are presented in Figure 28.

**Definition of parameters.** The eight parameters are defined as follows.

1. *Displacement of 1 through 2½ cm.* The block is otherwise correctly placed, including its orientation and location, but is displaced by at least 1 cm but no more than 2½ cm (see Example A in Figure 28). If one portion of a block is more displaced than another, it is scored on its greatest displacement.
2. *Displacement more than 2½ cm.* The block is otherwise correctly placed, including its orientation and location, but is displaced more than 2½ cm (see Example A in Figure 28). A block scored on this parameter is not scored on Parameter 1. However, if there is a more highly erroneous error, the block is scored on that parameter and not this parameter.
3. *Rotation more than 15 degrees.* The block is rotated around a vertical axis more than 15 degrees but not so much that it would be considered a right-left or front-back reversal (see Example A in Figure 28). To facilitate measuring, place one side of the 15-degree angle guide at or parallel to the line where the edge of the block should be and see if the block edge is at an angle greater than 15 degrees.
4. *Upside down, right for left, front for back, or end for end.* The block is rotated (usually 180 degrees) around a horizontal or vertical axis in such a

*Example A*

*Rotation and Displacement*

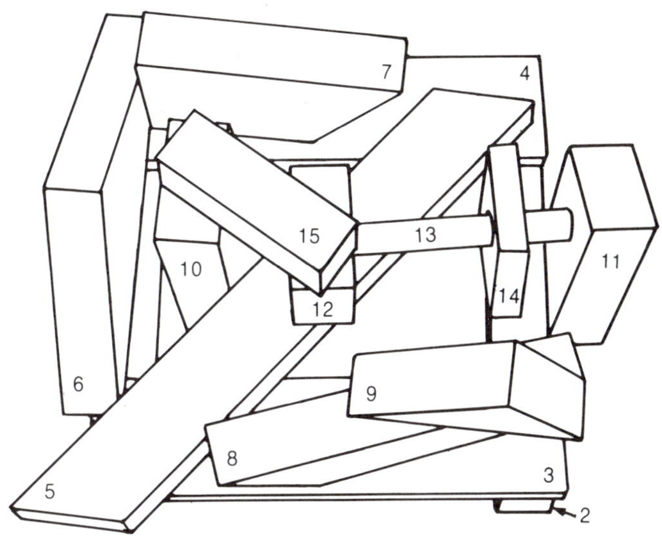


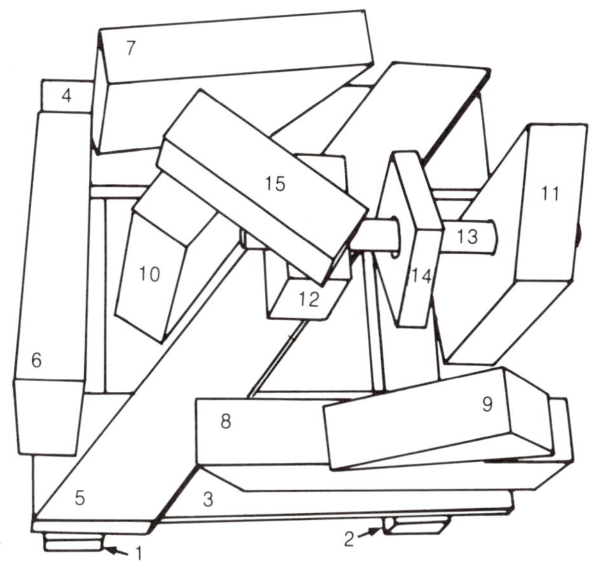


**Figure 28**
**Constructional Praxis: Scoring Examples for Structure II**

***figure continued on next page . . .***

*Example B*

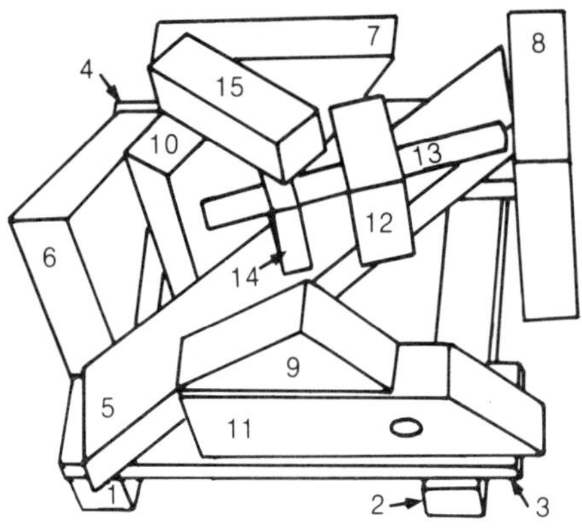

CHILD'S STRUCTURE

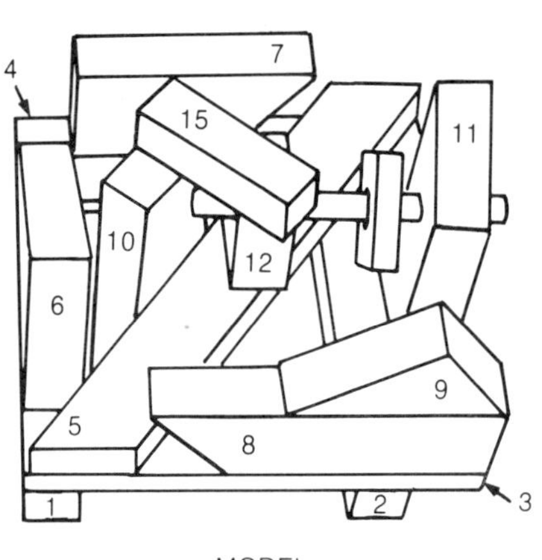

MODEL

| Block | Parameter | Reason |
|---|---|---|
| 1 | 8 | O.K. |
| 5 | 4 | Upside down |
| 8 | 5 | Incorrect but logical: substituted for block with similar shape |
| 9 | 1 or 2 | Depending on measurement; not scored down for previous error |
| 11 | 5 | Incorrect but logical: substituted for block with similar shape. Also upside down and R/L; however, scored on higher parameter |
| 12 and 14 | 5 | Incorrect but logical: similarly shaped blocks switched |
| 13 | 5 | Incorrect but logical: only through 2 blocks |

Blocks 2, 3, 4, 6, 7, 10, and 15 are displaced, rotated, or O.K., depending on results of measurement.

**Figure 28 (Continued)**
**Constructional Praxis: Scoring Examples for Structure II**

*figure continued on next page . . .*

*Example C*

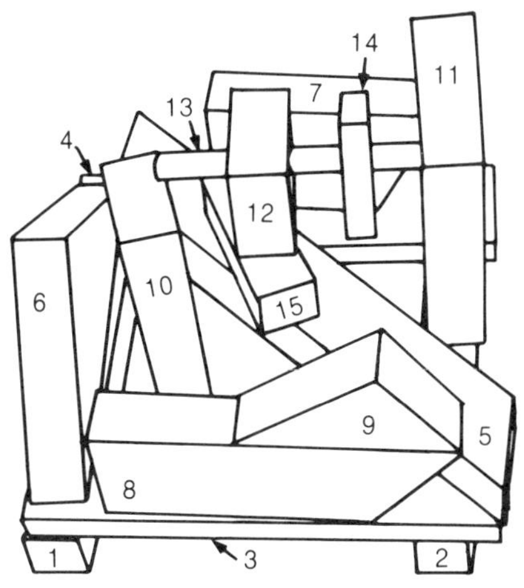

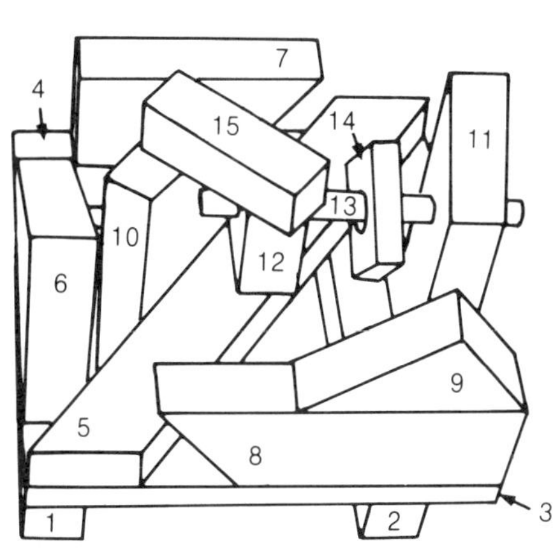


MODEL

| Block | Parameter | Reason |
|---|---|---|
| 1 | 8 | O.K. |
| 5 | 4 | R/L |
| 6 | 4 | Upside down |
| 7 | 1, 2, or 3 | Depending on measurements |
| 8 | 4 | R/L |
| 9 | 8 | O.K. |
| 11 | 5 | Incorrect but logical: on Block 2 rather than on table |
| 13 and 14 | 8 | O.K. |
| 15 | 6 | Gross mislocation |

Blocks 2, 3, 4, 10, and 12 are displaced, rotated, or O.K., depending on results of measurements.

**Figure 28 (Continued)**
**Constructional Praxis: Scoring Examples for Structure II**

***figure continued on next page . . .***

*Example D*

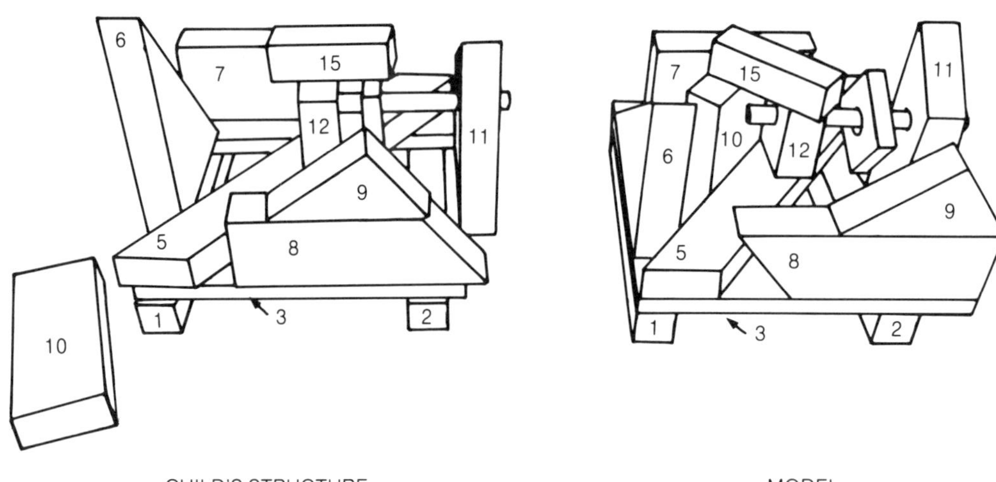


| Block | Parameter | Reason |
|---|---|---|
| 1 | 8 | O.K. |
| 6 | 4 | Rotated on horizontal axis |
| 8 | 4 | Upside down |
| 9 | 8 | O.K. |
| 10 | 7 | Omission |
| 11 | 4 | Reversed |
| 13 | 8 | O.K. |
| 14 | 5 | Incorrect but logical |
| 15 | 5 | Incorrect but logical: placed on 12 and 14 instead of 10 and 12 |

Blocks 2, 3, 4, 5, 7, and 12 are displaced, rotated, or O.K., depending on results of measurement.

**Figure 28 (Continued)**
**Constructional Praxis: Scoring Examples for Structure II**

*figure continued on next page . . .*

*Example E*

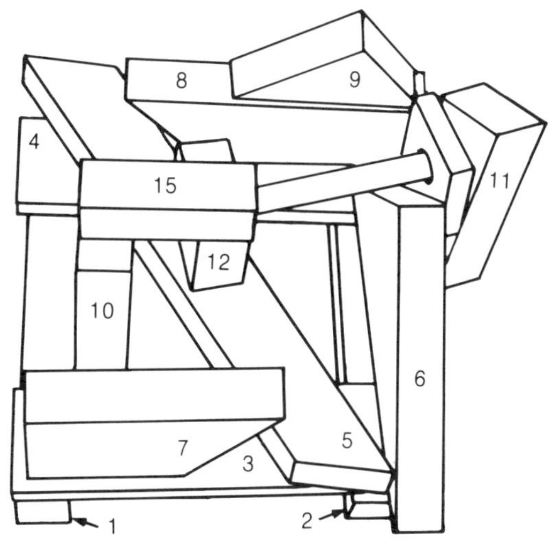


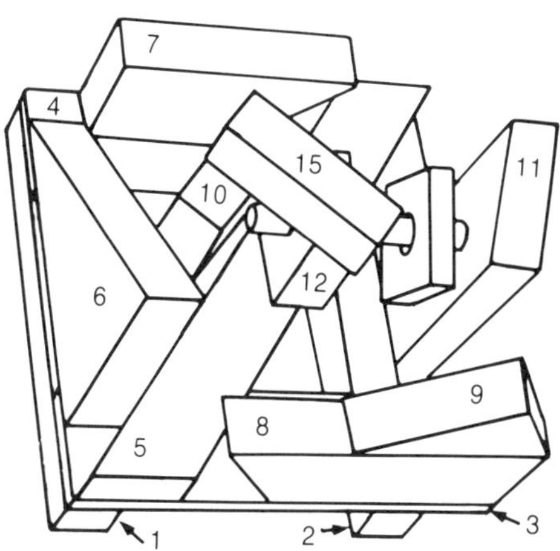


| Block | Parameter | Reason |
|---|---|---|
| 1 | 8 | O.K. |
| 5 | 4 | R/L and upside down |
| 6 | 5 | Incorrect but logical: placement of blocks contributes logically to the structure as a whole; model replication appears attempted |
| 7 | 5 | |
| 8 | 5 | |
| 11 | 4 | Rotated 180° around vertical axis |
| 13 | 8 | O.K.: horizontal and goes through Blocks 11, 12, and 14 |
| 14 | 8 | O.K.: appropriately on Block 13 |

Blocks 2, 3, 4, 9, 10, 12, and 15 are displaced, rotated, or O.K., depending on results of measurement.

**Figure 28 (Continued)**
**Constructional Praxis: Scoring Examples for Structure II**

***figure continued on next page . . .***

*Example F*

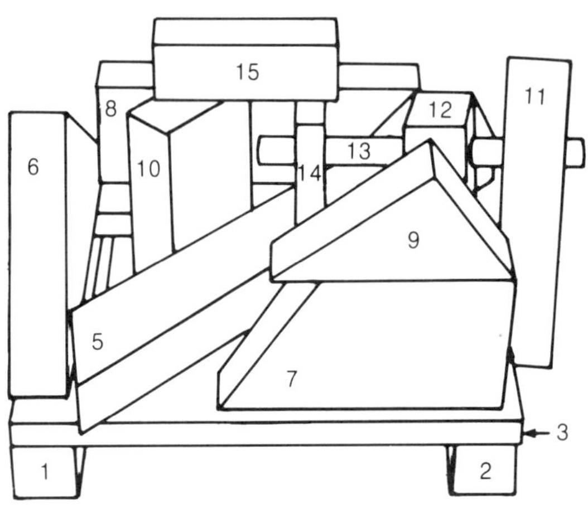


| Block | Parameter | Reason |
|---|---|---|
| 1 | 8 | O.K. |
| 5 | 4 | Upside down |
| 6 | 4 | Upside down |
| 7 | 5 | Incorrect but logical |
| 8 | 5 | Incorrect but logical |
| 9 | 8 | O.K. |
| 11 | 4 | Reversed |
| 12 | 5 | Switched with similar block |
| 13 | 8 | O.K. |
| 14 | 5 | Incorrect but logical |

Blocks 2, 3, 4, 10, and 15 are displaced, rotated, or O.K., depending on results of measurement.

**Figure 28 (Continued)**
**Constructional Praxis: Scoring Examples for Structure II**

*figure continued on next page . . .*

*Example G*

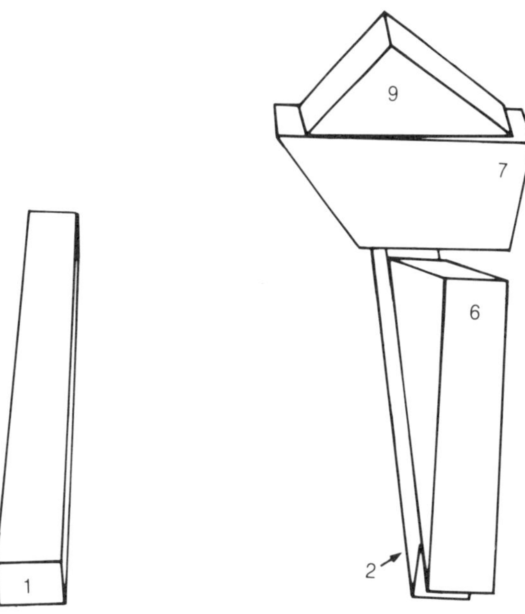


| Block | Parameter | Reason |
|---|---|---|
| 1 | 8 | O.K. |
| 2 | 1, 2, or 8 | Depending on measurement |
| 6 | 5 | Incorrect but logical |
| 7 | 5 | Incorrect but logical |
| 9 | 4 | Reversed end for end |

All other blocks would be scored as omitted.

**Figure 28 (Continued)**
**Constructional Praxis: Scoring Examples for Structure II**

*figure continued on next page . . .*

*Example H*

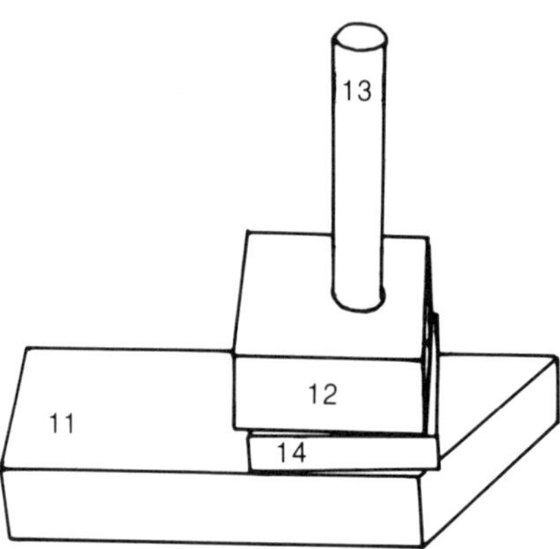


| Block | Parameter | Reason |
|---|---|---|
| 11 | 6 | Gross mislocation |
| 12 | 5 | Incorrect but logical |
| 13 | 6 | Vertical, therefore gross mislocation |
| 14 | 5 | Incorrect but logical |

All other blocks would be scored as omitted.

**Figure 28 (Continued)**
**Constructional Praxis: Scoring Examples for Structure II**

***figure continued on next page . . .***

*Example I*

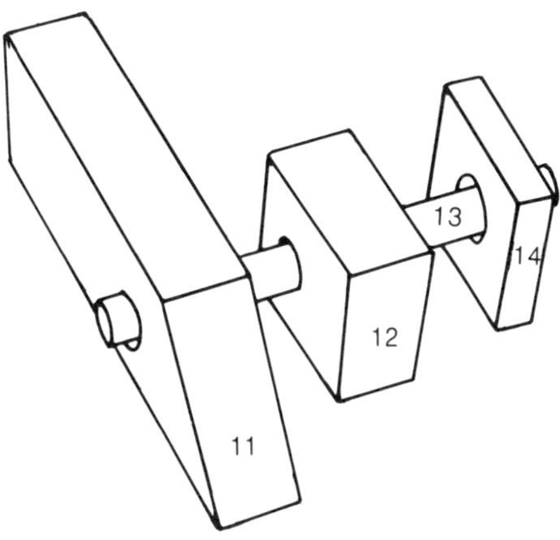


| Block | Parameter | Reason |
|---|---|---|
| 11 | 6 | Gross mislocation |
| 12 | 5 | Incorrect but logical: similar blocks switched |
| 13 | 8 | O.K.: horizontal and through all 3 blocks |
| 14 | 5 | Incorrect but logical: similar blocks switched |

All other blocks would be scored as omitted.

**Figure 28 (Continued)**
**Constructional Praxis: Scoring Examples for Structure II**

***figure continued on next page . . .***

*Example J*

| Block | Parameter | Reason |
|---|---|---|
| 1 | 8 | O.K. |
| 2 | 1,2,3, or 8 | Depending on measurement |
| 3 | 1 or 2 | Depending on measurement |
| 4 | 1 or 2 | Depending on measurement |
| 12 | 5 | Incorrect but logical: the relationship of the blocks to each other is similar to their relationship in the structure, although positioning within the structure is incorrect. |
| 13 | 5 | |
| 14 | 5 | |

All other blocks used in structure are grossly mislocated.
All blocks not used are "omitted."

**Figure 28 (Continued)**
**Constructional Praxis: Scoring Examples for Structure II**

manner that, although resting on proper block(s) or resting in the proper place, it is positioned upside down, in right-left reversal, in front-back reversal, end-to-end, or vertical instead of horizontal. Blocks 5, 6, 7, 8, 9, 11, and 12 are commonly scored on this parameter. The rotation of the block around the horizontal axis need not be 180 degrees for the block to be scored on this parameter; for example, Block 6 may lie on its longest side.

5. *Placement incorrect but logical.* An incorrect placement of a block which nonetheless makes a logical and recognizable contribution to the gestalt of the structure and is consistent with the model is scored on this parameter. The child is obviously trying to replicate the model. The most common incorrect but logical errors occur when a block is substituted for a similar block (e.g., switching placement of Blocks 7 and 8) or when incorrect placement of a block due to a previous error occurs (e.g., placing Block 15 on Block 12 only because of the omission of Block 10). When considering a score of incorrect but logical, analyze the block's placement in relation to the general vicinity in which the block should have been placed, the other blocks or block formations to which it should be related, and/or the block's similarity to other blocks. Observation of the child during construction may assist you in differentiating between scoring a block as incorrect but logical versus gross mislocation.
6. *Gross mislocation.* If a block has been placed on the structure without any logical resemblance to its or another block's correct placement in the model, it is scored on this parameter. Blocks simply lying on the table but pushed against the main structure or those that are added in a hit-or-miss manner fall into this category. Sometimes a child gives up and simply piles the blocks on top of each other, indicating that he or she knows they should be used but does not know how to use them.
7. *Omission.* When the child indicates that he or she has completed the structure, any block not used by the child as part of the structure is counted as an omission. The child may play with the block but not relate it to the structure.
8. *O.K.* If a block is not scored on any of the first seven parameters, it is considered correct.

**Scoring guidelines for each block.** The following guidelines pertaining to each block are provided to aid in scoring. In using these general guidelines, the assumption is that the lower numbered blocks listed as reference points are present. If reference block(s) are displaced, score in reference to the lowest numbered block, back to Block 1 if necessary. If the child has omitted any of the lower-numbered blocks, then proceed as previously directed. Score the lower-numbered block appropriately and then use it as the reference point for the scoring of other blocks which were used. Scoring examples for Structure II are presented in Figure 28.

*Block 1:* Since Block 1 is placed by you, it can be scored only as O.K. or omitted (i.e., the child built a structure which did not include Block 1).

*Block 2:* On the parameters of displacement and rotation, Block 2 is scored in relation to Block 1 (or block substituted for Block 1).

*Blocks 3 and 4:* On the parameters of displacement and rotation, score Blocks 3 and 4 in relation to Blocks 1 and 2. Blocks 3 and 4 may be scored as displaced in any direction in relation to Block 1, but are scored only for front-to-back displacement in relation to Block 2.

*Block 5:* On the parameters of displacement and rotation, score Block 5 in relation to Block 1 and/or Blocks 3 and 4. A common error is for Block 5 to be placed upside down (see Example B in Figure 28), right for left (see Example C in Figure 28), or both upside down and right for left (see Example E in Figure 28). Each of these would be scored on Parameter 4.

*Block 6:* On the parameters of displacement and rotation, score Block 6 in relation to Block 1 and/or Blocks 3 and 4. Note that Block 6 is *not* aligned with the front left corner of the structure when scoring the child's placement of this block. A common error is for Block 6 to be placed on its longest or shortest side (see Example D in Figure 28). This error would be scored on Parameter 4. If Block 6 is not on Blocks 1, 3, and/or 4 (e.g., is on the table) but still makes a recognizable contribution to the structure, score as incorrect but logical (Parameter 5).

*Block 7:* On the parameters of displacement and rotation, score Block 7 in relation to Block 4, even if Block 4 is displaced or rotated. Note that Block 7 is *not* aligned with the back left corner of the structure when scoring the child's placement of this block. A common error is to switch the position of Blocks 7 and 8 (see Example F in Figure 28), in which case both blocks are scored as incorrect but logical (Parameter 5).

*Block 8:* On the parameters of displacement and rotation, score Block 8 in relation to Block 3, even if Block 3 is displaced or rotated. (See Examples C and D in Figure 28 for examples of Parameter 4 errors on Block 8, and F for an example of a Parameter 5 error on Block 8.)

*Block 9:* On the parameters of displacement and rotation, score Block 9 in relation to Block 8 (or block substituted for Block 8), even if Block 8 is displaced or rotated. Note that in Example F in Figure 28, Block 9 is scored as O.K. even though it is not placed on Block 8.

*Block 10:* On the parameters of displacement, Block 10 is scored in relation to Blocks 1 and 5. When measuring for displacement, be sure to assess both side-to-side and front-to-back placement in relation to Blocks 1 and 5. Block 10 is measured for rotation only in relation to Block 5. If Block 5 is placed so that its front end is more toward the right than the left, score Block 10 for displacement only. For example, in Example C in Figure 28, Block 10 would be scored only for displacement due to the right-to-left placement of Block 5.

*Block 11:* On the parameters of displacement, score Block 11 for front-to-back and side-to-side displacement in relation to Blocks 2, 3, and 4, and on the parameter of rotation in relation to Block 2. If Block 11 is not on the table (e.g., on Blocks 2, 3, and/or 4) or is to the left of Block 2 (e.g., inside the structure), score Block 11 as incorrect but logical (see Example C in Figure 28).

*Block 12:* Score Block 12 in relation to Block 11 for displacement, unless Block 11 is displaced away from Block 2, in which case score in relation to Block 2. Do not score on rotation. Note that the hole in Block 12 is off center. If it is placed such that the hole is not in the upper middle position, score Block 12 on Parameter 4. If Block 12 is suspended rather than on Block 5, it is scored as incorrect but logical (see Example F in Figure 28).

*Block 13:* Do not score Block 13 (rod) on Parameters 1 through 4. Score as O.K. if it is horizontal (i.e., within 45 degrees to table edge) and goes through Blocks 11, 12, and 14, regardless of the arrangement of Blocks 11, 12, and 14 on the rod or the relationship of these blocks to the structure (see Example I in Figure 28). Score as incorrect but logical if it is horizontal and goes through one or two of Blocks 11, 12, and 14 (regardless of the arrangement of these blocks on the rod or their relationship to the structure). Score as grossly mislocated if Block 13 is vertical (i.e., greater than 45 degrees to tabletop) and goes through any of Blocks 11, 12, and 14 (see Example H in Figure 28), or is horizontal and does not go through any of the blocks it should go through, but is used.

*Block 14:* Do not score on Parameters 1 through 4. If placement of Blocks 12 and 14 is reversed (see Example B in Figure 28), both are scored as incorrect but logical. If Block 14 is used as support for another block (see Example D in Figure 28), score as incorrect but logical.

*Block 15:* On parameters of displacement and rotation, score in relation to Block 10, even if Block 10 is displaced or rotated. If Block 15 is only on either Block 10 or Block 12 (or blocks substituted for these) and not on both, it is scored as incorrect but logical. Another example of incorrect but logical placement of Block 15 is demonstrated in Example D in Figure 28 with Block 15 placed on Blocks 12 and 14 rather than on Blocks 10 and 12.

## Key Points to Remember

1. Seat the child to your left at the table, with adequate tabletop space for construction of "buildings."
2. You should be familiar with correct placement of blocks and structures in relation to the child and tabletop.
3. Complete the verbal directions before the child begins building.
4. The blocks which the child is to use are placed on the tabletop in the same orientation as they were in the box.
5. On both structures, place Block 1 in the appropriate place. The child is expected to build on it and not move it. Hold Block 1 in place if necessary.
6. On Structure I, the building is demonstrated in two parts.
7. Structure II is preassembled.
8. Four-year-olds do not build Structure II. You may elect not to ask certain other children to build Structure II.
9. The child cannot move either the Structure I or II models, but can stand up to look at them.
10. Start the stopwatch when the child picks up the first block.
11. The child is not allowed to build the structure next to yours or to add to your structure.
12. There is a time limit on both structures (3 minutes on each part of Structure I, 15 minutes on Structure II), although time is not recorded.
13. Stop the test at the time limit, after a period of nonproductive behavior (1 minute on Structure I, 2 minutes on Structure II), or when the child answers yes to "Are you finished?"
14. Discourage premature destruction of structure or conscious errors.
15. If the child destroys either of his or her structures, score from memory.
16. You should be familiar enough with scoring to be able to score quickly as the child builds, changing the score if child changes the structure.
17. To score Structure I, first review each scoring criterion. If the criterion is not met, score as 0; if the criterion is met, score as 1.
18. It is possible for the child to receive a score of 0 on the early items of Structure I and a score of 1 on later items.
19. To score Structure II, first review all scoring criteria.
20. Each block on Structure II is scored on only one parameter.
21. Score the block on the parameter that carries the heaviest penalty (the parameter with the highest number).

## 9. Postrotary Nystagmus (PRN)

**Materials**

- nystagmus board
- 30-degree angle guide
- protocol sheet
- stopwatch

The ball bearing mechanism of the nystagmus board must be kept clean and lubricated so that the board will turn freely. Use with persons weighing more than 70 pounds may damage the board.

In this test, the child is rotated both clockwise and counterclockwise on a board and the duration of postrotary nystagmus is timed.

The nystagmus board is placed on the floor about one meter from a blank wall and away from objects. Lighting should be even and of no greater or lesser intensity than usual. There should be no bright spots such as windows.

The child should not be allowed to play with the board before testing nor should he or she have engaged in any activity which would tend to produce extravestibular stimulation which could invalidate the test. Should such activity inadvertently occur, a rest of one minute should occur before PRN is readministered.

**Positioning the Child's Head**

Ask the child to sit cross-legged in the center of the nystagmus board and to hold on to the edge of the board in front of him or her. The child's head should be over the axis of the board and tilted forward 30 degrees. To determine the correct degree of neck flexion to tilt the head, the child is first asked to look straight ahead so that the head is upright. While in this position, the tip of the angle guide is placed so that it touches the shoulder and the left edge of the card is both vertical and lined up with the concha (visible cavity) of the ear. With the card held against the ear and pivoting on its point, the head is tilted until the right edge of the card is vertical.

Use of the card to position the head sometimes emphasizes head position so much that the child later flexes his or her head too much. If this is the case, a more accurate head position may be obtained without giving directions. Usually, when the child grasps the edge of the board, the head is positioned automatically and no further instruction on head position need be given other than in the following directions.

**Rotation to Left**

Stationing yourself on the left side of the child, hold the stopwatch in your left hand and place your right hand on the child's left knee, saying:

***"I am going to turn you around 10 times. While I'm turning you, hold your head like this. Don't move your head while you're turning. When you stop, look up and look at the wall. I will look at you, but don't you look at me; you look at the wall."***

Turn the child to his or her left (counterclockwise), maintaining a constant velocity of one rotation in 2 seconds. Each time the child's left knee comes around to the original position, push it again, keeping your hand on the child's knee as long as possible to keep a constant speed of rotation. To help maintain the correct speed, watch the stopwatch at least intermittently. The child's eyes should remain open during rotation.

It is important that the child not only maintain his or her head in the slightly flexed position, but also not rotate the head. Rotating the head in the direction of turning will decrease the duration of nystagmus, and rotating in the other direction will increase the duration of nystagmus.

On the completion of 10 rotations in 20 seconds, stop the child abruptly so that he or she is facing the wall. Switch the stopwatch back to zero and allow it to continue running. Say:

***"Look at the wall. Don't look at me.*** (Repeat after a few seconds:) ***Keep looking at the wall."***

Keep your hand on the child's knee when stopping the child to prevent his or her falling from the board. After stopping the board, watch the child's eyes carefully, noting the duration of the postrotary nystagmus. To avoid the child's focusing on you, observe from the side. Measure and record the duration of nystagmus as described in the section, "Recording Duration of Nystagmus," (see p. 86).

If you accidentally allow the board to go past the stopping point, do not move the board in the opposite direction to correct the child's position. Nystagmus is observed from that position.

Beware of providing unnecessary stimuli during the procedure, especially during the observation period, since extra stimuli such as noise may affect the nystagmatic response. However, if additional directions are required to maintain standard procedure, they should be given. It may be necessary to instruct the child to keep looking at the wall and not at you to reduce the inclination to turn the head and focus the eyes on you. If the child appears to be focusing on a spot on the wall, he or she should be advised to not look at anything. The child particularly should not look at a well-lighted area.

**Rotation to Right**

Allow at least a 30-second rest after the rotations to

the child's left (counterclockwise) before beginning the rotations to the child's right (clockwise). During this period, ask the child how he or she feels and, if necessary, say something reassuring. The child should stay seated on the board while you hold the board stable. After the 30 seconds have passed, say:

***"Now we'll go the other way. Keep your head this way*** (position head if necessary) ***and when you stop, look at the wall."***

The child is rotated to his or her right in the same manner as to the left, again pushing on the left knee, and the duration of nystagmus is recorded as described in the following section.

It is normal to be slightly dizzy after the 10 rotations to the left and even more after the additional rotations to the right, but the dizziness normally leaves quickly. If, after 20 rotations, the child is uncomfortably dizzy, you can modulate the effects of vestibular input by providing extra proprioceptive input from heavy work such as asking the child to pull hard on your arms.

**Recording Duration of Nystagmus**

Begin timing as soon as the board is stopped. Watch the child's eyes and note the duration of the nystagmus. When the nystagmus stops, stop the watch but continue to observe the child's eyes for at least 2 seconds. If the watch has been terminated prematurely, the appropriate amount of time is added to the clocked time. The duration is recorded on the protocol sheet to the nearest whole second. Be sure to enter the time as two digits—if less than 10 seconds, enter 0 as the first digit (e.g., enter 6 seconds as "06"). Be careful to record only the duration of the rhythmic, back-and-forth movements of primary nystagmus and not random eye movements (which frequently follow postrotary nystagmus) or secondary nystagmus. (In secondary nystagmus, the fast beat is in the opposite direction. It is not frequently seen in 4- through 8-year-olds.)

If the test is administered only one time, resulting in only one clockwise and one counterclockwise score, use those same scores when filling in bubbles for the second administration.

**Retest**

This test is administered a second time after the second half of the SIPT is completed, and the PRN scores for the retest are entered at that time.

## Key Points to Remember

1. Place nystagmus board away from objects.
2. Do not allow the child to play with the board prior to testing.
3. Position the child in 30-degree neck flexion (measure to be sure) and be sure the child knows to keep his or her head tilted throughout the rotation.
4. Rotate to the child's left first, then right.
5. Push on the child's left knee for both directions.
6. Maintain steady, even rotation by keeping your hand on the child's knee as long as possible.
7. Time rotations at 1 per 2 seconds, 10 revolutions in 20 seconds.
8. Restart the stopwatch as soon as the child is stopped. Stop the stopwatch when the nystagmus stops, but continue to watch the child's eyes for at least 2 seconds.
9. There should be no distracting objects in the line of vision after stopping the board.
10. The child should not focus on anything following rotation.
11. Allow at least 30 seconds of rest before the next rotation.
12. Get two readings in each direction: one set at each testing session.
13. Record the number of seconds of nystagmus to the nearest full second for both the left and right directions in both the initial test and the retest. Enter the seconds as two digits (e.g., enter 6 seconds as "06").
14. If the test is not administered a second time, record the scores obtained for the first administration as scores for the second administration as well.

## 10. Motor Accuracy (MAc)

**Materials**

- test sheet
- 2 red nylon-tipped pens
- tape
- line measure
- protocol sheet
- stopwatch

In this test, the child attempts to draw a line on top of a printed line with both the left and right hands.

Sit on the other side of the table facing the child. Smoothing the crease in the test sheet so that it is as flat as possible, fasten the open test sheet to the table with tape at all four corners to prevent its movement. The lower edge of the open test sheet is parallel to and one inch away from the front edge of the child's side of the table, with the middle of the test sheet directly in front of the midline of the child.

**Practice Line**

To begin the test, make sure the ink in the pen flows adequately. Give the pen to the child's previously determined writing hand and record that hand preference. If in doubt about the writing hand, place a pen at the midline of the child's body and ask the child to pick it up and write the first initial of his or her name. If still in doubt, ask the child to do the same thing with the other hand. Then say:

***"Watch me. I'm drawing a line on top of this black line."***

Using the second red ink pen, draw a line over the short black line provided for practice in the left center of the sheet (see Figure 29). Do not intentionally go off the line. Then say:

***"Now you draw a line on top of this black line."***

Point to the longer black line provided for practice. If the child draws the line so that he or she will take either less than 30 seconds or more than 120 seconds on the test proper, ask the child to go either faster or slower. Sixty seconds is ideal. If the child tries to improve his or her line by going back over the printed line, discourage this practice.

**Test Lines**

When the right hand is tested, the child starts his or her line just left of the middle of the test sheet. When the left hand is tested, the child's drawn line begins on the right side of the printed black line. In either case, the child draws a line over the entire heavy black line, or as close to it as possible.

Begin by saying:

***"Now draw a line on top of this black line beginning here*** (point to test proper) ***and going around to here.*** (Move finger indicating path and end of line.) ***Draw carefully."***

Start the stopwatch when the child's line passes the first horizontal short-dashed line one inch above the starting point. If the child draws the line so slowly that he or she is apt to take longer than 120 seconds, ask the child to go faster; tell the child that it is not necessary to stay exactly on top of the line but to do the best he or she can at the required speed. If the child draws so rapidly that he or she is apt to finish before 30 seconds, ask the child to go more slowly and to be careful to stay on the line. Again, 60 seconds is ideal. Encourage consistency of speed, but prompt the child's speed not more than a few times after he or she begins.

Stop the watch when the child's line passes the horizontal short-dashed line near the completion of the line. In the space provided on the test sheet, record the time to the nearest whole second. The accuracy of the child's line is scored later (see "Measuring the Accuracy of the Lines," p. 89).

On completion of the test with the preferred hand, the nonpreferred hand is tested. Turn the test sheet over, tape it, and indicate that the other hand is being used and that it is the second test. Make sure the child has an adequate grasp of the pen with the nonpreferred hand, and say:

***"Now you will draw a line with the other hand. You will draw on this black line beginning here*** (point to beginning of the line) ***and go around to here."*** (Sweep hand around over line and point to other end of black line.)

Use the same procedure as above to ensure that the child will maintain a fairly constant speed of drawing, requiring not less than 30 nor more than 90 seconds. Record the actual time taken by the child.

Some children tend to stop and go back over the printed black line to produce a better line when their drawn line deviates from the printed line. This practice is discouraged; it increases test time and does not improve scores since the most erroneous line enters into the scoring. Try to control the tendency to go back over a line with comments such as:

***"Do not stop and go back over the line. Keep going, but get back on the black line."***

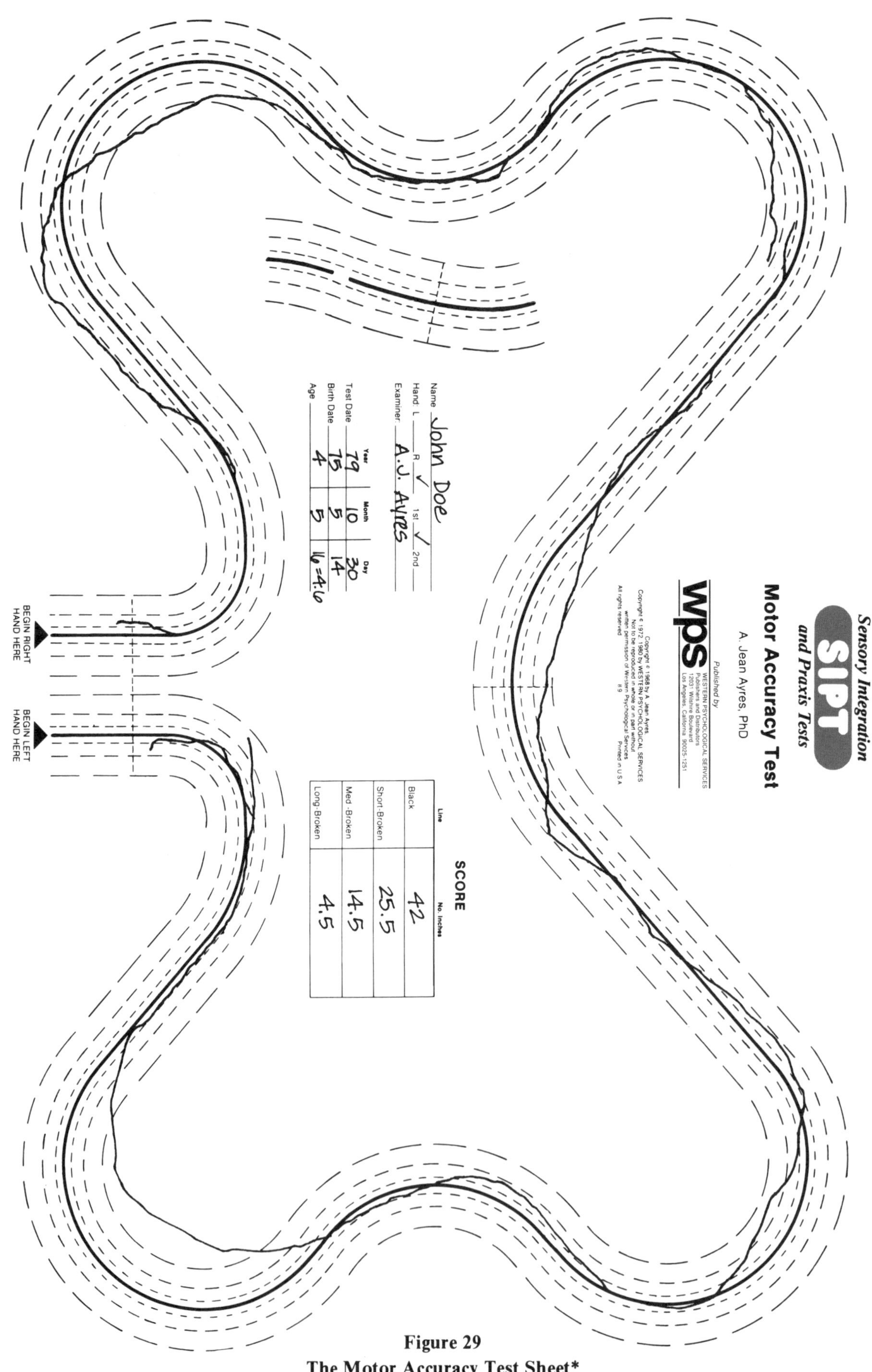


**Figure 29**
**The Motor Accuracy Test Sheet***

*Figure not shown at full size.

Drawing the line in the opposite direction intended is not allowed.

If the child skips a portion of the printed black line, quickly place the child's hand back to where he or she stopped, saying:

***"Begin where you left off."***

If skipping a portion of the task is not caught until after the child completes the task, tell the child to go back and draw in the part of the line that was skipped. *There should be no undrawn space on the printed black line or its adjacent areas.* Should a skip be overlooked, draw a straight line to join the two ends of the child's drawn lines to establish a basis for scoring all portions of the child's drawn line.

**Recording Time**

During the test, the number of seconds taken to draw the line with both the left and right hands is written on the test sheet. After the test is completed, fill in the circles corresponding to the time score on the protocol sheet. Be sure to enter the number of seconds as three digits; if less than 100, enter 0 in the first column (e.g., enter 52 seconds as "052").

**Measuring the Accuracy of the Lines**

Before measuring, check the reliability of the line measure on a 6-inch line. Then use the line measure to compute the total length of the child's drawn line that is off the solid black line or outside areas bounded by the short-, medium-, and long-dashed lines. If any part of the child's red line is not fully on the printed solid line, that length of the child's line is considered fully off the black line. In essence, you should not be able to see any part of the red line if it is fully on the black printed line. Similarly, the child's drawn line is considered outside the broken line area if any part of the red line is seen beyond the broken lines, even if the child's line is mainly within the stated boundaries of the broken lines.

To begin, the line-measure inches scale is set at 0 (or 39, which is the same place). Run the wheel of the line measure on the solid black line or the broken lines along those sections where the child's drawn line is off the printed line. The line measure is not run over the child's line. When the child's red line returns to the printed black line or within the area bounded by broken lines, lift the line measure from the printed line so that portion will not be measured. If the child has drawn two lines in the same area in an attempt to correct his or her line, the line *farthest* from the printed line is used for measurement. Measuring begins and ends at the horizontal broken line about one inch above where the child starts to draw. All measurements are made to the nearest half inch.

First run the line measure along the solid black line wherever the child's red line is not fully on it, and record the distance indicated on the line measure on the protocol sheet to the nearest half inch. The distance is entered as three digits—two digits for whole inches plus one digit for tenths (e.g., 6½ inches is entered as "06.5," 23 inches as "23.0," and if the child's line is not outside an area, "00.0").

Then reset the line measure at 0, and run the line measure along the short-dashed line wherever the child's red line is outside the short-dashed lines. When the child's line is to the right of that area, run the line measure along the right short-dashed line; when the child's line is to the left of the short-dashed line, run the line measure along the left short-dashed line. When the total length of the child's line outside the short-dashed lines is measured, the number of inches to the nearest half inch is recorded on the protocol sheet, as before.

Next, measure the distance that the child's red line falls outside the medium-dashed lines and record it on the protocol sheet.

Finally, if the line goes outside the long-dashed lines, run the line measure along the long-dashed line and record the total distance the child drew his or her line outside that line.

Thus, parts of a child's line may be measured up to four times—once each for solid, short-, medium-, and long-dashed lines. For example, any line outside the long-dashed lines would also be outside the solid, short-, and medium-dashed lines.

Remember that the child's line may be off the solid line (or even other areas) more than 39 inches, in which case, 39 must be added to the number indicated on the line measure.

For accurate results, the line measure must be kept vertical when run along a line. Lubricating with graphite will help prevent the mechanism from sticking. The line measure is a delicate instrument and has a limited life span. It will need to be replaced after considerable use. Thus, it is important to check the line measure before each use.

When the child makes a continuous line, scoring is a simple and easy process, but when the lines are irregular because of conditions such as neuromotor incoordination, scoring is complicated. Even when a line is made accidentally, due to incoordination, it is considered part of the test effort and is scored accordingly.

## Key Points to Remember

1. Sit across the table from the child.
2. During the demonstration, do not intentionally go off the line.
3. The child uses a red nylon-tipped pen.
4. The hand preferred for writing is the first hand tested.
5. Start timing when the child's line crosses the horizontal broken line about one inch above the starting point.
6. Remind the child not to stop or go back over the line.
7. The child is not allowed to draw the line opposite the intended direction.
8. Stop timing when the child crosses the horizontal broken line near the end of the test.
9. If the child skips a portion of the line, quickly place the child's hand where he or she stopped and say, "Begin where you left off." If the skipped portion is not caught until after the test, tell the child to go back and fill in the skipped portion. If the skipped portion is overlooked, draw a straight line to join the ends of the child's lines. If the child goes back to fill in the skipped portion, keep the stopwatch running to include time necessary to fill in the skipped portion.
10. Before measuring the accuracy of the line, check the reliability of the line measure on a 6-inch line.
11. Make sure that the line measure is set at 0 when beginning measurement.
12. Score any portion of the child's line that is not on the printed line, even if most is on the line.
13. Run the line measure along the printed line, not the child's line.
14. Run the line measure in the same direction; do not back up.
15. For each hand, record the number of seconds (to the nearest whole second) and the number of inches (to the nearest half inch) that the child's line is off the solid, short-, medium-, and long-dashed lines.
16. The dominant hand should complete the test between 30 and 120 seconds. The nondominant hand should complete the test between 30 and 90 seconds.

---

## 11. Sequencing Praxis (SPr)

### Materials

- protocol sheet

In this test, the child repeats a series of hand or finger tapping movements following your demonstration. There are three trial items and nine test items of sequenced motions. Each test item consists of five or six subitems. Within a given item, the first subitem sequence of positions is relatively simple; movements are added to each successive subitem, making them progressively more complex.

Sit across the table from and facing the child. Remove rings and other jewelry that make sounds when tapping the table. If possible, the child's jewelry should also be removed.

Give the motor sequences by tapping the table (or other designated location) as indicated on the protocol sheet. Be sure to start with the correct hand. Tap uniformly at the rate of one every half second (two taps per second) and firmly enough to make a sound.

The trial and test items are illustrated on the protocol sheet (see examples in Figure 30). The sequence of positions for each subitem on the protocol sheet is read across, left to right. Vertical lines separate parts of a sequence. If two hands are pictured with no dividing vertical lines, both hands tap simultaneously in the position pictured.

### Trial Items

To begin, Say:

***"I am going to move my hands. When I stop moving, you do the same thing. If I use this hand,*** (hold up left hand) ***you use this hand.*** (Touch child's right hand.) ***If I use this hand,*** (hold up right hand) ***you use this one.*** (Touch child's left hand.) ***If I use both hands,*** (hold up both hands) ***you use both hands.*** (Touch both of child's hands.) ***Now do this."***

Begin with both hands on the table near your midline, with fingers slightly flexed in a natural resting position. This is the starting position for each subitem sequence.

Execute Trial I by tapping your right hand on the table twice in 1 second. Tap firmly enough to make a sound. Audibility of taps should be consistent throughout the sequences. After performing the motor sequence, casually return your hands to the original starting position. The child is encouraged to do the same.

If the child completes the correct number of sequences in the correct position, commend the child and give the next trial item. If the child uses an incorrect hand position or does too few or too many motions (taps), repeat Trial I, saying:

***"Watch me again."***

During the three trial items, encourage the child to start with the correct hand, to wait until you finish before starting a subitem, and to use the correct number of motions or taps in the sequence. Give as much instruction as necessary to teach the child the correct method of performing the three trial items. When the child has mastered the concept of the test as much as possible, give the test items.

**Test Items**

Each item is demonstrated only once, so make certain that the child is paying attention before demonstrating.

The hand items (Items 1–6) are administered to all children, beginning on Item 1 and stopping when the discontinuance criterion is met (see "Discontinuing Items and the Test," on this page). The finger items (Items 7–9) are not administered to 4-year-olds.

Hand positions are shown in pictures on the protocol sheet. The code for hand items is: R = right hand, L = left hand, B = both hands. Some pictures provide a front view and others a top view, depending on which perspective provides the clearest representation. The items pictured from a front view (Items 3, 5, and 6) are to be mirrored in the demonstration. For the finger items, numbers indicate which finger is to be used: 1 = thumb, 2 = index, 3 = middle, 4 = ring, 5 = little. These symbols and the pictures of hands refer to *your* hands, not the child's.

**Sample Hand Items**

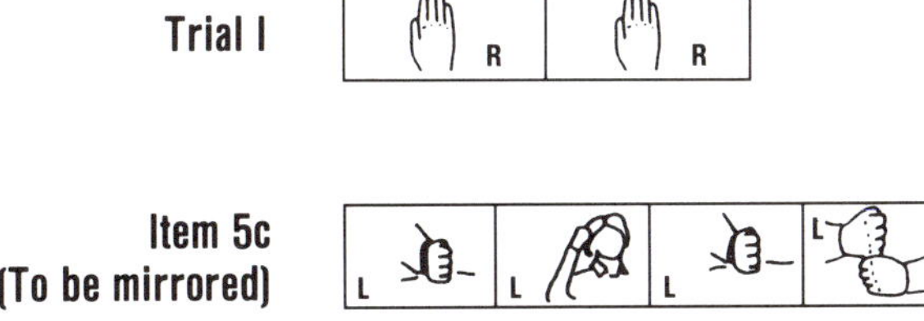

**Sample Finger Items**

Item 7a L 1 5 5

Item 8f L 1 2 3 1 3

**Figure 30**
**Sequencing Praxis: Sample Item Illustrations**

On Items 5 and 6, which include sequences in which one fist taps the other fist, do not tap the table with the bottom hand. Place the bottom hand on the table and tap it with the other hand. On Item 5, the head is also tapped with the fist.

After each sequence, record the child's score by filling in the 0, 1, or 2 circle to the right of the sequence on the protocol sheet. Criteria for determining the scores are described in the following section, "Scoring Sequences."

Be careful not to give the child cues by beginning to score before he or she finishes moving or by directing your eyes in a way that cues him or her to the correct position.

Do not give any instruction or feedback regarding the child's performance of the test items, except when it is necessary to ask the child to wait until you complete demonstration of a subitem. If a child begins before you finish, say:

***"Wait until I finish; then you do it."***

Then give the subitem again.

**Scoring Sequences**

Trial items are not scored. On the test items, score each subitem as 0, 1, or 2 by filling in the corresponding circle on the protocol sheet. Smoothness or speed of execution does not enter into scoring.

Score as 0 if the sequence is executed with the incorrect hand position or finger position, the movement is incorrect, or there are too few or too many motions in the sequence.

Score as 1 if the sequence was started or completed incorrectly (i.e., wrong hand position or finger position, or wrong motion) but error is recognized (such as verbally indicating error or by waving hands as if to "erase" movement) and the sequence is started over and completed correctly. Note that corrections scored as 1 pertain only to those for inaccurate position or motion. Corrections for starting with the wrong hand do not lower a score of 1 or 2.

Score as 2 if the sequence is completed with the correct hand positions or finger positions in the correct sequence with the correct number of taps. Also score as 2 if the hand or finger positions are correct in number and sequence but the sequence was started with the wrong hand (nonmirrored response).

**Discontinuing Items and the Test**

Discontinue each item when two consecutive subitems are incorrect (scored as 0) and go on to the next item. Discontinue the test when the first two subitems of two consecutive items are scored as 0.

## Key Points to Remember

1. Sit across the table from the child.
2. Remove all wrist and finger jewelry.
3. Administer the hand items (Items 1–6) to all children.
4. Do not administer finger items (Items 7–9) to 4-year-olds.
5. Perform the demonstration smoothly and uniformly, at two taps per second.
6. Be sure to start with the correct hand.
7. Test items are demonstrated only once. The child should be attending to your demonstration.
8. If you make an error, stop and repeat the item immediately.
9. If the child starts before you finish, stop and ask the child to wait until you have completed the item; then repeat the demonstration.
10. Do not give cues with your eyes or hands.
11. Your hands return to a midline resting position after each sequence. The child is encouraged to do the same.
12. Record the score after every subitem.
13. Score the sequence as 0 if the child uses an incorrect hand/finger position or completes too few or too many movements.
14. Score the sequence as 1 if the child begins the sequence incorrectly (in terms of hand/finger position or movement) and then starts over and performs the sequence correctly.
15. Score the sequence as 2 if it is completed with correct hand/finger positions and the correct number of movements.
16. Use of the incorrect hand (nonmirrored response) does not affect the child's score. Corrections for starting with the incorrect hand do not lower the score.
17. Discontinue each item when two consecutive subitems are scored as 0 and go on to next item.
18. Discontinue the test when the first two subitems of two consecutive items are scored as 0.

---

## 12. Oral Praxis (OPr)

**Materials**

- protocol sheet

In this test, you will be demonstrating a series of movements or positions of the lips, tongue, and mouth, which the child is asked to imitate. Do not wear brightly colored lipstick. Sit facing the child across the table.

**Test Items**

There are no trial items on this test. The movements or positions to be demonstrated in the test items are described on the protocol sheet. The test begins with Item 1. Say:

***"Watch my mouth and do this."***

Then demonstrate Item 1 by sticking out your tongue as far as possible, exaggerating the action. Score the child's performance as 0, 1, or 2, depending upon the skill of the response (see "Recording Item Scores," p. 93).

If the child receives a score of 2 (performs well), commend the child. If the child receives a score of 1 or 0 on Item 1, demonstrate the item again, saying:

***"Try to do this just as I do it."***

Reinstruct the child until he or she appears to understand the test requirements.

Prior to demonstrating Item 2, say:

***"Wait until I finish before you do these. Do it as many times as I do it."***

Administer Item 2 as described on the protocol sheet (click teeth three times in 1 second). Then reinforce the necessity of completing the item with a sufficient number of actions.

All items are administered to all children. Execute the movements in a precise manner, with separate parts of sequences clearly differentiated. Sequenced items are given within the approximate time indicated; however, smoothness is more important than timing. No verbal description of the item is given to the child except on Item 12. Keep the child's hands away from his or her mouth at all times.

Before each item, make sure the child is watching your face. Directions such as "watch me" or "watch this one" may be used before every item, if necessary, to gain the child's attention. If the child does not watch the demonstration, repeat the item. However, if the child watches any portion of the item, the item is not readministered.

If the child begins execution of an item before you have finished demonstrating, stop the demonstration. Before repeating the demonstration, say:

***"Wait until I finish; then you do it."***

**Recording Item Scores**

After each item, fill in the 0, 1, or 2 circle corresponding to the item score. If the score cannot be immediately determined, note the child's response and score later using the criteria listed below.

The response is scored as 0 if the action is incorrectly performed or not completed with a sufficient number of movements.

The response is scored as 1 if the action is executed with poor quality, or adequately with a sufficient number of movements but with poor sequencing. More, but not less, than the correct number of sequences may be completed, and the full sequence must be completed consecutively. Poor sequencing may be expressed as lack of rhythmic timing of a series of movements. A score of 1 is also given if the action is begun incorrectly but then a full response is completed accurately.

The response is scored as 2 if the action is executed well and with good sequencing where required. Good sequencing is defined as smooth and well-timed movement from one position to another. More, but not less, than the correct number of sequences may be completed.

Most children will mirror your action. If the child makes a nonmirrored response or reverses the action, such as starting on the right instead of the left on Item 9 or touching the bottom lip first instead of the top lip on Item 15, the item is scored as correct as long as the criteria listed below are met. The one exception to this rule is on Items 13 and 14. If the child's tongue moves to the right on Item 13, the tongue must move to the left on Item 14, and vice versa. If the child's tongue moves in one direction only, score Item 13 as correct (2 or 1) and Item 14 as incorrect (0).

The following are specific criteria for each item.

*Item 1:* Stick out tongue.

Score as 0 if incorrect.

Score as 1 if the tongue is visible but not protruding past lips.

Score as 2 if the child protrudes tongue past lips.

*Item 2:* Click teeth—three times (1 second).

Score as 0 if incorrect or insufficient number of movements.

Score as 1 if upper and lower teeth move up and down toward each other but do not touch and/or do not make an audible sound three times.

Score as 2 if upper and lower teeth make visible contact and make any audible sound three times.

*Item 3:* Pucker lips with obvious protrusion.

Score as 0 if incorrect.

Score as 1 if lips are pursed but not visibly protruding.

Score as 2 if lips are protracted as if kissing and are visibly protruding from surface of face.

*Item 4:* With lips together, puff out cheeks.

Score as 0 if incorrect.

Score as 1 if only slight cheek expansion is evident or only one cheek is expanded.

Score as 2 if both cheeks are filled with air so that definite and wide expansion is evident.

*Item 5:* Smack lips.

Score as 0 if incorrect.

Score as 1 if lips are pulled apart without turning inward first so that a lesser sound results.

Score as 2 if upper and lower lips turn inward and are pulled apart to make a smacking noise.

*Item 6:* With mouth open so that lips do not touch, cover teeth with lips.

Score as 0 if incorrect.

Score as 1 if lips cover teeth completely but upper and lower lips touch and/or lips do not turn inward.

Score as 2 if, with mouth open, upper and lower lips turn inward to cover teeth so that front teeth are not visible and mouth is open.

*Item 7:* Make "pkt" with audible, but not voiced, sound—two times (2 seconds).

Score as 0 if incorrect or insufficient number of movements.

Score as 1 if "p," "k," and "t" sounds are drawn out with vowel sounds attached as in "pa ka ta" and/or the sequence is completed in a sufficient number of times but poorly done.

Score as 2 if "p," "k," and "t" sounds are made in rapid succession two times without vowel sound attached and sequencing is good.

*Item 8:* Stick out tongue, retract, close mouth—three times (3 seconds).

Score as 0 if incorrect or insufficient number of movements.

Score as 1 if full sequence is completed but quality of sequence is poor, and/or tongue is visible but does not protrude past lips, and/or lips do not close completely.

Score as 2 if there is obvious protrusion of tongue followed by tongue retraction and lip closure three times.

*Item 9:* Put tongue in right cheek, then put tongue in left cheek—two times (2 seconds).

Score as 0 if incorrect or insufficient number of movements.

Score as 1 if tongue makes only slight protrusion of cheeks and/or lack of smooth, well-sequenced transition from one movement to another.

Score as 2 if tongue makes visible protrusion in one cheek then the other, then repeats, with smooth transition from one position to the other.

*Item 10:* Move jaw to right side, then move jaw to left side—two times (2 seconds).

Score as 0 if incorrect or insufficient number of movements.

Score as 1 if there is obvious shifting of jaw in both directions a sufficient number of times, but movement from one position to another is not smooth.

Score as 2 if lower jaw shifts any distance from one side to other, moving smoothly from one position to the other.

*Item 11:* Put tongue in left cheek.

Score as 0 if incorrect.

Score as 1 if tongue makes slight visible protrusion of cheek.

Score as 2 if tongue pushes cheek outward and makes definite obvious visible protrusion of cheek.

*Item 12:* Put tongue in your cheek and push on tongue momentarily. Then say, "Now hold it there," and push on child's tongue.

Score as 0 if incorrect.

Score as 1 if tongue takes initial or minimal resistance.

Score as 2 if tongue takes full resistance and maintains position in cheek.

*Item 13:* Start tongue midline upper lip, move tongue to right, and lick lips all the way around.

Score as 0 if incorrect.

Score as 1 if tongue licks lip (maintaining contact on lip) moving less than 360 degrees but at least 180 degrees and mouth is open.

Score as 2 if tongue licks lip (maintaining contact on lip) in complete 360-degree arc with mouth open.

*Item 14:* Start tongue midline upper lip, move tongue to left, and lick lips all the way around.

Score as 0 if incorrect.

Score as 1 if tongue licks lip in opposite direction of Item 13, moving less than 360 degrees but at least 180 degrees, with mouth open.

Score as 2 if tongue licks lip in *opposite direction* of Item 13 (maintaining contact on lip) in complete 360-degree arc with mouth open.

*Item 15:* Touch tongue to upper lip, then to lower lip—two times (2 seconds).

Score as 0 if incorrect or insufficient number of movements.

Score as 1 if child completes action a sufficient number of movements but is awkward in going from one position to another and/or does not open mouth so that there is a space between tongue and lips.

Score as 2 if tongue alternately touches upper lip (or above), then lower lip, then repeats, moving smoothly from one position to another with mouth open.

*Item 16:* Pucker (or purse) lips to right side of face, then to left side—two times (2 seconds).

Score as 0 if incorrect or insufficient number of movements.

Score as 1 if lips touch and move side to side but pucker or purse is not maintained and/or sequencing is sufficient in number but not in quality.

Score as 2 if puckered or pursed lips move to one side, then the other, then sequence is repeated. Lips remain puckered together. Sequencing is good.

*Item 17:* Bite lower lip, then bite upper lip—two times (2 seconds).

Score as 0 if incorrect or insufficient number of movements.

Score as 1 if above action is completed with a sufficient number of movements but movement transitions are not smoothly executed.

Score as 2 if upper teeth touch lower lip, then lower teeth touch upper lip, repeated with smooth transition.

*Item 18:* Stick out lower lip, close lips—two times (2 seconds).

Score as 0 if incorrect or insufficient number of movements.

Score as 1 if lower lip obviously protrudes past upper lip with full number of sequences, but sequencing is poor.

Score as 2 if lower lip obviously protrudes past upper lip, then mouth closes and action is repeated.

*Item 19:* Protrude jaw forward with open mouth, then retract jaw closing lips—two times (2 seconds).

Score as 0 if incorrect or insufficient number of movements.

Score as 1 if jaw protrudes forward two times but lips do not part or quality of movement is poor.

Score as 2 if jaw protrudes forward any distance with upper and lower lips parting, then jaw retracts and lips close; movement is repeated with a smooth transition.

## Key Points to Remember

1. Sit facing the child.
2. Do not wear brightly colored lipstick.
3. Administer all items to all children.
4. Execute movements in a precise manner, with separate parts of sequences clearly differentiated.
5. Items involving a sequence of movements should be completed within the established time frames.
6. Make sure that the child watches each item demonstration since items are not repeated.
7. Encourage the child to wait until you finish the demonstration before beginning.
8. Keep the child's hands away from his or her mouth.
9. Once the child begins to imitate the action, begin scoring.
10. Nonmirrored responses are acceptable, as long as the child executes motions in opposite directions on items requiring this.
11. The child is not penalized for performing more than the required number of movements.
12. The child must complete the full sequence to receive a score of 2 or 1.
13. The item is scored as 0 if the sequence is insufficient and/or the movement is incorrect.
14. The item is scored as 1 if the movement is completed with the full number of motions, but rhythm or sequencing is poor, or if the child begins the movement poorly but corrects self. Performance must also fulfill the specific item criteria as indicated for a score of 1.
15. The item is scored as 2 if the movement is executed well and meets scoring criteria.

---

## 13. Manual Form Perception (MFP)

**Materials**

- shield
- protocol sheet
- stopwatch

In addition, Part I requires:

- 8 white plastic forms of various shapes (e.g., heart, star, circle), about ⅛″ thick and 2″ in diameter, which are contained in a shallow box
- 9″ × 12″ response card with pictures of 4 forms on one side and 10 forms on the other

Additional materials for Part II include:

- 2 black plastic bases, one with 2 attached white forms and one with 5 attached white forms
- duplicates of the 7 forms

Remove mold marks from edges of forms with razor blade to eliminate tactile distraction.

In Part I of this test, the child is asked to feel a form, with eyes shielded, and simultaneously point to the picture of the form on a card. In Part II, the child feels a form with one hand and finds another form like it with the other hand.

**Part I Items**

Sit at the table with the child to your left. Open the box containing the eight forms and briefly, for about half a second, show the open box to the child. Say:

***"I'm going to put one of these blocks in your hand."***

With your left hand, hold the shield against the child's chest with consistent pressure throughout the entire test. Place the response card in front of the child with the four-form side exposed. You may attach the response card to the shield with a paper clip. Say:

***"You point to the picture here*** (point to the card) ***of the block in your hand. I'll put the first block in this hand*** (touch the child's right hand and place it under the shield palm up) ***and you point to its picture here*** (point to the card) ***with this hand."*** (Touch the child's left hand.)

There are no trial items on Part I. The forms to use on the test items are illustrated on the protocol sheet. The letter by each item number indicates whether the stimulus form should be placed in the child's right (R) or left (L) hand. Note that, in Part I, the hands are alternated so that a right-hand item is always followed by a left-hand item. All forms are placed in the same orientation as they appear on the protocol sheet, with the child's wrist forming the baseline.

To begin Item 1, the child's right hand should be

under the shield. Encourage the child to have his or her "pointer finger" ready. With your right hand, pick up and place the circle form in the child's right hand. Immediately prior to placement, say:

***"Which one is this?"***

Immediately switch the stopwatch to 0 after placing the form in the child's hand. Encourage the child to feel the block. If necessary, show the child how to feel it on Item 1. The child is allowed to change the orientation of the form while manipulating it.

As soon as the child points to a printed form, note the number of seconds, but leave the stopwatch running between items. Remove the block from the child's hand, record the time (in whole seconds), and score the accuracy of the child's response (0 = incorrect, 1 = correct).

A maximum of 30 seconds is allowed for each item. It is possible for the child to respond in 0 seconds, since it takes some time to move your hand from the child's hand to the stopwatch. If no choice is made within 25 seconds, say:

***"Point to one."***

If no choice is made within 30 seconds, record 30 seconds for the response time and score the accuracy as 0. The child may change his or her mind before the next form is placed if the 30-second time limit has not been reached. Score the last choice.

If the child responds correctly on Item 1, proceed to Item 2. If the child responds incorrectly on Item 1, record the response time and accuracy, then bring the child's right hand, with the form, from under the shield and reexplain the test to the child. For incorrect responses (including no response), provide more instruction and then readminister Item 1. *Do not record the accuracy and time for this second attempt.* More explanation can be provided before proceeding to Item 2.

For Item 2, remove the circle and ask the child to trade hands so the left hand is under the shield and the right hand is in front of the shield ready to point. You may need to position the child's hands for pointing and/or receiving the forms. Then place the star in the child's left hand and switch the stopwatch back to 0. Instructions may be repeated if necessary.

If the child responds correctly on Item 2, continue the test as described below. If the child does not answer correctly on Item 2, explain the procedure again and readminister the item. Again, do not record the time or accuracy of the second attempt. Teaching takes place during Items 1 and 2 only. Be sure the child understands the task before continuing.

Following the administration of Items 1 and 2, turn over the response card and show the pictures of the 10 geometric forms to the child. Say:

***"Now the block will be one of these."***

Place the form for Item 3 in the child's right hand. Proceed with the remaining test items for Part I as indicated on the protocol sheet, recording both response time and accuracy of response. If needed, encourage the child to look at all the pictures of the forms prior to choosing one.

During testing, the forms are kept in the box out of the child's sight, with the forms already used placed in the box lid. If the child accidentally sees a form before identifying it, do not record a response to that item at that time. Repeat the item later and record that response.

If the child wants to compare the plastic form with its picture, explain that he or she cannot do it then but will be allowed to do it when this game (or test) is through. If the child persists in pulling his or her hand from under the shield to see the block, form a "tunnel" with your fingers over the wrist of the child. The tunnel should be no tighter than necessary to prevent the child's hand from being pulled out from under the shield.

Manipulation of each form is encouraged, but only unilateral manipulation is allowed. Young children may need to be shown how to close their fingers around the form to obtain maximal tactile-kinesthetic cues.

The child is to indicate by pointing. Do not accept naming, except the first time it occurs. Then, explain that the child must point to his or her choice.

Discontinue Part I when four consecutive items are scored as 0 on Part I.

**Part II Items**

Part II is not administered to 4-year-olds. It is also not administered if five or more items on Part I were scored as 0 for accuracy (i.e., less than six items on Part I are scored as correct for accuracy). If the child is over age 4 and at least six items were answered correctly, test administration moves quickly from Part I to Part II. Remove the response card from Part I and turn over the protocol sheet. The shield remains in place, held against the child's chest with consistent pressure.

**Trial items.** There are two trial items for Part II. For Trial I, place the plastic base containing two forms on the table at the child's midline, parallel with and about 4 inches from the table edge. The hole in the base should be toward the child's right (see Figure 31). Place the stopwatch on the tabletop near the test form board so that you can quickly look from the child's choice to the stopwatch.

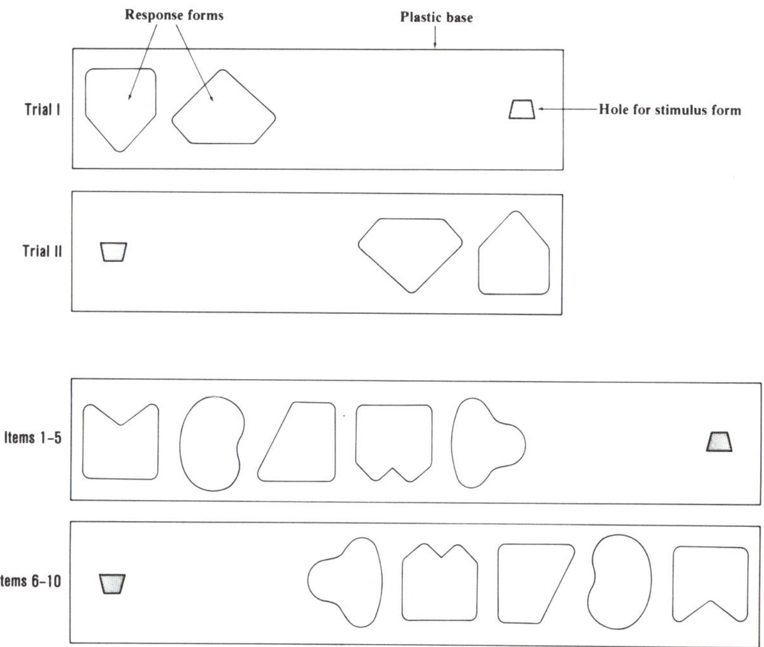


**Figure 31**
**Manual Form Perception: Orientation of the Formboard for Part II**

Set the first stimulus form in the hole so that its orientation is the same as that on the protocol sheet and the matching form on the base. The letter next to each item number on the protocol sheet indicates the hand that feels the (stimulus) form. The other hand feels the response forms. To begin, say:

***"Now both hands will feel blocks at the same time. This hand will feel this block*** (place child's right hand on stimulus form) ***and the other hand will find one like it over here."***

Lead the child's left hand across the two forms attached to the base, resting for a fraction of a second on each, saying:

***"Here's one, and here's one."***

This will provide the child with a tactile orientation to the forms. Hold the child's hand in such a way as to insure extension of the child's fingers and thumb. This will make it easier for you to lead the child's hand quickly across the forms. If necessary, you may show the child how to feel the forms. End with the child's hand resting back on the table. Then say:

***"Keep this hand on this block*** (tap child's right hand) ***and find a block just like it with this hand.*** (Tap child's left hand.) ***Tell me when you find the one that's the same. Be sure to feel each block. Begin here."***

Place the child's left hand on the first response form (the one closest to the stimulus form) at the completion of the verbal directions. Be sure the child understands that a verbal indication of his or her choice is needed; however, if he or she is unable to give one, pointing is acceptable.

After the child has made his or her choice, say:

***"This time we can look."***

Remove the shield to reinforce the nature of the task. Then move on to the second trial item, placing the shield back in place, removing the first form, and turning the base so that the empty hole is on the child's left (see Figure 31). This way, the stimulus item is introduced on the left and the response forms are reversed in sequence and top-to-bottom orientation. Insert the Trial II stimulus form, saying:

***"Now your other hand will feel this block*** (place child's left hand on the stimulus form) ***and this hand will find one like it over here.*** (Rub child's right hand lightly over response forms.) ***Be sure to feel each block. Begin here."***

If the child responds correctly, remove the shield to again reinforce the nature of the task. If the child chooses the wrong form, before removing the shield, ask the child to try feeling the other form. Then remove the shield and reexplain the task to the child. When it appears that the child understands the test procedure, move on to the test items. If the child does not seem to understand, repeat the trial items without the shield. Be sure that the child does understand the test procedure before going on to the test items.

**Test items.** Once Part II is begun, all Part II items are given, regardless of the number of errors made. To begin, place the plastic base with the five forms in front of the child, positioned the same as the trial base. The hole is on the child's right (see Figure 31) for Part II Items 1 through 5. The child should not see this form board until the test is over. Place the first stimulus form in the hole so that its orientation is the same as the matching form on the base and Item 1 on the protocol sheet. The shield is still in place. Say:

***"Now we'll do another set. Feel this block with this hand*** (place child's right hand on the stimulus form) ***and find one like it over here. Now there are more.*** (Rub child's left hand lightly across the five response forms.) ***Be sure to feel all the blocks. Begin here."***

Place the child's left hand on the first response form. Start the stopwatch as you finish saying the directions. The child can be told that the form he or she is trying to find is in the same orientation as the stimulus form (i.e., not reversed or upside down) if he or she asks or seems to need the information to respond.

A maximum of 45 seconds is allowed per item on Part II. If no choice is made within 40 seconds, say:

***"Choose one."***

If no choice is made within 45 seconds, record the time as 45 seconds and mark the item as incorrect. It is possible for the child to respond in 0 seconds.

Record response time (in whole seconds) and response accuracy (0 = incorrect, 1 = correct) after the child has made his or her choice.

If the child chooses the first response form without feeling any of the other forms on Item 1, ignore this choice and say:

***"Feel all the blocks before choosing one."***

If the child then chooses the second block, record the time and accuracy and go on to the next item. However, before giving the next item, remind the child to feel all the blocks. If needed, take the child's hand across the forms again before Item 2.

For Item 2, say:

***"Now we will do another one."***

Remove the first stimulus form and insert the second form, saying:

***"Feel this block with this hand*** (place child's right hand on the stimulus form) ***and find one like it with this hand*** (indicate child's left hand). ***Be sure to feel all the blocks. Begin here."***

Place the child's left hand on the first response form and start the stopwatch.

The rest of the items are administered in a similar manner. For each item, place the child's hands on the stimulus form and the first response form. The child must keep one hand on the stimulus form while the other hand finds the matching form. No bilateral manipulation of the response forms is allowed. You may give the above verbal directions and may orient the child's hands to the blocks between items as often as necessary.

Remind the child, as needed, to start with the first response form and to be sure to feel every form. Say, for example:

***"Keep this hand on this block (stimulus) and find one like it with this hand. Be sure to feel all the blocks. Begin here."***

If the child appears to understand the process, however, it is not necessary to repeat these directions each time. In this case, start the stopwatch as soon as the child touches the first response form.

After administering all the right-hand items, test the left hand by turning the base around so that the stimulus items are introduced on the left (see Figure 31). The response forms are also reversed in sequence and in top-to-bottom orientation. Insert the form for Item 6 while saying:

***"Now your other hand will feel this block*** (place child's left hand on stimulus form) ***and this hand will find one like it over here.*** (Rub child's right hand lightly over the response blocks.) ***Be sure to feel all the blocks. Begin here."***

The rest of the test proceeds as for the right hand. While recording the time and accuracy for an item, leave the stopwatch running. The child may change his or her mind up to the time the stimulus form is replaced with the next one.

After the test, fill in the circles corresponding to the time on each item. Be sure to enter the number of seconds as two digits (e.g., 6 seconds as "06").

## Key Points to Remember

1. Sit to the right of the child at the table.
2. The shield is held against the child's chest with constant and consistent pressure throughout the test.
3. On Part I, place the stimulus form in the child's hand in the samc orientation as it is on the protocol sheet, with the child's wrist forming the baseline.
4. The stopwatch runs continuously between items and the switchback is used to return to 0 immediately after the next form is placed in the child's hand.
5. Encourage the child to feel the form. If necessary, show him or her how (during first item). However, don't allow bilateral manipulation.
6. You can remind the child to look at all pictures of blocks before choosing.
7. The child is to indicate by pointing. Don't accept naming, except the first time the child does this.
8. Teaching takes place during Items 1 and 2 only. Be sure the child understands before proceeding.
9. Remember to flip the response card over after Item 2.
10. You may need to ask the child to trade or change hands between items.
11. The child is not allowed to see whether his or her choice is correct or incorrect during the test items.
12. If the child accidentally sees the form before identifying it, don't score. Repeat the item later.
13. Time in whole seconds.
14. The time limit per Part I item is 30 seconds.
15. If the child has not responded within 25 seconds, encourage making a choice.
16. The child may change his or her mind before the form or the next item is placed in his or her hand (if within 30 seconds). Score the child's last choice.
17. If no choice is made at the end of 30 seconds, record the time as 30 seconds.
18. It is possible for the child to respond in 0 seconds.
19. Score accuracy as 0 if the choice is incorrect or if no choice is made within 30 seconds.
20. Score accuracy as 1 if the correct choice is made within 30 seconds.
21. Discontinue Part I after 4 consecutive items are scored as 0 for accuracy.
22. Complete all verbal directions before placing block in hand.
23. Do not administer Part II to 4-year-olds, or if five or more items on Part I were scored as 0 (incorrect) for accuracy (less than six items correct).
24. If Part II is attempted, give all items of Part II.
25. Teaching takes place during the trial items only.
26. Orient the child's hand to the position of the response forms on each trial item, as well as before Items 1 and 6 and as often as needed. While orienting the child, hold the child's hands with fingers extended so that you can quickly move the hand across the forms.
27. Place the child's hand on the first response form *after* the verbal directions are completed.
28. If necessary, show the child how to manipulate the forms. No bilateral manipulation of the response forms is allowed.
29. You may need to remind the child to feel all the forms before choosing.
30. You may need to remind the child to keep his or her hand on the stimulus form while the other hand finds the matching form.
31. The child is encouraged to verbally indicate choice, but may respond with pointing if response is unequivocal.
32. The time limit per Part II item is 45 seconds. If the child has not responded in 40 seconds, encourage choice.
33. If no choice is made at the end of 45 seconds, record the time as 45 seconds.
34. The child can change his or her mind before the form for the next item is placed (if within 45 seconds).
35. Score accuracy as 0 if the choice is incorrect, or if no choice is made within 45 seconds.
36. Score accuracy as 1 if the correct choice is made within 45 seconds.

## 14. Kinesthesia (KIN)

**Materials**
- test sheet
- red pen
- shield
- masking tape
- centimeter ruler
- protocol sheet

Kinesthesia is measured by the accuracy with which the child moves his or her finger from one point on the test sheet to another point. To simplify this task, the test situation is described to the child as a game of "going visiting," which the child's hand will play without the help of his or her eyes.

The test has one trial item for each hand and five test items for each hand. The two test items with the most erroneous responses are readministered.

Sit directly across the table from the child. Tape the test sheet to the table with the trial items facing up. The edge of the sheet should be 2½ centimeters (one inch) from the table edge on the child's side of the table, with the center mark positioned at the child's midline. Wording and numbers are oriented so as to be read by you. Right-hand items, labeled "R," are designated by solid lines; left-hand items, labeled "L," are designated by broken lines.

Directions should always be given while the child's hand is stationary, not while in motion, since listening detracts from attending to the kinesthetic stimuli. Because it is easy for the child's mind to wander during test administration, giving the test rapidly will help to maintain the child's attention.

**Trial Items**

The first trial item is administered to the child's right hand with vision. The second trial item is then administered to the left hand with vision occluded. This allows the child to experience the task both visually and kinesthetically under trial conditions. Begin with Trial A (see Figure 32). Do not use the shield. Say:

***"We are going to play a game called 'going visiting.' I will take your finger to different 'pretend' houses. Point your finger like this."***

Demonstrate by holding your own hand up with the index finger in a pointing position; the other fingers are curled under. If the child's left hand is on the table, place it in the child's lap, saying:

***"We'll put this hand down here so we won't run into it."***

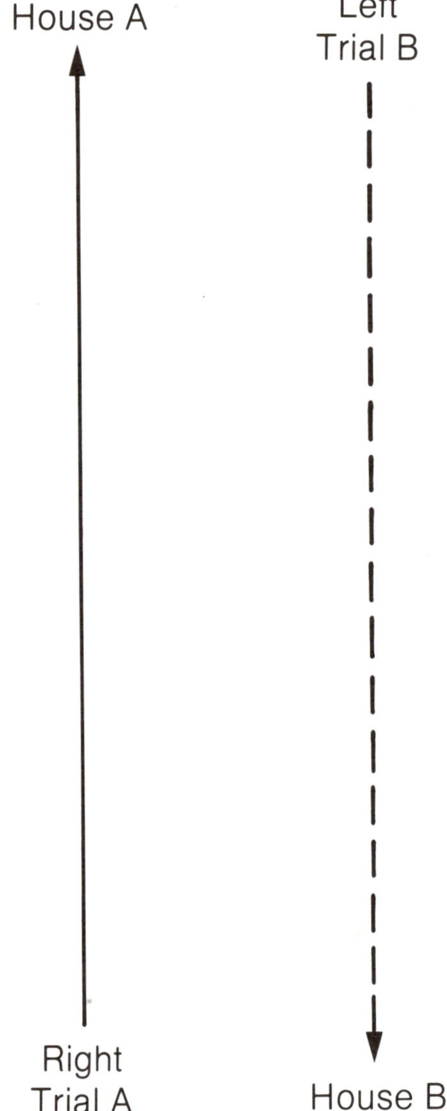


**Figure 32**
**Kinesthesia: Trial Items**

Hold the child's right index finger on the lateral borders of the distal joint and place the finger so the middle of the fingertip is at the beginning of the line for Trial A, saying:

***"This is where you live. I'm going to take you to House A. Think how it feels to go there so you can come back to House A by yourself."***

Still holding the child's index finger by its distal joint, lift the child's hand and place it so the middle of the child's index fingertip is at the arrow end of Trial A. Say:

***"This is where House A is. Remember where House A is so you can come back to it. Leave your finger here awhile."***

Release the child's finger and allow 3 seconds of silence (while counting to yourself 1001, 1002, 1003) for the child to concentrate on feeling where his or her finger is. Then say:

***"I'll take you home."***

Grasp the child's finger again and lift the child's hand back to the beginning of Trial A. Say:

***"This is where you live. Now put your finger on House A."***

The child should be able to visually locate the arrow end of Trial A and place his or her finger on it. If he or she is off target, assist the child with more exact placement, saying:

***"Place the tip of your finger exactly on the arrow."***

If it seems that the child understands what is expected up to this point and has responded appropriately, begin Trial B by saying:

***"Now let's see if your other hand can play the game without your eyes helping it. To do that, I will hold this shield here.*** (Position shield.) ***It will be easier for you to feel where your finger is if you close your eyes."***

Hold the shield just vertical enough to prevent eye contact with the child while administering the items. Position the shield firmly against the child's chest just below the clavicle and hold it consistently throughout administration of the test to promote habituation to the tactile stimuli. Use of the shield will reduce the distraction of being able to see your face and the tendency of some children to become preoccupied with either looking for signs of approval or disapproval, or attempting to detect correct item locations by watching your eye movements. To reduce discomfort to the child, there should be clothing covering the skin where the shield comes in contact. A layer or two of masking tape around the shield edge may also help to reduce scratching.

It may be necessary to redirect the child's right hand to his or her lap before continuing. If the child does not close his or her eyes voluntarily, suggest that he or she do so. Some children are unable to keep their eyes closed for long periods. In such cases, they are encouraged to close their eyes during the time the finger is at "the house." In any case, the shield is always held at a height which excludes most visual stimuli.

Administer Trial B to the left hand with the child's vision occluded, saying:

***"This is where you live. I'm going to take you to House B. Think how it feels to go there so you can come back to House B by yourself."***

Move the child's left index finger to the arrow end of Trial B in the same manner as for Trial A. While continuing to hold the child's finger on the arrow, say:

***"This is where House B is. Remember where it is so you can come back to it. Leave your finger here awhile."***

Release the child's finger for 3 seconds of silence, and then say:

***"I'll take you home."***

Return the child's finger to the beginning of Trial B and continue holding the child's finger until verbal directions are completed. Then say:

***"This is where you live. Now put your finger on House B."***

After the child has put his or her finger down, say:

***"Leave your finger on the spot until I finish measuring."***

Use the red pen to record the response on the test sheet by placing an inverted "V" with the point at the exact middle of the child's index fingernail, and writing the item number inside it (e.g., /1\). Quickly draw a line connecting the arrowhead of the item with the inverted V. Put the shield aside, saying:

***"On this one we can look to see how close you came to the house."***

Help the child examine his or her finger placement relative to the end of Trial B.

If you are satisfied that the child understands the expectations of the test, proceed with the test items.

**Test Items**

Untape the test sheet, turn it over, reposition it correctly, and tape it to the table. Cover the test sheet with the shield during this repositioning so that the child does not see the test sheet in advance.

Help the child to focus attention on kinesthetic sensation by avoiding other stimuli, especially cutaneous, auditory, and visual stimuli. Firm, steady positioning of the shield as described previously is suggested to prevent inconsistent tactual stimuli which might cause distraction or discomfort to the child. While positioning the shield to occlude the child's vision, say:

***"That part of the game was practice for you to learn how. Now, for the rest of the game, your hands will play without your eyes helping them. We will go to different houses."***

Ask for the child's right index finger to begin Item 1. Using the procedures described above for the trial items, say:

***"This is where you live. I'm going to take you to the first house. Think how it feels to go there so you can come back to the first house by yourself. This is where the first house is. Leave your finger here for awhile.*** (Wait 3 seconds.) ***I'll take you home. This is where you live. Now put your finger on the first house. Leave your finger on the spot until I finish measuring how close you are."***

Houses are referred to as first, second, third, fourth, and so on. However, if the child can better understand "House One" and so on, that wording should be used.

It is important that you carry the child's finger smoothly from house to house, and that you not overshoot or undershoot the destination. The movement from one point to the other should be in an arc, with the highest point of the arc being about 5 cm (2 inches) above the test sheet on the longer items, and about 2½ cm (1 inch) above the sheet on the shorter items. This is a general guideline and should not be adhered to at the expense of smoothness. You may note the ease or difficulty in moving the child's arm (postural adjustments).

Discourage the child from moving his or her finger to pick up tactile clues. If the child does not lift his or her finger but moves it on the surface of the sheet, stop the child and point out that his or her finger should be lifted from point to point. Readminister that item later. However, the child may move the finger until he or she feels it is in the correct position. The point where the child finally stops is the one that is scored.

Use the red pen to record the child's response on the test sheet. Draw an inverted V with the point at the exact middle of the child's index fingernail tip. Write the item number inside the inverted V (e.g., /3\). Draw a line between the inverted V and the arrowhead. Draw these lines for each item. Measure the length of the lines later (see "Recording Item Scores," on this page).

If the child insists upon seeing how close he or she came, say:

***"You may look at all your answers at the end of the game when we are finished."***

Based on the child's level of understanding and ability to cooperate with the test procedures, use your own judgment to determine which of the previous instructions need to be repeated for the rest of the test items. For example, the child may not need to be reminded to leave his or her finger in position while you mark the response. You may need to reiterate that checking responses is not allowed until the test is over, and that it is better if the child keeps his or her eyes closed. The more sophisticated and cooperative child may need only very basic verbal cues, such as:

***"This is your house. This is the second*** (third, etc.) ***house. This is your house. Now go back to the second*** (third, etc.) ***house."***

**Readministration of Two Items**

Following the administration of Item 10, the two items on which the child performed most poorly (i.e., his or her finger was farthest from the goal) are readministered. The two items to be readministered are determined by quickly measuring the most erroneous-appearing items and selecting the appropriate two. If there is a tie for the second most erroneous response, readminister the most recently administered of the tied items. When readministering the items, refer to them as "House 11" and "House 12" when speaking to the child, but record the actual number of the item. Place a small "2" outside the inverted V to indicate the second attempt and write the number of the retested item inside the inverted V (e.g., /5\ with 2).

**Recording Item Scores**

After the testing session has ended, measure the distance between the tip of the item arrowhead and the tip of the inverted V for each item. Use the centimeter ruler and record scores to the nearest tenth of a centimeter in two digits (e.g., "0.2" or "1.0"). Item scores may be, but need not be, entered on the test sheet. They must be entered on the protocol sheet. For the readministered items, enter the item numbers as well as the item scores of the second administration. If the child's score is greater than 9.9 cm for any single item, record that score as 9.9 cm.

## Key Points to Remember

1. Sit across the table from the child.
2. Tape the test sheet 2½ cm (1 inch) from edge of table on all four corners. It should be centered at the child's midline.
3. On the test sheet, right-hand items are labeled "R" and indicated by solid lines; left-hand items are labeled "L" and indicated by broken lines. (Indicates *child's* right or left.)
4. Two trial items (one right-hand and one left-hand) are presented on one side of the test sheet. On the second trial, vision is occluded.
5. During the test items, the child is encouraged to close his or her eyes.
6. The shield is used whether or not the child closes his or her eyes.
7. The shield is positioned firmly against the child's chest just below the clavicle and held consistently.
8. Give all directions while the child's hand is stationary, not in motion.
9. Pick up the child's index finger by the lateral borders of the distal joint, so that all other tactile stimuli (e.g., shield touching arm) are avoided.
10. Place the fingertip at the house, lift it, and place middle of fingertip on apex of arrow.
11. Don't lift the child's finger too high.
12. Let go of child's finger while at the house (arrow end) and count 3 seconds (1001, 1002, 1003).
13. If the child moves his or her finger while you are counting, remind the child to leave the finger in place and think what it's like to be there.
14. If the child slides his or her finger, stop the child and point out that the finger should be lifted from point to point.
15. Discourage the child from moving his or her finger to pick up tactile clues; however, the child may move finger until he or she feels that it is in the correct position. Measure from where the child finally stops.
16. Record the response directly on the test sheet by placing an inverted V with point touching the exact middle of the child's index fingertip. Write the item number inside it.
17. Between items, you can remind the child to trade or change hands.
18. Readminister the two items on which the child performed most poorly.
19. On the readministered items, record the item number inside the inverted V and write a small "2" outside the inverted V to indicate the second try.
20. After the testing session, measure in tenths of a centimeter from the point of the inverted V to apex of arrow.
21. On the protocol sheet, record the distances to the nearest tenth of a centimeter. The distances should be entered as two digits (e.g., "0.2" or "1.0").
22. For the two readministered items, enter the item numbers with their item scores (centimeters).
23. If the child's score is greater than 9.9 cm for any single item, record that score as 9.9 cm.

---

## 15. Finger Identification (FI)

**Materials**

- shield
- protocol sheet

In this test, one or two of the child's fingers are touched while the child's vision is occluded by the shield. The shield is then removed while the child points to the finger or fingers that were touched.

Sit across the table from the child, holding the shield in your left hand. The child's hands should be palms down on the tabletop. You may need to assist the child in spreading his or her fingers on the table.

The fingers to be touched are listed on the protocol sheet. When applying *one stimulus to one finger*, use your index finger to touch the dorsal surface of the middle phalanx. When applying *two stimuli to one finger*, simultaneously use your index finger and thumb to touch the proximal and middle phalanxes. Use the same two fingers to apply *two stimuli to two fingers*, and touch the middle phalanxes of both fingers. The touch stimulus should be given rapidly with one ounce of pressure (practice with a postal scale).

Your fingers should be at normal body temperature and you should avoid having cold fingers or long fingernails. Your fingernails should not touch the child. The child should not be able to obtain cues as to where he or she was touched by noting a blanched or dented spot on the finger touched. Also avoid unintentional tactile stimuli, such as allowing a sleeve to touch one of the child's hands.

**Trial Items**

The test items are preceded by four trial items. Vision

is *not* occluded during the trial items. To administer Trial I, the child's hands are placed palms down on the table. Touch the dorsal surface of the middle phalanx of the child's left middle finger with your index finger, saying:

***"Touch that finger."***

If the child does not touch the same finger with a finger of his or her right hand, pick up the child's right index finger and touch it to the left middle finger, saying:

***"Touch it this way."***

To administer Trial II, touch the dorsal surface of the middle phalanx of the child's right ring finger, saying:

***"Touch that finger."***

If the child does not touch the correct finger with a finger of his or her left hand, take the child's left index finger and place it on the child's right ring finger, indicating that the response should be made in that manner.

To administer Trial III, simultaneously touch the middle phalanxes of the child's right middle and right index fingers. No verbal directions are given unless the child fails to respond, in which case you say:

***"Touch those."***

Regardless of whether the child responds correctly or incorrectly, point out that two different fingers were touched at the same time:

***"(That's right.) That time I touched two of your fingers at the same time."***

To administer Trial IV, use your index finger and thumb to touch the child's left index finger on the proximal and middle phalanxes. No verbal directions are given unless the child fails to respond, in which case you say:

***"Touch that one."***

Regardless of whether the child responds correctly or incorrectly, point out that only one finger was touched, although it was touched by two of your fingers:

***"(That's right.) That time I touched one of your fingers in two different places."***

**Test Items**

Proceed to give the test items as indicated on the protocol sheet. Note that the items are arranged horizontally on the protocol sheet and that each right-hand item should be followed by a left-hand item.

Place the shield to occlude vision while giving the stimuli. The shield is held against the child's chest with consistent pressure, and just vertical enough that the child cannot make eye contact with you. Remove the shield for the child's response after the application of the tactile stimulus and after all indication of the stimulus has disappeared.

To begin, say:

***"Now let's see if you can tell which finger I touch when you cannot see the finger that I touch."***

As the child responds to each item, record the item score as described in the following section, "Recording Item Scores."

Sometime early in administering the test, if the child seems uncertain about a response, tell the child that he or she can change his or her mind. The comment is not made during the trial items unless the child's response is such that it is appropriate.

The attention of some children must be obtained before each item by such comments as:

***"Here's another."***

Any spoken words should precede application of the tactile stimulus to avoid possible occlusion of perception of the tactile input through simultaneous auditory input.

If an extra tactile stimulus is inadvertently applied, both stimuli are briefly "rubbed away" and the item is repeated later. Also, if an erroneous tactile stimulus is applied, it is briefly rubbed away and the correct item is given later.

**Recording Item Scores**

During test administration, the child's response to each item is scored as 0 (incorrect) or 1 (correct). Fill in the circle on the protocol sheet corresponding to the score on each item.

If the child points to the same finger that you touched but to a different phalanx, the response is considered correct. There are no partially correct responses. If a child first chooses an incorrect finger, then a correct one, give credit for the correct response. It is not necessary for the child to indicate that, in some instances, two stimuli were applied. He or she need only correctly indicate the finger that was touched. If two different fingers were touched, the child must indicate both fingers in order to receive credit for the item.

## Key Points to Remember

1. Sit across the table from the child.
2. You may need to assist the child in spreading his or her fingers on the tabletop.
3. Avoid cold hands and sharp, long nails.
4. The child's vision is not occluded during the trial items.
5. During the test items, the shield is held in place (with constant pressure against the child's chest) as the items are administered, and removed during the child's responses.
6. Administer all items to all children.
7. Touch the child's fingers in the locations specified on the protocol sheet.
8. Touch very rapidly with one ounce of pressure (practice with a postal scale).
9. Apply the stimulus to the middle phalanx of the child's finger by touching the area with your index fingertip.
10. If two stimuli are given, use your thumb and index finger to apply.
11. If two stimuli are applied to the same finger, touch the proximal and middle phalanxes.
12. During the test items, if the child appears uncertain, tell the child that he or she can change his or her mind. This should be done early in the test items.
13. If extra or incorrect tactile stimuli are inadvertently applied, the stimuli are "rubbed away" and the test item is repeated later.
14. Score the item as 0 if incorrect or 1 if correct.
15. There are no partially correct items.
16. When two stimuli are applied to one finger, the child need only indicate the finger touched (not both locations).
17. If the child points to the correct finger, but wrong phalanx, the response is considered correct.
18. If two fingers are touched, the child must correctly identify both to receive credit.

---

## 16. Graphesthesia (GRA)

### Materials

- shield
- protocol sheet

In this test, you will be drawing a series of designs on the dorsum of the child's hands while the child's view is occluded. After each design is drawn, the shield is removed so that the child sees his or her hand while attempting to make the same design on the same hand with the index finger of the other hand.

Sit across the table from the child, holding the shield in your left hand. It is important that your finger be of normal temperature and that your fingernail be short so that it does not touch the child. The child's hands are on the table with fingers parallel to each other and pointing away from his or her body.

The designs and the hand on which each design is drawn are shown on the protocol sheet. The designs are drawn in the same orientation as on the protocol sheet (i.e., from the child's position, the design will be upside down relative to the protocol sheet) and in the direction indicated. With your index finger, draw each design sufficiently large to occupy most of the space between the carpal and metacarpal joints. During the drawing, apply about one-half ounce of pressure. Avoid pulling the skin or providing cues to the child by allowing him or her to see blanched or dented skin. When possible, draw the design in one stroke; however, the child may use more than one stroke. Speed of administration should be about 1 second for simple items and 2 seconds for more complex items.

Encourage the child to draw carefully and accurately. The child draws the designs on the same hand on which you drew to avoid a tendency to make a mirror image of the design. However, if the child draws on the wrong hand, the response is accepted and scored as though it were the correct hand.

### Trial Item

The test items are preceded by a trial item. Note that the shield is *not* used on the trial item. To begin, say:

***"I am going to draw some designs (pictures) on the back of your hand*** (point to child's left hand) ***with my finger. You draw the same thing in the same place.*** (Point.) ***Draw with your finger.*** (Point to child's right index finger.) ***I will show you how with this one."***

Administer the trial item by drawing a straight horizontal line on the dorsum of the child's left hand. Then continue:

***"Draw what I drew. Draw it here.*** (Point to left hand.) ***Draw it carefully."***

If the child does not grasp the concept, hold the child's right index finger and draw the line appropriately. Then ask the child to draw the line.

**Test Items**

To administer the test items, say:

***"Now I'll draw something different on your other hand. I'll draw without you watching me."***

Position the shield just vertical enough to prevent eye contact during application of the stimulus. The shield is held against the child with constant and consistent pressure so that the child is not distracted from attending to the stimulus application by intermittent contact of the shield with the chest.

Items are arranged horizontally on the protocol sheet so that each right-hand item is followed by a left-hand item. The right and left columns on the protocol sheet refer to the *child's* hands, not yours.

For Item 1, draw a vertical line on the child's right hand. On completion of your drawing, remove the shield so the child can see his or her hands. (The shield should continue to occlude the child's view of the protocol sheet.) If the child does not automatically draw the line, say:

***"You draw the same thing. Draw it here.*** (Point.) ***Draw carefully."***

After the child's response, replicate the child's drawing in the blank box next to the item on the protocol sheet. The direction in which the child draws the design is not recorded. Because scoring is based on your replications of the drawings, draw as accurately as possible.

If, at any time during administration, you feel that the child is being careless, say:

***"Draw carefully. Make yours exactly like mine."***

If the child draws partially on the finger or hand, in the air, or entirely off the hand, it is acceptable if it can be scored. If not, advise the child to draw the figure again on his or her hand. When the child draws any place other than on the back of the hand, he or she is advised where to draw. If the child corrects a response, the more accurate response is recorded, but the item is not readministered.

Although the drawings may be scored later according to the criteria described in the following section, try to determine during administration whether each item would receive a score of 0 (very inaccurate). *The test is discontinued after four consecutive items are scored as 0.* If unsure about an item score, continue to administer items. The computer program will ignore any item score entered after the discontinuance criterion has been met.

**Recording Item Scores**

After the test session is completed, refer to your replications of the drawings and score each item as 0, 1, or 2, based on the specific item criteria listed below. On the protocol sheet, fill in the circle corresponding to each item score.

The item score is based on the accuracy with which the child replicated the design. The direction in which the child drew the design and the size of the child's drawing do not enter into scoring. Lines may be of any length unless otherwise specified. Also, the child is not penalized for drawing on the wrong hand.

Specific criteria for each design for scores of 0, 1, and 2 are defined below and illustrated in Figure 33. Any design less accurate than those illustrated for a score of 0 is scored as 0. If a drawing meets the criteria for a score of 1 or 2 with the exception that it is reversed or rotated more than 45 degrees and up to 180 degrees, the score is 1. A figure fitting several different criteria for a score of 1 is still scored as 1. A 10-degree allowance should be made when gauging whether lines are horizontal or vertical to the wrist.

*Items 1 and 4:*

Score as 0 if the design is very inaccurate (i.e., does not meet the criteria for a score of 1 or 2).

Score as 1 if the drawing is: (a) exactly one straight or slightly curved line of any orientation, including horizontal to the wrist, but not qualifying for a score of 2; (b) a straight or slightly curved line drawn back on itself or superimposed (you will need to indicate during administration whether a line was drawn back on itself or two lines were drawn); or (c) exactly one straight or slightly curved line with a slight hook on the end.

Score as 2 if the child drew exactly one straight line within 45 degrees of being vertical to the wrist.

*Item 2:*

Score as 0 if the design is very inaccurate (i.e., does not meet the criteria for a score of 1 or 2).

Score as 1 if the drawing is: (a) exactly one straight or slightly curved line of any orientation, including vertical to the wrist, but not qualifying for a score of 2; (b) a straight or slightly curved line drawn back on top of itself (superimposed); or (c) exactly one straight or slightly curved line with a slight hook on the end.

Score as 2 if the child drew exactly one straight line within 45 degrees of being horizontal to the wrist.

*Item 3:*

Score as 0 if the design is very inaccurate (i.e., does

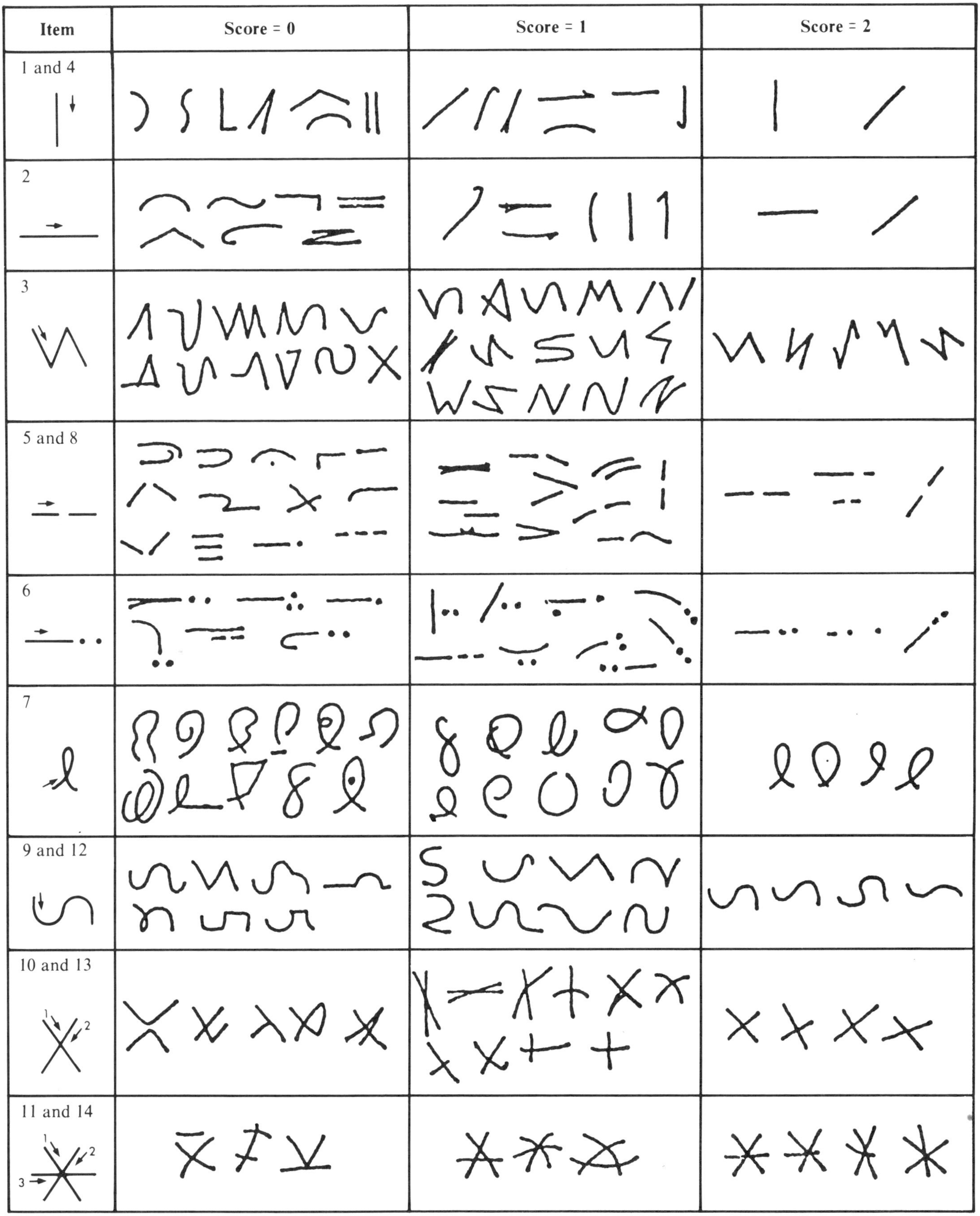

| Item | Score = 0 | Score = 1 | Score = 2 |
|---|---|---|---|
| 1 and 4 | | | |
| 2 | | | |
| 3 | | | |
| 5 and 8 | | | |
| 6 | | | |
| 7 | | | |
| 9 and 12 | | | |
| 10 and 13 | | | |
| 11 and 14 | | | |

**Figure 33**
**Graphesthesia: Scoring Examples**

not meet the criteria for a score of 1 or 2).

Score as 1 if the drawing has: (a) exactly three continuous straight or slightly curved lines, one line of which may cross another, forming a figure with two sharp or one or two slightly rounded angles; (b) exactly three continuous straight or slightly curved lines superimposed on each other; or (c) exactly four straight or slightly curved, continuous lines forming three sharp angles, with the ends of all lines within 10 degrees of parallel to each other.

Score as 2 if the child drew exactly three straight, continuous, nonsuperimposed lines forming a figure with two sharp angles and not rotated more than 45 degrees relative to the wrist.

*Items 5 and 8:*

Score as 0 if the design is very inaccurate (i.e., does not meet the criteria for a score of 1 or 2).

Score as 1 if the drawing has: (a) exactly two straight or slightly curved lines partly, entirely, or not at all superimposed on each other, or (b) lines are within 45 degrees of being parallel to each other or in line with each other, but not qualifying for a score of 2.

Score as 2 if the child drew exactly two straight, nontouching lines of any length, in line with each other and within 45 degrees of being horizontal to wrist.

*Item 6:*

Score as 0 if the design is very inaccurate (i.e., does not meet the criteria for a score of 1 or 2).

Score as 1 if the drawing has: (a) exactly one straight or slightly curved line with two dots at any location on the hand, including superimposed on each other, but not qualifying for a score of 2; or (b) exactly one straight or slightly curved long line with two very short lines, all in line with each other.

Score as 2 if the child drew exactly one straight line within 45 degrees of being horizontal to the wrist and two dots in line with each other. (A dot is made by touching the finger to the hand and not moving the finger along the surface of the hand.)

*Item 7:*

Score as 0 if the design is very inaccurate (i.e., does not meet the criteria for a score of 1 or 2).

Score as 1 if the drawing has: (a) one continuous curved line which turns back to meet itself but does not cross itself; rather, it forms a closed loop or almost closed loop (less than one-eighth the circumference); or (b) one continuous curved line crossing back over itself forming a self-contained loop with one or both ends of the line curved back around the loop but extending not more than halfway around the loop; or (c) one continuous curved line crossing back over itself, forming a closed loop with a short irregular extension of the end of one line.

Score as 2 if the child drew one line crossing back over itself, forming a self-contained loop within 45 degrees of being vertical.

*Items 9 and 12:*

Score as 0 if the design is very inaccurate (i.e., does not meet criteria for a score of 1 or 2).

Score as 1 if the drawing has: (a) one continuous line forming exactly two curves of definitely unequal size or one curve and one angle; or (b) one continuous line forming exactly two curves, or one curve and one angle and having an extension that does not form a third angle or curve.

Score as 2 if the child drew one continuous line forming exactly two curves, in opposite directions, of approximately equal size.

*Items 10 and 13:*

Score as 0 if the design is very inaccurate (i.e., does not meet criteria for a score of 1 or 2).

Score as 1 if the drawing has: (a) exactly two straight or curved lines of any length which intersect, but not necessarily within the middle third of each line, and form angles of any size; or (b) two intersecting lines, one of which is drawn back on itself.

Score as 2 if the child drew exactly two straight lines diagonal to the wrist (any number of degrees) and neither of which is greater than twice as long as the other, intersecting at the middle third of each line and forming four angles not grossly unequal in degrees.

*Items 11 and 14:*

Score as 0 if the design is very inaccurate (i.e., does not meet criteria for a score of 1 or 2).

Score as 1 if the drawing has: (a) exactly three straight lines, each of which intersects the other two but not in the same place; or (b) exactly three lines, one or more of which is slightly curved and each of which intersects the other two.

Score as 2 if the child drew exactly three straight lines which intersect each other in the same place; may be rotated.

## Key Points to Remember

1. Sit across the table from the child.
2. Avoid using cold index fingertip or sharp fingernail.
3. Designs are drawn in the same orientation as on the protocol sheet.
4. Designs are made on the dorsum of the child's hand with your index fingertip.
5. Use a half-ounce of pressure and avoid pulling the skin when drawing.
6. Avoid leaving marks on the child's skin.
7. Drawings should occupy most of the space between the child's carpal and metacarpal joints.
8. When possible, draw designs in one stroke. The child may use more than one stroke.
9. Draw simple items in about 1 second and complex items in about 2 seconds.
10. Trial item is given without the shield.
11. During the test items, the shield is held in place (with constant pressure against the child's chest) as the items are administered, and removed as the child responds.
12. Encourage the child to draw carefully and accurately.
13. If the child draws on the wrong hand, in the air, or anywhere else, the response is still accepted if scorable.
14. Copy the child's design on the protocol sheet and score later. However, during administration, try to determine whether each item would receive a score of 0.
15. Discontinue test after four consecutive items are scored as 0.
16. Use scoring criteria and examples to determine score of 0, 1, or 2 for each item.
17. If the child corrects his or her response, score the more accurate response.
18. Size of drawing and directionality are irrelevant to scoring.
19. Mirror images or rotated designs, if they otherwise meet the criteria for a score of 2, are scored as 1.

---

## 17. Localization of Tactile Stimuli (LTS)

**Materials**

- LTS pen
- centimeter ruler
- shield
- protocol sheet

In this test, stimuli are applied to various locations on the child's arms and hands using a special pen, and the child is asked to put his or her finger on the locations touched by the pen.

Sit across the table from the child, holding the shield in your nondominant hand. The child sits with forearms on the table, hands palms down with fingers parallel to each other and pointing away from the body.

Hold the LTS pen by grasping the sliding sheath between the thumb and index finger of your dominant hand and position it about 1 centimeter above the child's hand or arm. The device is held perpendicular to the table on which the child's arms rest. To apply the stimulus, quickly lower the pen tip to the child's arm or hand and continue lowering the sheath to about 1 centimeter from the bottom stop, then lift the sliding sheath to its original position (i.e., in contact with the upper stop and off the child's arm or hand). The total time from the time the sheath leaves contact with the upper stop to the time it returns contact with the upper stop should be one-half second. Note that the lower edge of the bottom stop should be in contact with the cap of the pen when the cap is on fully.

**Trial Item**

The trial item is given without the shield. To administer, say:

***"I am going to touch you lightly with this pen. Put your finger where I touch you. Put your finger here."***

Touch the back of the child's left hand with the tip of the pen as described above. Then, if necessary, say:

***"Put your finger exactly on the spot I touched."***

If the child does not put his or her finger on the place touched, show the child how to do it.

**Test Items**

During the test items, vision is occluded during both administration of the tactile stimulus and the child's response. Once the test items are started, the shield remains. Hold the shield against the child's chest with constant and consistent pressure, and just vertical enough to prevent eye contact with you during application of the stimuli.

The location at which each stimulus is to be applied is indicated by an *X* on the protocol sheet diagram and the sequence of items is numbered (see Figure 34). The right and left columns refer to the *child's* right and left, not yours. Note that the stimuli are administered in the order indicated by the item numbers on the protocol sheet, which is neither horizontally nor vertically in this case. Also, the stimuli are administered alternately to the right and left hands.

Points A and C on each diagram indicate elbow crease and the wrist, respectively. Point B, which is halfway between Points A and C, is *not* halfway between the distal and proximal stimuli. Also note that the stimuli near the elbow are applied at a location that enables vertical application of the LTS pen to the highest portion of the arm at that location.

To begin Item 1, say:

***"Now let's see how close you can put your finger when you can't see where I touch you. Put your finger here."***

Immediately after the word *here,* touch the back of the child's right hand as shown on the protocol sheet. When applying the stimuli, take care to avoid touching callused areas or providing other stimuli (e.g., auditory, other tactile, or visual stimuli). If an unintentional tactile stimulus occurs and confuses the child, repeat the test items. Also, an item may be repeated if the child appears distracted.

After the child responds, say:

***"If your finger doesn't land on the right spot, move it until it is on the right spot."***

When the child has stopped moving his or her finger and feels that it is on the right spot, say:

***"I'm going to measure you with this ruler. Leave your finger there until I finish."***

With the centimeter ruler, measure the distance from the child's fingernail to the periphery of the pen mark. Measure to the nearest tenth of a centimeter and record the score as two digits (e.g., "0.4") in the first space on the protocol sheet. If the child's score is greater than 9.9 centimeters for any single item, record that score as 9.9 cm. After the test, fill in the circles corresponding to each item score. Because the child is not asked to put his or her fingernail on the spot but rather the pad of the finger, the distance between the child's fingernail and pad of the finger will be considered a constant error.

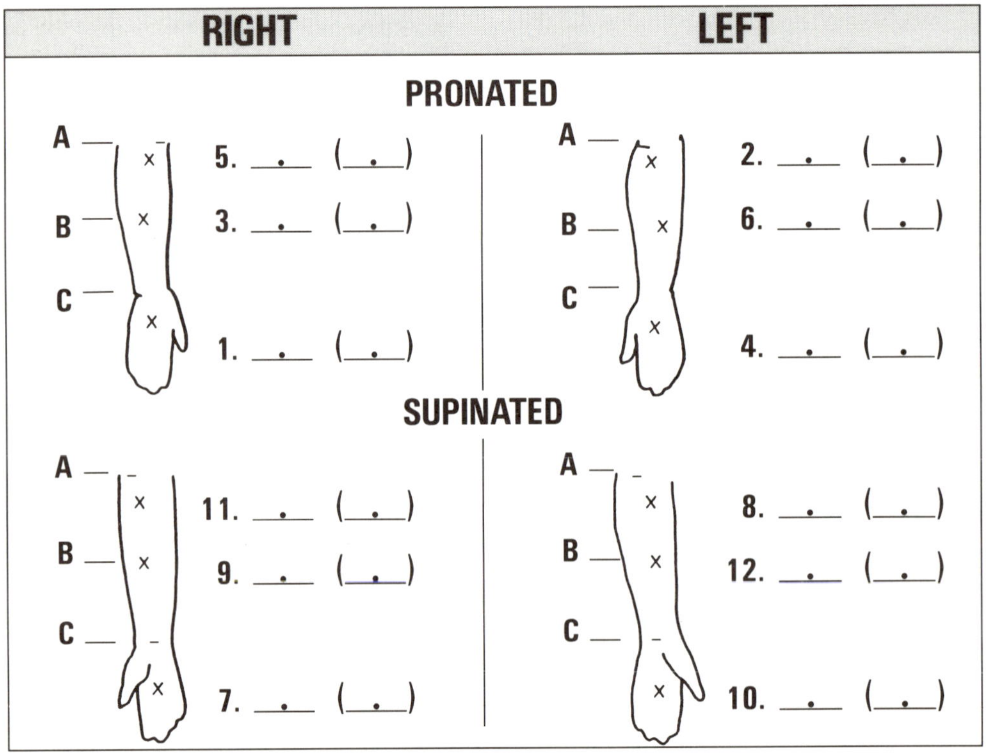

| RIGHT | | LEFT | |
|---|---|---|---|
| **PRONATED** | | | |
| A | 5. ___.___ (___.___) | A | 2. ___.___ (___.___) |
| B | 3. ___.___ (___.___) | B | 6. ___.___ (___.___) |
| C | 1. ___.___ (___.___) | C | 4. ___.___ (___.___) |
| **SUPINATED** | | | |
| A | 11. ___.___ (___.___) | A | 8. ___.___ (___.___) |
| B | 9. ___.___ (___.___) | B | 12. ___.___ (___.___) |
| C | 7. ___.___ (___.___) | C | 10. ___.___ (___.___) |

**Figure 34**
**Localization of Tactile Stimuli: Item Illustrations on the Protocol Sheet**

During measurement of distance, tell the child that he or she is feeling the ruler and at this time is not expected to put his or her finger on the spot touched, unless you specifically say that he or she is to do so.

Use the same procedure for the remaining items. For each item, call the child's attention to the coming stimulus by saying immediately *before* touching:

***"Put your finger here,"*** or ***"here's another."***

Frequently, children need to be reminded to point with only one finger and move the finger until it feels it is on the correct spot.

**Readministration of Two Items**

Following administration of Item 12, readminister the two items with the most erroneous scores (i.e., highest scores). These two items with the most erroneous scores may both be for one hand or one for each hand. If there is a tie for the second most erroneous score, retest the most recent item of the two. Record the scores from the second administration in the second spaces (in parentheses) on the protocol sheet. After the test, fill in the item numbers and their second scores in the circles at the bottom of the protocol sheet.

## Key Points to Remember

1. Sit across the table from the child.
2. The LTS pen is used to touch designated spots on the child's hand or arm.
3. Vision is not occluded during the trial item.
4. If necessary, you can show the child how to point during the trial item.
5. Test items are administered in sequence as numbered on the protocol sheet.
6. On the protocol sheet, right and left columns refer to the child's right and left hands, not yours.
7. The shield is held with consistent pressure against the child's chest throughout the test items.
8. Before each item, say, "Put your finger here" before touching.
9. To administer the stimulus, hold the LTS pen 1 cm above the spot and allow about half a second from leaving to returning to this position.
10. Avoid touching callused areas.
11. Guard against unintentional tactile stimuli. (Readminister item later if this occurs.)
12. If child appears distracted, you can repeat the item later.
13. Remember to say, "If your finger doesn't land on the right spot, move it until it is on the right spot."
14. When appropriate while you are measuring, you can remind the child that he or she is feeling the ruler.
15. Record measurements as two digits to the nearest tenth of a centimeter (e.g., "0.4").
16. Readminister the two most erroneous items.
17. If the child's score is greater than 9.9 cm for any single item, record that score as 9.9 cm.

# CHAPTER 4
## COMPUTERIZED SCORING AND INTERPRETATION

WPS TEST REPORT provides a computerized scoring and interpretation service for the SIPT which supplies users with a multiple-page report. *The WPS TEST REPORT for the SIPT is the only way to score the SIPT and compare the obtained SIPT scores to the results from the normative sample.* WPS is the sole provider of SIPT scoring.

One of the many benefits SIPT offers is computer scoring and interpretation. Users simply record the child's performance on WPS TEST REPORT Answer Sheets (WPS Catalog No. W-260A) and then send the completed Answer Sheets to WPS. On the same day they are received, WPS will score the SIPT Answer Sheets and mail a complete interpretive report to the user. Computer scoring saves hours of valuable time by performing all calculations for the user. It ensures quick, accurate scoring of all 17 tests. WPS TEST REPORT generates not only the total score for each test, but also specific content scores for many of the tests. It should be noted that WPS is always improving and updating its reports. The report shown in this chapter may not therefore be an exact replication of the newest version of the WPS TEST REPORT.

This chapter serves three functions. First, it provides an introduction to the SIPT computer report for the novice user. Second, it provides technical details that guide the methods of the computer report. These details are presented here in greater detail than they are in the report itself. For this reason, it is recommended that experienced users regularly review this chapter to ensure the continued correct use of the SIPT computer report. Third, this chapter provides guidelines for detailed clinical interpretation of the SIPT computer report.

### Technical Definitions of Terms Used in the Report

Both *SD* scores and percentile scores are provided in the SIPT computer report. *SD* scores are used as a standardized way of presenting test results relative to the normative sample. (*SD* scores, which always have a mean or average of 0 and a standard deviation of 1, are also commonly referred to as *z*-scores.) Average *SD* scores are usually considered to fall between –1 and +1. In a population of normal children, approximately two thirds of all scores will fall between –1.0 and +1.0 (i.e., no more than one standard deviation above or below the mean). Scores below –2 generally indicate definite problems, while scores between –2 and –1 are indicative of possible problems. Scores greater than 2 indicate generally superior performance, while scores between 1 and 2 indicate somewhat advanced performance. *SD* scores are linear scores, which means that the difference between a *SD* score of 0 and a *SD* score of 1 is equivalent to the difference between a *SD* score of 2 and a *SD* score of 3. For this reason, it makes the most sense to compare differences between individual subscale scores using *SD* scores, rather than raw scores or percentile scores.

The percentile scores indicate the percentage of individuals in the standardization sample whose scores were equal to or lower than the examinee's score. For example, if a child has a percentile score of 63, then that child's score equals or exceeds the scores of 63% of the children of the same age and sex in the standardization sample.* Average scores are usually considered to be between the 15th and 85th percentiles. Thus, a percentile score between 15 and 85 would be considered "average" relative to children of the same age and sex. A percentile score less than 15 would be considered to be appreciably more deviant than the "typical" child in the normative sample. Percentile scores are not linear. Therefore, the difference between scores of 50% and 60% is not the same as the difference between scores of 10% and 20%. This is why the percentile score anchor points are not evenly spaced on the graphs in the SIPT computer report. Percentile scores are provided because most people find them very easy to understand. It probably will be easiest to explain the child's performance to the parents in terms of percentile scores.

In addition to *SD* scores and percentiles, it is necessary to discuss the norms from which these scores are derived. Norms were obtained by giving the test to a large number of individuals who form a suitable comparison

*The percentiles presented in the SIPT report are expected values obtained from a normal distribution curve for *SD* scores of different values. The justification for this procedure is presented in chapter 6.

group. All the tests were then scored, and the distribution of test scores was examined statistically. The SIPT computer report uses separate norms for boys and girls, with a further differentiation of norms based upon age groupings into four-month intervals for younger children and six-month intervals for older children (see chapter 6). Each child's *SD* scores and percentiles are based upon these norms and indicate where the child stands relative to other children of the same age and sex.

**Limitations of Norms**

Norms do have some limitations that should be considered when interpreting *SD* scores. It is particularly important to consider how similar the child undergoing evaluation is to the children of the normative sample. If the child is very different from these comparison children, then the norms might not provide useful data. For example, the SIPT norms are based on 4- to 8-year-old children who have *not* been identified as needing clinical treatment. Therefore, if the SIPT was given to a 35-year-old, or a 2-year-old, the *SD* scores and percentiles provided would be very difficult to interpret because they would not show where the 35-year-old adult or 2-year-old child stands relative to his or her peers.

Note that if the child administered the test is younger than the youngest normative group (which is 4 years, 0 months to 4 years, 3 months), then the child must be compared to the norms for the youngest group, by coding an age of 4 years, 0 months on the Transmittal Sheet. If the respondent is older than the oldest normative group of 8 years, 6 months to 8 years, 11 months, then the client is automatically compared to the oldest group with an appropriate warning printed in the computer report. Based upon the data available at the time this Manual was printed, there seems to be little justification for giving the SIPT to children younger than the youngest normative group or more than two years older than the oldest normative group. Should the SIPT be normed on a wider range of age groups in the future, documentation released with the actual reports will indicate the extension.

*The narrative statements produced in the SIPT computerized interpretive report are based on the* SD *scores from the normative sample.* The clinician should bear this in mind when interpreting SIPT results. If the SIPT was administered to an individual who is not representative of the normative group, the narrative summaries may not apply or may be too extreme. For example, the children in the SIPT norming groups had no known brain damage nor other handicapping conditions. Children who have handicaps of different kinds may simply not be able to do certain SIPT tests for reasons other than low levels on the characteristic being assessed. For instance, a hearing impairment can cause poor performance on several tests that measure the ability to follow instructions, and could lower performance on most tests. Similarly, visual difficulties may cause poor performance even in the absence of any sensory integration difficulty.

A final caution addresses interpretation of scale scores. In interpreting any scale score, it is important to keep in mind that the obtained score is only an estimate of the individual's theoretical "true" score. As such, the obtained scale score should not be considered as an absolute. Scale scores are likely to vary as a function of the reliability and variance associated with the scale. As a result, scale scores should only be interpreted in the context of other convergent information about the client, and with consideration for the validity of the test protocol and the conditions under which the test was taken. A normal psychometric assumption is that if the child is retested a second time on the same test, the score will not be quite as extreme: children identified as low would be expected to evince somewhat better performance on retest, while children identified as high would be expected to have somewhat less superior performance on retest. *In general, those test scores that have the highest reliability will be the most stable from one occasion to another*. Therefore, the major summary scores for the 17 tests will generally be far more stable and reliable than brief subscale indices such as right versus left differences or summaries of smaller numbers of items.

**Necessary Information**

In order to process the SIPT, it is necessary to know the client's age and sex, because the SIPT norms are based on these variables. Answer Sheets submitted for processing without necessary information will be returned to the user. Note also that every SIPT Answer Sheet submitted for the child *must* have the same Transmittal Number that is given on the Transmittal Sheet. Without this number, the computer program will be unable to accurately sort the tests for the child.

It is extremely important that the individual protocol sheets submitted to WPS are completely filled out, as they are machine processed and cannot be corrected manually if they will not scan correctly. Use only a soft (No. 2), black-leaded pencil to mark responses. Sheets marked with ball-point pens, nylon- or felt-tipped pens, or pencils with harder lead cannot be machine scored. In order to ensure proper scoring, response circles must be dark and filled in completely. In changing a response, erase the first mark fully so that there is no "smudge" left on the page that might be interpreted as a response.

Responses, *as scanned,* are reported at the back of the computer report. If differences are found between the marks that were put on the protocol sheets and the marks printed out in the report, it should be noted that the scores and interpretive statements are based on the *responses as scanned.* If the report seems incomplete because some scores believed to be coded are not shown on the test report, immediately check the protocol sheet and the echoing of the responses by the computer program, to ensure that the two are consistent.

## Content of the WPS TEST REPORT for the SIPT

Each WPS TEST REPORT for the SIPT includes: (a) a listing of background information; (b) a summary graphic panel showing whether each of the 17 SIPT was administered to the child; (c) a table that shows the major result from the single "best" summary score for each of the 17 SIPT; (d) a table that shows the estimated "true" score for each major SIPT score, a standard error of measurement, and a frequency interval band of plus or minus two standard errors of measurement from the estimated true score; (e) the complete table of all the major *SD* scores for each of the 17 SIPT administered; (f) the *D*-square index of statistical fit of the child's profile to that obtained in key diagnostic groups. In addition, limited interpretive text is provided in some cases when the child's performance is exceptional; (g) an introduction to the SIPT with discussion of its limitations; (h) a listing of raw scored test data as recorded on the scan sheets; and (i) a description of the SIPT.

It should be noted that the SIPT computer report is classified as a professional-to-professional communication. Therefore, the SIPT computer report should not be shown directly to the child, or to the child's family without professional interpretation by an individual who is experienced in the administration and interpretation of the SIPT. The test administrator may wish to give this report directly to the child's family or to rewrite the report into a form judged more suitable for the purposes of the evaluation.

The SIPT computer report is intended to be, for the most part, self-documenting. Most of the information and results presented should be easily understood by a trained examiner with the aid of the narration printed throughout the report.

### Sample Report

A complete, computer-generated WPS TEST REPORT sample for the SIPT is presented as Figure 35. In this sample, all 17 SIPT were administered to the child; when a partial set of the tests is administered, then only relevant parts of the report are given. Figure 35 will be used to illustrate the following discussion of the report's content.

### Initial Material

After the report title is printed, the date of testing, client identifying information, and demographic data are given (see (1) in Figure 35). The next part of the first page is a master table of all 17 SIPT, which shows whether each test was administered to the child or not (see (2) in Figure 35). The table also includes a brief statement of the general purpose of each test. Norms used for the client are printed next (see (3) in Figure 35). If the age of the child was not within the age range associated with the normative group, a general caution is given.

### Summary Graph

The second page of the SIPT report is a summary graphic which shows the major score from each of the tests (see (4) in Figure 35). The major scores selected for the 17 tests are as follows: (1) the time-adjusted accuracy score on Space Visualization (SV), (2) the total accuracy score on Figure-Ground Perception (FG), (3) the Manual Form Perception (MFP) total score, (4) the Kinesthesia (KIN) total score, (5) the Finger Identification (FI) total score, (6) the Graphesthesia (GRA) total score, (7) the Localization of Tactile Stimuli (LTS) total score, (8) the total accuracy score on Praxis on Verbal Command (PrVC), (9) the Design Copying (DC) total accuracy score, (10) the total score on Constructional Praxis (CPr), (11) the Postural Praxis (PPr) total score, (12) the Oral Praxis (OPr) total score, (13) the Sequencing Praxis (SPr) total score, (14) the Bilateral Motor Coordination (BMC) total score, (15) the total score on Standing and Walking Balance (SWB), (16) the weighted Motor Accuracy (MAc) total score, and (17) the total Postrotary Nystagmus (PRN) score.

On the next page, there is a table that lists the actual *SD* score for each test as well as an estimated true score, a standard error of measurement, and a band of plus or minus two standard errors of measurement around the *estimated true score** (see (5) in Figure 35). The estimated true score is a "best guess" estimate of the child's latent ability correcting for the likely error of measurement in the testing. Note that, in general, this score will tend to be less extreme than the actual scores; the estimated true *SD* scores will tend to be closer to the theoretical average *SD* score of 0. If the child were to be tested a second time, it is about 95% likely that his or her score would be in the band of plus or minus two standard errors of measurement around the *estimated true score;* it is about 65% likely that his or her score on retest would be in the band of plus or minus one standard error of measurement around the estimated true score. (Note that it is possible for the actual *SD* score to fall outside the band of plus or minus two standard errors of measurement if the score is very extreme; this indicates that a score this extreme may have been obtained in part because of chance factors such as an inattentive child, environmental

*Stanley's method for constructing true score confidence bands was used for these calculations.

distractions, administrator error, or inappropriate use of the test.)

**Table of All Scorable SIPT Indices**

Page 4 of the SIPT report is a table that lists all SIPT scores that can be calculated from the data given for the child (see (6) in Figure 35). Many of the SIPT have several subscores. All scorable indices are summarized in this table in the metric of *SD* scores. (Note that these scores are later presented graphically in the next sections.)

**Comparison with Diagnostic Prototypes**

The next section of the SIPT report consists of a comparison of the child's scores with key diagnostic prototypes (see (7) in Figure 35). For each prototype, a D-square is given. D-square is an index of similarity of the degree of fit between the child's profile and the empirically-derived profile of another diagnostic group. A small D-squared value indicates a close fit, while a large value represents a poor fit.

**WPS ChromaGraph™ for the SIPT**

The last part of the professional section of the SIPT computer report is the WPS ChromaGraph™, which is only provided when the mail-in scoring service is utilized. A sample SIPT ChromaGraph™ is presented as Figure 36. Although the actual ChromaGraph™ included with each report is in full color, the sample ChromaGraph™ in this Manual is reproduced in black and white. The ChromaGraph™ provides a full-color graphic summary of the client's SIPT results. The information contained in the WPS ChromaGraph™ is taken directly from the regular SIPT computer report.

The WPS ChromaGraph™ for the SIPT is a single-page summary, which serves as a visual aid to convey major results from the testing to therapists, parents, and associated professionals. Drawn on high-quality paper, it is suitable for incorporation into the permanent chart of the child.

The WPS ChromaGraph™ for the SIPT graphically portrays three kinds of information: (a) deviance of the child's scores—identifying which individual test scores might be considered significant in a negative or positive direction; (b) patterns of score results expected for different kinds of children—displaying typical results that can be expected for six prototypic diagnostic groups, including both dysfunctional children and children who show average and superior patterns of sensory integration; and (c) comparison of scores—showing which, if any, of the six prototypic diagnostic groups had score profiles similar to the profile for this child.

At the top and bottom of the WPS ChromaGraph™ for the SIPT is a color "thermometer" corresponding to the scale of *SD* scores. Scores are compared to an unselected, national normative sample of the same age and sex as the child being tested. Scores on the left side of the plot show poor performance, those in the middle show average performance, and scores on the right side of the plot show better than average performance.

The magenta zone, containing *SD* scores from −3.0 to −2.5, is used to represent results that are clinically significant and of major import. Red is used for the zone from −2.5 to −2.0. Scores appearing here indicate a definite and real clinical problem. Scores from −2.0 to −1.0, represented in the orange zone, may be indicative of a clinical problem. *SD* scores between −1.0 and 1.0 appear in the green, or typical zone. Scores in the blue zone, from 1.0 to 2.0, indicate above-average sensory integration or praxis. Finally, the violet zone, from 2.0 to 3.0, indicates highly superior performance.

The color zones can be used to grade the individual scale scores of the child, which are presented as solid black squares in the body of the WPS ChromaGraph™. These black squares are connected when consecutive scales were given.

Six empirically derived comparison groups are named at the bottom of the WPS ChromaGraph™ for SIPT. Each of these prototypic diagnostic groups is plotted in a different color and in a different geometric shape, as indicated in the key on the ChromaGraph™. The child's score profile on the 17 major tests is superimposed on this graph. If the shapes within a color group are not connected, the profile of the child is not sufficiently enough like that of the group to suspect that he or she may be a typical member of that group. If the shapes for a color group are connected, it indicates that the child may be a member of that group. Shapes that are connected are drawn in a larger size. Note that although the colors of the *SD* zones and the group profiles follow the same order, a profile does not necessarily fall directly below its corresponding color zone. The groups are described in great detail in chapter 7.

The following method was used to determine whether or not the child's score profile matched these group profiles. For each of the six groups, a D-squared index was calculated. This index is defined as the sum of the squared differences between a child's scores and those of the group on those 17 major SIPT scores on which data are present. The resulting sum is then divided by the number of present tests. The child's D-squared values corresponding to each of the six SIPT groups are presented in the WPS TEST REPORT for SIPT, at the end of the section describing the ChromaGraph™ (see (7) in Figure 35). On the ChromaGraph™, groups are connected if the D-squared value is less than 1.0.

**Display of Item Responses**

In this part of the report (see (8) in Figure 35), the

client's responses to the individual questions on the SIPT are printed. The responses printed on this page are those used to generate the computer report. If the SIPT computer report results seem inconsistent with characteristics of the client already known to the user, it is often helpful, as a first step, to examine the item responses. If the item responses are inconsistent with prior knowledge of the client, the clinician should check to make sure that the client identification number is accurate. Next, the user should examine the response bubbles of the Answer Sheet. If the bubbles are too light, they cannot be read (or they may be misread, resulting in erroneous data) by the optical scanning equipment used to score the test. If the identification number is correct and the bubbles appear to be properly darkened, the clinician may wish to carefully weight the SIPT results with clinical observations made on the client during the administration, and information obtained through other sources, in order to reveal the source of the discrepancy. Be sure to darken leading zeros, otherwise responses will be misread by the computer.

## General Issues in Interpretation

### Validity of the Protocol

**Examination of the Answer Sheet.** Several sources of invalidity can be detected by examination of the Answer Sheet prior to sending it to WPS for computer scoring. One source of invalidity that can be detected in such a manner involves unscorable responses. Unscorable items are those that cannot be scored because the examiner has either entered no response or provided multiple responses. Under the standard scoring procedure, a given test may not be scored for all possible indices if some portions of the necessary information are omitted.

### Computer-Generated Narrative Interpretation of the SIPT Results

Interpretive summaries of results for each of the SIPT are generated by the SIPT computerized scoring and interpretation program. Narrative summaries are provided for each subsection of the SIPT using all possible indices for that SIPT. The summaries are derived in the following way: For each of the individual subscales, a number of specific clinical interpretations have been stored in the computer program. These statements are triggered by the level of the *SD* score on the particular index.

The narrative summaries are intended to aid the professional in interpreting the results of the SIPT. They are *not* intended to be shown directly to the child's family. Such presentation is inconsistent with accepted principles of psychological practice. These summaries, as well as all sections of the report are considered to be professional-to-professional consultations. Note that these computer-generated narratives are not designed to substitute for skilled clinical judgment. Computerized expert systems, such as the current one, can only respond to the information coded into the Answer Sheets. The computer program has no knowledge of additional information gathered from the SIPT administration, parent or school interviews, and other test results. Therefore, the computer program is only responding in an "average" way to the "typical" meaning of such test results; other interpretations may be appropriate, given the additional knowledge of a skilled administrator/interpreter of the SIPT and other sensory integration tests.

The individual narrative statements are generated by dividing the *SD* scores into four ranges. The lowest range is below –2.00 and is characterized by such phrases as "very low" or "much below average," depending upon the context. These scores are in the range most therapists would consider as definite indicators of dysfunction. The second lowest range is from –2.00 to –1.01. These *SD* scores are labeled as "low" and are in a range indicative of possible or probable dysfunction. *SD* scores from –1.00 to 1.00 are characterized as "average," while *SD* scores above 1.00 are characterized as "above average." Because the primary purpose of the SIPT is to identify children with various forms of dysfunction, scores in the above average range are not further differentiated.

### Specific Guidelines for Interpretation

Although the WPS TEST REPORT for the SIPT automatically provides interpretive information, presentation of a basic strategy for interpreting the SIPT should prove helpful to many users. Several basic considerations are worthy of discussion. As mentioned earlier, *SD* scores have a mean of 0 and a standard deviation of 1. Because *SD* scores are linear (see p. 113), it is easier to compare individual scales for the same client using *SD* scores rather than percentiles, because the difference between different *SD* scores is standard, regardless of the extremity of the scores. A general rule of thumb that works in most cases is that individual scale scores are considered to be significantly different from one another when they differ by at least one standard deviation (1 *SD* score point). A significantly deviant score is generally considered to be at least two standard deviations above or below the mean.

### Conclusion

This chapter has covered the individual components of the WPS computerized interpretive report for the SIPT. Important information for proper use and understanding of the computer report has been discussed. Specific guidelines to be used in the clinical interpretation and use of the SIPT computer report have also been presented.

Sensory Integration and Praxis Tests (SIPT)
by A. Jean Ayres, Ph.D.
A WPS TEST REPORT by Western Psychological Services
12031 Wilshire Boulevard
Los Angeles, California 90025-1251

Version: S800-004

Client Name: Tess T.
Age at Testing: 7 yrs. 08 mos.
Sex: Female
Grade: 2
Ethnic Background: White
Processing Date: 10/10/91

Transmittal Number: 0000wps
Preferred Hand: Right
No. of Tests Administered: 17
No. of Tests Scored: 17
No. of Unscorable Tests: 0

***** SIPT TEST REPORT *****

This WPS TEST REPORT for the SIPT provides detailed information on Tess's sensory processing and practic abilities. A summary of the tests that were scored for Tess is provided below:

| Test | Number of Subscores | Brief Description of Function(s) Measured | Was Test Administered? |
|---|---|---|---|
| SV | 5 | Motor-free visual perception; mental rotation | Yes |
| FG | 2 | Motor-free visual figure-ground perception | Yes |
| MFP | 14 | Recognition of forms held in hands; visualization | Yes |
| KIN | 3 | Somatic perception of arm position and movement | Yes |
| FI | 3 | Tactile perception of individual fingers | Yes |
| GRA | 3 | Tactile perception of simple designs; praxis | Yes |
| LTS | 3 | Identification of place on arm or hand touched | Yes |
| PrVC | 2 | Translation of verbal directions into action | Yes |
| DC | 4 | Visuopraxis; two-dimensional constructions | Yes |
| CPr | 3 | Three-dimensional visual space management | Yes |
| PPr | 1 | Planning and executing bodily movements | Yes |
| OPr | 1 | Imitating tongue\lip\jaw movements; somatopraxis | Yes |
| SPr | 3 | Sequencing movements, bilateral integration | Yes |
| BMC | 3 | Functional integration of the two sides of body | Yes |
| SWB | 5 | CNS processing of muscle, joint, gravity input | Yes |
| MAc | 6 | Eye-hand coordination; somatopraxis | Yes |
| PRN | 7 | CNS processing of vestibular (cupular) input | Yes |

This report presents score profiles and narrative summaries for the completed tests on the pages indicated in the above table. Page 2 contains a summary graph of the major SIPT scores, and Page 3 shows the estimated true scores on each of the major scales. A listing of Tess's scores on all of the SIPT subscales begins on Page 4. The final page of this report is a color plot, which contains a profile of this child's major SIPT scores, and shows how closely the child's profile matches the profiles of the six SIPT groups described in the SIPT Manual (WPS Catalog No. W-260M).

************************************************************
* Normative age group: 7 yrs., 6 mos. to 7 yrs., 11 mos. *
************************************************************

**Figure 35**
**Sample WPS TEST REPORT™ for the SIPT**

SIPT Test Report Transmittal Number: 0000wps Page: 2

SUMMARY GRAPH OF SIPT RESULTS

This graph shows the major scores for the 17 tests in the SIPT. No score is shown if the test was not administered, or if the test was partially administered in such a way that the major summary score could not be computed.

The SD scores shown correspond to a metric usually associated with the normal curve, and are also known as z-scores. In a normal distribution, SD scores have an average or mean value of 0 and a standard deviation of 1.

The percentile scores shown on the bottom of this graph (and on the graphs for each of the individual SIPT tests) indicate the percentage of children of this age in the general population who would be expected to score at or below a given value. For example, an SD score of 0 corresponds to the 50th percentile, which means that half of the children would be expected to obtain SD scores at or below 0. (Note that these are theoretical percentile scores, based upon the assumption that the test scores are normally distributed; such an assumption is warranted for most of the major SIPT scores, as discussed in chapter 5 of the SIPT Manual.)

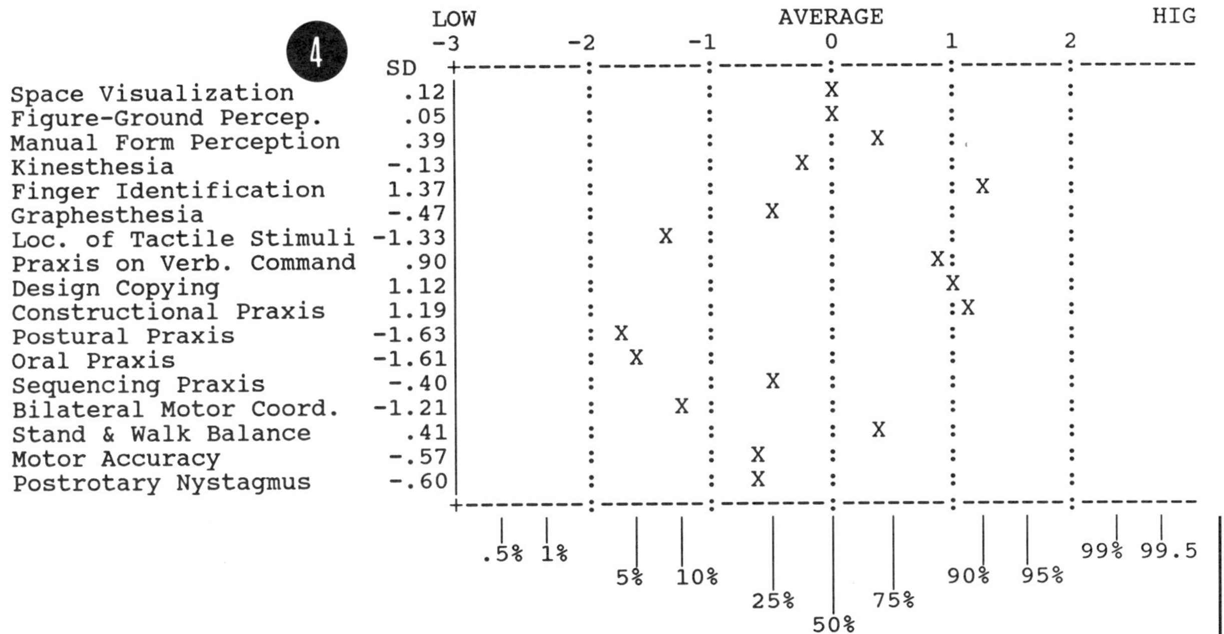


Western Psychological Services • 12031 Wilshire Boulevard • Los Angeles, California 90025-1251

WPS TEST REPORT™

**Figure 35 (Continued)**
**Sample WPS TEST REPORT™ for the SIPT**

SIPT Test Report Transmittal Number: 0000wps Page: 3

ESTIMATED TRUE SCORES

Below is a table which lists the major score for each test, the child's estimated true score on each test, the standard error of measurement, and a band of plus or minus two standard errors of measurement around the estimated true score. The estimated true scores are "best guess" estimates of the child's latent ability, correcting for the likely error of measurement in the testing. In general, estimated true scores will be less extreme than the actual scores.

If the child were to be tested again, it is about 95% likely that his or her score would be in the band of plus or minus two standard errors of measurement around the estimated true score; it is about 65% likely that his or her score on retest would be in the band of plus or minus one standard error of measurement around the estimated true score.

It is possible for the actual SD score to fall outside the band of plus or minus two standard errors of measurement if the score is very extreme; this means that a score this extreme may have been obtained in part because of chance factors such as an inattentive child, environmental distractions, administrator error, or inappropriate use of the test.

5

| Test | SD Score | Est. True Score | SEM | 2 SEM Band Lower | 2 SEM Band Upper |
|---|---|---|---|---|---|
| Space Visualization (SV) | .12 | .08 | .55 | -.99 | 1.16 |
| Figure-Ground Perception (FG) | .05 | .03 | .66 | -1.27 | 1.33 |
| Manual Form Perception (MFP) | .39 | .27 | .55 | -.80 | 1.35 |
| Kinesthesia (KIN) | -.13 | -.06 | .71 | -1.45 | 1.32 |
| Finger Identification (FI) | 1.37 | 1.02 | .51 | .02 | 2.02 |
| Graphesthesia (GRA) | -.47 | -.34 | .52 | -1.36 | .68 |
| Loc. of Tactile Stimuli (LTS) | -1.33 | -.71 | .69 | -2.05 | .64 |
| Praxis on Verb. Command (PrVC) | .90 | .81 | .33 | .15 | 1.46 |
| Design Copying (DC) | 1.12 | 1.04 | .26 | .52 | 1.56 |
| Constructional Praxis (CPr) | 1.19 | .83 | .55 | -.24 | 1.91 |
| Postural Praxis (PPr) | -1.63 | -1.40 | .37 | -2.13 | -.67 |
| Oral Praxis (OPr) | -1.61 | -1.45 | .32 | -2.07 | -.83 |
| Sequencing Praxis (SPr) | -.40 | -.33 | .40 | -1.12 | .45 |
| Bilateral Motor Coord. (BMC) | -1.21 | -.99 | .42 | -1.82 | -.16 |
| Standing & Walking Balance (SWB) | .41 | .35 | .39 | -.41 | 1.10 |
| Motor Accuracy (MAc) | -.57 | -.49 | .39 | -1.25 | .27 |
| Postrotary Nystagmus (PRN) | -.60 | -.29 | .72 | -1.70 | 1.13 |

N/A: Test not administered U/S: Test unscorable

*: SD scores below -3.00 are reported as -3.00
**: SD scores above 3.00 are reported as 3.00





**Figure 35 (Continued)**
**Sample WPS TEST REPORT™ for the SIPT**

SIPT Test Report Transmittal Number: 0000wps Page: 4

SIPT SCORES

The following table is a complete list of this child's SIPT test scores. Major scores for each of the 17 tests are listed with an asterisk (*). Test scores outside the average range of -1.00 to 1.00 are interpreted as: Much Below Average (-3.00 to -2.01), Below Average (-2.00 to -1.01), Above Average (1.01 to 2.00), and Much Above Average (2.01 to 3.00). The DC atypical approach parameters and CPr Part II errors are interpreted as Possible Problem (-3.00 to -1.50) and Possible Strength (1.50 to 3.00).

6

| Test | SD Score | SD Score Interpretation (if not in average range) |
|---|---|---|
| **Space Visualization (SV)** | | |
| * Time-adjusted accuracy | .12 | |
| Accuracy | .47 | |
| Time | -.95 | |
| Contralateral use | -1.35 | Below Average |
| Preferred hand use | -.13 | |
| **Figure-Ground Perception (FG)** | | |
| * Accuracy | .05 | |
| Time | -1.75 | Below Average |
| **Manual Form Perception (MFP)** | | |
| * Total accuracy | .39 | |
| Total time | -2.15 | Much Below Average |
| Part I accuracy | .53 | |
| Part I right accuracy | -.15 | |
| Part I left accuracy | 1.11 | Above Average |
| Part I time | -1.11 | Below Average |
| Part I right time | -1.14 | Below Average |
| Part I left time | -.85 | |
| Part II accuracy | .17 | |
| Part II right accuracy | .08 | |
| Part II left accuracy | .21 | |
| Part II time | -2.04 | Much Below Average |
| Part II right time | -2.11 | Much Below Average |
| Part II left time | -1.49 | Below Average |
| **Kinesthesia (KIN)** | | |
| * Total accuracy | -.13 | |
| Right hand accuracy | -.23 | |
| Left hand accuracy | .00 | |
| **Finger Identification (FI)** | | |
| * Total accuracy | 1.37 | Above Average |
| Right hand accuracy | .74 | |
| Left hand accuracy | 1.52 | Above Average |

(Continued on Next Page)

*: Major score used in ChromaGraph
N/A: Test not administered U/S: Test unscorable

Western Psychological Services • 12031 Wilshire Boulevard • Los Angeles, California 90025-1251

WPS TEST REPORT™

**Figure 35 (Continued)**
**Sample WPS TEST REPORT™ for the SIPT**

SIPT Test Report    Transmittal Number: 0000wps    Page: 5

SIPT SCORES (Continued)

| Test | SD Score | SD Score Interpretation (if not in average range) |
|---|---|---|
| **Graphesthesia (GRA)** | | |
| * Total accuracy............... | -.47 | |
| Right hand accuracy.......... | -.44 | |
| Left hand accuracy........... | -.37 | |
| **Localization of Tactile Stimuli (LTS)** | | |
| * Total accuracy............... | -1.33 | Below Average |
| Right hand accuracy.......... | -1.75 | Below Average |
| Left hand accuracy........... | -.29 | |
| **Praxis on Verbal Command (PrVC)** | | |
| * Total accuracy............... | .90 | |
| Total time................... | .74 | |
| **Design Copying (DC)** | | |
| * Total accuracy.............. | 1.12 | Above Average |
| Adjusted accuracy........... | 1.14 | Above Average |
| Part I accuracy............. | .64 | |
| Part II accuracy............ | 1.21 | Above Average |
| Atypical approach parameters: | | |
| - boundaries | .92 | |
| - additions | .29 | |
| - segmentations | .02 | |
| - reversals | -1.16 | |
| - right-to-left | .78 | |
| - inversions | .05 | |
| - jogs | -.93 | |
| - distortions | .31 | |
| **Constructional Praxis (CPr)** | | |
| * Total accuracy.............. | 1.19 | Above Average |
| Part I accuracy............. | .87 | |
| Part II accuracy............ | 1.51 | Above Average |
| Part II error: | | |
| Displacement 1-2.5 cm | 1.41 | |
| Displacement > 2.5 cm | -.69 | |
| Rotation > 15 degrees | .99 | |
| Reversals | .73 | |
| Incorrect but logical | .58 | |
| Gross mislocations | .12 | |
| Omissions | .04 | |

(Continued on Next Page)

*: Major score used in ChromaGraph
N/A: Test not administered    U/S: Test unscorable



**Figure 35 (Continued)**
**Sample WPS TEST REPORT™ for the SIPT**

SIPT Test Report          Transmittal Number: 0000wps          Page: 6

SIPT SCORES (Continued)

| Test | SD Score | SD Score Interpretation (if not in average range) |
|---|---|---|
| **Postural Praxis (PPr)** | | |
| * Total accuracy | -1.63 | Below Average |
| **Oral Praxis (OPr)** | | |
| * Total accuracy | -1.61 | Below Average |
| **Sequencing Praxis (SPr)** | | |
| * Total accuracy | -.40 | |
| Hand accuracy | .13 | |
| Finger accuracy | -.88 | |
| **Bilateral Motor Coordination (BMC)** | | |
| * Total accuracy | -1.21 | Below Average |
| Arm accuracy | -1.01 | Below Average |
| Feet accuracy | -1.13 | Below Average |
| **Standing and Walking Balance (SWB)** | | |
| * Total score | .41 | |
| Eyes open | -.24 | |
| Eyes closed | .96 | |
| Right foot | -.18 | |
| Left foot | -.20 | |
| **Motor Accuracy (MAc)** | | |
| * Weighted total acc | -.57 | |
| Unweighted total acc | -.63 | |
| Pref hand weight acc | -.51 | |
| Pref hand unweight acc | -.61 | |
| Nonpref hand weight acc | -.64 | |
| Nonpref hand unwght acc | -.65 | |
| **Postrotary Nystagmus (PRN)** | | |
| * Average nystagmus | -.60 | |
| Average clockwise | .27 | |
| Average cnt clockwise | -1.46 | Below Average |
| Time 1 clockwise | .26 | |
| Time 1 cnt clockwise | -2.13 | Much Below Average |
| Time 2 clockwise | .26 | |
| Time 2 cnt clockwise | -.47 | |

*: Major score used in ChromaGraph
N/A: Test not administered          U/S: Test unscorable



**Figure 35 (Continued)**
**Sample WPS TEST REPORT™ for the SIPT**

SIPT Test Report Transmittal Number: 0000wps Page: 7

COMPARISON WITH DIAGNOSTIC PROTOTYPES

Six prototypic diagnostic groups of children have been identified, including both dysfunctional children and children who show average and superior patterns of sensory integration. The major test scores for each of these groups are compared to this child's scores, and the resulting D-squared values estimate similarity. The diagnostic groups and the statistical method used in the comparisons are described in the SIPT Manual.

When the estimate indicates that a child may be a typical member of a diagnostic group (D-squared value is less than or equal to 1.00), the pattern of fit is shown graphically on the WPS ChromaGraph with the symbols indicated below in brackets.

For technical and research purposes, D-squared vaules are listed below. Note that a small D-squared value indicates a close fit, while a large value represents a poor fit.

| Group | D-squared |
|---|---|
| 1. Low Average Bilateral Integration and Sequencing [green plus sign] | .61 |
| 2. Low Average Sensory Integration and Praxis [blue circle] | .92 |
| 3. Generalized Sensory Integration Dysfunction [magenta square] | 3.58 |
| 4. Dyspraxia on Verbal Command [orange asterisk] | 1.86 |
| 5. Visuo- and Somatodyspraxia [red triangle] | 1.69 |
| 6. High Average Sensory Integration [violet diamond] | 1.14 |

7

Western Psychological Services • 12031 Wilshire Boulevard • Los Angeles, California 90025-1251

WPS TEST REPORT™

**Figure 35 (Continued)**
**Sample WPS TEST REPORT™ for the SIPT**

SIPT Test Report Transmittal Number: 0000wps Page: 8

CAUTIONS

The Sensory Integration and Praxis Tests are a complicated set of tests designed to be administered to children between the ages of 4 and 9 years old. The SIPT should be administered by a trained examiner who has been specifically trained in the use of the SIPT or related psychological and educational assessment methods.

This WPS TEST REPORT for SIPT has been written as a professional-to-professional consultation device. The results in this WPS TEST REPORT should be interpreted for the child's parents by an individual trained in the interpretation of the SIPT. Test result interpretations will be most valid when the individual interpreting these results has either directly tested the child using SIPT or has observed the child in similar tasks. These test results should always be used in conjunction with other information about the child including a social and medical history, direct observation of the child's behavior, and interviews with parents and school officials.

Individuals interpreting the SIPT should be familiar with the reliability and validity evidence presented in the SIPT Manual (WPS Catalog No. W-260M). This WPS TEST REPORT and the SIPT results cannot be understood fully without reading the SIPT Manual and understanding the strengths and limitations of the instrument.

A special caution is in order about the use of a selected subset of SIPT exercises. The 17 SIPT procedures form a comprehensive assessment battery. While a subset of the SIPT can be used for specialized purposes, the overall results will be most accurate when SIPT is employed in its totality. See the SIPT Manual for a discussion of the predicted effects of eliminating subtests.

DESCRIPTION OF THE CHROMAGRAPH FOR THE SIPT

The WPS ChromaGraph for the SIPT is a single-page summary, drawn on a graphics plotter, that presents major testing and statistical results in six colors, plus black and white. It was designed as a visual aid to convey major results from the testing to therapists, parents, and associated professionals. Drawn on high-quality paper, it is suitable for incorporation into the permanent chart of the child.

The WPS ChromaGraph for the SIPT graphically portrays three kinds of information: a) deviance of the child's scores--which individual test scores might be considered significant in a negative or positive direction; b) patterns of score results expected for different kinds of children--typical results that can be expected for six prototypic diagnostic groups, including both dysfunctional children and children who show average and superior patterns of sensory integration; and c) comparison of scores--the extent to which the child's pattern of test scores matches the pattern shown by each of the six prototypic diagnostic groups.

At the top and bottom of the WPS ChromaGraph for the SIPT is a color "thermometer" corresponding to the scale of SD scores. Scores are compared to an unselected national normative sample of the same age and sex. Scores on the left side of the plot show poor performance, those in the middle show average performance, and scores on the right side of the plot show better than average performance.



**Figure 35 (Continued)**
**Sample WPS TEST REPORT™ for the SIPT**

SIPT Test Report Transmittal Number: 0000wps Page: 9

Sensory Integration and Praxis Tests (SIPT)
by A. Jean Ayres, Ph.D.
A WPS TEST REPORT by Western Psychological Services
12031 Wilshire Boulevard
Los Angeles, California 90025-1251

Version: S800-004

***** LISTING OF SIPT TEST DATA RECORDED ON SCAN SHEETS *****

Client Name: Tess T.
Age at Testing: 7 yrs. 08 mos.
Sex: Female
Grade: 2
Ethnic Background: White
Processing Date: 10/10/91
Father's occupation: Not employed outside the home
Mother's occupation: Clerical, sales, technical
Child's Home zip code: 90025
Testing Date: 10/05/91
IQ Test Administered: Not provided
Tactile Defensiveness: Mild defensiveness
Obs. 1: Prone Extension: Not Provided
Obs. 2: Supine Flexion: Not Provided
Obs. 3: Ocular Pursuits: Not Provided
Special Education Classes: No

Transmittal Number: 0000wps
Preferred Hand: Right
No. of Tests Administered: 17
No. of Tests Scored: 17
No. of Unscorable Tests: 0
Neurological Impairment: No
IQ Score: 97

8

1. SPACE VISUALIZATION (SV) Tr:0000wps
Accuracy: 1=right, 0=wrong Hand: 0=left, 1=both, 2=right

| Item: | 1 | 2 | 3 | 4 | 5 | 6 | 7 | 8 | 9 | 10 | 11 | 12 | 13 | 14 | 15 |
|---|---|---|---|---|---|---|---|---|---|---|---|---|---|---|---|
| Acc : | 1 | 1 | 1 | 1 | 1 | 1 | 1 | 1 | 1 | 0 | 1 | 1 | 1 | 1 | 1 |
| Hand: | 0 | 1 | 0 | 2 | 1 | 1 | 0 | 2 | 2 | 0 | 2 | 0 | 2 | 0 | 2 |
| Time: | 01 | 03 | 02 | 02 | 04 | 01 | 03 | 01 | 01 | 01 | 01 | 02 | 01 | 03 | 02 |

| Item: | 16 | 17 | 18 | 19 | 20 | 21 | 22 | 23 | 24 | 25 | 26 | 27 | 28 | 29 | 30 |
|---|---|---|---|---|---|---|---|---|---|---|---|---|---|---|---|
| Acc : | 1 | 1 | 1 | 1 | 0 | 1 | 1 | 1 | 1 | 1 | 1 | 0 | 1 | 1 | 0 |
| Hand: | 2 | 0 | 2 | 2 | 2 | 0 | 1 | 2 | 2 | 2 | 2 | 2 | 2 | 2 | 2 |
| Time: | 03 | 01 | 01 | 04 | 04 | 01 | 02 | 05 | 04 | 03 | 03 | 04 | 04 | 02 | 24 |

2. FIGURE-GROUND PERCEPTION (FG) Tr:0000wps
1 = selected, 0 = not selected

| Item: | 1 | 2 | 3 | 4 | 5 | 6 | 7 | 8 |
|---|---|---|---|---|---|---|---|---|
| Select: | 010110 | 011010 | 101001 | 011010 | 010011 | 100110 | 100110 | 010110 |
| Time: | 15 | 10 | 30 | 27 | 35 | 32 | 55 | 53 |

| Item: | 9 | 10 | 11 | 12 | 13 | 14 | 15 | 16 |
|---|---|---|---|---|---|---|---|---|
| Select: | 000111 | 000110 | 000000 | 000000 | 000000 | 000000 | 000000 | 000000 |
| Time: | 40 | 50 | ** | ** | ** | ** | ** | ** |

3. STANDING AND WALKING BALANCE (SWB) Tr:0000wps

| Item: | 1 | 2 | 3 | 4 | 5 | 6 | 7 | 8 | 9 | 10 | 11 | 12 | 13 | 14 | 15 | 16 |
|---|---|---|---|---|---|---|---|---|---|---|---|---|---|---|---|---|
| Best response: | | | | 10 | 10 | 10 | 10 | 10 | 15 | 15 | 03 | 03 | 01 | 01 | 06 | 03 |



**Figure 35 (Continued)**
**Sample WPS TEST REPORT™ for the SIPT**

```
SIPT Test Report             Transmittal Number: 0000wps           Page: 10

 4. DESIGN COPYING (DC)
    Hand Used: Right
                        Part I    SNH:                              Tr:0000wps
      1. Acc :  2     2. Acc :  2    3. Acc :  1    4. Acc :  2
                         SNH PARMS:     SNH PARMS:     SNH PARMS:
                          R :N           R :N           R :N
                          L :N
                          I :N
      5. Acc :  2     6. Acc :  2    7. Acc :  0    8. Acc :  2
         SNH PARMS:      SNH PARMS:     SNH PARMS:     SNH PARMS:
          I :N            R :N           R :Y           R :N
          L :N            I :N           L :N           I :N
                                         I :N
      9. Acc :  2    10. Acc :  2   11. Acc :  2   12. Acc :  0
         SNH PARMS:
          R :N
          I :N
     13. Acc :  0
                        Part II:  *=error                           Tr:0000wps
     14.  SH   SNH  15.  SH   SNH  16.  SH   SNH  17.  SH   SNH
          1:Y  B:N       1:Y  B:N       1:Y  B:N       1:Y  B:N
          2:Y  L:N       2:Y  A:N       2:Y  A:N       2:Y  A:N
          3:Y            3:Y  R:N       3:Y  R:N       3:Y  S:N
          4:Y            4:Y  L:N            L:N            R:N
                         5:Y  I:N            D:N            I:N
                                                            J:N
     18.  SH   SNH  19.  SH   SNH  20.  SH   SNH  21.  SH   SNH
          1:Y  B:N       1:Y  B:N       1:Y  B:N       1:Y  B:N
          2:Y  A:N       2:Y  A:N       2:Y  A:N       2:Y  A:N
          3:Y  I:N       3:Y  S:N       3:N  S:N       3:Y  S:N
          4:Y  L:N            R:N            R:N       4:N  L:N
               J:N            L:N            J:N            I:N
     22.  SH   SNH  23.  SH   SNH  24.  SH   SNH  25.  SH   SNH
          1:Y  B:N       1:Y  B:N       1:Y  B:N       1:Y  B:N
          2:Y  A:N       2:Y  A:N       2:Y  A:N       2:Y  A:N
          3:Y  R:N       3:N  S:N       3:Y  S:N       3:Y  S:Y
               L:N                      4:Y  R:N       4:Y  R:N
               I:N                      5:Y  D:N       5:N
               J:Y

 5. POSTURAL PRAXIS (PPr)                                           Tr:0000wps
    0=incorrect or exceeded 7 seconds
    1=correct in 4-7 seconds partially correct
    2=correct in 3 seconds
    Item:   1  2  3  4  5  6  7  8  9 10 11 12 13 14 15 16 17
    Score:  2  2  1  2  2  1  1  1  1  2  1  0  0  1  0  1  1

 6. BILATERAL MOTOR COORDINATION (BMC)                              Tr:0000wps
    0 = incorrect
    1 = approximately correct
    2 = correct
    Item:   1  2  3  4  5  6  7  8  9 10 11 12 13 14
    Score:  2  1  1  0  1  0  1  0  1  1  1  0  0  0
```



**Figure 35 (Continued)**
**Sample WPS TEST REPORT™ for the SIPT**

SIPT Test Report Transmittal Number: 0000wps Page: 11

**7. PRAXIS ON VERBAL COMMAND (PrVC)** Tr:0000wps

Accuracy: 0=incorrect, 1=correct

| Item: | 1 | 2 | 3 | 4 | 5 | 6 | 7 | 8 | 9 | 10 | 11 | 12 |
|---|---|---|---|---|---|---|---|---|---|---|---|---|
| Accuracy: | 1 | 1 | 1 | 1 | 1 | 1 | 1 | 1 | 1 | 1 | 1 | 1 |
| Time: | 01 | 00 | 01 | 02 | 03 | 02 | 03 | 03 | 01 | 01 | 01 | 01 |

| Item: | 13 | 14 | 15 | 16 | 17 | 18 | 19 | 20 | 21 | 22 | 23 | 24 |
|---|---|---|---|---|---|---|---|---|---|---|---|---|
| Accuracy: | 1 | 1 | 1 | 1 | 1 | 1 | 1 | 1 | 1 | 1 | 1 | 1 |
| Time: | 00 | 02 | 01 | 02 | 01 | 02 | 01 | 01 | 04 | 02 | 03 | 05 |

**8. CONSTRUCTIONAL PRAXIS (CPr)** Tr:0000wps

Structure I

| Item: | 1 | 2a | 2b | 3a | 3b | 3c | 3d | 4a | 4b | 4c | 5a | 5b | 6a | 6b | 7a | 7b |
|---|---|---|---|---|---|---|---|---|---|---|---|---|---|---|---|---|
| Score: | Y | Y | Y | Y | Y | Y | Y | Y | Y | Y | Y | Y | Y | Y | Y | Y |

Structure II

| blkno: | 1 | 2 | 3 | 4 | 5 |
|---|---|---|---|---|---|
| Parms: | ------01 | 00000001 | 00000001 | 00000001 | 01000000 |
| blkno: | 6 | 7 | 8 | 9 | 10 |
| Parms: | 00010000 | 01000000 | 00000001 | 00000001 | 00000001 |
| blkno: | 11 | 12 | 13 | 14 | 15 |
| Parms: | 00000001 | 00-00001 | ----0001 | ----0001 | 00000001 |

**9. POSTROTARY NYSTAGMUS (PRN)** Tr:0000wps

| | First Test | Retest |
|---|---|---|
| Counterclockwise (to the left) | 00 | 07 |
| Clockwise (to the right) | 10 | 10 |

**10. MOTOR ACCURACY (MAc)** Tr:0000wps

Preferred Hand: Right

| | Right Hand | Left Hand |
|---|---|---|
| Time in Seconds | 120 | 067 |
| Solid Line | 35.0 | 43.0 |
| Short Broken Line | 01.5 | 08.5 |
| Medium Broken Line | 00.0 | 00.0 |
| Long Broken Line | 00.0 | 00.0 |



**Figure 35 (Continued)**
**Sample WPS TEST REPORT™ for the SIPT**

SIPT Test Report Transmittal Number: 0000wps Page: 12

11. SEQUENCING PRAXIS (SPr) Tr:0000wps
    0 = executed with wrong hand position or movement, or too few or too many motions in sequence
    1 = started with wrong hand position or movement but started over and completed correctly
    2 = completed with correct hand positions in correct sequence and correct number of motions and/or taps

| | Item 1 | | | | | |
|---|---|---|---|---|---|---|
| Part: | a | b | c | d | e | |
| Score: | 2 | 2 | 2 | 2 | 2 | |

| | Item 2 | | | | | |
|---|---|---|---|---|---|---|
| Part: | a | b | c | d | e | f |
| Score: | 2 | 2 | 2 | 0 | 2 | 2 |

| | Item 3 | | | | | |
|---|---|---|---|---|---|---|
| Part: | a | b | c | d | e | f |
| Score: | 2 | 2 | 2 | 2 | 2 | 2 |

| | Item 4 | | | | | |
|---|---|---|---|---|---|---|
| Part: | a | b | c | d | e | f |
| Score: | 2 | 2 | 2 | 2 | 2 | 2 |

| | Item 5 | | | | | |
|---|---|---|---|---|---|---|
| Part: | a | b | c | d | e | f |
| Score: | 2 | 2 | 2 | 0 | 0 | 0 |

| | Item 6 | | | | | |
|---|---|---|---|---|---|---|
| Part: | a | b | c | d | e | f |
| Score: | 1 | 2 | 1 | 1 | 0 | 0 |

| | Item 7 | | | | | |
|---|---|---|---|---|---|---|
| Part: | a | b | c | d | e | f |
| Score: | 2 | 0 | 0 | 0 | 0 | 1 |

| | Item 8 | | | | | |
|---|---|---|---|---|---|---|
| Part: | a | b | c | d | e | f |
| Score: | 1 | 2 | 0 | 0 | 0 | 1 |

| | Item 9 | | | | | |
|---|---|---|---|---|---|---|
| Part: | a | b | c | d | e | f |
| Score: | 0 | 0 | 0 | 0 | 0 | 0 |

12. ORAL PRAXIS (OPr) Tr:0000wps
    0=incorrect, 1=poor quality, 2=well executed

| Item: | 1 | 2 | 3 | 4 | 5 | 6 | 7 | 8 | 9 | 10 | 11 | 12 | 13 | 14 | 15 | 16 | 17 | 18 | 19 |
|---|---|---|---|---|---|---|---|---|---|---|---|---|---|---|---|---|---|---|---|
| Score: | 2 | 2 | 2 | 0 | 2 | 1 | 2 | 1 | 1 | 1 | 2 | 1 | 1 | 1 | 2 | 1 | 0 | 0 | 0 |

13. MANUAL FORM PERCEPTION (MFP) Tr:0000wps

Part I

| Item: | 1 | 2 | 3 | 4 | 5 | 6 | 7 | 8 | 9 | 10 |
|---|---|---|---|---|---|---|---|---|---|---|
| Accuracy: | 1 | 1 | 1 | 1 | 0 | 1 | 1 | 1 | 1 | 1 |
| Time: | 02 | 02 | 05 | 05 | 06 | 08 | 08 | 04 | 12 | 07 |

Part II

| Item: | 1 | 2 | 3 | 4 | 5 | 6 | 7 | 8 | 9 | 10 |
|---|---|---|---|---|---|---|---|---|---|---|
| Accuracy: | 1 | 1 | 1 | 0 | 1 | 1 | 1 | 0 | 1 | 1 |
| Time: | 28 | 22 | 40 | 45 | 25 | 25 | 15 | 44 | 15 | 30 |

14. KINESTHESIA (KIN) Tr:0000wps

Item Scores (cm)

| Item: | 1 | 2 | 3 | 4 | 5 | 6 | 7 | 8 | 9 | 10 |
|---|---|---|---|---|---|---|---|---|---|---|
| Score: | 1.2 | 4.5 | 4.0 | 0.4 | 3.2 | 5.8 | 3.0 | 4.9 | 1.6 | 9.0 |

Readministered Items (2 most erroneous items)

06. Score: 1.3 10. Score: 1.2

15. FINGER IDENTIFICATION (FI) Tr:0000wps
    (0 = incorrect 1 = correct)

| Item: | 1 | 2 | 3 | 4 | 5 | 6 | 7 | 8 | 9 | 10 | 11 | 12 | 13 | 14 | 15 | 16 |
|---|---|---|---|---|---|---|---|---|---|---|---|---|---|---|---|---|
| Score: | 1 | 1 | 1 | 1 | 1 | 1 | 1 | 1 | 1 | 1 | 0 | 1 | 1 | 1 | 1 | 1 |

16. GRAPHESTHESIA (GRA) Tr:0000wps

| Item: | 1 | 2 | 3 | 4 | 5 | 6 | 7 | 8 | 9 | 10 | 11 | 12 | 13 | 14 |
|---|---|---|---|---|---|---|---|---|---|---|---|---|---|---|
| Score: | 2 | 1 | 1 | 2 | 2 | 2 | 0 | 2 | 0 | 0 | 1 | 0 | 2 | 2 |

Western Psychological Services • 12031 Wilshire Boulevard • Los Angeles, California 90025-1251

WPS TEST REPORT™

**Figure 35 (Continued)**
**Sample WPS TEST REPORT™ for the SIPT**

SIPT Test Report Transmittal Number: 0000wps Page: 13

17. LOCALIZATION OF TACTILE STIMULI (LTS) Tr:0000wps

Item Scores (cm)

| Item: | 1 | 2 | 3 | 4 | 5 | 6 | 7 | 8 | 9 | 10 | 11 | 12 |
|---|---|---|---|---|---|---|---|---|---|---|---|---|
| Score: | 1.0 | 1.4 | 0.5 | 0.7 | 3.8 | 0.5 | 0.6 | 5.6 | 2.5 | 1.5 | 4.5 | 2.5 |

Readministered Items (2 most erroneous items)

08. Score: 1.5 11. Score: 6.5

Clinical Observations: Possible tactile defensiveness

NOTE:

TRANSMITTAL NUMBERS

'Tr:' indicates the transmittal number recorded on the SIPT test sheet.

INVALID OR MISSING RECORDED RESPONSES

'Not Provided', 'N/g', '-', '*', and blanks all indicate missing or multiply-marked responses.



Figure 35 (Continued)
Sample WPS TEST REPORT™ for the SIPT

SIPT Test Report Transmittal Number: 0000wps Page: 14

## DESCRIPTION OF THE SENSORY INTEGRATION AND PRAXIS TESTS FOR PARENTS

A. Jean Ayres, Ph.D.

The Sensory Integration and Praxis Tests (SIPT) help us to understand why some children have difficulty learning or behaving as we expected. The SIPT do not measure intelligence in the usual sense of the word, but they do evaluate some important abilities needed to get along in the world. They do not measure language development, academic achievement, or social behavior, but they assess certain aspects of sensory processing or perception that are related to those functions. They also evaluate praxis or the child's ability to cope with the tangible, physical, two- and three-dimensional world.

Sensory integration is that neurological process by which sensations (such as from the skin, eyes, joints, gravity, and movement sensory receptors) are organized for use. Praxis is that ability by which we figure out how to use our hands and body in skilled tasks like playing with toys, using a pencil or fork, building a structure, straightening up a room, or engaging in many occupations. Practic ability includes knowing what to do as well as how to do it. Practic skill is one of the essential aptitudes that enables us "to do" in the world.

"Dys" means "difficult" or "disordered." Sensory integrative dysfunction may result in difficulty with visual perception tasks or in inefficiency in the interpretation of sensations from the body. A dyspraxic child has difficulty using his or her body, including relating to some objects in the environment. A dyspraxic child often has trouble with simply organizing his or her own behavior.

There are 17 SIPT tests. They fall, roughly, into four overlapping types: (1) motor-free visual perception, (2) somatosensory, (3) praxis, and (4) sensorimotor.

(1) Motor-free visual perception

These tests evaluate the ability to visually perceive and discriminate form and space without involving motor coordination. The Space Visualization is a puzzle-like test in which the child indicates which of two forms will fit a formboard. Although the child is invited to place the form in the hollow of the formboard, the motor aspect of the test does not enter into scoring the test. The examiner does keep track of whether the child used the right or left hand in picking up the blocks and, in doing so, whether he or she crossed the body's midline or tended to use each hand on its own side of the body. In the Figure-Ground Perception, the child points to pictures that are hidden among other pictures. The test measures how well a child visually perceives a figure against a confusing background.

(2) Somatosensory

These tests assess tactile, muscle, and joint perception. ("Soma" means "body.") During somatosensory testing the child is encouraged to "feel" rather than "see." A large piece of cardboard held over the area where the arms and hands are working helps the child concentrate on what is felt. Being touched where the child cannot see the touching often makes the child feel uncomfortable even though none of the tactile stimuli really hurt the child. If the child's negative reaction to the testing is strong, the response is referred to as "tactile defensiveness."



**Figure 35 (Continued)**
**Sample WPS TEST REPORT™ for the SIPT**

SIPT Test Report Transmittal Number: 0000wps Page: 15

On the Manual Form Perception, the child identifies through the tactile and kinesthetic senses unusual shapes held in the hand. On the Kinesthesia, the conscious sense of joint position and movement is evaluated by the child's attempt to put his or her finger at the same place the therapist had previously put it. Tactile perception is measured with three tests: a) the Finger Identification, in which the child points to his or her finger that the therapist touched; b) the Graphesthesia, in which the child draws with a finger the same simple design the therapist drew on the back of the child's hand; and c) the Localization of Tactile Stimuli, in which the child points to the spot where the therapist had lightly touched the child's arm or hand with a pen. This last test leaves 14 tiny, washable spots on the child's arm and hand.

(3) Praxis

Practic skill is evaluated six different ways: 1) Praxis on Verbal Command assesses the ability to interpret verbally given instructions to assume certain positions and to then assume them. A typical test item might be "Put your hands on top of your head." 2) Design Copying evaluates the ability to copy simple designs. 3) Constructional Praxis evaluates the child's ability to build with blocks, using structures built by the therapist as models. Both the Design Copying and the Constructional Praxis require visual form and space perception, in addition to practic abilities. 4) Postural Praxis requires the child to imitate the unusual body postures assumed by the therapist. 5) Oral Praxis asks the child to imitate movements and positions of the tongue, lips, and jaw. 6) Sequencing Praxis asks the child to imitate a series of simple arm and hand positions.

(4) Sensorimotor

Four sensorimotor tests are included in the SIPT because their tasks require sensory integration. Bilateral Motor Coordination evaluates the ability to coordinate the two sides of the body in a series of arm movements. Standing and Walking Balance assesses the degree of sensory integration of the proprioceptive (muscle and joint) and vestibular (gravity and head movement) senses. On the Motor Accuracy, eye-hand coordination is measured by how well a child draws a line on top of a printed line. Executing the task requires eye muscle control, practic ability, visual perception, and motor coordination. Finally, the Postrotary Nystagmus measures the duration of the reflexive back and forth eye movements following rotation of the body (10 times in 20 seconds). This observation is one way of telling how well the nervous system is integrating the sensations from the vestibular system.



Figure 35 (Continued)
Sample WPS TEST REPORT™ for the SIPT

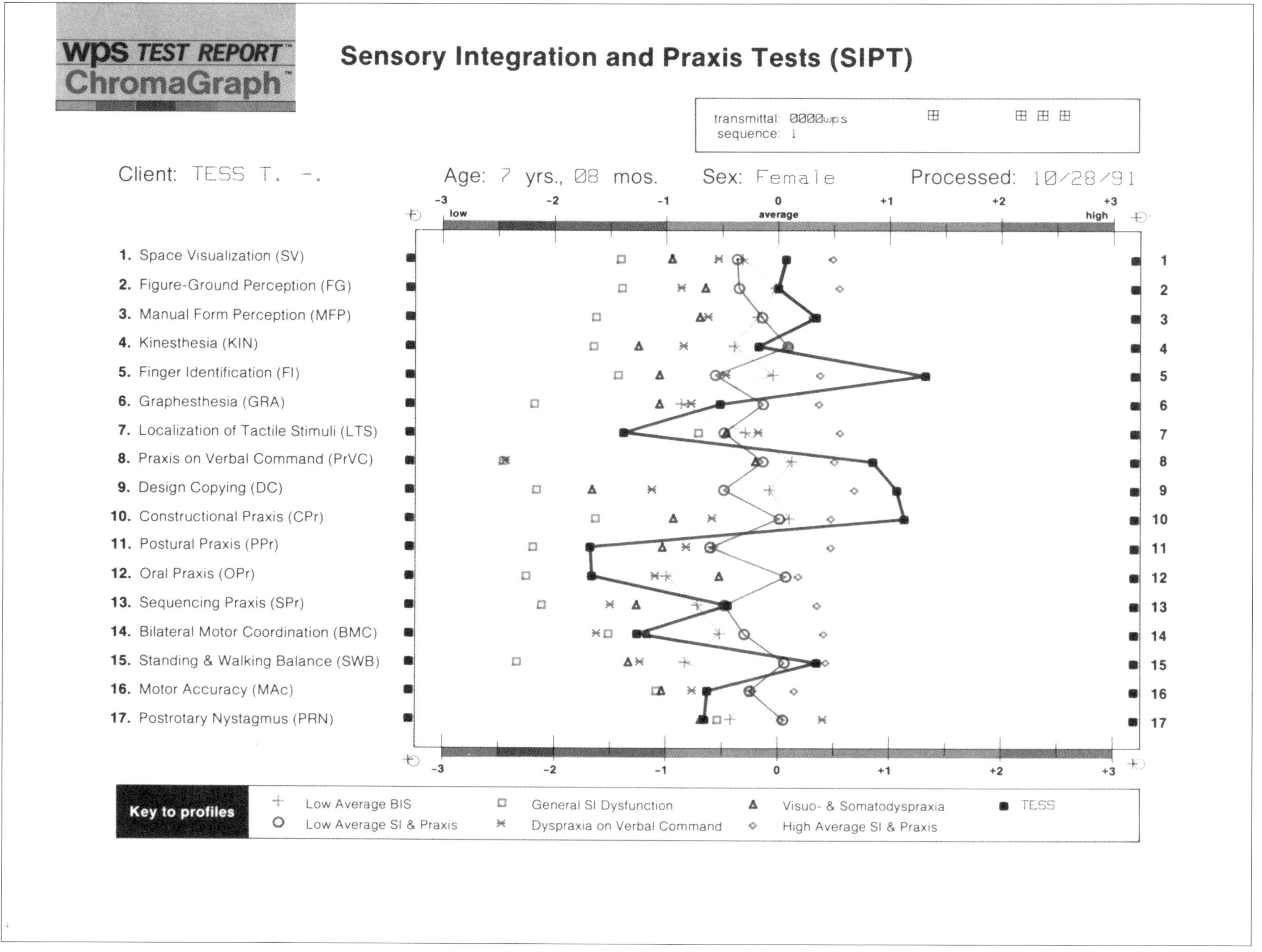


**Figure 36**
**Sample WPS ChromaGraph™ for the SIPT**

# CHAPTER 5
## INTERPRETATION OF THE SIPT SCORES

### General Considerations

The SIPT was designed to assist in the clinical understanding of children with mild to moderate irregularities in learning and behavior. The WPS TEST REPORT for the SIPT provides comprehensive descriptive information about the child's performance on each of the 17 SIPT. In addition, the WPS ChromaGraph™ for the SIPT compares the child's overall pattern of SIPT scores with the patterns that characterize six different diagnostic clusters of children. (For a detailed description of the WPS TEST REPORT and WPS ChromaGraph™ for the SIPT, see chapter 4 of this Manual; an empirical study deriving the six SIPT groups is presented in chapter 7.) When combined with other information, the WPS TEST REPORT and ChromaGraph™ can aid in the diagnosis of sensory integrative dysfunction, including dyspraxia, and bilateral integration and sequencing (BIS) deficit.

Other information and clinical observations should always be considered when making diagnostic judgments. In particular, the clinician should have: (a) a clear and comprehensive description of the presenting problem; (b) a relevant history of the child; (c) a general knowledge of the child's intellectual capacity, language development, and academic achievement; (d) any pertinent psychological and medical diagnoses; and (e) clinical observations, including those of ocular and postural responses, defensive reactions to tactile stimuli, and degree of gravitational security. The SIPT scores should be interpreted in light of all of these additional sources of information, and should not be interpreted in isolation from other sources of information.

For example, a child who is mentally retarded, or one whose medical diagnosis is inconsistent with dyspraxia, should not be diagnosed as dyspraxic on the basis of his or her SIPT scores. Although a mentally retarded child or a child with chromosomal disorder may have poor practic ability as evidenced by low praxis scores, that child does not have dyspraxia in the usual sense of the word. Similarly, although autistic children usually obtain low scores on tests of praxis, their scores should not be interpreted as evidence of dyspraxia, per se. On the other hand, a child with a diagnosed brain disorder such as cerebral palsy or traumatic brain injury might be diagnosed as dyspraxic.

In general, the diagnosis of bilateral integration and sequencing (BIS) deficit (or a variant of that diagnosis) should be made only when the major characteristics of such a deficit are present and when there are no other conditions that might account for those characteristics. When symptoms characterizing a BIS deficit appear along with other sensory integrative or practic deficits, scores on the tests that are associated with BIS functions may be described as part of the total picture, or they may be described separately if a BIS deficit is clearly present.

### SIPT Interpretation Strategy

At least three sources of information are needed by the test user when interpreting SIPT scores: (a) a relevant professional knowledge base, (b) familiarity with the SIPT validity data, and (c) a WPS TEST REPORT and ChromaGraph™. For children who fall into one of the exclusion categories mentioned above (e.g., autistic children, mentally retarded children, and children with significant medical or psychological diagnoses), the ChromaGraph™ results should be interpreted with caution.

### Interpretation With WPS ChromaGraph™ Assistance

The WPS ChromaGraph™ plots the scores of the tested child (as solid black squares) and also plots the profiles for any SIPT groups whose score profiles resemble the child's profile on the SIPT. (The six SIPT groups are described in detail in chapters 4 and 7 of this Manual.) The score profiles for the six SIPT groups are shown in Figure 37.

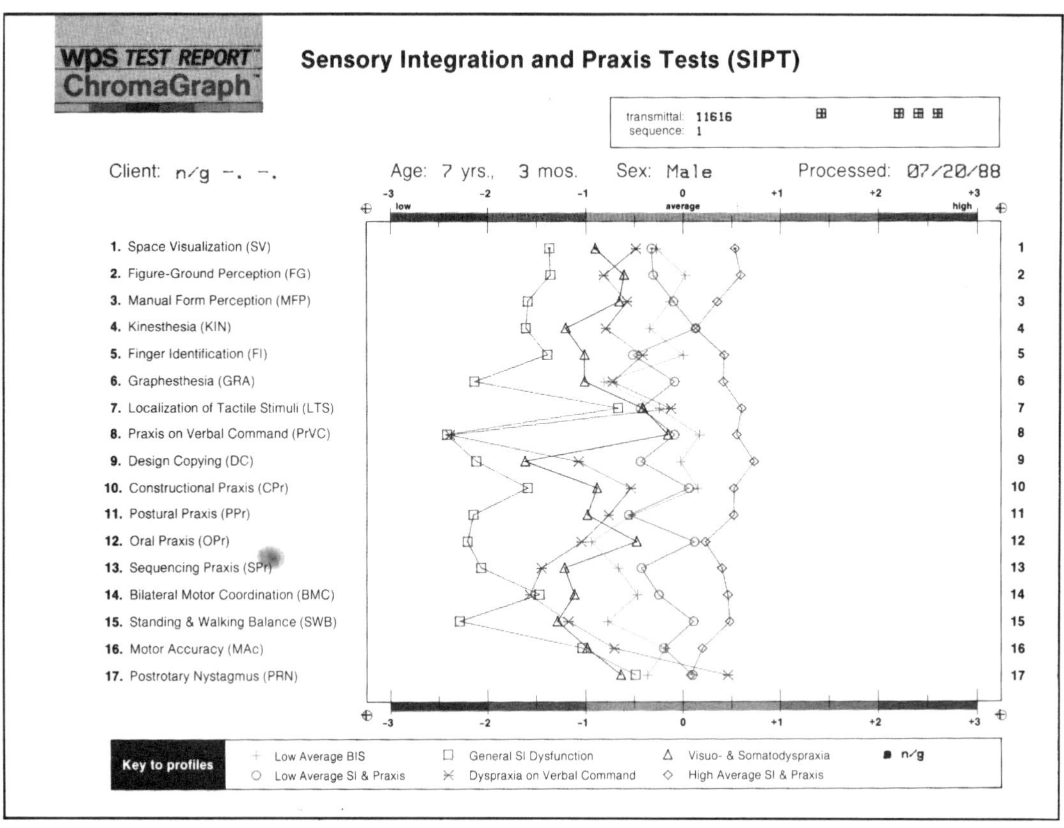


**Figure 37**
**ChromaGraph™ of Means for 6 SIPT Groups**

On the WPS ChromaGraph™, each SIPT group is identified by a distinctive geometric shape and color. The colors and shapes for the six SIPT groups are as follows:

| **Color and Shape** | **SIPT Group** |
|---|---|
| Green plus sign | Group 1: Low Average Bilateral Integration and Sequencing |
| Magenta open square | Group 2: Generalized Sensory Integrative Dysfunction |
| Red triangle | Group 3: Visuo- and Somatodyspraxia |
| Blue circle | Group 4: Low Average Sensory Integration and Praxis |
| Orange asterisk | Group 5: Dyspraxia on Verbal Command |
| Violet diamond | Group 6: High Average Sensory Integration and Praxis |

The clinician can interpret a child's profile of scores by comparing the scores to those of each of the SIPT groups against which the computer compared the scores. This method requires referring to the data on cluster analysis in chapter 7.

An alternative method of interpretation is presented here. This method is considered supplemental to that which relies on chapter 7 data. It differs mainly in an orientation of detecting and diagnosing dysfunctions as opposed to primarily matching profiles to one of six clusters. This method was developed on a wider data base than the SIPT clusters alone and provides greater flexibility in interpretation. In studying a large number of factor analyses, a search was made for those tests that were not only the strongest but the most consistent representatives of a function.

Of the six SIPT groups generated by cluster analysis, only three represent different types of dysfunction. The three types are: (a) deficit in the bilateral integration and sequencing (BIS) function; (b) visuo- and somatodyspraxia, which can be further divided into visuodyspraxia and somatodyspraxia; and (c) dyspraxia on verbal command. SIPT Groups 4 and 6 represent low and high average function and are not types of dysfunction. Their score patterns do not assist in diagnosing problems. SIPT Group 2—Generalized Sensory Integrative Dysfunction—is not so much qualitatively as quantitatively different from the other dysfunctional groups. In addition to the three major types of dysfunction, there are a number of partial but meaningful score patterns.

## Description of Major Dysfunctional Types

### Deficit in Bilateral Integration and Sequencing

It is assumed that the score constellation of SIPT Group 1 presents the purest and clearest picture of the BIS deficit. Although the SIPT Group 1 scores are all within average range, when the lowest scores fall well below others, or when there is a significant contrast in scores in the presence of a presenting problem, a deficit is justifiably presumed. The cluster analysis (Figure 37) shows the prototypic pattern of deficit in the BIS function: low scores (from lowest to better) on Oral Praxis (OPr), Standing and Walking Balance (SWB), Sequencing Praxis (SPr), Bilateral Motor Coordination (BMC), Graphesthesia (GRA), and Postural Praxis (PPr). In contrast, scores on the following tests fell between 0.2 and –0.2 *SD:* Space Visualization (SV), Figure-Ground Perception (FG), Manual Form Perception (MFP), Finger Identification (FI), Localization of Tactile Stimuli (LTS), Praxis on Verbal Command (PrVC), Design Copying (DC), Constructional Praxis (CPr), and Motor Accuracy (MAc). Highest scores were on Praxis on Verbal Command and Constructional Praxis. Five of the six lowest scoring tests, excluding Postural Praxis, were consistently the best representatives of the BIS function in all analyses, but the relative magnitude of the test identity with the BIS function differed among populations studied. Sequencing Praxis (SPr), Oral Praxis (OPr), and Bilateral Motor Coordination (BMC) often had the highest loading on factor analyses. A BIS deficit often appeared without low Postural Praxis (see Table 9). Contrast between scores on the BIS tests and the rest of the tests is an important criterion in diagnosis.

The three major statistical procedures (cluster, discriminant, and factor analyses) that defined the BIS function indicated that: (a) the major BIS deficit-identifying tests are Oral Praxis (OPr), Sequencing Praxis (SPr), Bilateral Motor Coordination (BMC), Standing and Walking Balance (SWB), and Graphesthesia (GRA), not necessarily in that order; (b) no one of these five identifying tests is much more likely than another to have a lower score, although Bilateral Motor Coordination is probably the purest indicator of bilateral integration; and (c) Postural Praxis may or may not be low in cases of BIS deficit.

### Visuo- and Somatodyspraxia

As a dysfunctional type, visuo- and somatodyspraxia is, of course, based on SIPT Group 3. Cluster analysis groups symptoms of visuodyspraxia and somatodyspraxia into one category (see Figure 37) with members of this group scoring lowest on Design Copying (DC), Standing and Walking Balance (SWB), Sequencing Praxis (SPr), Kinesthesia (KIN), Bilateral Motor Coordination (BMC), Postural Praxis (PPr), Motor Accuracy (MAc), Graphesthesia (GRA), and Finger Identification (FI), in that order. The rest of the scores fall between –0.4 and –0.9 *SD.* The highest score is Praxis on Verbal Command, and that score is in contrast to the rest of the scores. The Postrotary Nystagmus (PRN) score is –0.6 *SD,* but either high or low Postrotary Nystagmus scores can be associated with visuo- or somatodyspraxia.

Visuodyspraxia and somatodyspraxia often can be identified separately, increasing diagnostic precision. Analysis of scores of 1,750 normative children (see Table 8) show Constructional Praxis (CPr), Space Visualization (SV), and Design Copying loaded highest on the visuopraxis factor. The fourth factor of the analysis of scores of 125 dysfunctional children (see Table 9) shows that tests with highest loadings on the factor were (in order of high to lower loadings) Motor Accuracy, Design Copying, Space Visualization, Figure-Ground Perception (FG), Constructional Praxis, Finger Identification, Standing and Walking Balance, and Manual Form Perception. Scores on any two of the following tests probably indicate visuodyspraxia: Motor Accuracy, Design Copying, Space Visualization, or Constructional Praxis. Low Finger Identification consistently associated with visuodyspraxia.

Tests identifying somatodyspraxia, as separate from visuodyspraxia, are set forth in Table 9 (Chapter 7). Postural Praxis is the highest loading test on Factor 5, (somatopraxis) in that table. Constructional Praxis and Oral Praxis (OPr) also load substantially. Constructional Praxis requires both visuo- and somatopractic skill. Graphesthesia and Finger Identification represent the somatosensory component in the factor. Other analyses strongly linked Postural Praxis with Localization of Tactile Stimuli (LTS). Table 9 identified Oral Praxis as a separate function, but it may be seen as part of somatodyspraxia. Both instances of low Oral Praxis scores as manifestations of dyspraxia are different from that indicated by low Oral Praxis as part of a BIS deficit (see Table 9).

Of the dyspraxia tests, Oral Praxis and Postural Praxis show the highest linkage with tactile perception and are among the best representatives of somatodyspraxia. This representation is not evident on the SIPT Group 3 profile, which reflects visuodyspraxia to a greater extent than somatodyspraxia, but it is clearly evident in SIPT Group 2 (generalized dysfunction). The three lowest scoring somatodyspraxia tests in the generalized dysfunction cluster are Oral Praxis, Postural Praxis, and Sequencing Praxis (see Figure 37). The most consistent somatosensory processing deficits apt to be associated with somatodyspraxia are Standing and Walking Balance, Localization of Tactile Stimuli, Kinesthesia, Graphesthesia, and Finger Identification.

**Dyspraxia on Verbal Command**

This type of dysfunction, identified primarily by SIPT Group 5, is the most discrete and least variable in its manifestation among children. The major identifying feature of the prototypic profile generated by cluster analysis is the contrast between a very low Praxis on Verbal Command (PrVC) score and a higher Postrotary Nystagmus (PRN) score (see Figure 37). The Praxis on Verbal Command accuracy—not time—score was used in all computations reported. Another identifying feature is moderately low (between -1.0 and -1.6 *SD*) scores on Design Copying (DC), Oral Praxis (OPr), Sequencing Praxis (SPr), Bilateral Motor Coordination (BMC), and Standing and Walking Balance (SWB). Equally important in characterizing the dyspraxia on verbal command profile are average to low average scores on the rest of the tests, which include the somatosensory, the visual form and space, and the visuopraxis tests, except Design Copying (DC), which fell at -1.1 *SD*. Children in this cluster scored highest, excluding Postrotary Nystagmus, on Localization of Tactile Stimuli (LTS), Finger Identification (FI), Space Visualization (SV), and Constructional Praxis (CPr). Because Praxis on Verbal Command performance taps postural praxis as well as linguistic interpretation, the Postural Praxis (PPr) score of -0.8 *SD* on the prototypic profile is important to note.

The Praxis on Verbal Command-Postrotary Nystagmus score relationship was consistent in several factor analyses of scores of different (but not usually independent) populations which included dysfunctional children. Notably, the low Praxis on Verbal Command-higher Postrotary Nystagmus pattern was not evident in the factor analysis of scores of 1,750 normative children (see Table 8), suggesting it is a dysfunctional pattern. Factor 2 of Table 9 gives one example of the pattern in dysfunctional children. Negative factor loadings on that factor by Bilateral Motor Coordination and Figure-Ground Perception (FG) support the idea that these tests are vulnerable to the same conditions that result in high Postrotary Nystagmus scores. Other untabled analyses supported the idea that scores on Sequencing Praxis, Oral Praxis and Graphesthesia (GRA) are also apt to be depressed when Postrotary Nystagmus has a high positive factor loading. The insignificant loading (-.03) of Postural Praxis on Factor 2 (see Table 9) indicates a strong independence of Postural Praxis from the essential features of Factor 2. It is strongly recommended that low scores on Bilateral Motor Coordination, Sequencing Praxis, Oral Praxis, and Graphesthesia not be interpreted as somatodyspraxia or as a BIS deficit, but as concomitants of dyspraxia on verbal command. Similarly, a single low score on a visuopraxis test should not be diagnosed as dyspraxia in a clear case of dyspraxia on verbal command.

## Interpretation of a SIPT Profile With Little or No ChromaGraph™ Assistance

Sometimes only part of the SIPT group profile fits the child's profile. Sometimes there is only a partial, but recognizably meaningful, score pattern. Occasionally there is no SIPT group profile, but only the child's profile plotted on the ChromaGraph™. These cases require a different interpretative approach.

**Partial Patterns**

Study of a large number of untabled factor analyses revealed natural linkages among tests. These partial patterns also appeared on some of the children's profiles. In one analysis, low Postrotary Nystagmus (PRN) shared variance with low Design Copying (DC), Constructional Praxis (CPr), Space Visualization (SV), Finger Identification (FI), Motor Accuracy (MAc), Manual Form Perception (MFP), and Figure-Ground Perception (FG). The pattern suggests inefficient CNS processing of vestibular input was associated with poor processing of visual input.

Another analysis showed a strong relationship among Space Visualization, Figure-Ground Perception, Design Copying, Motor Accuracy, and Constructional Praxis. Two or more of these tests could identify a visual form and space perception deficit, with or without dyspraxia. The diagnostic terms "visuoconstruction deficit" or "visual space management problem" could be used.

Postural Praxis (PPr) and Oral Praxis (OPr) appeared as essentially a doublet in several analyses, indicating a likely linkage. In a number of analyses, the somatosensory tests correlated with each other irrespective of praxis, suggesting a sensory integrative deficit without dyspraxia. In other analyses Postrotary Nystagmus correlated with the somatosensory tests, suggesting a vestibulosomatosensory processing deficit.

## Interpretation by Individual Test

Guidelines for interpreting individual SIPT are given in the following paragraphs. These guidelines incorporate clinical experience, theory, and statistical evidence for each of the 17 tests.

**Space Visualization (SV)**

Because Space Visualization shares many elements in common with Design Copying (DC), Constructional Praxis (CPr), Motor Accuracy (MAc), and Figure-Ground Perception (FG), the interpretation of a child's Space Visualization score should also consider the child's scores on these other tests. In dysfunctional children, the Design Copying scores are apt to be slightly lower than the Space Visualization scores. Although Space Visualization is basically a motor-free test, it requires practic ideation and motor planning and, therefore, may be interpreted as reflecting visuopraxis as well as space visualization and visual form and space perception. The Space Visualization total score is less helpful than Design Copying and Constructional Praxis in the identification and differential diagnosis of sensory integrative dysfunction. The Space Visualization score is also apt to be less reliable than the Design Copying score.

The scores derived from Space Visualization performance provide unique data on hand usage and may contribute to understanding a child's lateralization of skilled hand use. A low Space Visualization Contralateral Use (SVCU) score indicates a tendency to use each hand on its own side of the body, and may suggest slow development in establishing unilateral hand preference. In connection with other supportive data, a low SVCU score may be interpreted as a sign of poor functional integration of the two sides of the body. One of the results of poor bilateral integration may be a tendency for right-handed children to draw right to left or to make reversals.

Very low scores by right-handed children on Space Visualization Preferred Hand Use (PHU) may be associated with poor functional integration of the two sides of the body, with right-left confusion, and with diminished preferred-hand visuomotor coordination. These associations are not necessarily strong and may easily be obscured by other symptoms of dysfunction. Many different central nervous system inefficiencies may interfere with the normal development of preferred-hand use or the use of each hand in space on both sides of the body.

**Figure-Ground Perception (FG)**

Performance on Figure-Ground Perception may help to identify a primary visual perception deficit. The test is not directly related to somatosensory processing or to praxis. The Figure-Ground Perception accuracy score is preferred over the time score as an indication of figure-ground perception, and is the score appearing on the ChromaGraph™. A high time score indicates that the child spent less time on the test than most children of his or her age. Note, however, that the time score reflects the total time for all attempted items; thus, a high time score may result from *either* rapid performance (if an average number of items were attempted) *or* slow or average performance speed (if relatively few items were attempted).

A low accuracy score on Figure-Ground Perception in the presence of adequate scores on the other visual tests should not be interpreted as a sign of inadequate visual perception. Figure-Ground Perception is one of the less reliable measures in the SIPT, probably because the test is vulnerable to chance factors, such as impulsivity or distractibility. The Figure-Ground Perception time score may help the examiner evaluate the amount of diligence the child employed in taking the test.

The validity data suggest that Figure-Ground Perception is more sensitive than most of the SIPT to high-level central nervous system integrity. An interpretation of possible cerebral inefficiency associated with low Figure-Ground Perception accuracy should be supported by other evidence. A high score on Postrotary Nystagmus (PRN), accompanied by a low score on Praxis on Verbal Command (PrVC), would help provide such support.

**Manual Form Perception (MFP)**

Manual Form Perception helps identify problems with form and space perception. A low Manual Form Perception score accompanied by a low score on Constructional Praxis (CPr) or Design Copying (DC) would indicate a form and space perception deficit across different sensory systems. If a low score on Manual Form Perception is accompanied by low somatosensory test scores, including Standing and Walking Balance (SWB), this may suggest the presence of a sensory processing inefficiency, but visualization problems should also be considered as a more likely deficit. A low Manual Form Perception score in conjunction with evidence of visuodyspraxia should be interpreted as part of the visuodyspraxia and not separately as a form and space perception deficiency.

Part II of Manual Form Perception requires the functional integration of the two sides of the body. Thus, when a low score on Manual Form Perception is accompanied by low scores on other tests associated with bilateral integration and sequencing (BIS), the low Manual Form Perception score should not be taken as evidence of a form and space perception problem. Neither should the examiner rely on Manual Form Perception alone to identify a BIS deficit.

Manual Form Perception is particularly sensitive to the conditions that may produce high Postrotary Nystagmus (PRN) scores. In some cases a low Manual Form Perception score may reflect a difficulty with cognitive problem solving that is based on somatosensory input and visualization. When there is evidence of a more generalized problem-solving deficit, as in the mildly retarded, the Manual Form Perception score should be interpreted as indicating haptic perception status rather than form and space perception.

**Motor Accuracy (MAc)**

Both the right- and the left-hand Motor Accuracy scores are fairly good indicators of overall sensory integrative and practic status in children 4 to 6 years of age. However, Motor Accuracy is somewhat less effective at 7 and 8 years of age. Motor Accuracy and Design Copying (DC) correlate well, and Motor Accuracy can be used as a visuopraxis test with the young impaired child who cannot perform the Design Copying task. Motor Accuracy performance requires visual space management, vestibulosomatopractic ability, functional integration of the two sides of the body and extracorporeal space, and neuromotor coordination. Of these parameters, Motor Accuracy is best as an index of visual space management or eye-hand coordination. Before a low Motor Accuracy score can be interpreted as an index of any of the many functions it taps, however, it must be accompanied by scores on other tests assessing similar functions.

The left-hand score is as accurate a representation of these functions as the right hand is. Comparison of right- and left-hand scores can help in making judgments about hand preference. Both right- and left-hand Motor Accuracy scores appear to depend more heavily upon the right cerebral hemisphere than upon the left cerebral hemisphere. The validity data indicated no direct relationship between Motor Accuracy performance and Praxis on Verbal Command (PrVC). The simplest and most direct interpretation for a parent or teacher is that Motor Accuracy measures eye-hand coordination. Such a catchall phrase is insufficient for a real understanding of the test score, but it can be directly related to skill in use of a writing instrument. The Motor Accuracy task demands low-order skill across many areas, but it is not a good screening instrument for sensory integrative dysfunction in 7- and 8-year-old children. Overall, however, Motor Accuracy was the best single measure in the SIPT for discriminating between normal and dysfunctional children (in a matched sample of 176 normal children and 117 children with learning disabilities or sensory integrative deficits).

**Kinesthesia (KIN)**

This is one of the somatosensory tests that helps identify the sensory processing deficits associated with dyspraxia. However, Kinesthesia scores are not consistently low when somatodyspraxia is present. Both Kinesthesia and Standing and Walking Balance (SWB) assess proprioception, so the Standing and Walking Balance score should be checked when interpreting the Kinesthesia score. Kinesthesia also can be checked against Localization of Tactile Stimuli (LTS), as the two tests sometime correlate moderately. The correlation, however, may be due to the dependence of both tests on the ability to focus attention. If a low Kinesthesia score is seen in conjunction with a similarly low praxis score, especially in Oral Praxis (OPr) or Sequencing Praxis (SPr), the Kinesthesia score may be an indication that poor kinesthetic ability is contributing to poor somatopractic ability. A poor kinesthetic sense also may contribute to poor visuopractic skill, as measured by Design Copying (DC) or Motor Accuracy (MAc).

**Finger Identification (FI)**

Finger Identification is one of the most important measures of tactile perception in the SIPT. A low score on Finger Identification, seen in conjunction with low Postrotary Nystagmus (PRN) and/or low scores on another tactile test or praxis test, may indicate a deficit in processing tactile input. Low scores on Finger Identification are most apt to be found in cases of visuo- or somatodyspraxia. Finger Identification is usually not low in cases of dyspraxia on verbal command. When it is, another or additional diagnosis should be considered. Finger Identification scores are expected to be low in cases of generalized dysfunction. Finger Identification is vulnerable to the same conditions that may contribute to low Postrotary Nystagmus scores; Finger Identification also may be affected by the conditions associated with high Postrotary Nystagmus, but this is less often the case. Poor performance on Finger Identification may be part of a diagnosis of poor somatosensory processing without dyspraxia. A low Finger Identification score may or may not be seen in cases of BIS deficit. When it is, Finger Identification usually is not interpreted as evidence of a tactile perception deficit.

**Graphesthesia (GRA)**

A low Graphesthesia score most often indicates a deficit in the interpretation of complex tactile stimuli or a problem in translating complex tactile stimuli into practic action. The latter interpretation might accompany a diagnosis of visuo- or somatodyspraxia. A low score on Graphesthesia also may reflect the somatic component of somatodyspraxia or the tactile concomitant of oral dyspraxia.

A low Graphesthesia score also may contribute to the diagnosis of a deficit in bilateral integration and sequencing (BIS). In that case, the low Graphesthesia score should be accompanied by a low score on at least two of the following: Oral Praxis (OPr), Sequencing Praxis (SPr), Bilateral Motor Coordination (BMC), and Standing and Walking Balance (SWB). Graphesthesia scores may be affected by conditions that contribute to low Postrotary Nystagmus (PRN) scores, but they can also be affected by conditions associated with high Postrotary Nystagmus scores.

**Localization of Tactile Stimuli (LTS)**

Localization of Tactile Stimuli has the advantage of being a fairly simple test, making it easy for an impaired child to understand what is expected of him or her, and the advantage of requiring little or no practic ability. These advantages, however, must be balanced against the fact that Localization of Tactile Stimuli is one of the less reliable measures in the SIPT. Accordingly, a low score should be interpreted cautiously unless it is supported by similarly low scores on one or more of the other tactile tests. For example, low scores on Finger Identification (FI) or Kinesthesia (KIN), with which Localization of Tactile Stimuli correlates moderately, would provide support for the accuracy of the Localization of Tactile Stimuli score. When a low score on Localization of Tactile Stimuli accompanies a low score on Postural Praxis (PPr), Oral Praxis (OPr), or Constructional Praxis (CPr), the dyspraxia may result from inefficient processing of tactile input.

Localization of Tactile Stimuli is the most likely of the SIPT to elicit defensive or aversive responses when administered to a tactually defensive child. Tactually defensive children do not always have low tactile perception scores, but when tactile test scores are low, they help to define the nature of that dysfunction.

**Postrotary Nystagmus (PRN)**

Postrotary Nystagmus evaluates the integrity of a relatively discrete vestibular-ocular reflex. The interpretation of Postrotary Nystagmus scores discussed in this Manual assumes that the child will not have a peripheral vestibular problem; if he or she does have any peripheral vestibular problems, the following interpretation technique is not appropriate: If there is no peripheral vestibular problem, then the Postrotary Nystagmus score is seen as an indication of how efficiently the central nervous system (CNS) processes this type of vestibular sensory input. An atypically low Postrotary Nystagmus score is interpreted as a sign of low CNS responsivity to the vestibular input. An atypically high Postrotary Nystagmus score is interpreted as a sign of insufficient CNS inhibition of the lower nystagmatic reflex. Neither condition is favorable.

Short nystagmus duration is most apt to be associated with deficits in the somatosensory and visuopraxis tests, but similar deficits may be found in association with long nystagmus. Other tests whose scores are most apt to be low when the Postrotary Nystagmus score is low are Postural Praxis (PPr), Oral Praxis (OPr), Sequencing Praxis (SPr), and Motor Accuracy (MAc). These associations suggest that vestibular sensory input may contribute to various aspects of praxis and to functional integration of the two sides of the body. Low scores on tests of bilateral integration and sequencing (BIS) in conjunction with short nystagmus may mean something different from similarly low scores in conjunction with prolonged nystagmus. Thus, the low scores on Sequencing Praxis, Bilateral Motor Coordination (BMC), and Standing and Walking Balance (SWB), shown by SIPT Group 5 (see chapter 7), may reflect a condition that is quite different from the condition associated with the similarly low scores of SIPT Group 3.

Postrotary nystagmus is produced by stimulation of the semicircular canals. These semicircular canals, which are primarily concerned with ocular functions, represent just one of two major types of vestibular receptors. The other major receptor type consists of the gravity receptors in the macula of the utricle and saccule. The gravity receptors are primarily concerned with body balance. The only test in the SIPT expected to tap integration of sensory input from the gravity receptors is Standing and Walking Balance. Thus, performance on Standing and Walking Balance should be considered in the evaluation of vestibular function. However, low scores on Standing and Walking Balance are not necessarily indicative of poor vestibular function, as Standing and Walking Balance also depends upon proprioceptive, visual, and neuromotor function. Test users are advised to expand their data base for interpreting CNS integration of vestibular input by making related clinical observations.

**Standing and Walking Balance (SWB)**

Scores on Standing and Walking Balance are sufficiently reliable to be used in the absence of other test scores, but the scores cannot be interpreted meaningfully without other SIPT scores. Standing and Walking Balance scores may be interpreted in several different ways. First, and primarily, Standing and Walking Balance performance reflects vestibuloproprioceptive processing. Vestibular input is primarily from the macular receptors of the saccule and utricle. Judging by Standing and Walking Balance correlations, this input appears to contribute to performance on many of the other SIPT tests. A propriovestibular interpretation of a low Standing and Walking Balance score should be supported by a low score on one or more of the other tests tapping proprioception, namely Kinesthesia (KIN), Oral Praxis (OPr), Bilateral Motor Coordination (BMC), Sequencing Praxis (SPr), or Postural Praxis (PPr). A purely vestibular interpretation of Standing and Walking Balance would be supported by a low Postrotary Nystagmus (PRN) score.

When a low score on Standing and Walking Balance is seen in conjunction with low Postrotary Nystagmus and a low score on any of the praxis tests except Praxis on Verbal Command (PrVC), it is reasonable to interpret the Standing and Walking Balance score as an indication that there is insufficient propriovestibular sensory integration to support practic ability.

When a low score on Standing and Walking Balance is seen in connection with a high Postrotary Nystagmus score, a more likely interpretation is that the same conditions that produced high Postrotary Nystagmus also interfered with Standing and Walking Balance. A likely result is a neuromotor incoordination that interferes with body balance. A low score on Motor Accuracy (MAc) (which is also a sensorimotor test) and clinical observations should help verify the presence of a neuromotor deficit.

Standing and Walking Balance is often, but not invariably, associated with bilateral integration and sequencing (BIS). A low Standing and Walking Balance score may be interpreted as part of a BIS deficit when accompanied by low scores on two of the following: Sequencing Praxis, Bilateral Motor Coordination, Oral Praxis, or Graphesthesia (GRA). Of all the tests of BIS function, Standing and Walking Balance is the least consistently associated with the function. Until further research clarifies the meaning of low scores on the tests of BIS in children identified as belonging to SIPT Group 5 (the group characterized by dyspraxia on verbal command), a low Standing and Walking Balance score should not be treated as an indication of a BIS problem in that group.

### Praxis on Verbal Command (PrVC)

A low score on Praxis on Verbal Command indicates that the child has difficulty interpreting the test's verbal directions into planned postural positions. The problem may be in linguistic processing or in postural praxis or in both. Unless there is evidence of *severe* somatodyspraxia, the linguistic interpretation is also preferred when low Praxis on Verbal Command is accompanied by an average or high Postrotary Nystagmus (PRN) score and low Oral Praxis (OPr), Bilateral Motor Coordination (BMC), or Sequencing Praxis (SPr) scores. In this case, which replicates the pattern found in SIPT Group 5, most of the remaining SIPT scores, except Design Copying (DC), should be within normal limits. When low Praxis on Verbal Command is accompanied by low Postural Praxis (PPr) and low or low average Postrotary Nystagmus, the problem may be in the postural praxis component. If a low Praxis on Verbal Command score is accompanied by low scores on Figure-Ground Perception (FG), Manual Form Perception (MFP), and Design Copying, the possibility of a common cognitive limitation should be considered. When low Praxis on Verbal Command is accompanied by many other low SIPT scores, the linguistic and postural components cannot be distinguished, and the simple interpretation is that the child has difficulty in translating verbal directions into practic action. Until further research clarifies the meaning of low scores on Sequencing Praxis, Bilateral Motor Coordination, and Standing and Walking Balance (SWB), accompanied by low Praxis on Verbal Command and high Postrotary Nystagmus (as found in SIPT Group 5), a diagnosis of a BIS deficit in connection with dyspraxia on verbal command should not be made. The validity data do not indicate whether a low score on those tests reflects a bilateral integration or a sequencing problem. Sequencing of action is classically considered primarily a left cerebral hemisphere function, as is linguistic interpretation.

### Design Copying (DC)

Design Copying is the best single indicator of visuopractic ability in the SIPT. Assuming that the child is of at least average intelligence, his or her Design Copying score can be used independently of the SIPT, but the only conclusion that can be drawn from a low Design Copying score is that the child may have a visuopraxis problem. Design Copying will not necessarily reflect the status of the child's other practic abilities. It may suggest that the child is likely to have trouble with tasks requiring two-dimensional space management, such as writing or drawing. Design Copying correlates well with Motor Accuracy (MAc), and a low Motor Accuracy score may help to confirm the accuracy of the Design Copying score in younger children who have trouble performing the Design Copying task. The Motor Accuracy task, however, requires only simple practic ability and does not discriminate well at ages 7 or 8 years. A low Design Copying score in conjunction with a low Constructional Praxis (CPr) score increases the accuracy of the diagnosis of visuopraxis, and expands the interpretation to include the ability to use objects in the three-dimensional world. If the Design Copying and Praxis on Verbal Command (PrVC) scores vary widely, the examiner should consider the possibility of unilateral dysfunction and look for other differences that might indicate the presence of a unilateral problem.

Generally, the presence of more atypical approaches than characteristic of a child with the same score simply indicates visuodyspraxia. An exception to this generality is the right-to-left parameter. High right-to-left, reversals, inversions, and segmentation scores are associated with a leftward orientation, poor visual space management, and lack of well-established right-hand preference. McAtee

(1987) found reversals related to body balance, crossing the midline of the body, and Space Visualization (SV) scores. In her sample, right-to-left and reversals were the only parameters correlating significantly with body balance. Naturally, left-handed children may differ from the normative expectations. High scores on the additions, boundary, segmentation, distortion, and jogs parameters are most apt to suggest visuodyspraxia. Visuodyspraxia is usually present in conjunction with somatodyspraxia. A safe interpretation of a high atypical approach score is that it generally indicates dysfunction. Failure to establish firm unilateral hand preference or a good sense of directionality means the same thing. Many variations of CNS inefficiency can and do interfere with lateralization of hand use. When directionality problems are present, the more underlying and fundamental deficit should be addressed. The most likely deficits are visuodyspraxia, unilateral inefficiency, and inadequate bilateral integration.

### Constructional Praxis (CPr)

Performance on Constructional Praxis can be interpreted as an indication of the child's skill in relating objects to each other in an orderly arrangement and through systematic assembly. A child who performs poorly on the test is apt to have trouble organizing his or her behavior in general, and specifically, in relating to objects in a manipulative or organizational manner. Constructional behavior is easily observed during test administration, and can be described to illustrate how the child probably manages objects in daily life. The cognitive and conceptual demands characteristic of all visuo- and somatopractic ability is assessed well by Constructional Praxis. Spatial reasoning may be an important aspect of this task.

Constructional Praxis correlates most strongly with Design Copying (DC). It is particularly useful in diagnosing visuodyspraxia in those children who are unable to perform Design Copying well enough to provide a sufficient sample of behavior for evaluation. A low score on Motor Accuracy (MAc), which correlates well with both Constructional Praxis and Design Copying, can help verify visuodyspraxia in younger children with low Constructional Praxis scores. Constructional Praxis is also closely associated with Postural Praxis (PPr), supporting the idea that an adequate body percept is needed for body-environment interaction. Low scores on Constructional Praxis may be associated with either very low or very high scores on Postrotary Nystagmus (PRN).

Unless the type of error made in block placement is supported by other similar data, high parameter scores (representing many errors) on Part II are best interpreted as inefficient praxis. A high score on some parameters will be accompanied by low scores on other parameters. The low scores could indicate the child made other more serious errors. Dysfunctional children's scores vary considerably on Constructional Praxis. The test is more useful for differential diagnosis than it is for making general distinctions between normal and dysfunctional children, especially in older children.

### Postural Praxis (PPr)

Postural Praxis taps a central practic ability common to all of the praxis tests, but especially to Oral Praxis (OPr), Sequencing Praxis (SPr), and Constructional Praxis (CPr). A low score on Postural Praxis is often a major indicator of somatodyspraxia. However, a low score on Postural Praxis without a low score on another praxis test or on a somatosensory test probably should be questioned and possibly should not be diagnosed as dyspraxia. When a low somatosensory test score accompanies a low Postural Praxis score, the diagnosis of somatodyspraxia is appropriate. When low Postrotary Nystagmus (PRN), and sometimes a low score on Standing and Walking Balance (SWB), accompanies a low Postural Praxis score, it can be assumed that inadequate proprio-vestibular sensory integration contributes to the poor postural praxis. Postural Praxis is particularly vulnerable to the conditions that contribute to low Postrotary Nystagmus scores.

The frequent co-occurrence of visuoconstruction deficits (as evidenced by a low Constructional Praxis score) with a low Postural Praxis score illustrates the importance of an adequate body precept and praxis to visually directed, skilled upper-extremity use and manipulation of objects.

A child with a low Postural Praxis score, accompanied by some somatosensory deficit, will probably have trouble with skilled body-environment interaction.

### Oral Praxis (OPr)

Of all the praxis tests, Oral Praxis is most closely related to perception and interpretation of sensations from the body, especially tactile sensation. The SIPT does not test somatosensory perception in the oral area, and there is no guarantee that the somatosensory tests of the arms and hands accurately represent the oral area. However, the validity data do suggest that they are closely related. A low Oral Praxis score, in conjunction with a low score on Localization of Tactile Stimuli (LTS), is especially suggestive of poor tactile perception in the oral area. Oral Praxis scores also correlate moderately well with Graphesthesia (GRA) scores. Graphesthesia requires the interpretation of complex tactile stimuli from the hands, and requires practic planning based upon those sensations. Kinesthesia (KIN) and Finger Identification (FI) also correlate with Oral Praxis. Accordingly, a low score on Oral Praxis in conjunction with a low score on Localization of Tactile Stimuli, Graphesthesia, or Kinesthesia, can be interpreted as oral dyspraxia in conjunction with poor somatosensory integration in the oral area. Severe deficits would likely affect speaking, eating, and other related oral activity. The validity data (see factor analysis results in chapter 7) indicate that an oral praxis problem is frequently associated with a postural praxis problem. In such a case, the oral dyspraxia might be considered part of a more general practic ideation, planning, and execution problem. Poor somatosensory integration will likely accompany the problem.

Poor functional integration of the two sides of the body and poor sequencing also may contribute to a low Oral Praxis score. The relationship is identified by concomitant low scores on Sequencing Praxis (SPr), Bilateral Motor Coordination (BMC), and Graphesthesia. When a child is diagnosed as having a bilateral integration and sequencing (BIS) deficit, a low Oral Praxis score is not considered a function of oral dyspraxia, per se.

A low Oral Praxis score associated with a low Postrotary Nystagmus (PRN) score probably does not reflect the same practic condition as that associated with a high Postrotary Nystagmus score and may not respond to treatment similarly. In the presence of an average or high Postrotary Nystagmus score and other indications of possible left cerebral hemisphere inefficiency, a low Oral Praxis score may be interpreted as part of a more comprehensive deficit.

**Sequencing Praxis (SPr)**

Sequencing Praxis is often interpreted as measuring a quality that is central to praxis. A low Sequencing Praxis score in conjunction with a low score on another praxis test and one or more low somatosensory processing test scores may be interpreted as evidence of somatodyspraxia. A low Sequencing Praxis score in conjunction with a low score on either Oral Praxis (OPr), Bilateral Motor Coordination (BMC), or Graphesthesia (GRA) may be interpreted as part of a bilateral integration and sequencing (BIS) deficit. The bilateral nature of the Sequencing Praxis task is as important to interpretation of the test as is the sequencing aspect. The sequencing aspect should not be overemphasized. Finally, a low Sequencing Praxis score is not a sufficient basis for assuming a left cerebral hemisphere problem unless other test scores or clinical observations also support that diagnosis.

**Bilateral Motor Coordination (BMC)**

Bilateral Motor Coordination should not be interpreted in isolation from other SIPT scores. To be interpreted as evidence of poor functional integration of the two sides of the body, a low Bilateral Motor Coordination score should be accompanied by a low score on Oral Praxis (OPr), Sequencing Praxis (SPr), or Space Visualization Contralateral Use (SVCU). Bilateral Motor Coordination also can aid in the identification of a BIS deficit when a low score on Bilateral Motor Coordination is accompanied by a low score on either Oral Praxis, Standing and Walking Balance (SWB), Graphesthesia (GRA), or Sequencing Praxis.

Bilateral Motor Coordination appears to have a somatic foundation, and is easily affected by somatodyspraxia. A low Bilateral Motor Coordination score that does not indicate a BIS deficit and is accompanied by a low Postural Praxis (PPr) score may be a sign of somatodyspraxia. Performance on Bilateral Motor Coordination may be negatively affected by the conditions that contribute to high or low Postrotary Nystagmus (PRN). Bilateral Motor Coordination is also vulnerable to neuromotor incoordination and to unilateral dysfunction. When a unilateral problem is suspected, the interpretation of scores on Bilateral Motor Coordination should reflect that possibility.

## Examples of Profile Interpretation

A number of SIPT profiles from the dysfunctional and normative groups are presented and discussed in the remainder of this chapter to illustrate the interpretation process. Children are considered in groups according to SIPT group membership, but interpretation is based on the dysfunctional types approach. When interpreting profiles on the WPS ChromaGraph™ for SIPT, it is important to remember that: (a) the SIPT groups that resemble the child's pattern of responses are plotted if the overall fit between the child's scores and the SIPT group scores are sufficiently close, and (b) all 17 SIPT are given equal weight in those computations. For clinical and diagnostic purposes, the differential importance of individual tests to each of the major dysfunctional groups is critical.

The profile of Case A, a 6-year, 3-month-old girl selected from the normative group, shows how widely SIPT scores can vary in an apparently "normal" child (see Figure 38). Her profile meets some of the criteria for likeness to Group 1 (Low Average Bilateral Integration and Sequencing), Group 4 (Low Average Sensory Integration and Praxis), and Group 6 (High Average Sensory Integration and Praxis). It is noteworthy that no score fell below −1.0 *SD*. Overall, this appears to be a fairly normal child.

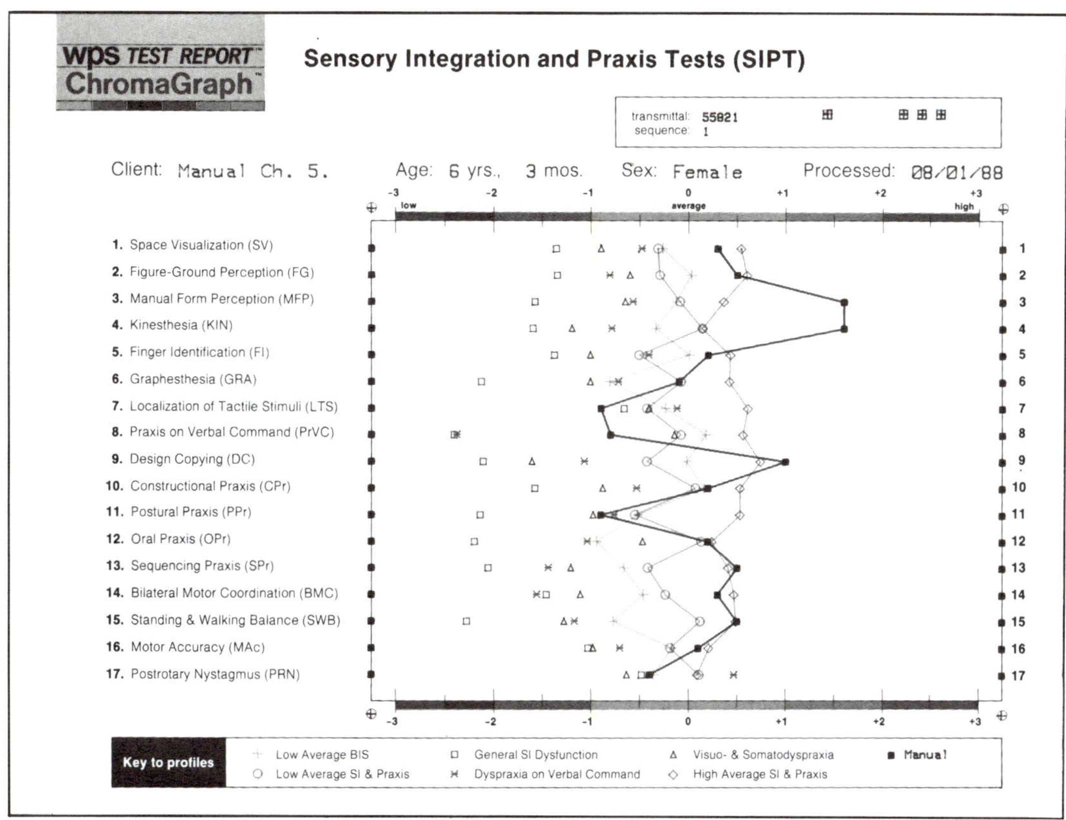


**Figure 38**
**Case A**

**SIPT Group 1:**
**Low Average Bilateral Integration and Sequencing**

All of the scores of Group 1 fall within the normal range. When the low scores that reflect bilateral integration and sequencing (BIS) are lower than those of Group 1, a BIS deficit may be present. Tests, low scores on which are critical to identifying a BIS deficit, were established earlier as Oral Praxis (OPr), Sequencing Praxis (SPr), Bilateral Motor Coordination (BMC), Standing and Walking Balance (SWB), and Graphesthesia (GRA). Postural Praxis (PPr) may or may not be low. An equally important criterion to diagnosis is score contrast between this group of scores and the rest of the scores.

The profile of Case B, a 6-year, 4-month-old boy with an occupational therapy diagnosis of sensory integrative dysfunction, generally follows the SIPT Group 1 profile (see Figure 39). Although many of Case B's scores are within normal limits on the tests that are most important to the diagnosis of BIS deficit, they are typically somewhat lower than his other scores. Also, he scored particularly well on Constructional Praxis (CPr) and on Praxis on Verbal Command (PrVC). Among his lowest scores are Graphesthesia (GRA), Standing and Walking Balance (SWB), and Oral Praxis (OPr), all of which are among the most important criteria for diagnosing BIS deficit. Thus, this child can confidently be given a diagnosis of BIS deficit. That deficit can be expected to make academic learning difficult. This child's strengths lie in visuopraxis and in implementing verbal commands for postural activity.

The SIPT profile of Case C, a 6-year, 3-month-old boy selected from the normative group (see Figure 40), also shows a pattern that is characteristic of children in SIPT Group 1. Three of the tests that identify the BIS function are quite low and in considerable contrast to most of his other scores. Of all the tests that identify the BIS function, only Graphesthesia (GRA) is inconsistently high, and it is considerably lower than his scores on the other tactile tests. The fact that this child's scores on Oral Praxis (OPr), Bilateral Motor Coordination (BMC), and Standing and Walking Balance (SWB) are considerably lower than the average scores for Group 1 increases the possibility of a more severe BIS deficit. Case C's scores on Localization of Tactile Stimuli (LTS) and Finger Identification (FI) are much higher than those of Group 1. However, because scores on these tests are expected to be within average limits in BIS cases, Case C's high scores do not detract from the BIS diagnosis. Case C has many sensory integrative strengths, and these are in contrast to what is probably a mild deficit in the BIS function.

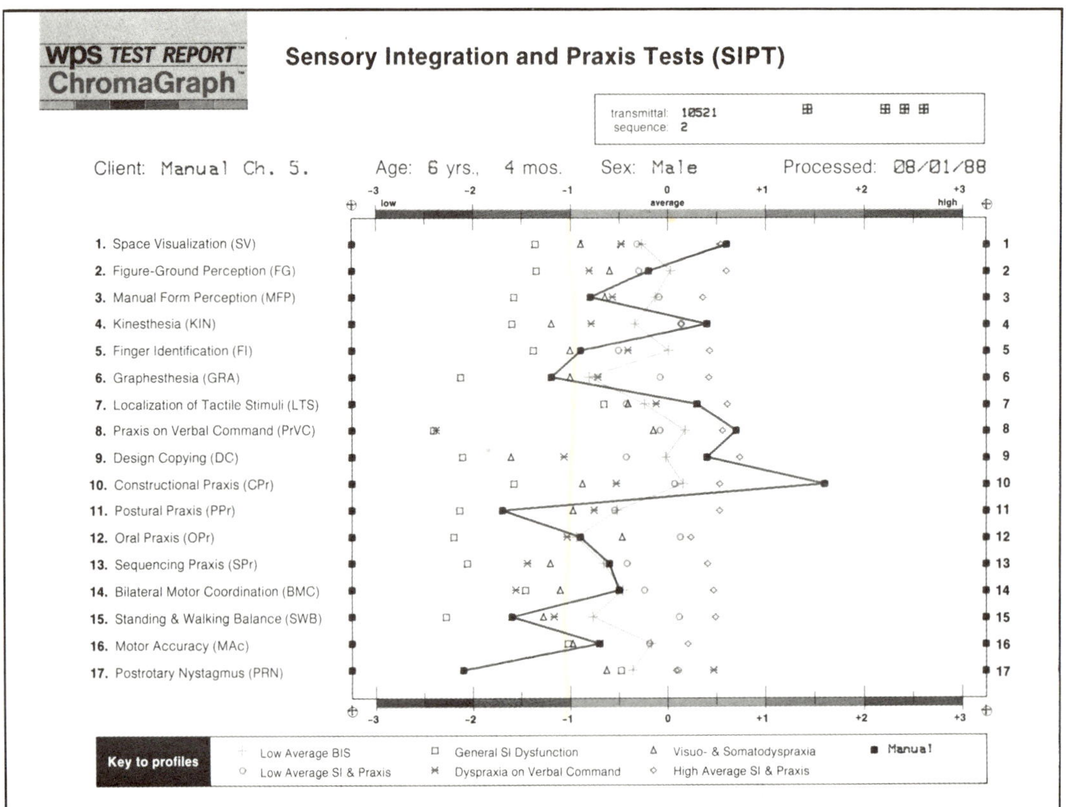


**Figure 39**
**Case B**

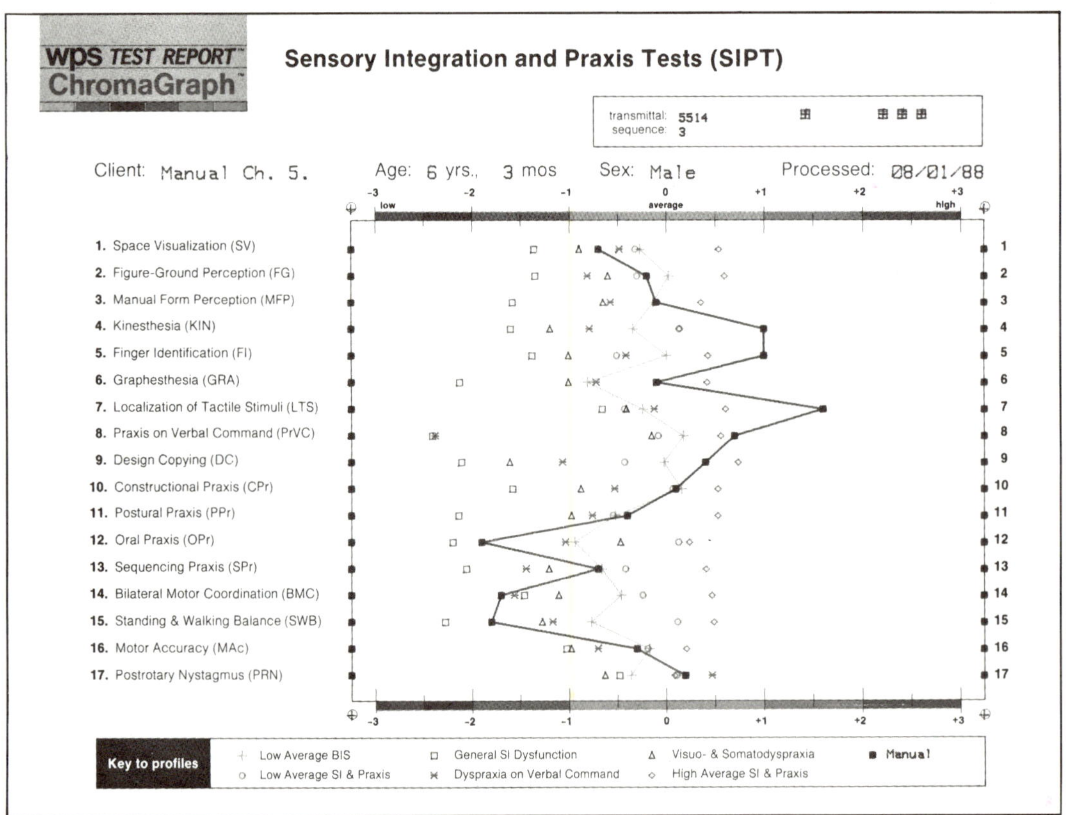


**Figure 40**
**Case C**

**SIPT Group 2:**
**Generalized Sensory Integrative Dysfunction**

Children whose profiles match SIPT Group 2 may be diagnosed "generalized dysfunction," but a more specific diagnosis is preferred. "Generalized" does not mean "severe," although some children with generalized dysfunction do have severe dysfunction. Generalized means that the child has difficulty (scores –1.0 *SD* or lower) in all of the major areas tested, with no identifiable or distinctive pattern to the scores. Occasionally, the ChromaGraph™ will indicate the Generalized Sensory Integrative Dysfunction when a few important scores fall within the average range. In such cases, the clinician is advised to supplement the diagnosis of possible generalized dysfunction, and concentrate on the child's individual pattern of deficits and nondeficit areas. The sensory processing areas that are most and least impaired should be considered and related to the other findings. Although strong performance on Bilateral Motor Coordination (BMC) or Standing and Walking Balance (SWB) may be irrelevant for most purposes, high scores on other tests such as Motor Accuracy (MAc) are apt to be quite important.

The profile of Case D, a 6-year, 7-month-old girl with an educational diagnosis of learning disability, illustrates a good fit to the SIPT Group 2 profile (see Figure 41). Scores are –1.0 *SD* or lower on all praxis tests and on Graphesthesia (GRA), Standing and Walking Balance (SWB), Manual Form Perception (MFP), Space Visualization (SV), and Motor Accuracy (MAc). She had only one score in the average to above-average range, and that was on Localization of Tactile Stimuli (LTS), which is relatively unimportant in the diagnosis of generalized dysfunction. Although Postrotary Nystagmus (PRN) is not an important criterion for group likeness, the fact that this child's Postrotary Nystagmus is very low should be noted. The low Postrotary Nystagmus, in conjunction with her low Praxis on Verbal Command (PrVC), suggests that the dyspraxia on verbal command may arise more from difficulty in planning postural responses than from problems with linguistic interpretation. In the sensory processing area, this child's most serious problem appears to involve proprioception, as evidenced by her low scores on Kinesthesia (KIN) and Standing and Walking Balance (SWB) and her high score on Localization of Tactile Stimuli (LTS). If she scored poorly on Kinesthesia simply because of failure to focus attention, she probably would have scored poorly on Localization of Tactile Stimuli.

Although Case D has generalized dysfunction, a more specific diagnosis is more helpful. Her SIPT scores indicate the presence of visuo- and somatodyspraxia. Visual and haptic form and space tasks not requiring

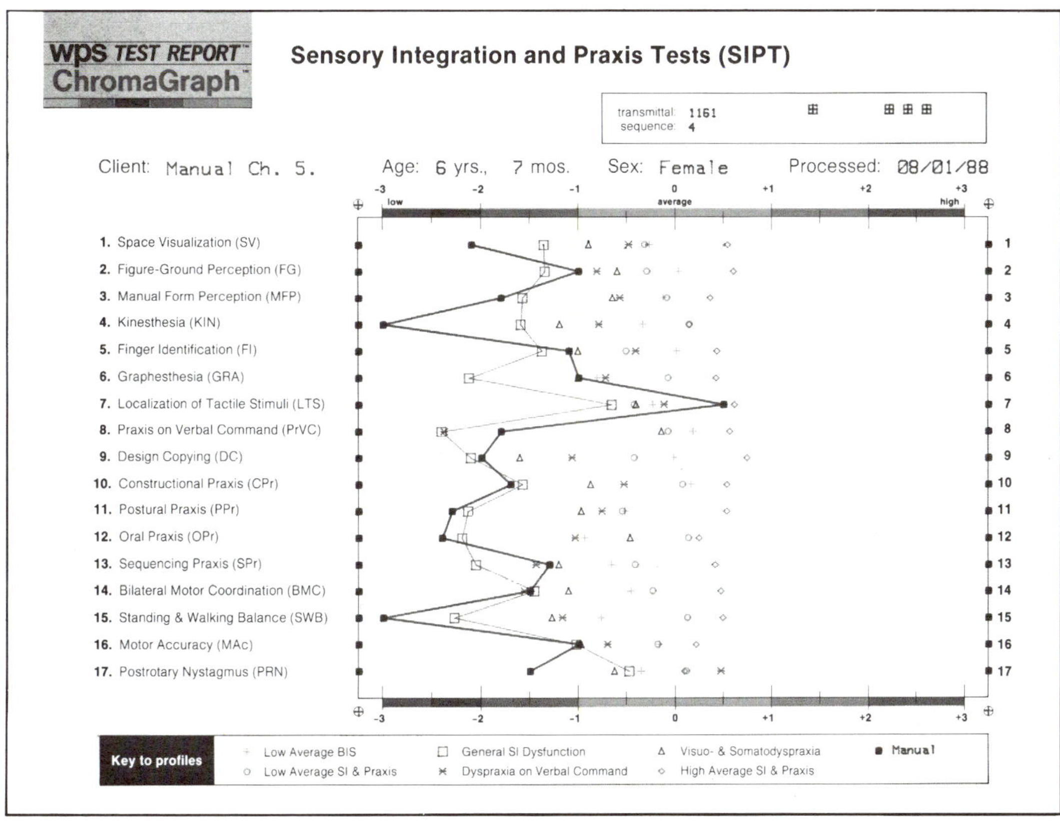


**Figure 41**
**Case D**

practic ability also fall in a problematic area. Moderate to severe propriovestibular processing deficits appear to accompany the dyspraxia. Complex tactile sensory processing is difficult, but simple localizing of tactile input is average. She may have trouble translating verbally given directions into bodily activity, and this possibility should be checked clinically.

The profile of Case E, an 8-year, 4-month-old boy with an occupational therapy diagnosis of sensory integrative dysfunction, also resembles the children in SIPT Group 2, but his score pattern (see Figure 42) is considerably different from that of Case D. Not all praxis test scores are low. His Constructional Praxis (CPr) falls well within the average range. Also, scores on Manual Form Perception (MFP), Space Visualization (SV), and Motor Accuracy (MAc) fall within normal limits. It appears that this child has adequate form and space perception and three-dimensional construction skill. With these strengths, a diagnosis of generalized dysfunction should not be made. Instead, the dysfunctional areas should be specifically named. Case E has low scores on four praxis tests (Oral Praxis [OPr], Postural Praxis [PPr], Praxis on Verbal Command [PrVC], and Sequencing Praxis [SPr]), but his profile does not show quite the degree of somatosensory perception deficit expected with such low praxis scores. Scores on all the tests that identify BIS are low, and it is noteworthy that Graphesthesia (GRA) is much lower than the other somatosensory test scores. Low Oral Praxis, Sequencing Praxis, Bilateral Motor Coordination (BMC), and Standing and Walking Balance (SWB) could be due to dyspraxia, or they could be part of a BIS deficit. Case E's scores can be interpreted as reflecting dyspraxic areas, especially in interacting adaptively with the physical world. He may have trouble translating verbal directions into appropriate actions, but this possibility should be verified clinically. His ability to construct and perceive form and space both visually and haptically will help him with behavioral organization in relation to the physical world.

**SIPT Group 3:**
**Visuo- and Somatodyspraxia**

Most children with visuodyspraxia also have somatodyspraxia, but that is not invariably the case. A low Design Copying (DC) score and a low score on either Constructional Praxis (CPr), Space Visualization (SV), or Motor Accuracy (MAc) are typical of visuodyspraxia. Low Finger Identification (FI) often accompanies visuodyspraxia. A low score on Postural Praxis (PPr), Sequencing Praxis (SPr), or Oral Praxis (OPr) and a low score on two tests of sensory processing—Finger Identification (FI), Graphesthesia (GRA), Standing and

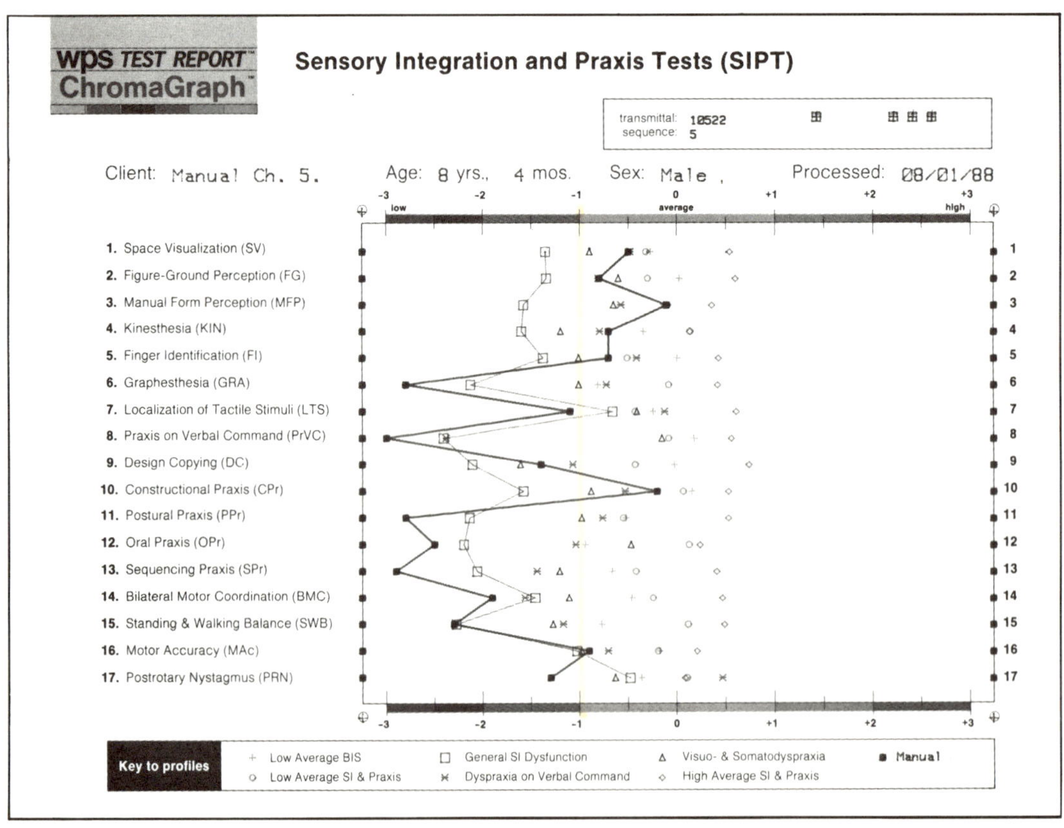


**Figure 42**
**Case E**

Walking Balance (SWB), Kinesthesia (KIN)—are characteristic of somatodyspraxia. In reporting visuo- or somatodyspraxia, it often is helpful to specify the areas of sensory processing inefficiency. Children with visuo- and somatodyspraxia are apt to have difficulty with practic ideation and concept formation. Interacting adaptively with the physical world usually presents problems. Disorganized behavior is often typical.

The profile of Case F, a 7-year, 2-month-old boy with an occupational therapy diagnosis of sensory integrative dysfunction, best fits the profile of SIPT Group 3 (see Figure 43). His low scores on Design Copying (DC), Graphesthesia (GRA), Space Visualization (SV), and Motor Accuracy (MAc) are consistent with many of the aspects of description of visuodyspraxia as a type of dysfunction. His low score on Sequencing Praxis (SPr) is also consistent with visuo- or somatodyspraxia. On an untabled factor analysis, the following tests loaded highest on the visuopraxis factor: Motor Accuracy, Design Copying (DC), Space Visualization, Figure-Ground Perception (FG), Finger Identification (FI), Standing and Walking Balance (SWB), and Manual Form Perception (MFP), in that order. Five of Case F's scores on these tests are low. The following tests loaded highest on the somatopraxis factor: Postural Praxis (PPr), Constructional Praxis (CPr), Oral Praxis (OPr), Graphesthesia (GRA), Finger Identification (FI), and Space Visualization (SV), in that order. Case F's Graphesthesia and Space Visualization scores are the only ones of this group that are low. The case for visuodyspraxia is stronger than for somatodyspraxia, but the latter cannot be excluded. In this case, visuodyspraxia is associated with poor somatosensory processing. To summarize, Case F appears to have mild visuodyspraxia and possibly some trouble with somatodyspraxia. It also appears that proprioceptive and tactile sensory inputs are not being processed well.

The profile of Case G, a 6-year, 7-month-old boy with an educational diagnosis of learning disability, provides another illustration of Group 3 (see Figure 44). The fit is only fair. A number of Case G's scores fall far below the mean scores of Group 3, and some fall above. The lower or higher scores do not necessarily invalidate the possible diagnosis of visuo- and somatodyspraxia as a dysfunctional type, but they do necessitate close analysis. Motor Accuracy (MAc), Finger Identification (FI), and Standing and Walking Balance (SWB) are the only low scores of this child that are also usually low on visuopraxis factors, and none of these is primarily a visuopraxis test. In contrast, his scores on Postural Praxis (PPr) and Sequencing Praxis (SPr) are both low, thus meeting two of the most important criteria for

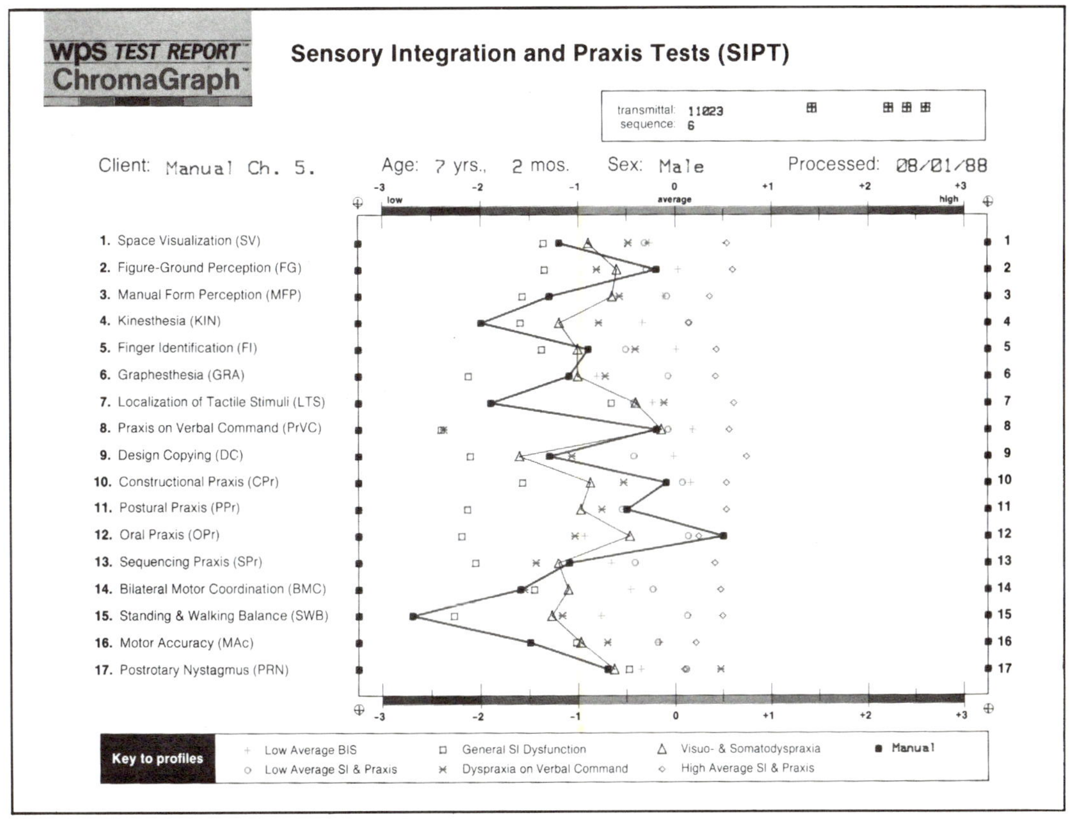


**Figure 43**
**Case F**

somatodyspraxia. His scores are also low on Standing and Walking Balance, Kinesthesia (KIN), and Bilateral Motor Coordination (BMC), all of which are associated with somatodyspraxia. Case G's profile also meets the Group 3 pattern of a higher score on Praxis on Verbal Command (PrVC) than on any of the other praxis tests. Visuodyspraxia is not a problem; somatodyspraxia *is* a problem. His oral, postural, and sequencing praxis are all affected. The dyspraxia appears to be accompanied by inefficient propriovestibular sensory processing. A deficit in localizing tactile stimuli is probably related to the oral dyspraxia. Case G's strengths lie in visual form and space perception.

Case H is a 7-year, 6-month-old learning-disabled girl. The WPS ChromaGraph™ plotted the scores of Case H against the profiles of both Group 2 and Group 3 (see Figure 45). There are likenesses to each. Case H has an average score on Praxis on Verbal Command (PrVC), but the other praxis test scores are all low. Also, scores on Manual Form Perception (MFP), Finger Identification (FI), and Localization of Tactile Stimuli (LTS) are within the normal range. Those normal scores, plus Praxis on Verbal Command invalidate a "general dysfunction" diagnosis. More specificity is needed. Case H's other praxis tests, along with her Graphesthesia (GRA), Kinesthesia (KIN), and Standing and Walking Balance (SWB) scores meet the most important criteria for a diagnosis of visuo- and somatodyspraxia. Low Space Visualization (SV) and Bilateral Motor Coordination (BMC), while less important to consider, do add further to the diagnosis. Case H's profile of SIPT scores meets the qualifications for a diagnosis of visuo- and somatodyspraxia. Her strongest area identified by the SIPT is translating verbal commands into postural responses.

The profile of Case I, a 6-year, 2-month-old boy was selected from the SIPT normative sample. His scores (see Figure 46), which span three standard deviations, are plotted against Group 1 (Low Average Bilateral Integration and Sequencing), Group 3 (Visuo- and Somatodyspraxia), and Group 4 (Low Average Sensory Integration and Praxis). Three of the low scores—Space Visualization (SV), Design Copying (DC), and Constructional Praxis (CPr)—indicate some difficulty with visuopraxis. In light of his low Postrotary Nystagmus (PRN) score, it is likely that this problem is related to inadequate CNS processing of vestibular sensory input. Case I's strength lies in somatosensory processing.

### SIPT Group 4: Low Average Sensory Integration and Praxis

The mean SIPT scores of Group 4 fall in a narrow range without much contrast or identifiable pattern. No

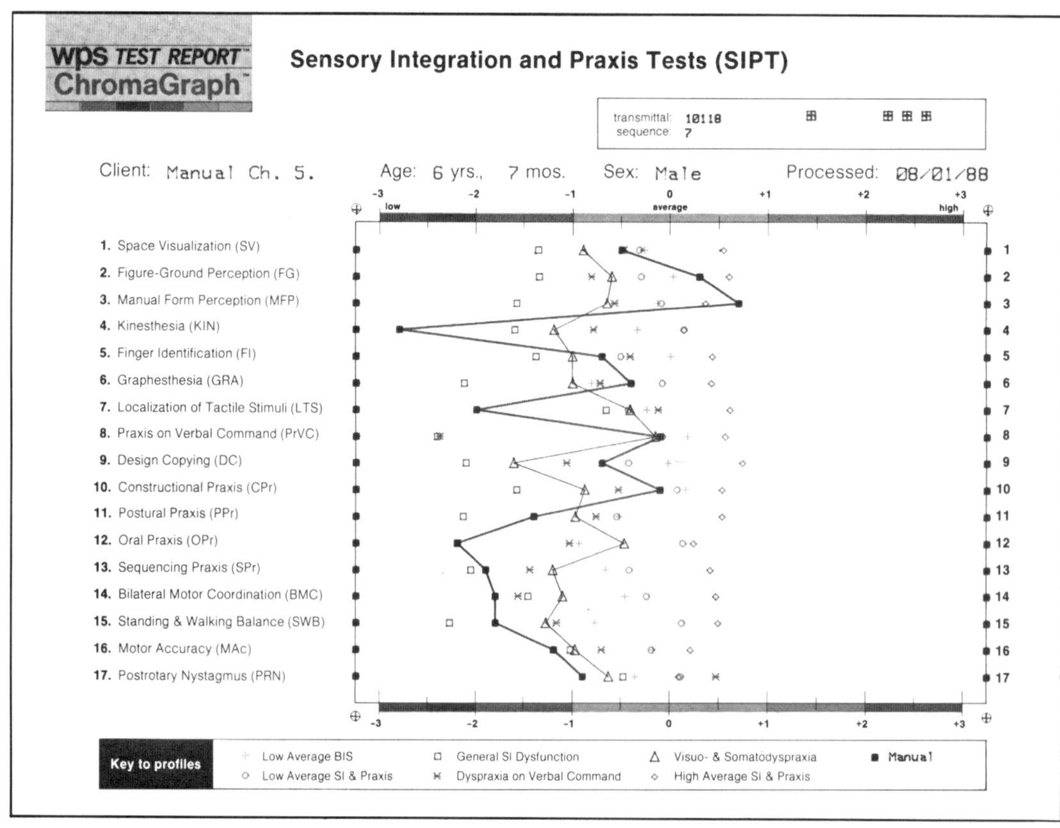


**Figure 44**
**Case G**

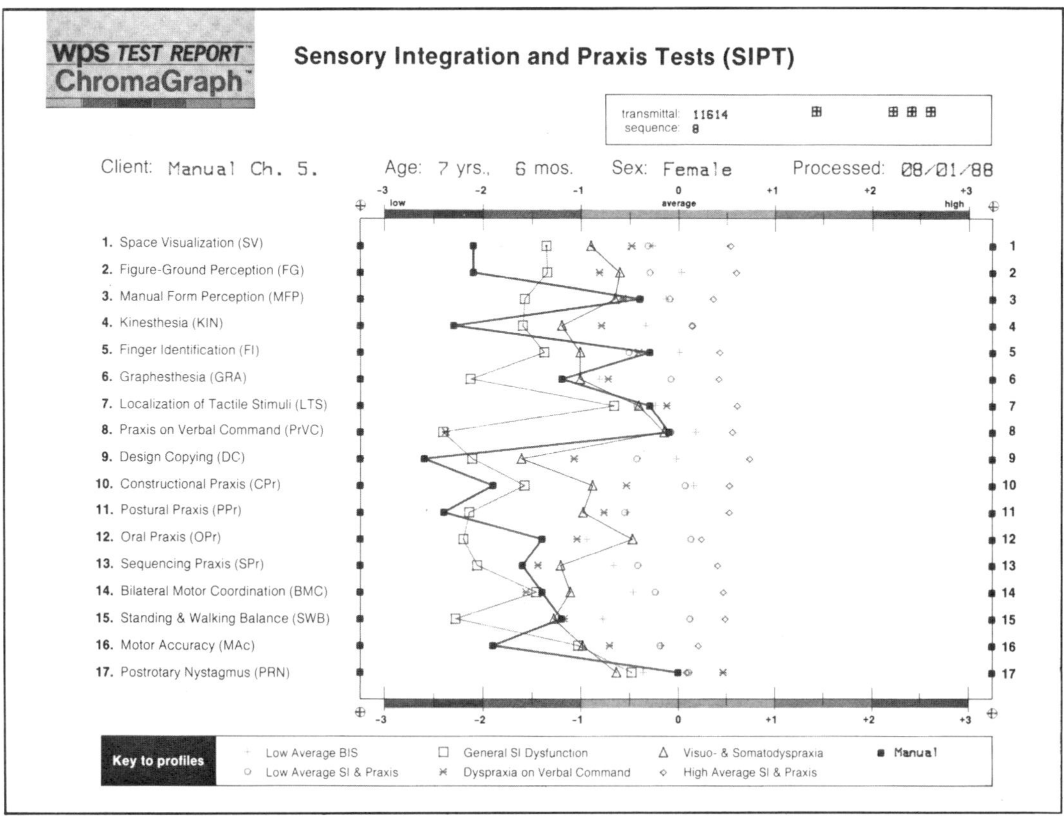


**Figure 45**
**Case H**

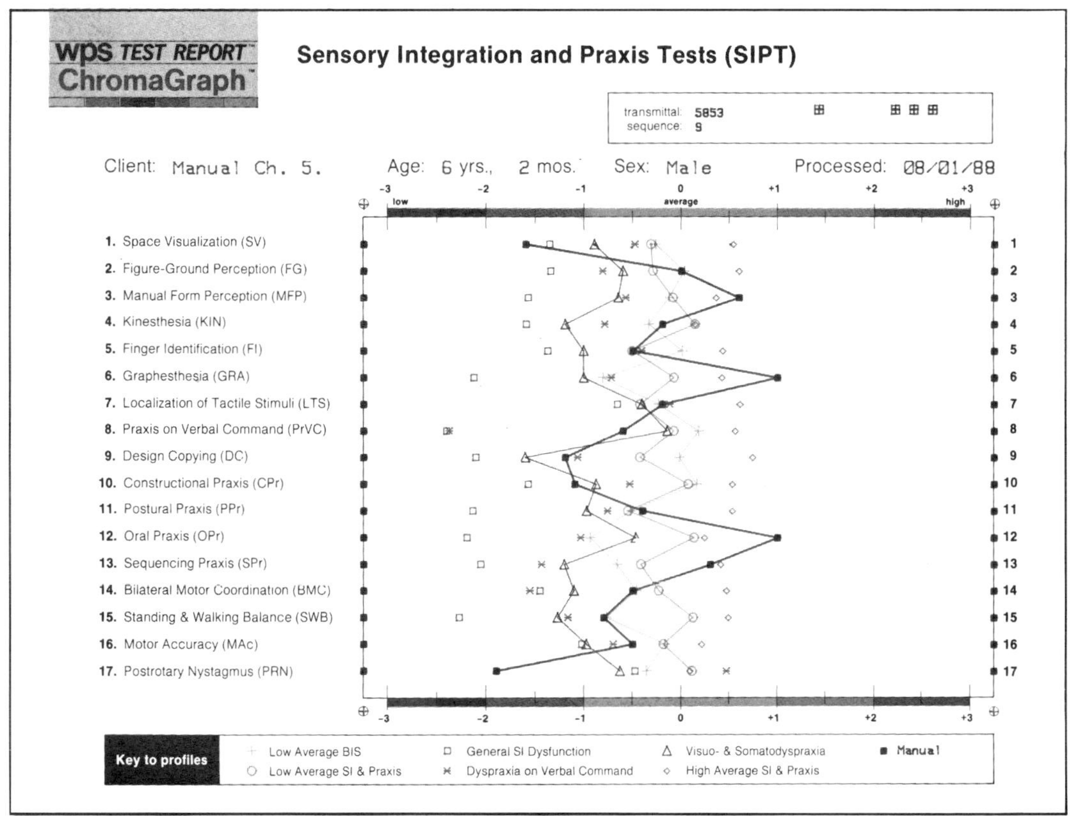


**Figure 46**
**Case I**

dysfunction is identified when a child's score falls within that range. However, some children's profiles are computer fitted to Group 4, yet show enough contrast to require an alternate interpretation.

The SIPT profile of Case J, a 7-year, 3-month-old learning-disabled boy, is shown in Figure 47. Most of his scores are within the average range, but several are low enough to suggest some possible explanations for the problems he is having with academic learning. His scores were -1.0 *SD* or lower on four of the tests—Finger Identification (FI), Kinesthesia (KIN), Motor Accuracy (MAc), and Design Copying (DC). These form a meaningful group, for they are all related to each other. The low scores take on extra significance because they are in contrast to the many scores falling above the mean. His overall performance is consistent with low average sensory integration and praxis (Group 4). More importantly, he appears to have mild visuodyspraxia or sensorimotor deficit related to a somatosensory processing deficit.

The profile of Case K, a 5-year, 9-month-old learning-disabled girl, was matched with Group 1 (Low Average BIS) and Group 4 (Low Average Sensory Integration and Praxis). Her profile (see Figure 48) has: (a) two scores lower than -1.5 *SD* (Praxis on Verbal Command [PrVC] and Oral Praxis [OPr]), (b) three at approximately -1.0 *SD* (Constructional Praxis [CPr], Standing and Walking Balance [SWB], and Postrotary Nystagmus [PRN]), and (c) four at 0.5 *SD* or above (Figure-Ground Perception [FG], Kinesthesia [KIN], and Graphesthesia [GRA]). This profile does not unambiguously meet the criteria for a BIS deficit or for a low average interpretation. Her lowest score is on Praxis on Verbal Command, which typically is one of the highest praxis scores in cases of BIS disorders. Her scores on Graphesthesia (GRA), Finger Identification (FI), and Bilateral Motor Coordination (BMC) are in the high average to superior range. There is a two-and-a-half *SD* spread in scores. That spread could reflect measurement error, but a contrast in scores is also one of the bases for interpreting low scores as dysfunctional. Case K's low scores do provide a meaningful grouping, and indicate that she may have some difficulty with oral praxis and with praxis on verbal command. Her relatively low Postrotary Nystagmus and Standing and Walking Balance scores suggest that this difficulty may be related to inefficient CNS processing of macular and cupular vestibular sensory processing. Her greatest strength appears to lie in somatosensory processing.

## SIPT Group 5: Dyspraxia on Verbal Command

To qualify for a diagnosis of dyspraxia on verbal

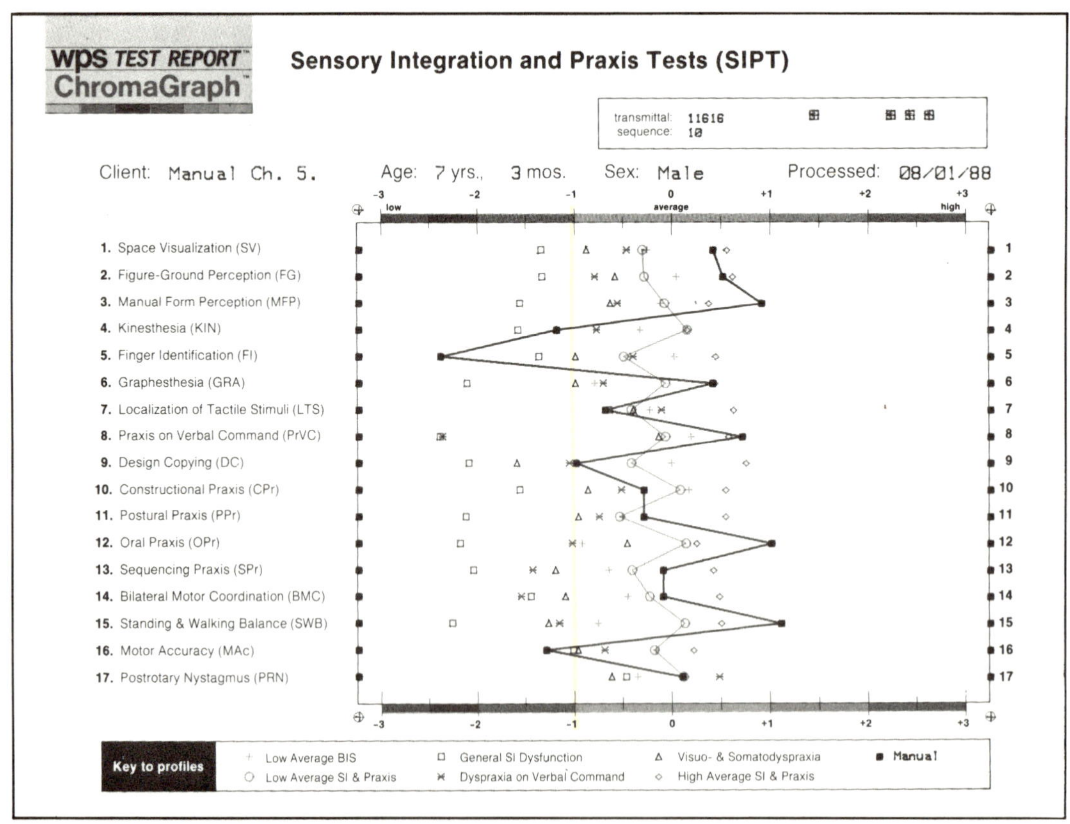


**Figure 47**
**Case J**

command, the child's profile must show a low score on Praxis on Verbal Command (PrVC) and should also show an average to high score on Postrotary Nystagmus (PRN). Scores on form and space perception, visuopraxis, somatosensory tests, Postural Praxis (PPr), and Constructional Praxis (CPr) are generally in the low average range or better. Four of the five tests that usually identify the BIS function—Bilateral Motor Coordination (BMC), Oral Praxis (OPr), Sequencing Praxis (SPr), and Standing and Walking Balance (SWB)—are expected to be below average. In Group 5, these low scores are interpreted as signs of cerebral inefficiency rather than a BIS deficit.

There is a linguistic and a postural aspect to Praxis on Verbal Command. A SIPT diagnosis of dyspraxia on verbal command generally presumes impairment of the linguistic component; if Postural Praxis is also low, the problem may lie in the postural component rather than the linguistic component. Sometimes the source of the problem must be checked clinically.

The SIPT scores of Case L, a 5-year, 3-month-old learning-disabled boy, illustrates a fairly typical Group 5 profile (see Figure 49). This profile meets the two most important descriptors of low Praxis on Verbal Command (PrVC) and average to high Postrotary Nystagmus (PRN). Additional low scores for Group 5 include those on Oral Praxis (OPr), Sequencing Praxis (SPr), Bilateral Motor Coordination (BMC), and Standing and Walking Balance (SWB). Case L's scores on these latter tests are not generally low but, with the exception of Oral Praxis, they tend to follow the Group 5 constellation, and they are lower than the majority of this child's scores. The form and space and somatosensory test scores meet the "low average or better" criterion. This child's relatively high score on Constructional Praxis (CPr) is also consistent with the Group 5 pattern. Overall, the SIPT data support the diagnosis of dyspraxia on verbal command.

The SIPT profile of Case M, a 7-year, 3-month-old learning-disabled boy, also fits the Group 5 profile (see Figure 50). The two most important criteria of low Praxis on Verbal Command (PrVC) and average to high Postrotary Nystagmus (PRN) are both present. Although scores on the Oral Praxis (OPr), Sequencing Praxis (SPr), Bilateral Motor Coordination (BMC), and Standing and Walking Balance (SWB) group are not consistently low, there are no apparent patterns in his scores to suggest an alternative diagnosis or to invalidate the diagnosis of dyspraxia on verbal command. However, this child's low Postural Praxis (PPr) score suggests that, unlike the typical Group 5 child, he may have trouble with the postural component of Praxis on Verbal Command as well as with the linguistic component.

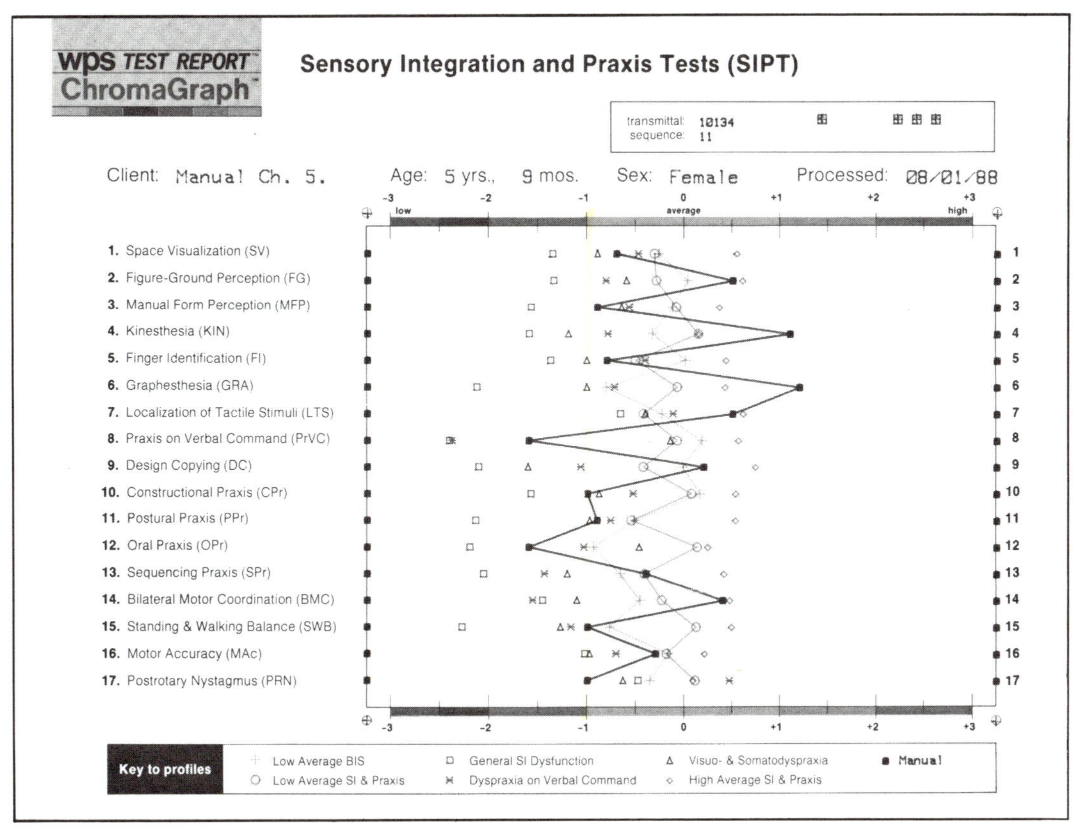


**Figure 48**
**Case K**

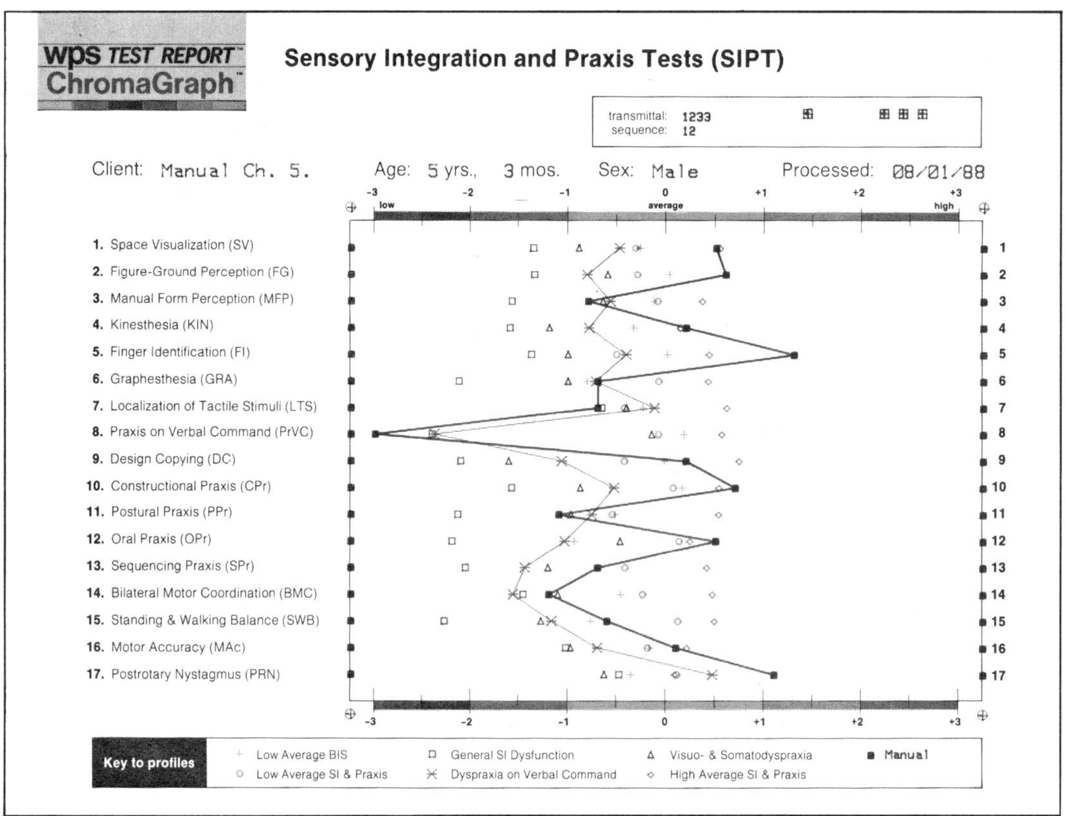


**Figure 49**
**Case L**

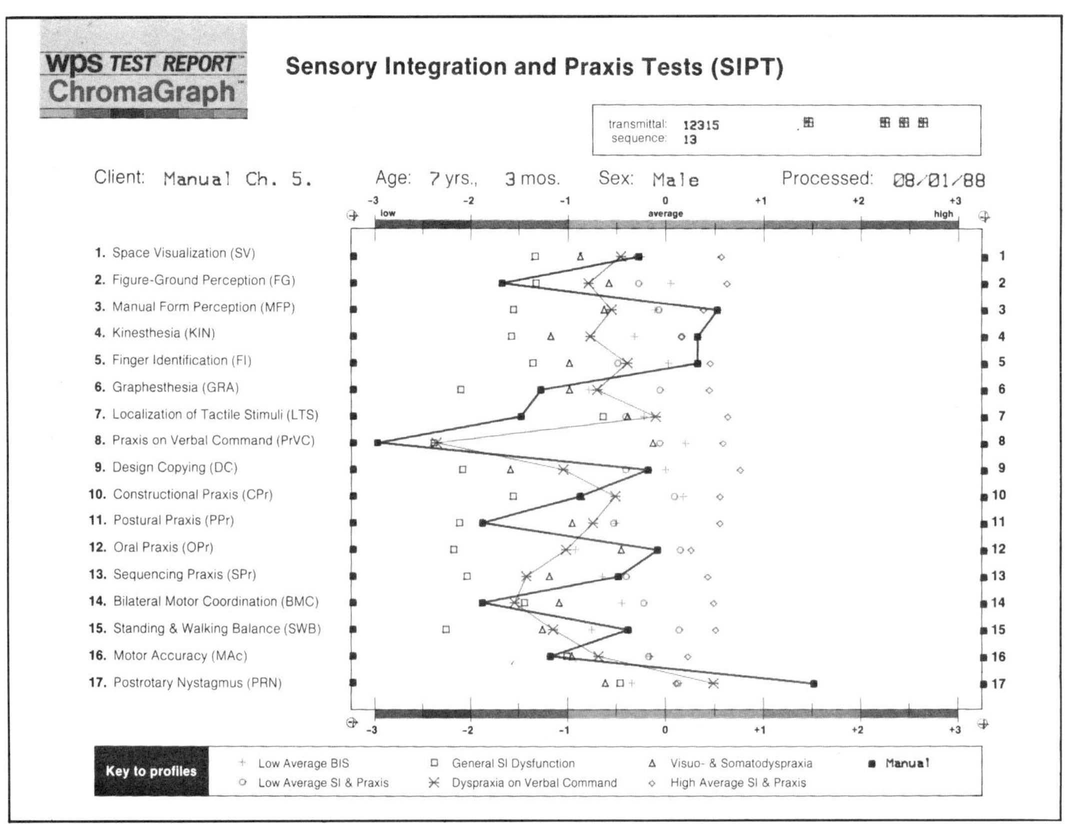


**Figure 50**
**Case M**

Case N is a 5-year, 10-month-old learning-disabled girl, whose SIPT profile (see Figure 51) shows some similarities to both Group 5 (Dyspraxia on Verbal Command) and Group 3 (Visuo- and Somatodyspraxia). Her scores on Praxis on Verbal Command (PrVC) and Postrotary Nystagmus (PRN) satisfy the most important description of dyspraxia on verbal command. Her profile also meets most of the less important description of low scores on Sequencing Praxis (SPr), Bilateral Motor Coordination (BMC), and Standing and Walking Balance (SWB). The low Design Copying (DC) and Motor Accuracy (MAc) are important to consider, but there is insufficient supporting evidence for a firm diagnosis of visuo- or somatodyspraxia. The best interpretation of Case N's profile is dyspraxia on verbal command. In addition, she shows some difficulty in two-dimensional space management.

The ChromaGraph™ of Case O (see Figure 52), a 6-year, 7-month-old learning-disabled boy, plots his profile against that of Group 1 (Low Average Bilateral Integration and Sequencing), Group 3 (Visuo- and Somatodyspraxia), and Group 5 (Dyspraxia on Verbal Command). The interpretation approach in this case consists of the careful examination of scores –1.0 *SD* or lower, looking for meaningful relationships. Those low scores, in order from lowest to higher scores, are: Oral Praxis (OPr), Sequencing Praxis (SPr), Standing and Walking Balance (SWB), Motor Accuracy (MAc), Localization of Tactile Stimuli (LTS), Space Visualization (SV), Praxis on Verbal Command (PrVC), Design Copying (DC), Postural Praxis (PPr), and Kinesthesia (KIN). Of the five tests on which low scores identify the BIS pattern, Case O scored low on three. Case O scored low average to average scores on Finger Identification (FI), and Constructional Praxis (CPr), but scored atypically low on Praxis on Verbal Command and on Oral Praxis. Overall, his pattern does not support a diagnosis of BIS deficit.

Checking next for dyspraxia on verbal command, Case O's profile only marginally meets the criteria of low Praxis on Verbal Command and average to high Postrotary Nystagmus (PRN). Of the four other tests expected to be low in Group 5 cases, Case O's scores are low on three (Oral Praxis, Sequencing Praxis, and Standing and Walking Balance. Most of his form and space and somatosensory test scores are in the low average range. There are certain resemblances to the dyspraxia on verbal command profile, but such a diagnosis cannot be made with confidence.

The most important criteria for visuo- or somatodyspraxia are low scores on a number of visuo- and somatopraxis tests and on the somatosensory tests. Case

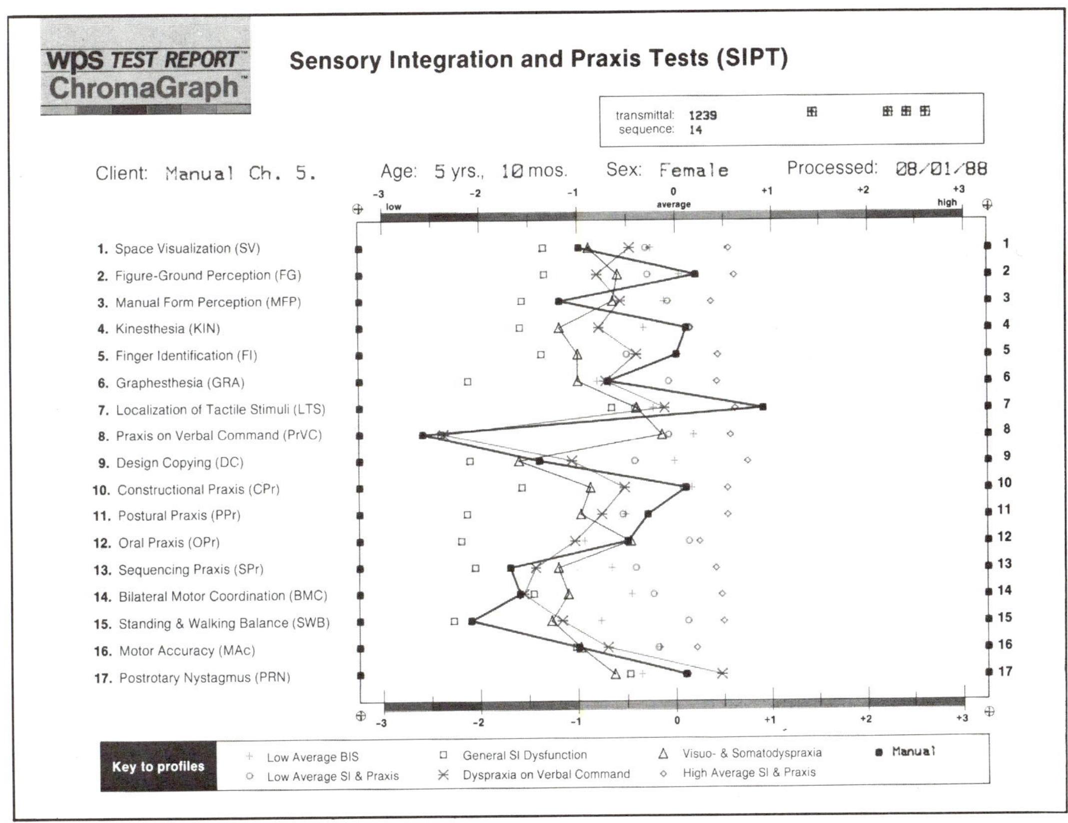


**Figure 51**
**Case N**

O has low scores on the following important tests: Oral Praxis, Sequencing Praxis, Design Copying, and Postural Praxis. Also, his scores are low on Space Visualization, Motor Accuracy, and Bilateral Motor Coordination (BMC). Of the important somatosensory tests, Case O's scores are low on Standing and Walking Balance and Kinesthesia. Although Case O's scores on Praxis on Verbal Command and on Oral Praxis are lower than one would expect in Group 3, this group nevertheless appears to provide the closest match. The best interpretation of Case O's profile is visuo- and somatodyspraxia accompanied by poor proprioception. There is also some difficulty in translating verbally given commands into postural positions. Whether the trouble lies in linguistic interpretation or postural praxis needs to be determined clinically.

**SIPT Group 6:**
**High Average Sensory Integration and Praxis**

Children whose scores fit the Group 6 profile clearly do not have a problem identified by the SIPT. There are occasions, however, when the child's profile is compared against Group 6, yet the scores show the possibility of areas of difficulty.

Case P, a 6-year, 1-month-old boy with an educational diagnosis of learning disability, is an example of a child whose overall scores provide an acceptable match to SIPT Group 6, but who shows some rather sharp contrasts in his scores (see Figure 53). The average of all his SIPT scores may place him in the high average range, but SIPT diagnosis should take into account the score contrasts. Two of Case P's scores—Postural Praxis (PPr) and Kinesthesia (KIN)—are –1.0 *SD* or lower, and his Oral Praxis (OPr) score is –0.9 *SD*. Although these scores are not exceptionally low, they are well below Case P's average and highest scores. Ordinarily, a test user would conclude this profile is that of a child with varying abilities, but probably without serious problems. Yet, because Case P has already been determined to have a learning disability, the SIPT task is to try to add to the clinical understanding of the child. The interpretation task is to determine whether there is a meaningful pattern among the lowest scores. Postural Praxis and Oral Praxis are often related. Postural Praxis and Kinesthesia correlate less frequently, but sufficiently often to warrant interpreting Case P's profile as showing difficulty with postural and oral praxis, perhaps associated with inefficient kinesthetic processing.

**Nonclassified**

The profile of Case Q, a 7-year, 2-month-old learning-disabled boy, illustrates a situation in which the

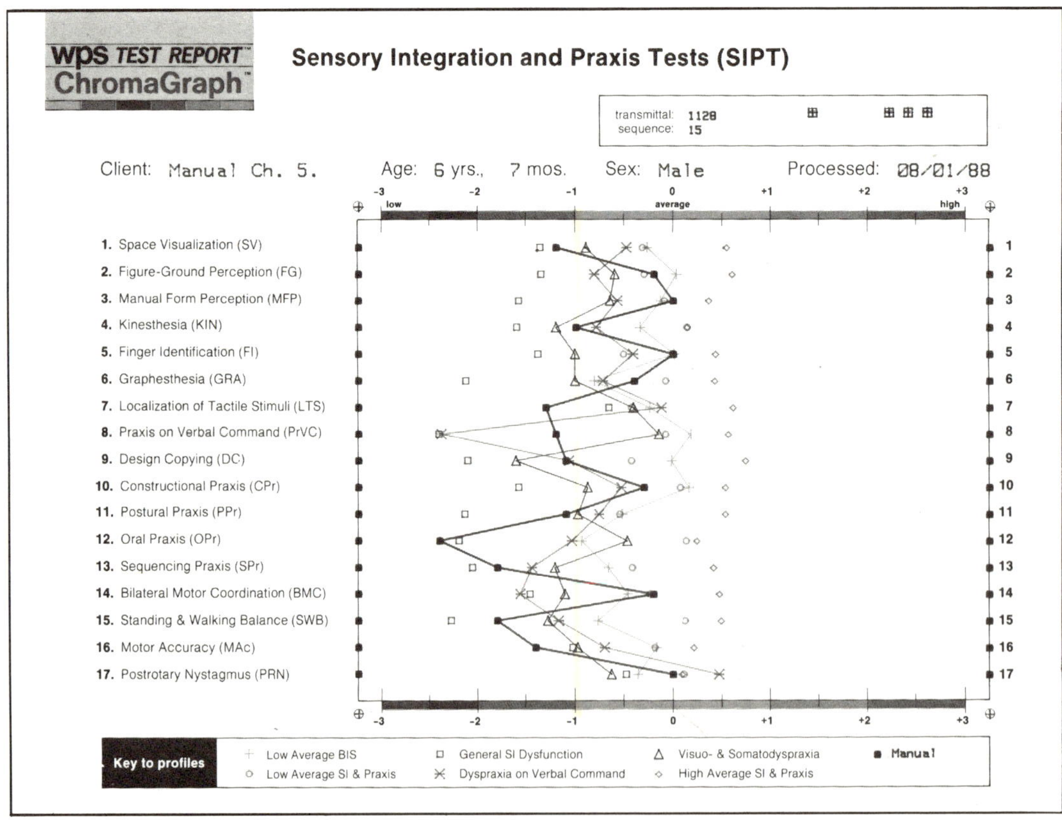


**Figure 52**
**Case O**

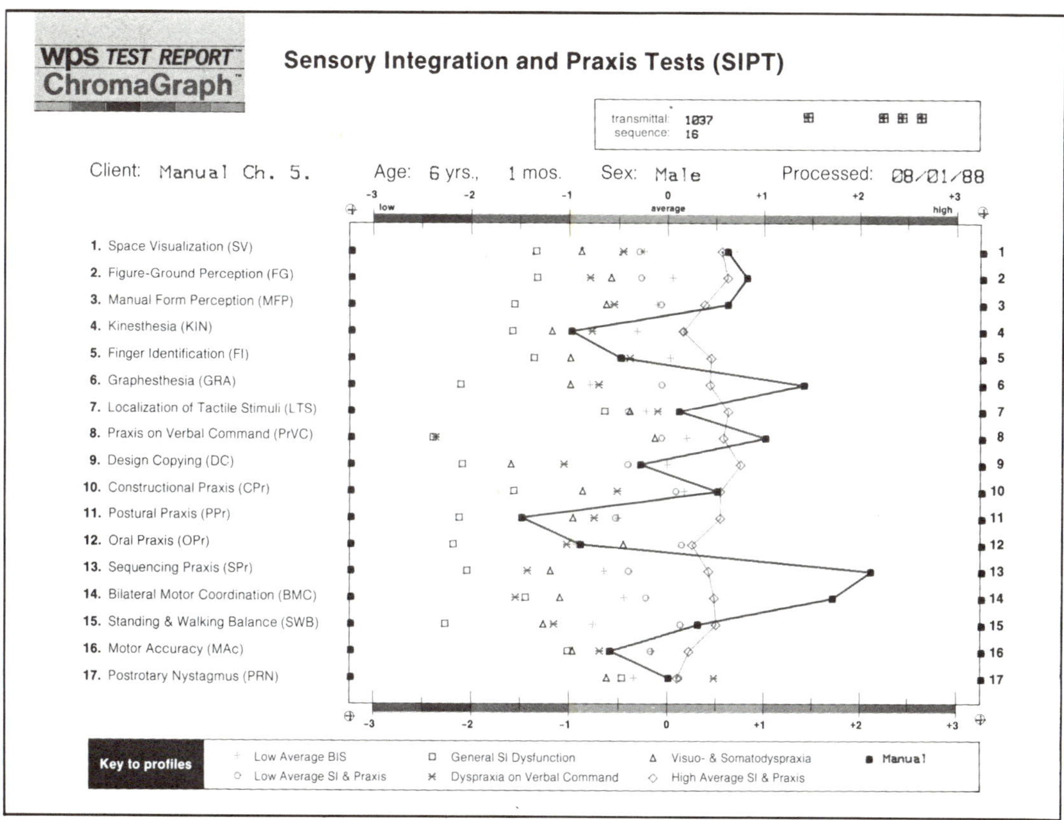


**Figure 53**
**Case P**

score constellation does not meet any of the computer-generated criteria for SIPT group likeness (see Figure 54). The scores are very widely dispersed with several strengths (especially Constructional Praxis [CPr] and Figure-Ground Perception [FG]) contrasted with a number of weaknesses. The objective is to find recognizable patterns among the low scores. The SIPT group's constellations may be of some use, but they cannot be expected to fit well. Two patterns are quite evident: (a) scores on the five tests of the BIS function are among the lowest, and (b) the major criteria for dyspraxia on verbal command are met. Of the five remaining praxis tests, Case Q scores poorly on four: Oral Praxis (OPr), Sequencing Praxis (SPr), Design Copying (DC), and Postural Praxis (PPr). Space Visualization (SV), Motor Accuracy (MAc), and Bilateral Motor Coordination (BMC) also are low. Scores on these tests and Praxis on Verbal Command (PrVC) provide support for an interpretation of dyspraxia. The somatosensory test scores are puzzling. The three low ones—Graphesthesia (GRA), Standing and Walking Balance (SWB), and Localization of Tactile Stimuli (LTS)—indicate a probable deficit that supports the dyspraxia conclusion. A diagnosis of dyspraxia is the most conventional and flexible of any SIPT diagnosis. It is concluded that Case Q's scores show dyspraxia in the following areas: (a) two-dimensional visual space management, (b) translating verbally given directions into postural positions, (c) skilled postural and oral movements, and (d) the sequencing of movement. There may be some tactoproprioceptive deficit. Case Q's strength lies in three-dimensional construction and in visual figure-ground perception.

As mentioned earlier in this chapter, interpretation of the WPS ChromaGraph™ for SIPT is problematic when the tested child has known medical or psychological problems that make him or her very different from most learning-disabled, sensory integration disordered, and normal children. Five such cases are discussed in the remainder of the chapter, without reference to the WPS ChromaGraph™ for SIPT. Their major SIPT scores are shown in Table 3.

The SIPT scores of Case R, a traumatically brain injured boy, 8 years, 3 months of age, are shown in Table 3. Interpretation of Case R's scores are based on relevant validity data and general knowledge of brain function. The possibility of a neuromotor problem resulting from brain injury must be addressed first in order to avoid misinterpretation of low scores. Tests most vulnerable to neuromotor coordination are Motor Accuracy (MAc), Standing and Walking Balance (SWB), and Bilateral Motor Coordination (BMC). Case R's scores on these three tests are all low: –2.7, –3.0, and –2.4 *SD,* respectively. Scores on Space Visualization (SV) and Figure-Ground Perception (FG) are the least likely influenced by

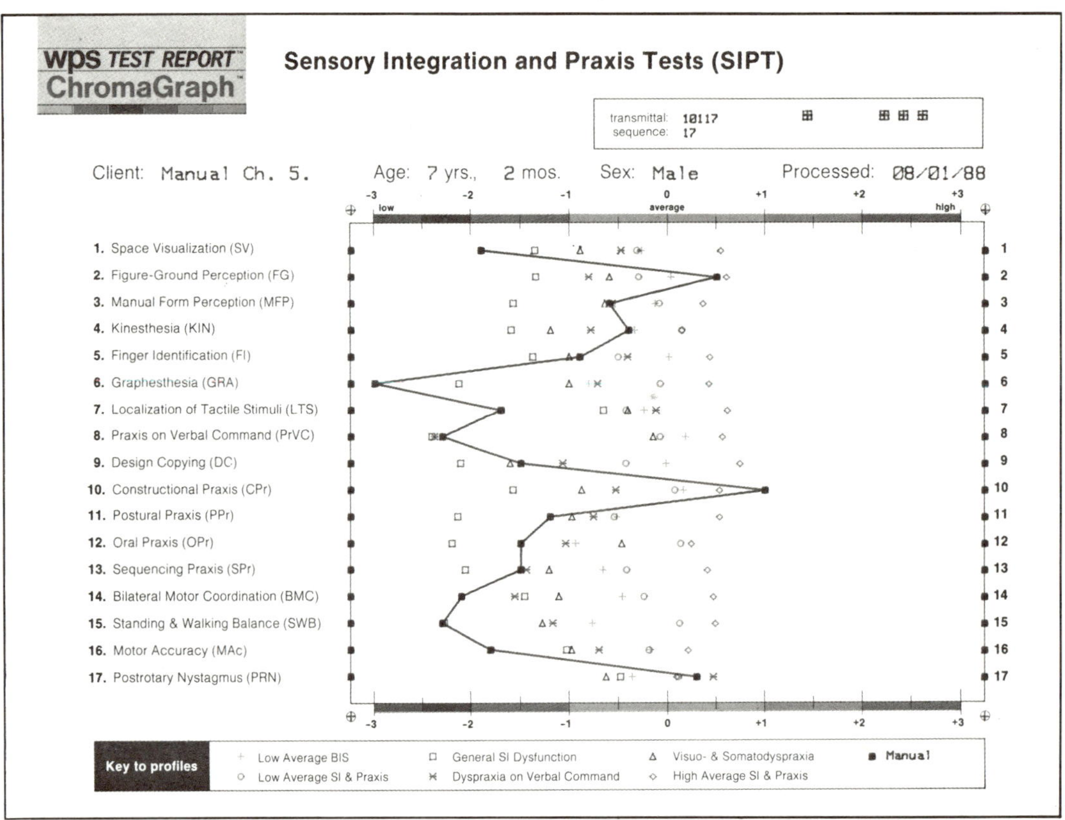


**Figure 54**
**Case Q**

a neuromotor condition. Case R scored within normal limits on these tests (0.3 *SD,* –0.8 *SD*, respectively). Vulnerable test scores may be depressed by a neuromotor problem. The strong contrast between scores of tests that group themselves logically aids in interpretation. Scores on almost all of the form and space and visuopraxis tests—Space Visualization, Figure-Ground Perception, Manual Form Perception (MFP), Constructional Praxis (CPr)—fall within the average range. The two exceptions are Design Copying (DC), with a –1.7 *SD* score, and Motor Accuracy, with a -2.7 *SD* score. In view of performance on the other tests of visual perception, the Design Copying and Motor Accuracy scores are probably slightly depressed by a motor problem, but dyspraxia should not be ruled out until other praxis test scores are considered. Case R scored poorly on five out of the six praxis tests, and of these, only Design Copying is likely to be appreciably depressed by motor incoordination. Dyspraxia (not developmental dyspraxia) was one of Case R's problems at the time of testing. Whether the Praxis on Verbal Command (PrVC) problem is one of linguistic interpretation or of postural praxis should be checked clinically. The somatosensory scores vary. Two low scores on Graphesthesia [GRA] (–1.4 *SD*) and Localization of Tactile Stimuli [LTS] (–1.9 *SD*) indicate some loss of tactile perception, but the sense of kinesthesia and finger identification have been preserved. The high Postrotary Nystagmus (PRN) score suggests loss of higher level CNS inhibition on the nystagmatic reflex. The possibility of inhibitory loss on behavior should not be overlooked. Although all of the scores of tests associated with the BIS function are low, that function has been identified only with developmental problems and cannot be applied to cases of traumatic brain injury. Case R's scores show dyspraxia in the following abilities tested: postural execution of verbal commands, copying designs, postural and oral motor planning, and executing skilled use of the body. Upper extremity kinesthetic sense and finger identification have been preserved. Visual and haptic form and space perception not requiring praxis is age adequate. Motor incoordination is suspected.

The SIPT scores of Case S, a 7-year, 2-month-old girl with an educational diagnosis of reading disorder, are shown in Table 3. Because a reading disorder is a learning disorder, it would be appropriate to utilize the computerized application of SIPT group analysis to identify Case S's sensory integrative status. The case also serves to illustrate diagnosis without computer assistance. Most of Case S's scores fall in the low average to difficult range. Two scores, Design Copying [DC] (–3.0 *SD*) and Bilateral Motor Coordination [BMC] (1.1 *SD*), are notably very different from the rest of the scores, and neither of these scores is well supported by comparably low scores on tests with which Design Copying or Bilateral Motor Coordination usually correlate. It is probable that both scores reflect some measurement error. Taking that likelihood into consideration, it is noted that Case S's form and space perception and management ability are below average, but not seriously so. Tasks requiring these abilities are difficult for Case S. Scores on the somatosensory tests fall mainly in a similarly low area. The inefficient somatosensory processing is probably related to the poor form and space perception and management. Case S scored in the low to low average range on four of the six praxis tests, indicating lack of well-developed visuo- and somatopraxis. The -1.8 *SD* score on Praxis on Verbal Command (PrVC) does suggest a problem, but the profile as a whole does not sufficiently resemble SIPT Group 5 to warrant a diagnosis of dyspraxia on verbal command. Case S shows mild deficits in: (a) visual and haptic form and space management, (b) translating verbally given directions into postural responses, and (c) somatosensory perception. These areas of difficulty are probably related to Case S's reading disorder.

The SIPT scores of Case T, a 5-year, 7-month-old boy considered to be "high risk" at birth, are shown in Table 3. The 4.1 *SD* range of scores raises the question of reliability of the Kinesthesia (KIN) score of -3.0 *SD*. The vulnerability of the Kinesthesia score to distraction contributes toward its being one of the less reliable of the SIPT. On the other hand, the presence of so many high scores suggests Case T was an adequate test taker. The Kinesthesia score is checked by noting the test scores with which it is most apt to correlate, namely Standing and Walking Balance (SWB) and Localization of Tactile Stimuli (LTS). Case T's scores on Standing and Walking Balance and Localization of Tactile Stimuli are slightly below average, but still within normal limits. Thus, his true Kinesthesia score is probably low, but not as low as -3.0 *SD*. The interpretation approach is to look for meaningful relationships among the highest scores and similarly among the low scores. Tests with scores at the mean or above show strengths in visual form and space perception, in visuopraxis, and in interpreting verbally given directions into planned motor action. Tests with scores -1.0 *SD* or lower are Kinesthesia, Oral Praxis (OPr), Postrotary Nystagmus (PRN), and Bilateral Motor Coordination (BMC). Kinesthesia and Oral Praxis often correlate well, and both can be influenced by the same conditions that lower Postrotary Nystagmus scores. Oral Praxis and Bilateral Motor Coordination also often correlate well. In summary, Case T has generally adequate sensory integration and praxis but appears to have some difficulty with oral praxis associated with vestibulokinesthetic sensory processing deficiency.

The SIPT scores of Case U, a 6-year, 4-month-old boy with a medical diagnosis of spina bifida, are given in Table 3. The score pattern is not similar to the SIPT groups, and the child's diagnosis precludes the use of that model. The scores cover a wide range, with most falling below the normative mean. There are a number of scores -1.0 *SD* or lower indicating dysfunction, especially for visuoconstruction and eye-hand coordination tests such as Motor Accuracy (MAc), Design Copying (DC), and Constructional Praxis (CPr). Interpretation of complex somatosensory input is poor for Manual Form Perception (MFP) and Graphesthesia (GRA), especially when it involves forms and shapes. However, Case U was able to identify fingers well. The kinesthetic sense is inefficient. Because a neuromotor deficit is apt to depress the Standing and Walking Balance score, that score is not analyzed. A motor problem could also depress Motor Accuracy performance. In contrast to these deficit areas, Case U was age adequate in translating verbally given directions into planned motor action. Visual perception tasks not requiring practic skill or motor coordination—Space Visualization (SV) and Figure-Ground Perception (FG)—were performed adequately. To summarize Case U's sensory integrative and praxis status, there is a definite problem with skilled visual-motor performance, but nonmotor visual perception tasks are not difficult. Similarly, simply localizing tactile input is age adequate, but interpretation of complex forms or shapes with the hands and relating the input to action is very difficult for Case U.

The SIPT scores of Case V, a 9-year, 4-month-old autistic girl of normal or above intelligence, are compared to the data from the normative sample for 8 years, 6–11 months (see Table 3). This profile is not represented as typical of autistic children, whose scores on the SIPT vary considerably. Case V's most severe deficit lies in development of practic ability involving the body (Praxis on Verbal Command [PrVC], Postural Praxis [PPr], Oral Praxis [OPr], Sequencing Praxis [SPr]). The problem was equally severe whether the examiner gave directions verbally or by demonstration. Low scores on Standing and Walking Balance (SWB), Postrotary Nystagmus (PRN), and Kinesthesia (KIN) indicate inadequate CNS processing of propriovestibular sensory input. The deficit is probably associated with the inadequate practic development. Tactile perception scores fall in the low (Finger Identification [FI]) to average (Graphesthesia [GRA], Localization of Tactile Stimuli [LTS]) range. Case V's strength lies in the visual domain, where most scores fall in the average range (Space Visualization [SV], Figure-Ground Perception [FG], Manual Form Perception [MFP], Motor Accuracy [MAc]). Design Copying (DC), however, is -1.3 *SD* and Constructional Praxis (CPr) 1.0 *SD*. Chance factors could favor a high Figure-Ground

Perception or Manual Form Perception score, but could not very well favor a high Constructional Praxis (CPr) score. Case V cannot be considered dyspraxic, but there is strong evidence of deficits in the conceptualization, planning, and execution of adaptive actions of the body and in CNS processing of vestibular sensory input.

**Table 3**
**SIPT Scores for Case Studies**

| | Case ID | | | |
|---|---|---|---|---|
| **Test** | **R** | **S** | **T** | **U** |
| Space Visualization (SV) | 0.3 | -1.3 | 1.1 | 0.1 |
| Figure-Ground Perception (FG) | -0.8 | -1.4 | 0.0 | 0.1 |
| Manual Form Perception (MFP) | -0.1 | -1.6 | -0.9 | -3.0 |
| Kinesthesia (KIN) | 0.0 | -1.4 | -3.0 | -1.0 |
| Finger Identification (FI) | 0.8 | -1.8 | 1.1 | 0.9 |
| Graphesthesia (GRA) | -1.4 | -1.1 | -0.1 | -2.3 |
| Localization of Tactile Stimuli (LTS) | -1.9 | -0.5 | 0.7 | -0.8 |
| Praxis on Verbal Command (PrVC) | -2.8 | -1.8 | 0.8 | 0.1 |
| Design Copying (DC) | -1.7 | -3.0 | 0.4 | -2.3 |
| Constructional Praxis (CPr) | 0.2 | -0.9 | 0.7 | -2.1 |
| Postural Praxis (PPr) | -2.8 | -0.8 | -0.8 | -0.8 |
| Oral Praxis (OPr) | -1.7 | -0.2 | -2.2 | -1.8 |
| Sequencing Praxis (SPr) | -3.0 | -0.9 | 0.9 | -0.2 |
| Bilateral Motor Coordination (BMC) | -2.4 | 1.1 | -1.0 | 0.5 |
| Standing and Walking Balance (SWB) | -3.0 | -1.1 | -0.3 | -3.0 |
| Motor Accuracy (MAc) | -2.7 | -1.1 | -0.6 | -2.5 |
| Postrotary Nystagmus (PRN) | 3.0 | -0.2 | -1.0 | — |

# CHAPTER 6
## DEVELOPMENT AND STANDARDIZATION OF THE SIPT

### Determining the Content Domain

The domain covered by the SIPT has been defined by several major sources. The perceptual, practic, and motor parameters, which leading investigators have found to be affected by adult-onset brain damage, were the original basis for devising a large number of procedures for assessing similar parameters in dysfunctional children. The major criterion for the selection of these procedures was the capacity of each measure to discriminate between dysfunctional and nondysfunctional children of normal intelligence. A guiding question was, "What aspects of sensory processing and related foundational skills are particularly vulnerable to disorder and are associated with learning and behavior problems in children?" A second criterion was evidence of a natural relationship among the measures, as determined through factor analysis. A third criterion was whether the parameter lent itself to reliable measurement and analysis by standard psychometric methods.

The preliminary content domain had been previously defined in the development of the SCSIT (Ayres, 1980b). However, some parts of the SCSIT yielded less meaningful and useful information than others, and further clinical experience with dysfunctional children revealed important aspects of dyspraxia that were not assessed by the SCSIT. Many years of collective experience by investigators of adult-onset brain disorder served as a model for selection of additional behavioral domains to be included in the SIPT. A final winnowing of the many evaluation procedures was made in connection with consultation with the experts in the field of sensory integrative dysfunction.

These stages of SIPT content determination are discussed in four sections of this chapter: (a) establishing the initial domain; (b) redefining the domain for the SCSIT; (c) eliminating the less helpful elements of the SCSIT; and (d) expanding the domain for the SIPT.

**Establishing the Initial Domain**

Although apraxia and developmental dyspraxia were recognized early in this century, it was not until the middle of the century that agnosia, apraxia, and aphasia in adults became subjects of extensive and intensive investigation. Study of comparable disorders of perception and dyspraxia in children has lagged behind, although visual perception problems in learning disorders has been a subject of considerable interest.

The concept of the *body schema* was introduced by Head (1920), who believed all sensations were brought into relationship in the internalized postural model of the body. If the schema was deficient through brain damage, Head particularly noted, touch localization was likely to be poor. The concept of body schema, as dependent upon organization of bodily sensation for a postural model, was further developed by Nielsen (1938, 1955). Expanding the construct to include psychic processes, Schilder (1942, 1950) pointed out that the representations of movement could not be separated from perceptions from skin, muscles, and the vestibular system. Brain disorders could result in agnosia of the body image, including finger agnosia.

An early idea of sensory integration was expressed by Bender (1938), who stated that perception could not be understood as a sum of single sensations, and that the organism related to constellations of stimuli by a total process. Halstead (1947) recognized that perception, memory, and the formation of concepts were important to ideation, and that the association of ideas was a necessary precondition to the most complex products of the intellect. He applied factor analysis to a number of perceptual-motor behaviors and other measures to isolate the basic factors of what he termed "biological intelligence." Denny-Brown (1958) also recognized that apraxia involved disturbance of conceptual function, including its relationship to ideation, and where this process was poor, there was sensory perceptual defect. The sense of kinesthesia was thought to be especially affected in apraxia.

The many manifestations of agnosia and apraxia in the mature but damaged brain were explored and extensively described by such experts as Benton (1959), Critchley (1953), Denny-Brown (1958), Halstead (1947), Goody and Reinhold (1952), Hecaen, Penfield, Bertrand, and Malmo (1956), and Nielsen (1946, 1955). Recognition of the close association between sensation and perception and movement was expressed well by Gooddy and Reinhold (1952) in their comment, *"Orientation and movement are inseparable. Both have their origin in sensation"* (p. 481). Human orientation in space, according to these investigators, is dependent upon all forms of perception. Somewhat later, Luria (1966) addressed the integrative activity of the central nervous apparatus and considered the role of cutaneokinesthetic reception as well as visual, auditory, and proprioceptive reception.

The association of somatosensory input and praxis has been reported throughout the decades of study of apraxia. Dee and Benton (1970) found patients had as much trouble with tactile and kinesthetic spatial perception as with visual perception. Hecaen and Albert (1978) suggested that defective sensory guidance of fine motor behavior might be a unitary basis for disorders of general behavior and that such a deficit might represent a necessary, but not sufficient, condition for apraxia. Gubbay, Ellis, Walton, and Court (1965) found that apraxia was associated with finger and topographical agnosia in 21 individuals from 9 to 21 years of age. Although Poeck and Orgass (1971) have declared the concept of the body schema useless, they have linked that disturbance quite frequently with apraxia and visuospatial disorientation. The close relationship between perception and praxis has been expressed by use of the term *apractognosia* (Critchley, 1953; Hecaen, Penfield, Bertrand, & Malmo, 1956). Lesny (1980) has used the term *dyspraxia-dysgnosia* in reference to developmental disorder in children.

Perceptual disorders in children were first studied by those concerned with overt brain damage. Most of these studies were focused on visual perception. Strauss and Lehtinen (1947) emphasized the space perception requirements for academic achievement. Cruickshank and Bice (1955) suggested that psychomotor perception difficulty may be typical of brain injury per se, but that it was not always present. They administered a figure-background test and the Bender-Gestalt (Bender, 1938), and on the latter test analyzed the approach on several parameters. Another figure-background perception study of cerebral palsied children (Cruickshank, Podosek, & Thomas, 1965) established variations of perceptual deficit in that population. Benton (1955) determined that finger localization skills develop progressively from 6 to 9 years of age.

These and similar sources provided the universe of content from which the SCSIT evolved, and from which the SIPT subsequently was developed.

**Refining the Domain for the SCSIT**

In an early study (Ayres, 1965), 29 perceptual-motor and related procedures were administered to 100 dysfunctional and 50 normal 6- and 7-year-old children. Of the 29 measures, those showing the strongest discriminating capacity and factor loadings were selected for further study. Those that were selected, plus 64 additional postural, auditory-language, and behavioral parameters, were included in a Q-technique factor analysis (Ayres, 1969b). Mainly from the result of these two studies, the tests that eventually became the SCSIT were selected. In this selection process, some behavioral measures that expert therapists found useful for the assessment of adults failed to meet expected psychometric standards when applied to young children, and therefore, were not included in the SCSIT.

For example, drawings of people have been used as indices of body percept. Bender (1956) found that in cases of organic brain damage, the disruption in the patterning of somatosensory impulses was reflected in drawings of a man. However, Ayres and Reid (1966) reported other findings. Self-drawings and scores on a number of perceptual-motor variables, including early forms of the SCSIT, were obtained from the 100 dysfunctional children and 50 normal children mentioned previously (Ayres, 1965). The drawings were highly discriminating, as expressed in a critical ratio of 5.72 ($p < .01$), but apparently were less dependent upon somatic than upon visual perception. Scores on self-drawings correlated significantly with somatosensory tests ($r = .39$), visual perception tests ($r = .53$), motor planning tests ($r = .44$), tests that required drawing with a pencil ($r = .56$), and tests that required integration of function of the two sides of the body ($r = .34$). The ability to draw with a pencil was of greatest importance to self-drawings, and that skill was more discretely measured by the Motor Accuracy Test. Therefore, self-drawings were not included in the SCSIT.

A number of other parameters, many of which proved to be clinically useful and closely linked to somatosensory integration, were deliberately omitted from the SCSIT. For example, tactile defensive responses, gravitational insecurity, and hyperactivity with distractibility, while all very significant behavioral observations, were not included because of the difficulty in obtaining reliable ratings of these behaviors. Similarly, important postural and ocular responses showed close association with the other sensory integrative measures (Ayres,

1972b, 1977), but were excluded for psychometric reasons. Choreoathetoid-like involuntary movement is frequently seen in children with sensory integrative dysfunction, but failing to correlate appreciably with scores on sensory integration tests (Ayres, 1976a), it was presumed to be a neuromotor, rather than a sensory integrative, deficit and, therefore, not included in the SCSIT.

**Eliminating the Less Helpful Elements of the SCSIT**

Some of the procedures evaluated by the SCSIT did not lend themselves well to conventional psychometric analysis, some were difficult to administer, and some tests yielded less valuable information than others. Before discarding some of the SCSIT in planning the SIPT, the opinion of the faculty members of Sensory Integration International (SII) was sought and considered. These experts provided experience and guidance that helped determine the revised content of the modified and retained tests that were standardized as the SIPT.

**Expanding the Domain for the SIPT**

Most of the SCSIT tests were substantially revised or redesigned for the SIPT. In addition, the Southern California Postrotary Nystagmus Test (SCPNT; Ayres, 1975) was added to the SIPT as the Postrotary Nystagmus (PRN) test, and four new praxis tests were developed specifically for the SIPT.

**Basis for the selection of the Postrotary Nystagmus test.** Although assessment of the adult's central and peripheral vestibular system has been common for many years, only relatively recently has it been recognized that inefficient central nervous system processing of vestibular sensory input is often associated with minimal brain dysfunction in children. The inclusion of measures of this processing is based more on the frequency with which integrative deficits are found than on the understanding of the way in which these deficits affect learning and behavior. Ayres (1978) tested 128 learning-disabled children 6–10 years of age (mean = 97.6 months, *SD* = 14.3 months), and found that approximately 50% of these children had SCPNT scores of −1.1 *SD* or lower (in a normal population, less than 14% of all children would be expected to obtain scores this low).

This finding is consistent with the results of a number of other studies. For example, in a study of 39 children between the ages of 3 and 6 (14 normal children and 25 children diagnosed as having minimal neurological impairment), Steinberg and Rendle-Short (1977) found that more than half of the impaired children demonstrated no postrotational nystagmus elicited through angular deceleration. Frank and Levinson (1973, 1975–1976) reported that in examining more than 1,000 dyslexic children, the vast majority of the problems were found to be of cerebellar-vestibular origin. Studies by deQuiros (1976) and by deQuiros and Schrager (1978) found that many learning-disabled children had vestibular problems with related proprioceptive and postural disturbances. These authors have suggested that beginning conceptualization is determined by knowledge of one's own body, which, in turn, is dependent upon sensory integration.

Although there are a number of ways in which disordered CNS processing of vestibular input is manifested and should be observed in the child, only the duration of elicited postrotary nystagmus and body balance were chosen as tests for inclusion in the SIPT. Other parameters were excluded because their properties did not lend themselves to standard psychometric analysis. The procedure chosen for measuring elicited postrotary nystagmus duration in the SIPT is similar to the procedure which, for many years, has been standard for testing adults (see, for example, Fukuda, 1975; Tibling, 1969; Tuohimaa, 1978).

**Basis for selection of praxis tests.** The praxis tests that were selected for the SIPT were based upon many of the procedures employed by investigators of adult-onset apraxia. Early classifications of apraxia into ideational and ideomotor types (see Poeck, 1982) have been replaced with a proposal of two stages or levels of complexity in the practic act (DeRenzi, 1985; Hecaen & Rondot, 1985, Poeck, 1985). The first stage is ideation or conceptualization, and the second stage is planning action. With this view, the various procedures conventionally used to test for adult-onset apraxia represent different behavioral goals. The tasks most frequently used with adults have been imitating or performing on verbal command: (a) acts of symbolic representation such as showing how to use an imaginary comb, (b) culturally meaningful gestures such as saluting, (c) nonsymbolic facial and manual positions or movements, (d) actual use of objects, (e) drawing or copying designs, and (f) block construction (for examples, see Aram & Horwitz, 1983; Benson & Barton, 1970; Benton & Fogel, 1970; DeRenzi, Motti, & Michelli, 1980; Ettlinger, 1969; Gainotti, Miceli, & Caltagirone, 1977; Kertesz & Hooper, 1982; Poeck, 1982). Sequencing or ordering a series of movements is an essential element in these practic tasks (see Kimura, 1977). While that quality can be subjectively observed as the person is performing the act, objectively measuring the sequencing skill requires a specially structured situation.

Those who have constructed instruments for assessing perceptual and practic disorders in children have found imitating postures and movements, copying designs, and manipulating objects most helpful (Beery, 1982; Berges & Lezine, 1975; Haworth, 1970; Kephart, 1971; Gubbay, Ellis, Walton, & Court, 1965).

When the selection of praxis tests for the SIPT was under consideration, several types of tests used with adult-onset apraxia were ruled out. The symbolic use of objects was not included because of possible cultural bias and the need for dependence upon visual perception of language comprehension for understanding the task requirement. Further support for excluding the task was obtained by Ayres, Mailloux, and Wendler (1987), who found that the Manual Expression subtest of the Illinois Test of Psycholinguistic Abilities (Kirk, McCarthy, & Kirk, 1968), a test requiring symbolic use of objects, tended to share less variance with other praxis tests than did the preliminary versions of the SIPT tests of praxis. The ITPA subtest carried loadings of only .29 and .33 on the major somatopractic factor of two separate factor analyses. The mean score of 182 dysfunctional children administered the test fell at age expectancy on the subtest, suggesting symbolic use required in the test was not a sensitive indicator of developmental dyspraxia. Furthermore, DeRenzi (1985) has stated that symbolic use of objects as used for testing adults is just a more complex form of object use and easier to use at bedside. The evaluative use of culturally meaningful gestures with young children is also inappropriate because of the factors of age and cultural dependence.

With these matters under consideration, selection of practic tasks for inclusion in the SIPT was made. The use of imitation of facial and manual positions and movements with adult-onset apraxia was translated into the involvement of the whole body in both imitating postures (Postural Praxis [PPr]) and translating verbal directions into practic movement (Praxis on Verbal Command [PrVC]). Because the child's performance on these two tests does not reflect sequencing ability, an important aspect of praxis, the Sequencing Praxis (SPr) test was designed specifically to assess the ability to plan and execute a series of movements or to make transitions from one upper extremity position to another. The items of the SIPT Oral Praxis (OPr) test also require some sequencing ability and are typical of those conventionally used in the assessment of both adults and children.

Both two- and three-dimensional construction have been used to assess constructional apraxia in adults. Accordingly, the SCSIT Design Copying test was expanded to require drawing capability more similar to that used in testing of adults. Because two- and three-dimensional construction tasks do not necessarily require the same practic skill, the three-dimensional Constructional Praxis (CPr) test was developed for the SIPT.

## Test Standardization

### Selection and Training of Examiners

Solicitation of examiners for the SIPT was on a nationwide basis. Announcements of the standardization project were placed in appropriate journals, newsletters, and other professional publications. A large number of applications were received, and each applicant was rated on a set of defined criteria, including past training and experience, previous exposure to sensory integration theory and assessment, and ability to gain access into the selected communities. The necessary number of potential examiners were selected for each region along with several alternates. Potential examiners and alternates attended 5-day workshops aimed at teaching the administrative and scoring procedures of the SIPT. All examiners were tested following training; any individual not meeting defined accuracy criteria was dropped as an examiner. These procedures resulted in a final sample of 100 normative examiners.

### Sampling Procedure

Once the characteristics of the ideal sample were determined, a three-stage sampling procedure was used: (a) selection of the communities, (b) selection of the schools, and (c) selection of the children.

To insure a nationally representative sample, a modified random sampling procedure was used, stratified on the (previously mentioned) variables to reflect the population distribution characteristics contained in the 1980 U.S. Census Bureau data. The total number of children tested in each region was based on the number of children between the ages of 4 to 14 living in the region, as provided by the census data. The percentage of urban versus rural children and the number of children belonging to each ethnic group for each region were based on total population figures provided by the census.

**Selection of communities.** All communities in the United States were divided into two types—urban (populations greater than 2,500) and rural (populations less than 2,500)—based on a newly published atlas. Urban cities were randomly selected for each region based on the population total for the region, and a list of urban cities was compiled for each region in which potential examiners could test. Sampling for the rural communities proved impossible for two reasons: (a) the large number of rural areas made it difficult to accurately sample from them, and (b) it was unlikely that an examiner would live near any of the rural areas selected. Each individual accepted as an examiner was designated as either an urban or rural examiner. Rural examiners were allowed to select the rural community nearest them. Urban examiners were to choose a community from the list of communities selected for their region. In a few instances urban examiners were unable to obtain permission to test in any schools in urban communities listed for their region. These examiners were allowed to replace the listed community with a nearby community having highly similar demographic characteristics.

**Selection of schools.** Once the communities were selected, each examiner was responsible for obtaining permission to test in the schools. Examiners were to meet with school officials to decide which schools were most representative of the entire community. If the community had a large number of elementary schools, examiners were encouraged to test in a minimum of three schools in order to obtain a good representation of the community. In the rural areas, or in communities that had only one or two schools, examiners were to test in all elementary schools.

**Selection of children.** Examiners were required to submit lists of approximately 150 children attending each school that agreed to be part of the standardization sample. If the school was small so that 150 children could not be obtained, the examiner was to test in other schools from the same community. Some rural examiners tested in several different rural communities in order to secure the required number of students on the list. Each list contained the following information: (a) a number for each child, (b) ethnic background, (c) sex, (d) age in years and months, and (e) indication if the child was receiving any special services. Student names were not to be included on the list.

Children on each examiner's list were categorized into age × sex × ethnic groupings. Each grouping was then randomly sampled to determine which children the examiner would test. Alternates were also provided for each age × sex × ethnic group; for each child selected, at least two alternates were identified. In the great majority of cases, the 150-child sampling proved sufficient.

Once children were selected, the examiner was responsible for obtaining parental permission. If the parents or child declined to participate, the examiner solicited the first alternate. Most parents and children approached agreed to be part of the study. In addition to the parental permission form, a background information sheet was also completed by the parents. The information sheet was used to collect various demographic data to insure that the standardization sample included children from a broad range of socioeconomic levels.

### Characteristics of the Standardization Sample

The children in the SIPT normative sample were chosen to be as representative as possible of the population of children ages 4-0 (4 years, 0 months) to 8-11 (8 years, 11 months) living in the United States. In order to insure a national representation, the following variables were considered as part of the sampling plan:

1. **Age Group.** The standardization sample contained children belonging to 12 age groups between the ages 4-0 to 8-11. Ages 4-0 to 5-11 were divided into 4-month intervals, while 6-month intervals were used for ages 6-0 to 8-11.
2. **Sex.** The sample consisted of approximately equal numbers of boys and girls at each age interval.
3. **Ethnicity.** Five categories were used: (a) Asian, (b) white, (c) black, (d) Hispanic, and (e) other ethnic groups. The number of children sampled from each ethnic group was based upon the 1980 U.S. Census data.
4. **Type of Community.** Communities were divided into two major groupings: those with populations less than 2,500 (rural) and those with populations greater than 2,500 (urban).
5. **Geographic Location.** The nine divisions specified by the U.S. Census were used—New England, Mid-Atlantic, East North Central, West North Central, South Atlantic, East South Central, West South Central, Mountain, and Pacific—in addition to a subsample from Canada.

A total of 1,997 children were included in the normative sample. The majority of testing was accomplished during the 1984–1985 school year (September through June), with some additional testing occurring during the summer of 1985. Children were drawn from regular classrooms in public and private preschools and elementary schools. Children identified as having motor impairment (e.g., cerebral palsy, apraxia) or severe visual handicap were not included in the standardization sample. However, children who were receiving other special services (e.g., remedial reading, speech therapy) were retained in the sample.

Table 4 shows the distribution of children from each census region. Approximately 4% of the U.S. sample came from the New England region, 12% from the Mid-Atlantic region, 16% from the South Atlantic region, 21% from the East North Central region, 9% from the East South Central region, 9% from the West North Central region, 11% from the West South Central region, 8% from the Mountain region, and approximately 10% from the Pacific region. Overall, approximately 73% of the children came from urban areas and 27% from rural areas.

Because sensory integration principles are employed throughout Canada, it was anticipated that the SIPT would be used there as a major assessment tool. Therefore, a number of Canadian children were also included in the standardization sample. Examiners chosen from Canada had to meet the same eligibility requirements and attended the same workshops as the examiners from the United States. Canadian children were randomly

**Table 4**
**Normative Sample by Age, Sex, and Geographic Region**

| | Geographic Region[a] | | | | | | | | | | | | | | | | | | | | |
|---|---|---|---|---|---|---|---|---|---|---|---|---|---|---|---|---|---|---|---|---|---|
| | New Eng | | Mid Atl | | South Atl | | East No. Cent | | East So. Cent | | West No. Cent | | West So. Cent | | Mtn | | Pac | | Can | | |
| **Age Group (yrs/mos)** | **M** | **F** | **M** | **F** | **M** | **F** | **M** | **F** | **M** | **F** | **M** | **F** | **M** | **F** | **M** | **F** | **M** | **F** | **M** | **F** | **Total** |
| 4/0–4/3 | 3 | 2 | 3 | 2 | 2 | 3 | 7 | 5 | 1 | 1 | 2 | 3 | 1 | 1 | 1 | 2 | 4 | 0 | 0 | 1 | 44 |
| 4/4–4/7 | 7 | 3 | 3 | 5 | 5 | 6 | 10 | 9 | 6 | 3 | 6 | 5 | 4 | 6 | 2 | 3 | 3 | 7 | 3 | 3 | 99 |
| 4/8–4/11 | 3 | 5 | 7 | 7 | 7 | 4 | 9 | 7 | 4 | 2 | 4 | 8 | 5 | 4 | 1 | 1 | 9 | 9 | 4 | 4 | 104 |
| 5/0–5/3 | 4 | 6 | 8 | 5 | 6 | 3 | 13 | 17 | 3 | 5 | 2 | 6 | 8 | 7 | 2 | 1 | 6 | 3 | 3 | 0 | 108 |
| 5/4–5/7 | 2 | 6 | 10 | 6 | 16 | 9 | 14 | 17 | 9 | 3 | 6 | 3 | 11 | 12 | 4 | 4 | 7 | 6 | 9 | 3 | 157 |
| 5/8–5/11 | 3 | 3 | 10 | 13 | 16 | 17 | 14 | 20 | 9 | 10 | 5 | 6 | 10 | 9 | 8 | 6 | 7 | 7 | 7 | 4 | 184 |
| 6/0–6/5 | 4 | 3 | 14 | 15 | 21 | 23 | 28 | 24 | 11 | 10 | 8 | 10 | 10 | 14 | 13 | 14 | 12 | 11 | 6 | 8 | 259 |
| 6/6–6/11 | 2 | 2 | 9 | 11 | 22 | 21 | 19 | 18 | 10 | 9 | 6 | 12 | 10 | 11 | 9 | 7 | 10 | 12 | 9 | 7 | 216 |
| 7/0–7/5 | 2 | 2 | 14 | 12 | 17 | 14 | 18 | 19 | 8 | 5 | 11 | 14 | 12 | 12 | 6 | 9 | 10 | 8 | 10 | 9 | 212 |
| 7/6–7/11 | 1 | 4 | 13 | 13 | 15 | 16 | 20 | 25 | 6 | 6 | 8 | 8 | 10 | 12 | 8 | 12 | 12 | 11 | 7 | 9 | 216 |
| 8/0–8/5 | 3 | 2 | 12 | 11 | 14 | 15 | 14 | 20 | 9 | 9 | 10 | 9 | 13 | 6 | 9 | 7 | 6 | 8 | 7 | 7 | 191 |
| 8/6–8/11 | 4 | 5 | 11 | 15 | 22 | 17 | 19 | 25 | 11 | 6 | 5 | 7 | 12 | 8 | 10 | 4 | 6 | 7 | 7 | 6 | 207 |
| **Subtotal** | 38 | 43 | 114 | 115 | 163 | 148 | 185 | 206 | 87 | 69 | 73 | 91 | 106 | 102 | 73 | 70 | 92 | 89 | 72 | 61 | |
| **Total** | 81 | | 229 | | 311 | | 391 | | 156 | | 164 | | 208 | | 143 | | 181 | | 133 | | 1,997 |

*Note.* Within each geographic region, M = males, F = females.

[a]Geographic Region Code:
New Eng = New England
Mid Atl = Mid Atlantic
South Atl = South Atlantic
East No. Cent = East North Central
East So. Cent = East South Central
West No. Cent = West North Central
West So. Cent = West South Central
Mtn = Mountain
Pac = Pacific
Can = Canada

sampled by age and sex, and urban and rural communities were chosen in the same manner as those in the United States.

Table 5 shows the ethnic distribution of children in the standardization sample. Of the 1,961 children for whom ethnic identification was provided, approximately 78% were white, 12% were black, 6% were of Hispanic origin, 2% were Asian, and 2% belonged to other ethnic groups. Table 6 shows the distribution of children by type of school attended.

### Analysis of the Normative Data

The normative data analyses for the SIPT were conducted in three stages: (a) *preliminary analyses,* which examined age- and gender-related differences in SIPT performance, and determined appropriate scoring and stopping rules for the SIPT tests; (b) *computation of means and standard deviations,* which included the examination of developmental trends on each of the 17 tests; and (c) *determination of major SIPT scores,* based upon the extent to which each of the various subscores was able to discriminate between normal and dysfunctional children.

**Preliminary analyses.** Multivariate analysis of variance (MANOVA) was used to examine age- and gender-related differences in SIPT performance on the SIPT tests. This analysis revealed significant sex differences on all except Manual Form Perception (MFP) and Postrotary Nystagmus (PRN). There also were significant age effects on all tests except Postrotary Nystagmus.

Item data were examined to determine appropriate stopping rules for the tests. These stopping rules were determined on the basis of two related criteria: (a) the difference between the score for the total test and the

**Table 5**
**Normative Sample by Age, Sex, and Ethnicity**

| Age Group (yrs/mos) | Ethnic Identification | | | | | | | | | | Total |
|---|---|---|---|---|---|---|---|---|---|---|---|
| | Asian | | Black | | Hispanic | | White | | Other | | |
| | M | F | M | F | M | F | M | F | M | F | |
| 4/0–4/3 | 0 | 1 | 3 | 2 | 0 | 0 | 19 | 17 | 2 | 0 | 44 |
| 4/4–4/7 | 0 | 0 | 1 | 4 | 2 | 4 | 46 | 39 | 0 | 2 | 98 |
| 4/8–4/11 | 3 | 2 | 5 | 6 | 3 | 0 | 41 | 38 | 0 | 2 | 100 |
| 5/0–5/3 | 0 | 3 | 5 | 4 | 3 | 1 | 45 | 43 | 0 | 1 | 105 |
| 5/4–5/7 | 3 | 2 | 11 | 8 | 5 | 1 | 67 | 56 | 1 | 0 | 154 |
| 5/8–5/11 | 1 | 2 | 10 | 10 | 4 | 4 | 71 | 76 | 3 | 3 | 184 |
| 6/0–6/5 | 1 | 5 | 14 | 17 | 9 | 14 | 98 | 93 | 2 | 0 | 253 |
| 6/6–6/11 | 3 | 4 | 17 | 13 | 6 | 4 | 79 | 84 | 0 | 2 | 212 |
| 7/0–7/5 | 2 | 1 | 12 | 12 | 9 | 6 | 80 | 82 | 3 | 1 | 208 |
| 7/6–7/11 | 0 | 3 | 10 | 14 | 10 | 8 | 77 | 88 | 0 | 1 | 211 |
| 8/0–8/5 | 3 | 0 | 12 | 15 | 8 | 7 | 69 | 70 | 3 | 1 | 188 |
| 8/6–8/11 | 2 | 1 | 14 | 15 | 3 | 2 | 85 | 76 | 1 | 5 | 204 |
| **Subtotal** | 18 | 24 | 114 | 120 | 62 | 51 | 777 | 762 | 15 | 18 | |
| **Total** | 42 | | 234 | | 113 | | 1,539 | | 33 | | 1,961 |

*Note.* Within each ethnic group, M = males, F = females.

**Table 6**
**Normative Sample by Age and by Type of School**

| Age Group (yrs/mos) | Type of School | | | | | Total |
|---|---|---|---|---|---|---|
| | Private Preschool | Public Preschool | Private Elementary | Public Elementary | Other | |
| 4/0–4/3 | 37 | 1 | 1 | 1 | 4 | 44 |
| 4/4–4/7 | 76 | 5 | 8 | 9 | 1 | 99 |
| 4/8–4/11 | 78 | 5 | 14 | 5 | 2 | 104 |
| 5/0–5/3 | 64 | 2 | 20 | 14 | 8 | 108 |
| 5/4–5/7 | 33 | 2 | 25 | 89 | 8 | 157 |
| 5/8–5/11 | 16 | 1 | 40 | 125 | 2 | 184 |
| 6/0–6/5 | 15 | 1 | 59 | 183 | 1 | 259 |
| 6/6–6/11 | 2 | 1 | 46 | 164 | 3 | 216 |
| 7/0–7/5 | 0 | 0 | 37 | 172 | 3 | 212 |
| 7/6–7/11 | 0 | 0 | 43 | 169 | 4 | 216 |
| 8/0–8/5 | 0 | 0 | 33 | 152 | 6 | 191 |
| 8/6–8/11 | 0 | 0 | 59 | 145 | 3 | 207 |
| **Total** | 321 | 18 | 385 | 1,228 | 45 | 1,997 |

score that would have been obtained using the stopping rule should be negligible; and (b) the predictive validity of the score that would have been obtained using the stopping rule should not be significantly lower than the predictive validity of the total (unstopped) score. Pearson product-moment correlations and multiple discriminant analyses were used to compare the predictive validity of test scores with and without stopping rules. On the basis of these criteria, stopping rules were included in 8 of the 17 tests: Space Visualization (SV), Figure-Ground Perception (FG), Standing and Walking Balance (SWB), Design Copying (DC), Bilateral Motor Coordination (BMC), Sequencing Praxis (SPr), Manual Form Perception (MFP), and Graphesthesia (GRA).

**Computation of group means and standard deviations.** Because the preliminary analyses indicated significant age and gender differences in SIPT scores, separate norms were developed for boys and girls in each of the 12 age groups. The age differences discussed earlier indicated that the developmental curve for each of the tests except Postrotary Nystagmus should be monotonically increasing (i.e., mean test scores should increase as age increases). To ensure monotonicity across age groups, and to reduce the effects of sampling error, polynomial regression was used to obtain smoothed means and standard deviations separately for boys and girls in each age group. Each child's *SD* scores on the SIPT reflect the degree to which the child's scores deviate from the smoothed means for children of the same age group and gender, expressed in standard deviation units.

**Determination of major SIPT scores.** Most of the individual tests yield several different subscores, often including subscores for time and accuracy of performance. For both theoretical and pragmatic reasons, it was expected that different diagnostic groups might exhibit different speed-accuracy trade-offs in performance on some of the tests. Thus, time-adjusted accuracy scores are important on a number of these tests. To determine the optimal statistical weights for time and accuracy, a sample of 176 children with sensory integrative dysfunction or other learning disabilities was matched (on age, gender, geographic region, parental occupation, and parental education) with a group of 176 normal children. Using this combined sample of 352 children, multiple discriminant analyses were used to determine the appropriate weights for test time and accuracy. A number of different monotonic transformations of accuracy and time measures were tried. The final weights were determined by selecting those weights that best discriminated between normal and dysfunctional children within and across the 12 age groups.

**Developmental Trends**

The SIPT are designed to be administered to boys and girls within an age range where some aspects of development can be rapid and pronounced. It is, therefore, important to examine the major developmental trends in the different measures. As noted earlier in the chapter, because there are significant sex differences on many of the variables, it is also important to examine the developmental trends separately for boys and girls.

In this section, the major trends in the SIPT variables are shown graphically. A series of graphs (Figures 55–71) shows both the smoothed age trends (used in the norms tables for the *SD* score calculations) and the raw mean scores from which the age trends were calculated. In order to put all of the figures into the same metric (so that each graph is directly comparable to one another in scale to show differences in maturation over time across variables), raw and smoothed mean scores were calculated for boys and girls on a variable at all ages. These mean scores were then converted to average *SD* scores using the norms for boys from 6 to 6½ years of age, so that all scores should be interpreted as results relative to those for boys in the middle of the age distribution in the norms table. Because each variable was standardized in this way, the age trends on one variable, such as Space Visualization (SV), can be compared to the age trends on another variable, such as Design Copying (DC). Also, when the line for the girls is higher than that of the boys, such as is frequently the case, it indicates that relative to boys, the girls score somewhat better, on the average. Again, because the metric has been equalized on the different figures, it is possible to compare the plots to one another.

Another piece of information supplied on the plots is the *goodness of fit* of the actual means to the smoothed curves. Note that in this curve fitting, the regressions were restricted to be monotonic (never-decreasing) ones, so that in a couple of cases, tighter statistical fit could have been obtained at the risk of theoretical validity. An examination of the developmental curves given in the following figures, however, supports well the conclusion that the smoothed curves represent the data very well.

Figure 55 shows the developmental curves on Space Visualization (SV). Note that the developmental curves for boys and girls are about the same, and that there is a steady, and approximately equal developmental trend throughout the span tested without obvious leveling of the curves. This is one of the few tests where the boys tend to do slightly better than the girls.

Figure 56 shows the developmental curves for Figure-Ground Perception (FG). Note that the curves for the boys and girls are almost identical. There is some leveling off in the rate of development at the older ages.

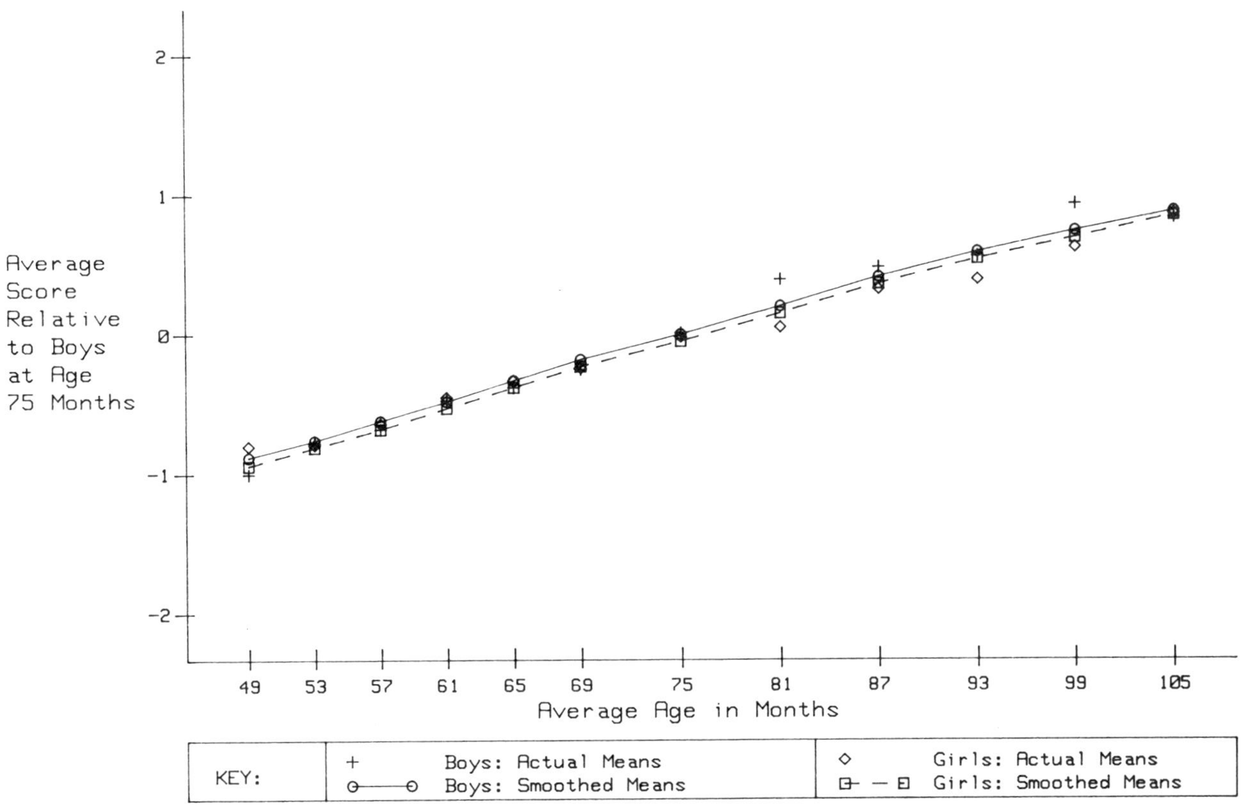


**Figure 55**
**Actual and Smoothed Age Trends for Boys and Girls on SV**

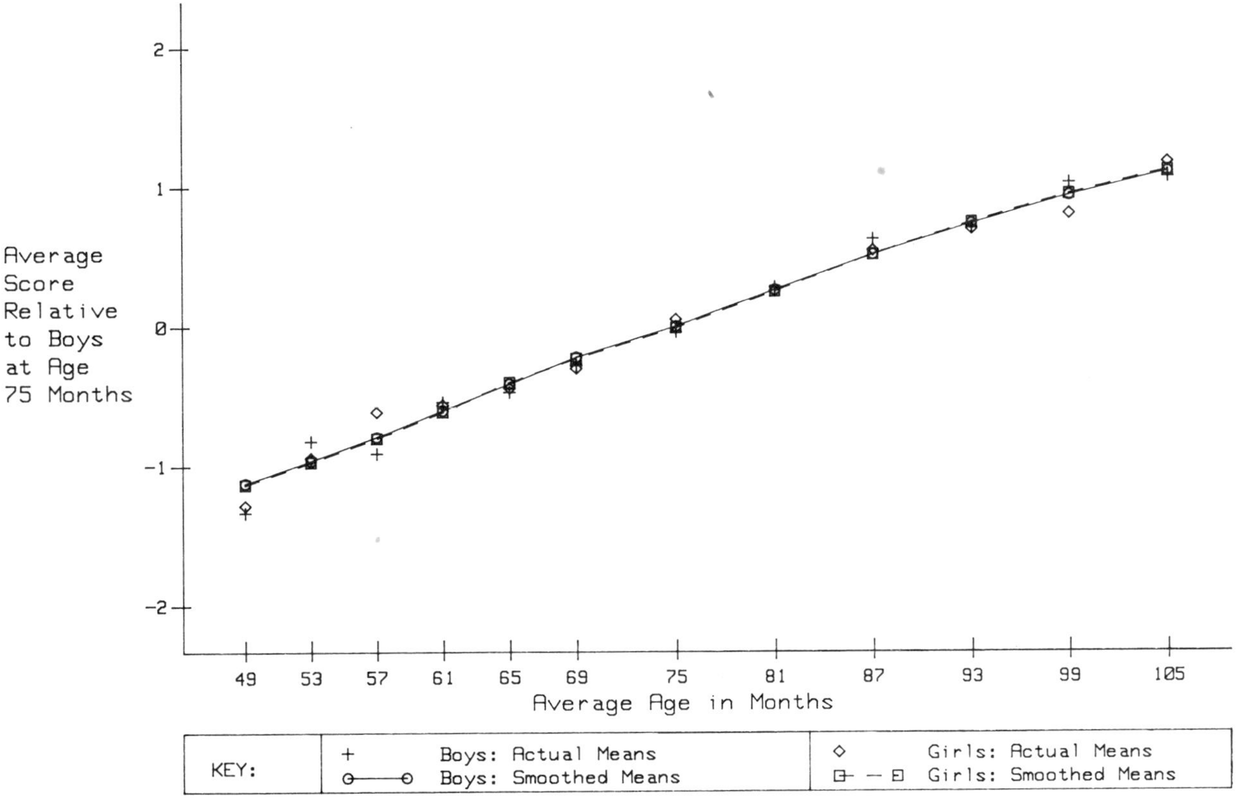


**Figure 56**
**Actual and Smoothed Age Trends for Boys and Girls on FG**

The Manual Form Perception (MFP) developmental trends are shown in Figure 57. Development is fairly pronounced in the earlier and middle ages while decreasing somewhat for the older children. Boys and girls have about the same developmental trends.

Kinesthesia (KIN) developmental trends are shown in Figure 58. Again, the curves for the two sexes are close to identical, and developmental rate starts to level off at the older ages.

Figure 59 shows the developmental curves for Finger Identification (FI). There is a noticeable slowing in the rate of development at the end of the age span for both boys and girls.

The developmental curves for Graphesthesia (GRA) are given in Figure 60. Boy and girl trends are almost identical, with observed sex differences in actual means only occurring at the younger ages. The curves seem to reach an asymptotic level at the upper ages.

Fairly dramatic sex differences in the developmental curves for boys and girls are observed for Localization of Tactile Stimuli (LTS) in Figure 61. Note that the curves for boys and girls are parallel, but fairly distinctly separated. The rate of development on this test is fairly small across the age span in the SIPT normative sample.

Relatively large developmental trends are shown in Figure 62 for Praxis on Verbal Command (PrVC). Girls consistently score better than boys.

Figure 63 shows the developmental trends for boys and girls on Design Copying (DC). Note that the developmental trend on this test is so pronounced compared to the other SIPT that the scale of this figure is different from the other figures showing developmental trends. Boys and girls score at approximately the same level.

Constructional Praxis (CPr) developmental trends are given in Figure 64. Boys score consistently better than girls, and development is pronounced through the age span, with some leveling off at the upper ranges.

As shown in Figure 65, girls are consistently better than boys on Postural Praxis (PPr). Note that the developmental rates for both sexes are pronounced, and that girls evince a leveling off at an age of about a year younger than boys.

The developmental trends for Oral Praxis (OPr) are given in Figure 66. Again, girls score better than boys in a consistent manner. The rate of development of this characteristic throughout the age span sampled is dramatic and fairly rapid.

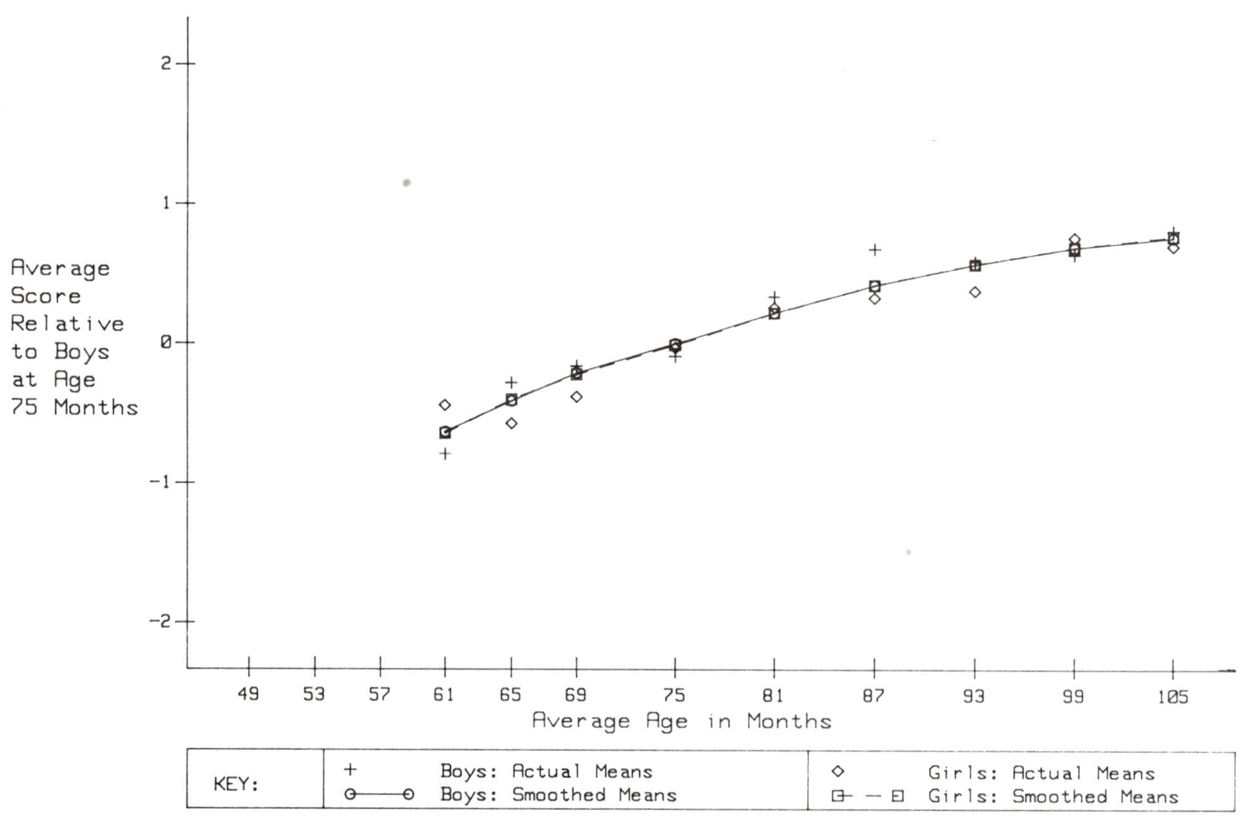


**Figure 57**
**Actual and Smoothed Age Trends for Boys and Girls on MFP**

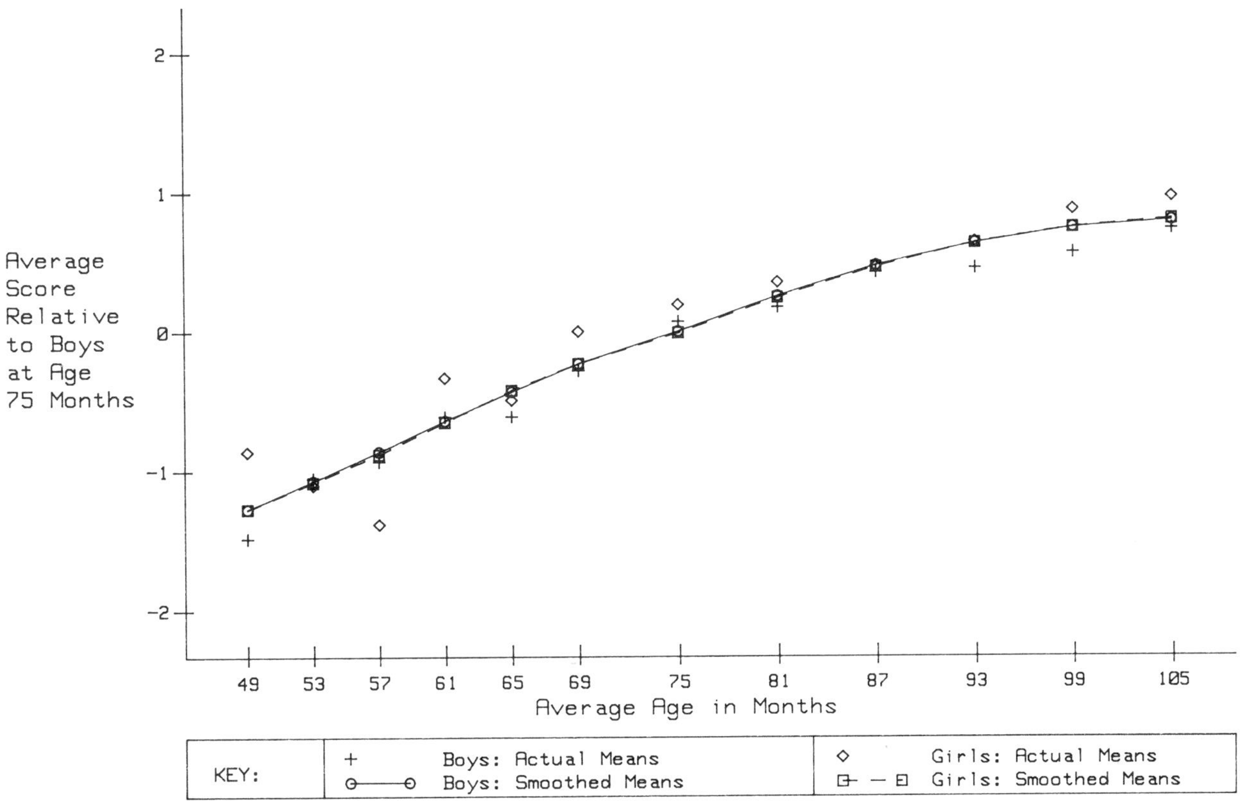


**Figure 58**
**Actual and Smoothed Age Trends for Boys and Girls on KIN**

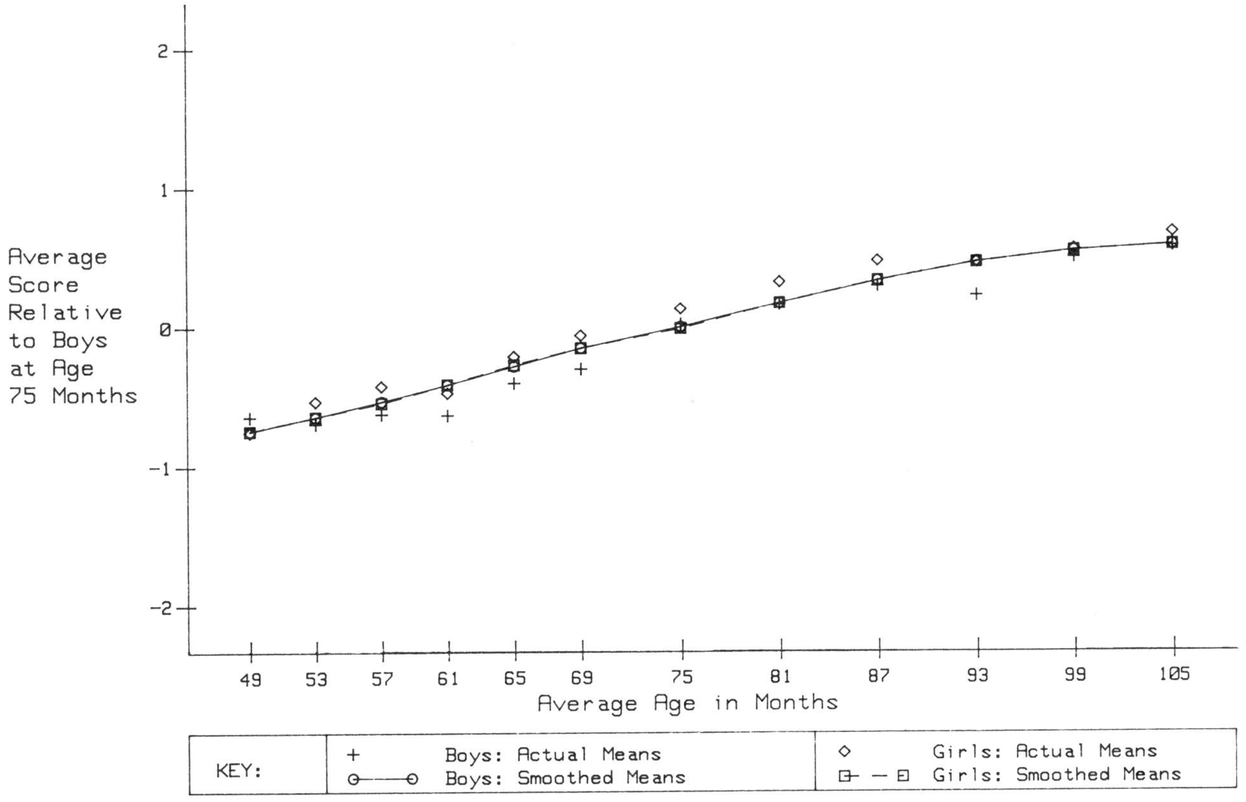


**Figure 59**
**Actual and Smoothed Age Trends for Boys and Girls on FI**

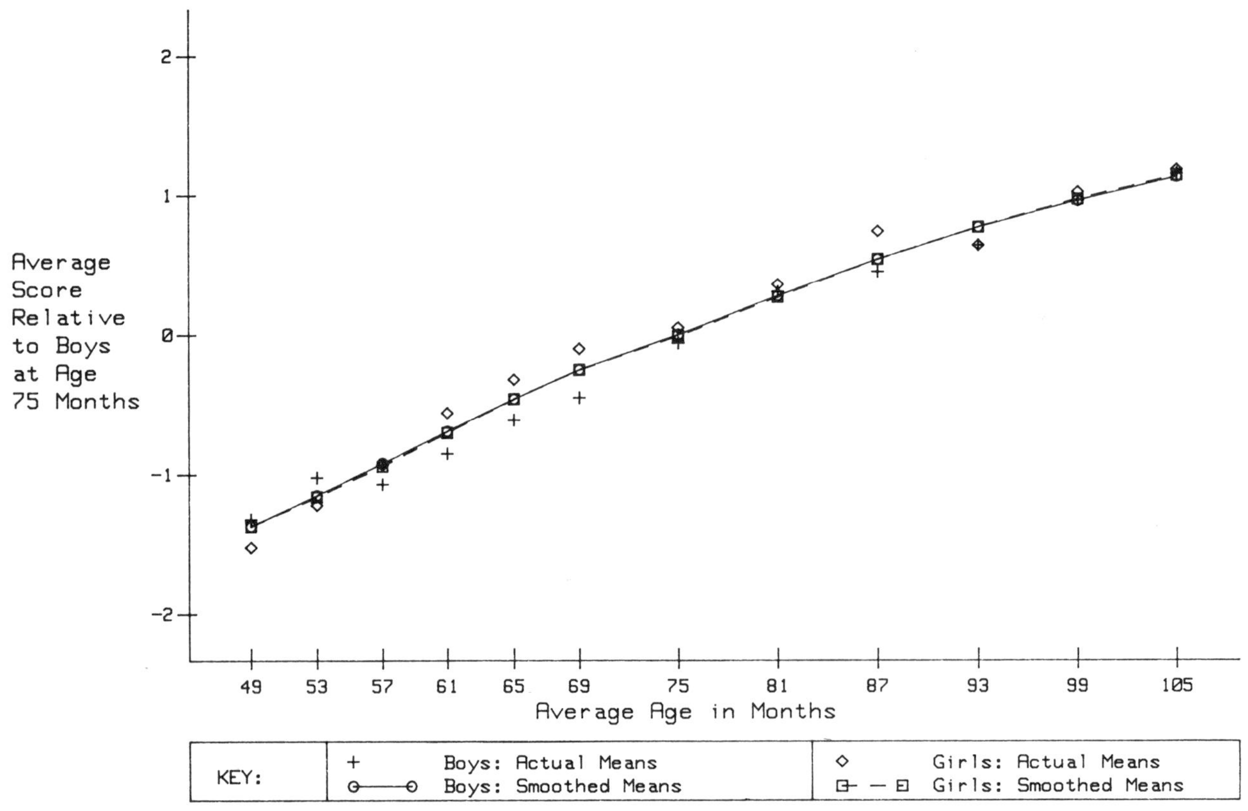


**Figure 60**
**Actual and Smoothed Age Trends for Boys and Girls on GRA**

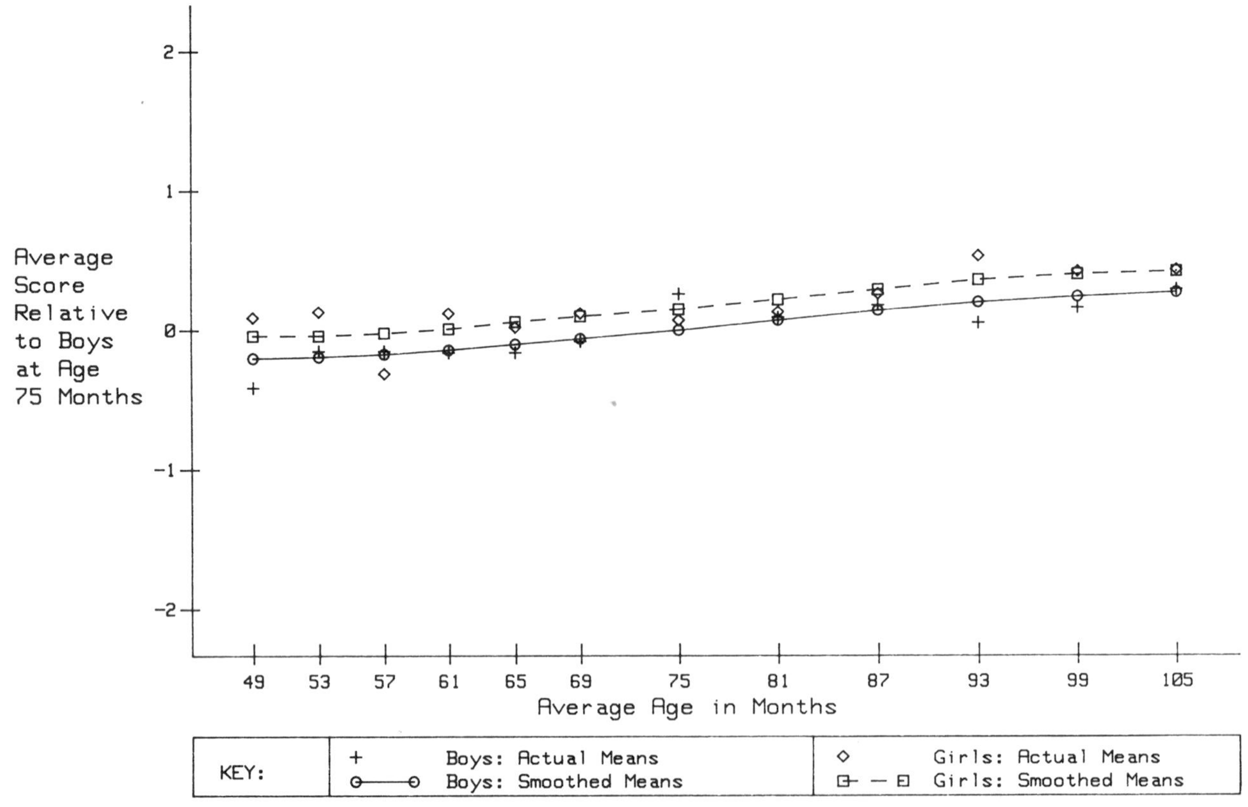


**Figure 61**
**Actual and Smoothed Age Trends for Boys and Girls on LTS**

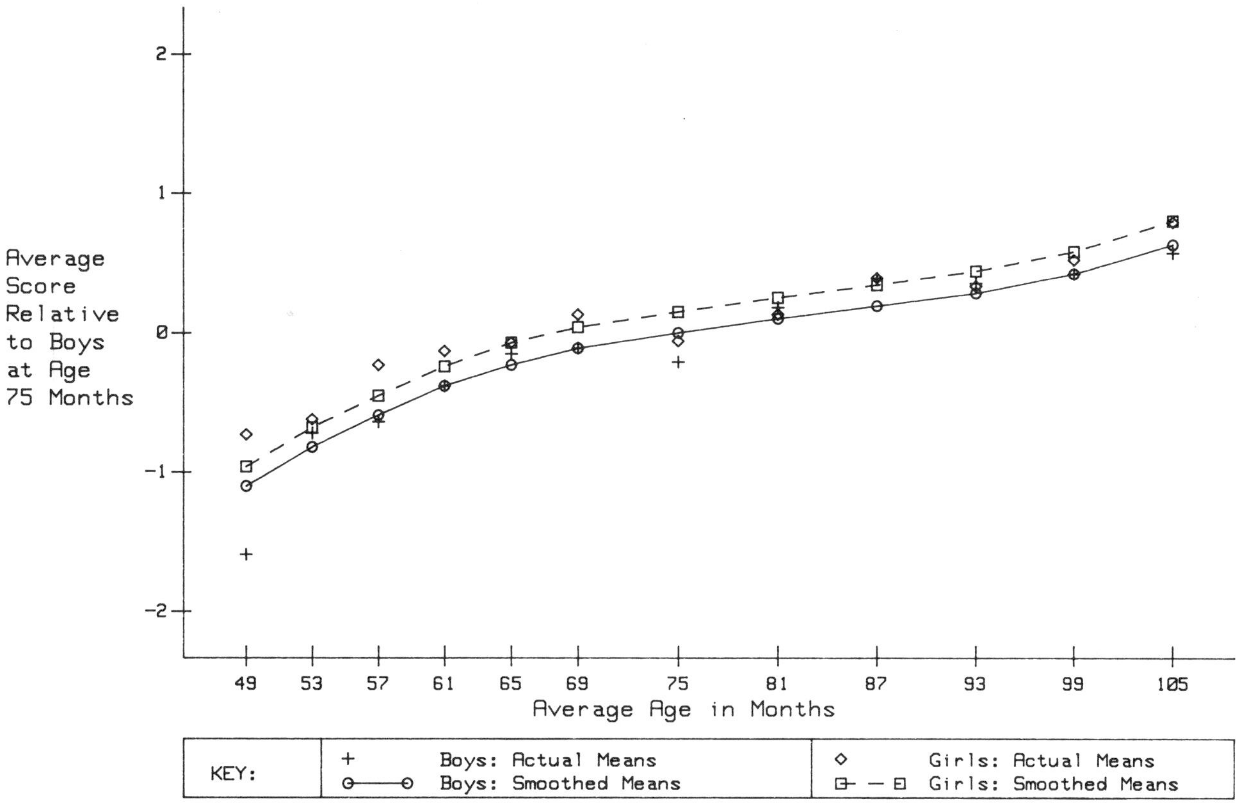


**Figure 62**
**Actual and Smoothed Age Trends for Boys and Girls on PrVC**

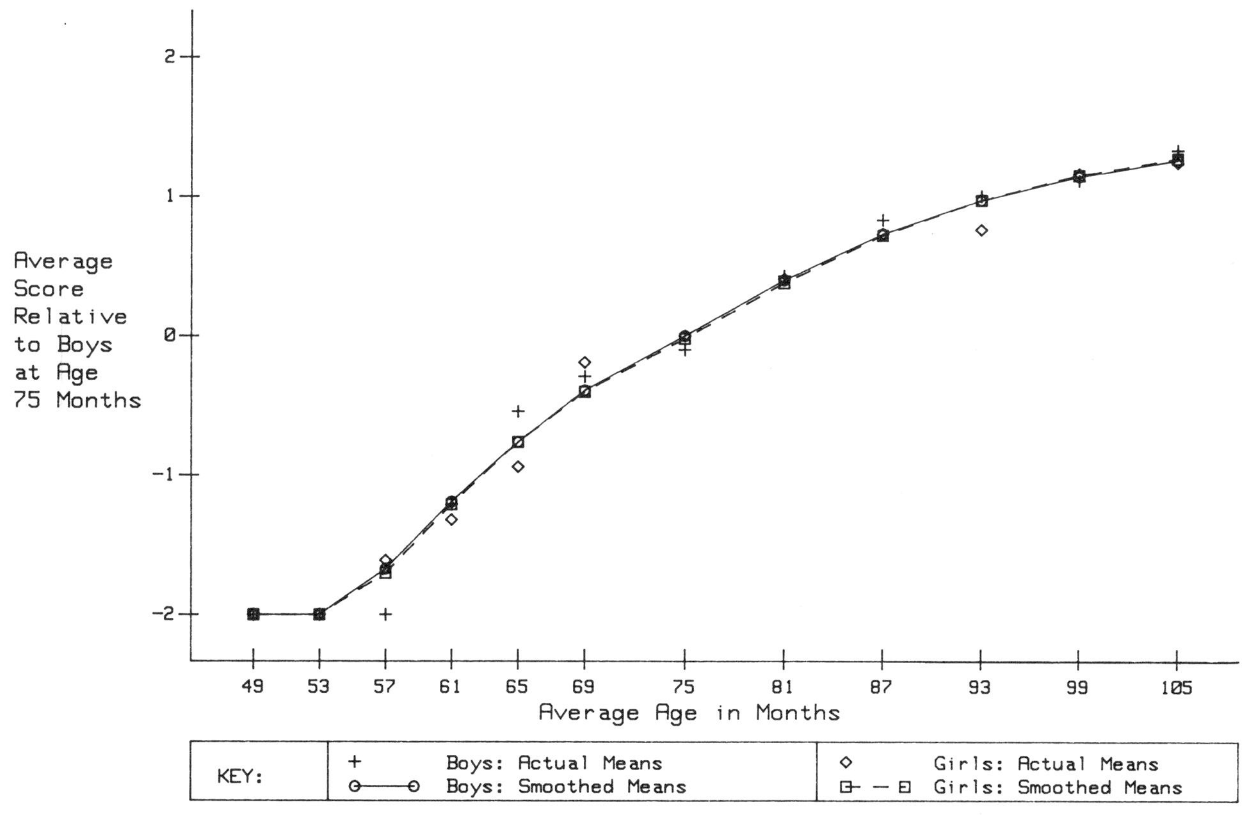


**Figure 63**
**Actual and Smoothed Age Trends for Boys and Girls on DC**

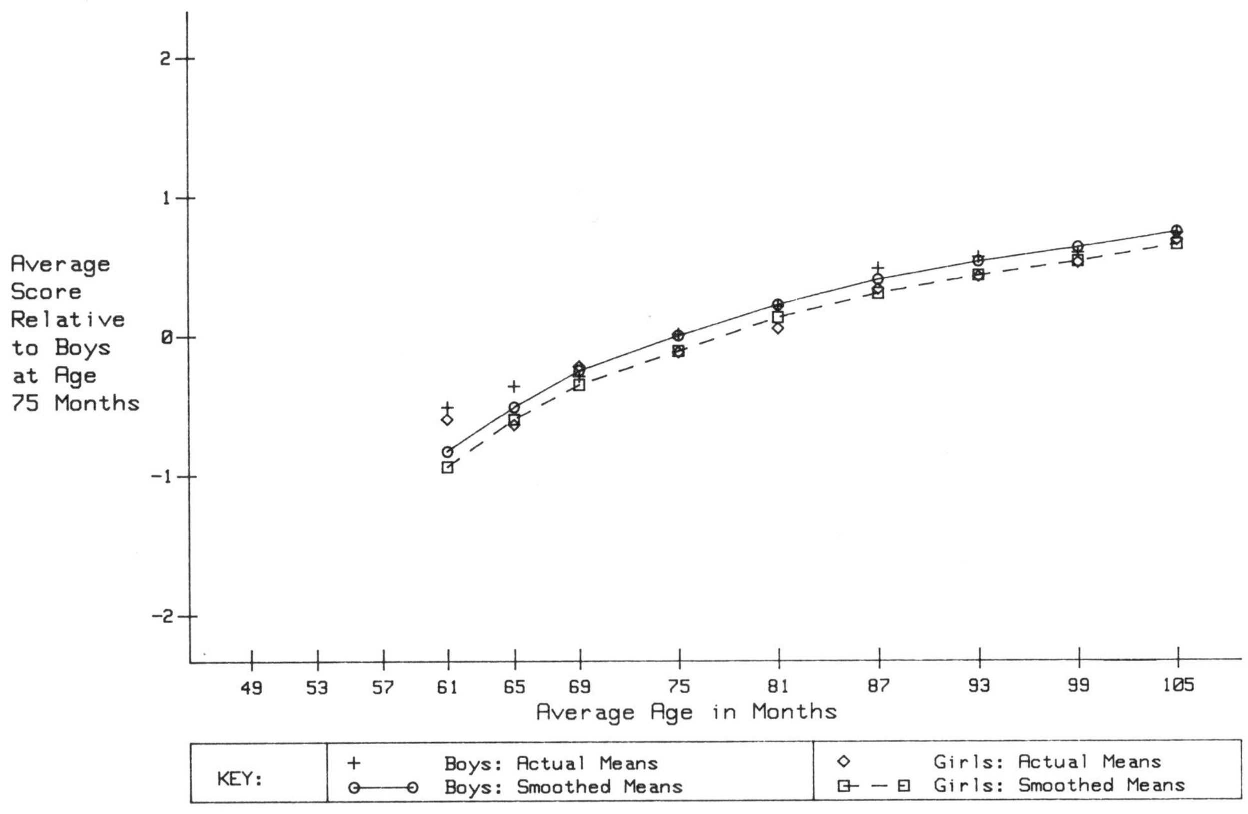


**Figure 64**
**Actual and Smoothed Age Trends for Boys and Girls on CPr**

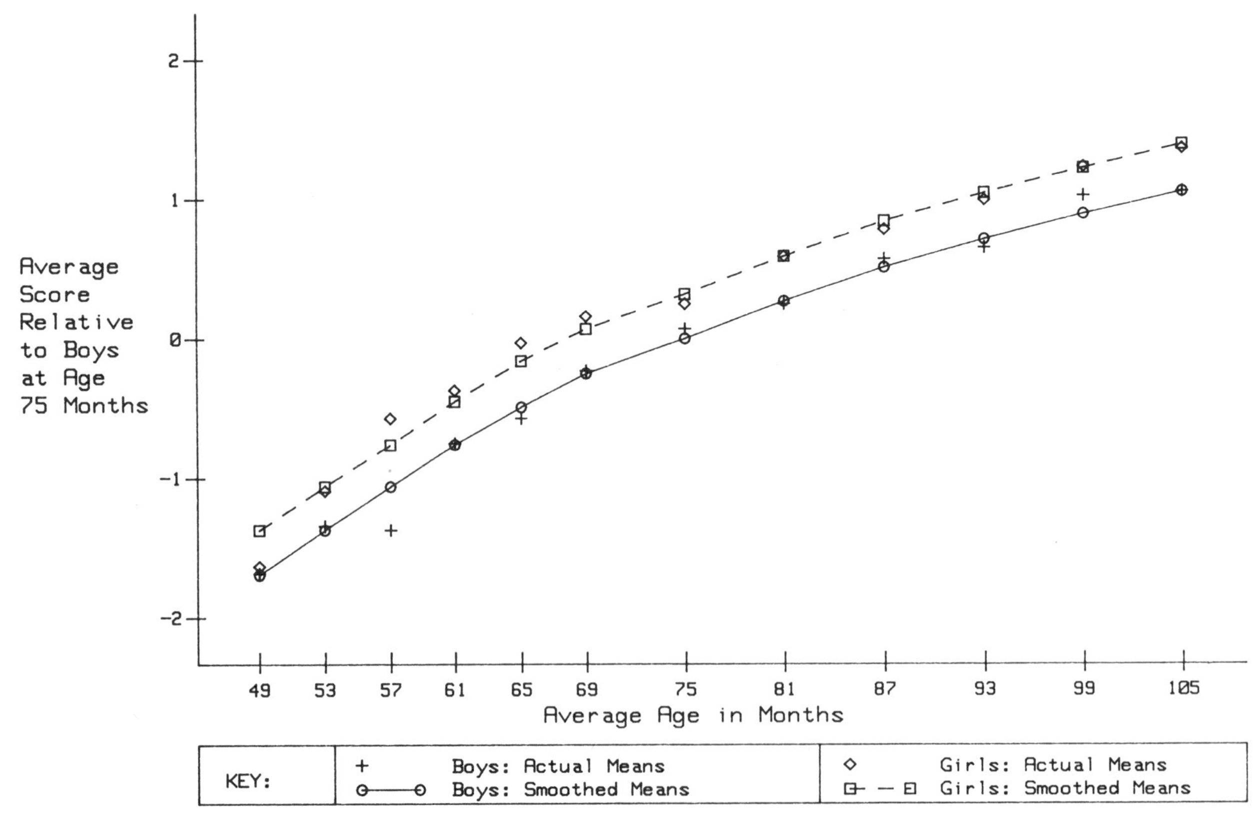


**Figure 65**
**Actual and Smoothed Age Trends for Boys and Girls on PPr**

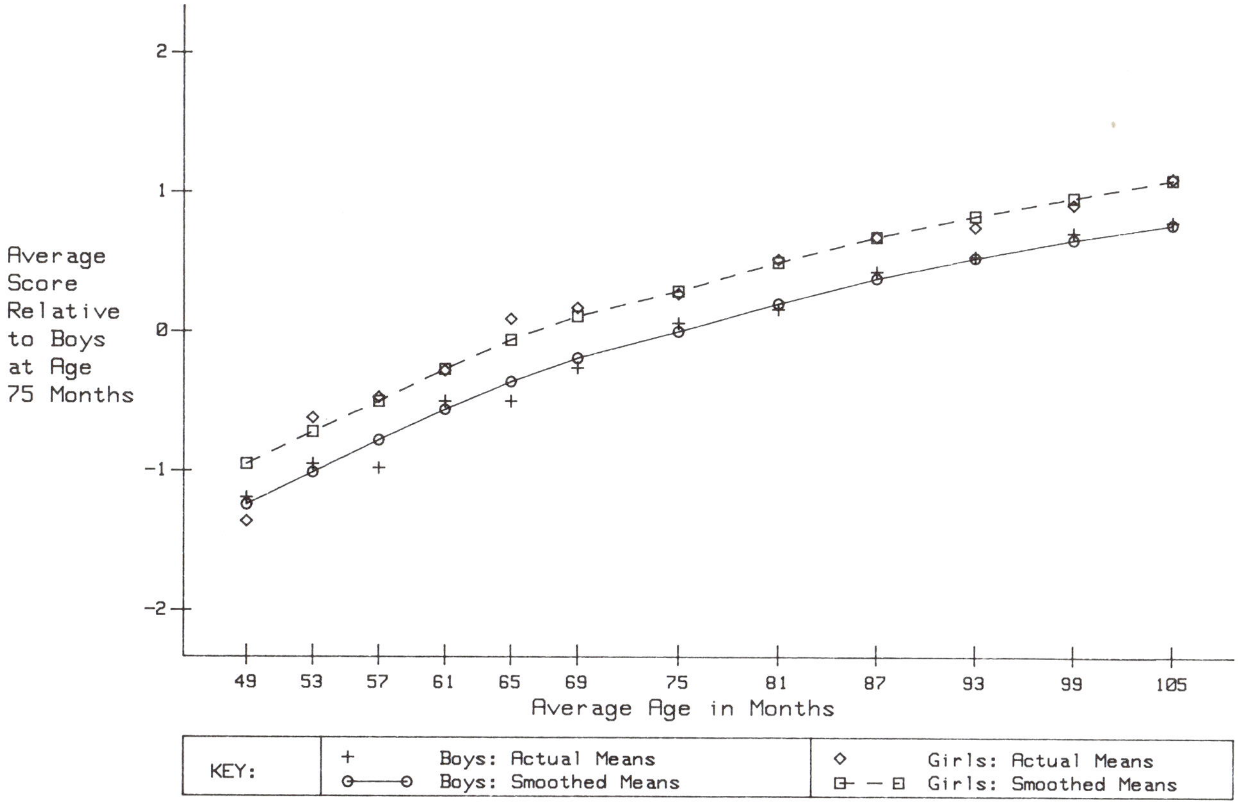


**Figure 66**
**Actual and Smoothed Age Trends for Boys and Girls on OPr**

Developmental curves of Sequencing Praxis (SPr) are given in Figure 67. Although there is pronounced developmental tendencies until the later ages, the curves for boys and girls are very similar.

Bilateral Motor Coordination (BMC) tendencies across the age span are given in Figure 68. Girls do consistently better than boys, although their curve flattens about a year earlier. There is pronounced development throughout the span.

Developmental curves for Standing and Walking Balance (SWB) are shown in Figure 69. Girls tend to do better than boys, and there is dramatic development, leveling for girls around 7½ years and for boys around 8 years.

Figure 70 shows the developmental trends for boys and girls on Motor Accuracy (MAc). Girls consistently do better than boys. There is pronounced development in this tendency.

Postrotary Nystagmus (PRN) did not show a significant developmental trend for either boys or girls. Therefore, all data were pooled for the norms calculations. Figure 71 shows that the small and unsystematic residuals from the baseline level suggest the lack of a systematic developmental curve.

**The Normality of the Distribution of the SIPT Variables**

In order to determine the degree to which the variables fit the normal distribution, the entire group of 1,997 children in the normative sample was used to calculate a cumulative frequency distribution for each variable calculated in the WPS TEST REPORT for the SIPT. These frequency distributions were calculated after converting the raw scores to *SD* scores using the norms for the appropriate age and sex. Also calculated were the mean score, standard deviation, skewness, and kurtosis for each frequency distribution.

The mean or average score in each distribution should theoretically be 0, and the standard deviation should be 1. Note that there will be slight departures from these values due to the fact that the variables are not perfectly continuous, and that some data smoothing was done in the calculation of the norms. Skewness is a measure of the degree to which the distributions are not symmetric. Values of 0 are obtained from perfectly normal distributions. In a sample, of course, there may be some departure from a perfect value of 0. Finally, kurtosis is a measure of whether the distribution is more peaked or flatter than a normal distribution. For a normal distribution, a value of 0 will be obtained, although in a sample there may be some departure from this value.

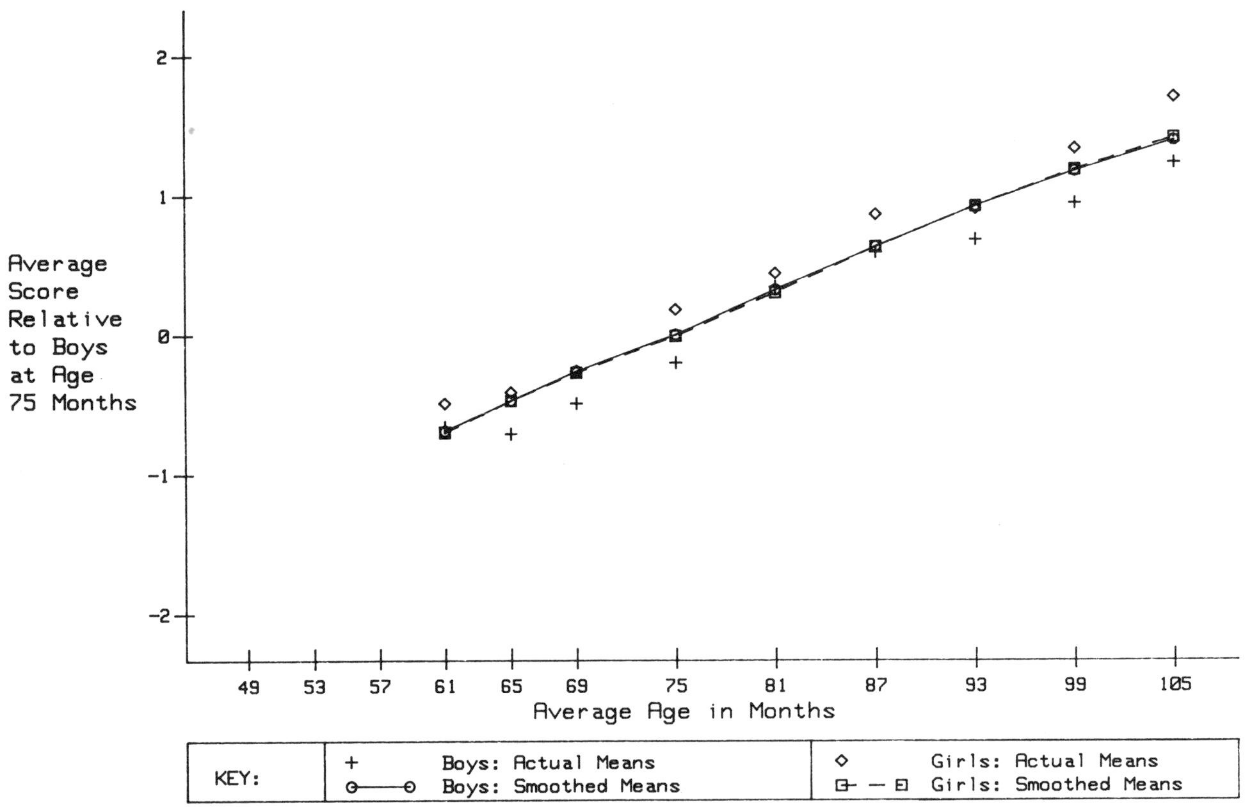


**Figure 67**
**Actual and Smoothed Age Trends for Boys and Girls on SPr**

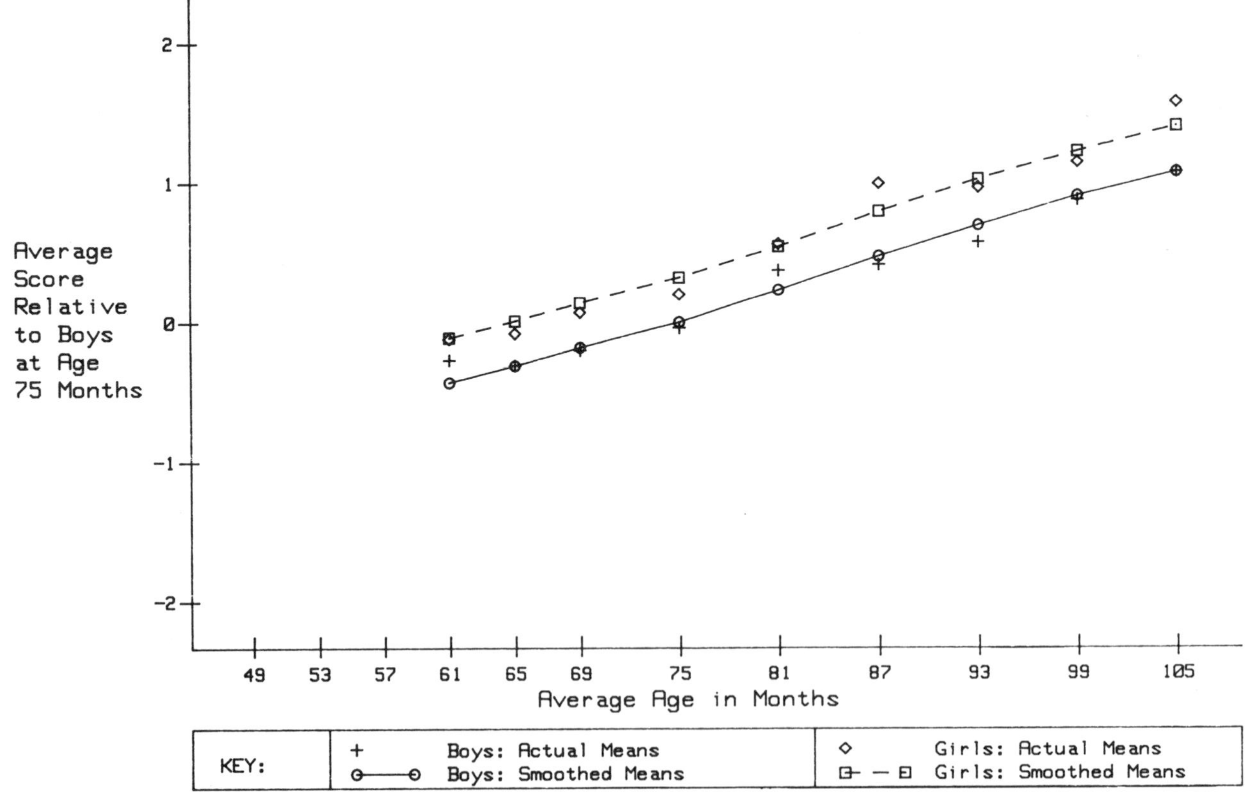


**Figure 68**
**Actual and Smoothed Age Trends for Boys and Girls on BMC**

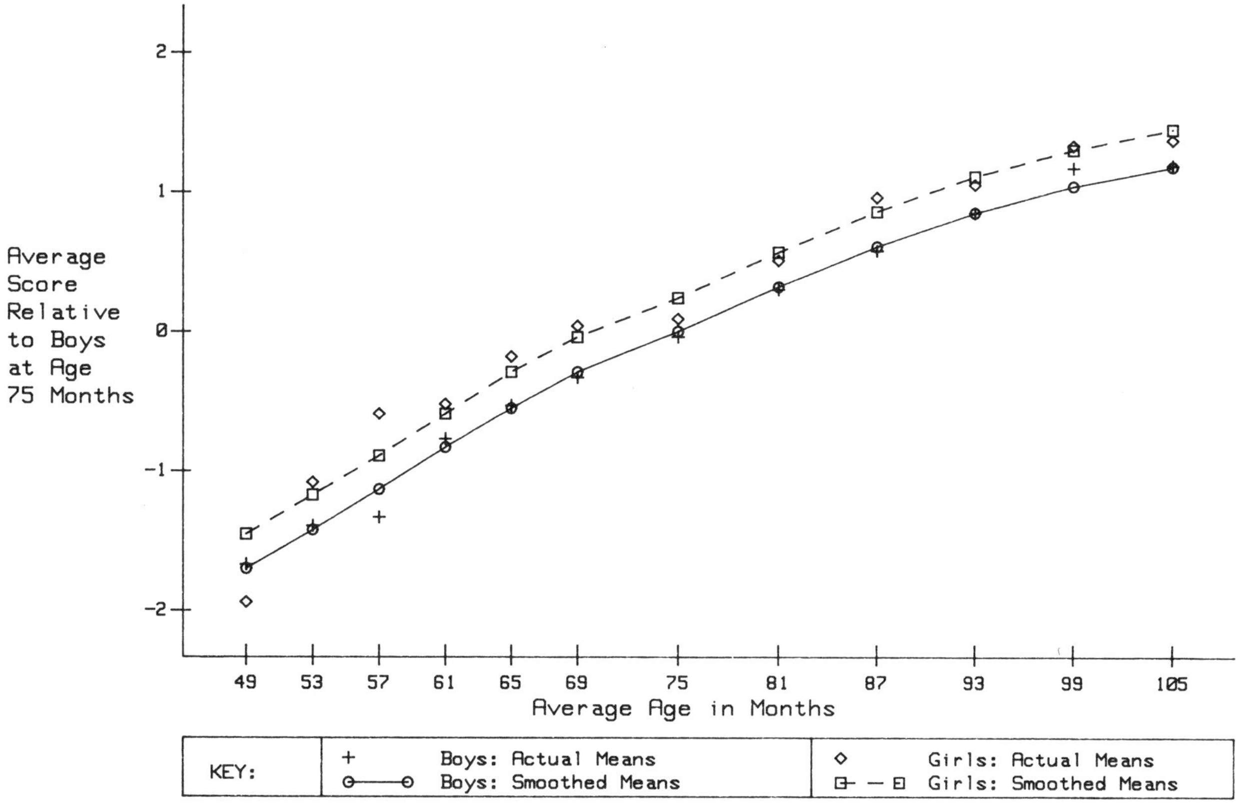


**Figure 69**
**Actual and Smoothed Age Trends for Boys and Girls on SWB**

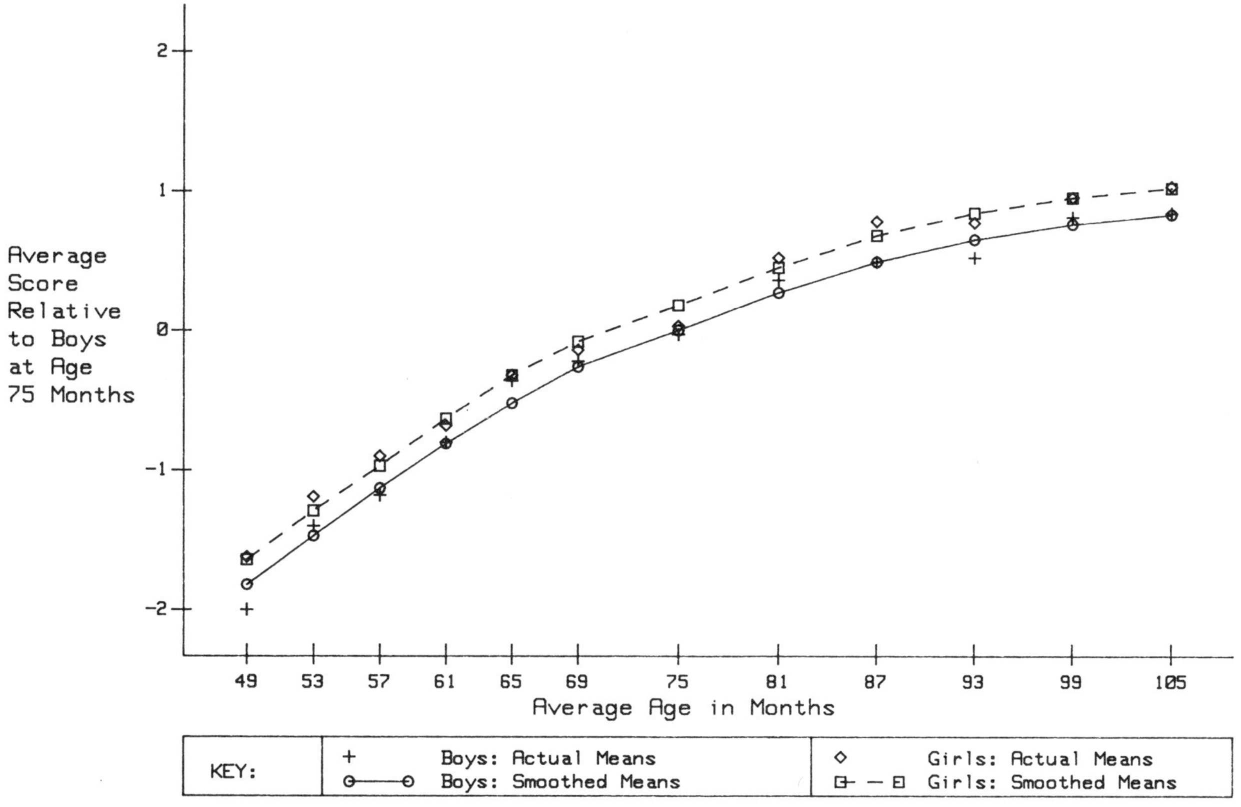


**Figure 70**
**Actual and Smoothed Age Trends for Boys and Girls on MAc**

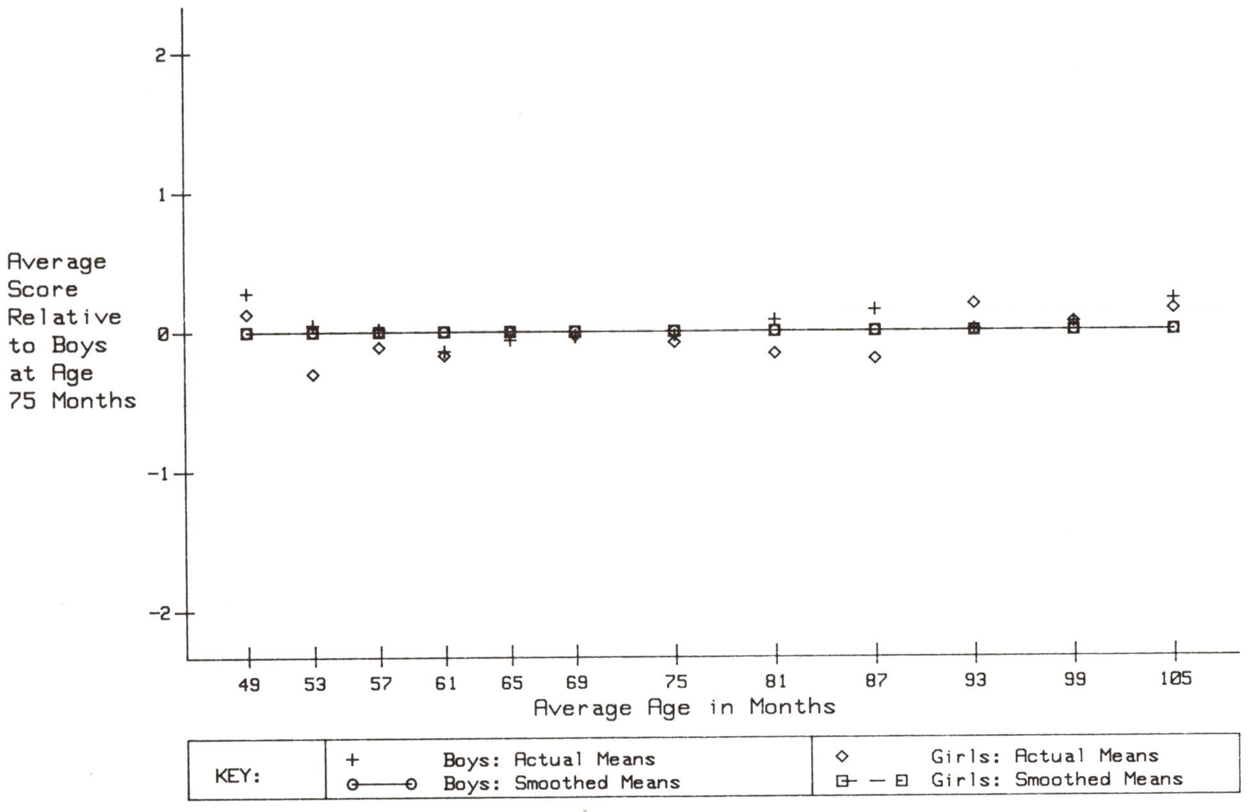


**Figure 71**
**Actual and Smoothed Age Trends for Boys and Girls on PRN**

Table 7 shows the mean, standard deviation, skew, and kurtosis for each SIPT variable. In the table, a notation is also included as to whether a graphic inspection of the distribution, comparing the obtained frequency curve to a plotted, theoretically expected normal curve, led three experienced data analysts to conclude that the variables were reasonably normally distributed and could be characterized as such.

An overall conclusion that can be drawn from Table 7 is that almost all of the *major* SIPT variables—that is, the single score for each test that is seen as the major summary of the performance on that test—are reasonably well normally distributed when assessed in the combined normative sample. That is not to say that the individual distributions do not have some minor statistical departures from normality; this is generally to be expected in clinical instruments with truncated ranges or a limited number of items. The normal distribution is not as good a summary for individual subscores or part scores. This result is not surprising, as many of the subscores are based on a very limited number of items or are untransformed measures. Certain data transformations are made on the major scores so as to help them to be more normally distributed.

The conclusion that the normal distribution fits the major scores relatively well is an important one for the use of the test scores. It has been a longstanding tradition in the use of precursor tests of the SIPT (e.g., the Southern California Sensory Integration Test) to think of the scores as being scaled to fit within a bell-shaped normal distribution. Because it has been shown that this sample is reasonably well distributed along the lines suggested by the normal distribution, it can be assumed that the population is in fact normally distributed. Hence, percentile scores, as shown in the WPS TEST REPORT for SIPT and which are directly related to the normal curve, are appropriate.

**Table 7**
**Normality of the Distribution of SIPT Variables**

| Test | Mean | *SD* | Skew | Kurtosis |
|---|---|---|---|---|
| Space Visualization (SV) | .00 | .99 | -.77 | 1.18 |
| Figure-Ground Perception (FG) | .00 | .97 | .13 | .07 |
| Manual Form Perception (MFP) | .00 | .98 | -.45 | -.03 |
| Kinesthesia (KIN) | .02 | .96 | -.80 | .87 |
| Finger Identification (FI) | .01 | .99 | -.23 | -.43 |
| Graphesthesia (GRA) | .00 | .98 | -.11 | .08 |
| Localization of Tactile Stimuli (LTS) | .00 | .98 | -.39 | -.08 |
| Praxis on Verbal Command (PrVC) | .01 | .94 | -1.22 | 1.23 |
| Design Copying (DC) | .01 | .95 | -.33 | .21 |
| Constructional Praxis (CPr) | .03 | .80 | -.42 | .39 |
| Postural Praxis (PPr) | .00 | 1.00 | -.30 | -.09 |
| Oral Praxis (OPr) | .00 | .98 | -.33 | -.15 |
| Sequencing Praxis (SPr) | -.01 | 1.00 | .05 | -.20 |
| Bilateral Motor Coordination (BMC) | -.01 | .98 | .26 | -.39 |
| Standing and Walking Balance (SWB) | .00 | .98 | .21 | .38 |
| Motor Accuracy (MAc) | .00 | .81 | -.14 | .63 |
| Postrotary Nystagmus (PRN) | .00 | .82 | .01 | .71 |

# CHAPTER 7
## VALIDITY AND RELIABILITY OF THE SIPT

### Validity

Test validity refers to the ability to draw meaningful inferences from test scores to meet an intended purpose. The primary purpose of the SIPT is the assessment of the sensory integrative and practic status of children with known or suspected problems. A secondary purpose of the SIPT is to provide a basis for treatment to ameliorate sensory integrative and practic deficits. Two types of validity are addressed in this section: (a) construct validity (the extent to which the test assesses the relevant theoretical construct); and (b) criterion-related validity (the extent to which performance on the test can be used to predict the child's current or future performance on related tasks). Content validity (i.e., the extent to which the test items provide a representative sampling of performance on important aspects of the construct) was addressed in chapter 6 of this Manual.

### Construct Validity

**The Construct**

As examined by the SIPT, sensory integration is defined as those neural processes that organize sensation from the body and from the environment. Spatiotemporal inputs from touch, kinesthetic, proprioceptive, vestibular, and visual sensory receptors allow the individual to perceive the body in its environment, then to conceptualize, organize, and execute a plan of action. Therefore, the SIPT includes not just measures of organization of inputs from the sensory systems, but also of the practic behaviors associated with them, which allow the individual to engage in adaptive interaction with the physical world.

The SIPT is intended to measure the efficiency of some of the neural substrate of learning and behavior. Identification of sensory integrative processes and the organization of related behavioral parameters into a meaningful theoretical construct have been accomplished through statistical analyses, from the study of clinical and basic brain research, and from several decades of clinical experience.

**The Evidence Prior to SIPT Publication**

Construct-related evidence has evolved through a series of factor analyses of the SCSIT, and of measures that were earlier forms of some of the SIPT tests. The first analysis (Ayres, 1965) was made of scores on 35 behavioral measures. Most of these measures were tests of perceptual and motor functions, but the test battery also included the Marianne Frostig Developmental Test of Visual Perception (Frostig, Lefever, & Whittlesey, 1961), the Ayres Space Test (Ayres, 1962a), and some clinical observations of hand usage. The entire battery was administered to 100 children with and 50 children without suspected perceptual deficits as judged by physicians, teachers, or test reports. The age range of the dysfunctional group was 70–96 months (mean = 84.1, $SD$ = 7.3). The IQ range was 71–139 (mean = 96.97, $SD$ = 14.05). A control group of 50 children was chosen to match the dysfunctional group on sex and on mean, variance, and range of mental age.

A principal components analysis of these measures yielded four factors for the children with suspected perceptual deficits. The first unrotated factor to emerge, with an eigenvalue of 13.9, was labeled developmental apraxia. The tactile and motor planning tests had loadings on this factor that ranged from the .40s to .70s. The second factor to emerge was a fairly strong form and space perception factor, which linked kinesthesia with visual perception. The third factor appeared to reflect the syndrome of tactile defensiveness and hyperactivity with distractibility, and shared variance with the tactile perception tests with loadings from .30 to .51. The fourth factor appeared to reflect a deficit in the functional integration of the two sides of the body, with high factor loadings for scores on right-left discrimination, crossing the body's midline, bilateral motor coordination, and number concepts. In a similar analysis of scores of the nondysfunctional children, the first factor was a visuomotor factor related to number concepts, but not to tactile functions. Neither the apraxia, tactile defensiveness, nor the bilateral integration factors emerged among the comparison group data, thus providing support for considering these score constellations to be indicative of patterns of dysfunction as opposed to types of normal developmental skills.

In another study (Ayres, 1966d), 19 of these measures were administered to 92 children between 4 and 8 years of age. Most of the children in this study were considered normal, but about 10% had been medically determined to have possible or probable central nervous system dysfunction. For this group of 92 children, the two major factors were a somatomotor factor and a visuomotor factor. In contrast, analysis of the scores of a subgroup of 64 children, selected from the 92 on the basis of having shown normal scores on the Gesell Developmental Schedules (Gesell & Amatruda, 1974) before one year of age, failed to find a somatomotor factor. Most of the variance in this subgroup was accounted for by a single visuomotor factor, with emphasis on the visual component. In other words, the distinctive association of somatosensory and motor planning parameters was dependent upon the inclusion of dysfunctional children in the sample.

In a later study, a Q-technique factor analysis (Ayres, 1969b) was used to examine 64 variables including the SCSIT, clinical observations of postural responses and ocular pursuits, measures of preferred-hand usage, language tests, and tests of intelligence and academic achievement. The measures were administered to 36 children (29 boys, 7 girls) in public school classes for children with educational handicaps. The age range was 73–118 months (mean = 97.7, *SD* = 11.9). The mean IQ was 93.5 (*SD* = 11.3). This study identified new markers for the pattern of scores previously referred to as *bilateral integration* and suggested expanding the name of the disorder to *postural and bilateral integration deficit*. Inadequate development of postural and ocular responses and poor bilateral integration were the identifying features of the children with this symptom constellation. These children were also more intelligent and had better language skills than many of the other dysfunctional children in the study. The postural and bilateral integration symptom complex that emerged from this factor analysis also has been noted in clinical evaluation and in treatment, where the test scores of certain learning-disabled children have reflected this pattern. Clinical experience also has indicated that short duration of postrotary nystagmus is often associated with this symptom constellation. The symptom complex was renamed vestibular-bilateral integration (VBI) deficit, thereby placing emphasis on central nervous system processing of vestibular input.

A subsequent factor analysis (Ayres, 1972b) examined scores on tests of sensory integration, psycholinguistic skill, academic achievement, intelligence, and postural and ocular responses obtained from 148 children (110 boys, 38 girls) determined by public schools as learning disabled. The mean age was 92.6 months (*SD* = 12.0); the mean IQ was 96.5 (*SD* = 12.3), and the IQ range was 70 to 132. In this analysis, the standardized scores on tactile, kinesthetic, and visual tests tended to share variance, with loadings of .32 for the SCSIT Kinesthesia test, .36 to .51 for four tactile tests, and .35 to .57 for visual tests. Motor planning, hyperactivity, and tactile defensiveness showed close association with each other, with loadings of .50 and .40 for motor planning, .59 for freedom from hyperactivity, and .50 for freedom from tactile defensiveness. Psycholinguistic tests shared variance with one another and with the intelligence tests, with factor loadings ranging from .34 to .70.

Similar relationships were found in a factor analysis (Ayres, 1977) of scores on the SCSIT, the Southern California Postrotary Nystagmus Test (SCPNT), several psycholinguistic and auditory perception tests, a dichotic listening test, a test of academic achievement, and clinical measures of postural and ocular function. The subjects of the study were 102 boys and 26 girls, 6–10 years of age (mean = 97.6 months, *SD* = 14.3), all diagnosed by schools as learning disabled. The close association between praxis and tactile function was demonstrated by their loadings of .64 and .69 on the same factor. Clinical tests of postural and ocular function shared variance on a single factor with loadings in the .70s. Auditory-language abilities all loaded on a single factor. Postrotary nystagmus carried a negative loading of –.37 on the auditory-language factor. In entering SCPNT scores into the analyses, no attempt was made to differentiate between prolonged or short nystagmus duration, neither of which is normal, yet each of which was overrepresented in the population. The children with abnormally long duration nystagmus tended to score lower on the rest of the measures than did those with abnormally short duration.

In another study, Silberzahn (1975) factor analyzed scores on a number of measures, including the SCSIT, in a sample of 66 boys and 21 girls between 5.0 and 9.3 years of age (mean = 7.6 years), all of whom had been referred to a child guidance clinic because of behavior disorders. In this analysis, the first factor to emerge was a combination of eye-hand coordination (as measured by the Motor Accuracy Test) and form and space perception (as measured by the SCSIT Manual Form Perception, Space Visualization, Design Copying, and Kinesthesia tests). A praxis factor also emerged, comprised of Bilateral Motor Coordination, Standing Balance Open and Closed, Imitation of Postures, and Graphesthesia. In addition, a visual perception factor emerged, defined by high factor loadings for Figure-Ground Perception, Design Copying, Position in Space, and Graphesthesia.

Some support for considering the sensory integrative functions tapped by the SCSIT as a single construct was found in a study of scores of 182 children with known or suspected sensory integrative dysfunction (Ayres, Mailloux, & Wendler, 1987). The age range of the children was 4 years, 0 months to 8 years, 11 months (mean = 6 years, 6 months, *SD* = 17.4 months). The majority (65.4%) of the children in this study were males. Four standardized instruments were administered: the SCSIT, the SCPNT, the Sentence Repetition Test (Spreen & Benton, 1963), and the Manual Expression and Auditory Sequential Memory subtests of the Illinois Test of Psycholinguistic Abilities (Kirk, McCarthy, & Kirk, 1968). Clinical observations of prone extension and supine flexion postures, ocular pursuits, a percentage of preferred-hand use, and a contralateral hand use score were recorded. In addition, preliminary forms of three of the new praxis tests of the SIPT (Sequencing Praxis [SPr], Oral Praxis [OPr], and Praxis on Verbal Command [PrVC]) were included, as well as a specially constructed block building test, a precursor to the SIPT Constructional Praxis (CPr) test. In the factor analysis of these measures, the SCSIT scores (with the sole exception of Bilateral Motor Coordination) all loaded on a major visual-somatosensory-praxis factor. The highest factor loadings were carried by the block building test (.73), the tactile tests as a group (.72), and the visual tests as a group (.55). The Sentence Repetition Test (.68) and Auditory Sequential Memory (.91) defined a second factor. A third factor reflected the functional integration of the two sides of the body, with significant loadings for Bilateral Motor Coordination (BMC; .79), Oral Praxis (.47), and Sequencing Praxis (.41). Finally, prone extension and supine flexion postures defined a fourth factor, with loadings of .86 and .37, respectively. Postrotary Nystagmus (PRN) loaded –.35 on this factor.

Investigators of adult-onset apraxia (DeRenzi, Faglioni, & Sorgato, 1982; Hecaen & Rondot, 1985; Poek, 1982) have been concerned about whether developmental dyspraxia is a unitary function or a constellation of several discrete types of dyspraxia. The aforementioned analyses seem to suggest the existence of both a general practic function and more specific differential practic skills defined by behavioral goals. In addition, a conceptual system common to praxis also appears to serve visual perception.

**The Evidence From Factor Analysis of the SIPT**

A number of factor analyses of SIPT scores from several different populations further clarify the nature of the construct assessed by the SIPT.

Table 8 summarizes the results of a principal-components analysis of the 17 major SIPT scores for 1,750 children from the SIPT standardization sample. The sample contained equal numbers of male and female children, between the ages of 5.0 and 8.9 years of age (mean = 6.92, *SD* = 1.13). This sample included white (76%), black (12%), Hispanic (6%), and Asian (2%) children. Approximately 90% were right-handed, and approximately 10% were left-handed. All of the children in this sample were presumed to be normal (i.e., had not been diagnosed as learning disabled or as suffering from sensory integrative dysfunction, had not been identified as having any major sensory or motor impairment, and were enrolled in regular education classes in their schools). In this group of presumably normal children, the first factor to emerge (Factor 1) was a visuopraxis factor, best represented by the haptic and visual form and space perception and visuoconstruction tests (i.e., Space Visualization [SV], Figure-Ground Perception [FG], Manual Form Perception [MFP], Design Copying [DC], Constructional Praxis [CPr]). Factor 2, a somatopraxis factor, includes such measures as Oral Praxis (OPr), Bilateral Motor Coordination (BMC), Postural Praxis (PPr), and Graphesthesia (GRA). Factor 3 appears to be a vestibular and somatosensory processing factor, defined by Postrotary Nystagmus (PRN), Localization of Tactile Stimuli (LTS), Finger Identification (FI), and Kinesthesia (KIN). The loadings on this factor indicate that, in the normative sample, deficiencies in somatosensory processing were more apt to be associated with low Postrotary Nystagmus scores than with high scores; this seems to be a natural relationship. Factor 4 is essentially a Kinesthesia and Motor Accuracy (MAc) doublet, although Postural Praxis loaded –.37 on the factor. Although Motor Accuracy has both visuopractic and somatopractic aspects, it tests something different from either of these abilities, as evidenced by the correlations of .16 and .12 between the fourth factor and the visuopraxis and somatopraxis factors.

There is some resemblance between these results and the factor structure of the SIPT scores of 125 children with learning or sensory integrative deficits (see Table 9). The learning-disabled children in this sample were identified by the school systems; the diagnoses of sensory integrative dysfunctions were determined by occupational therapists certified in use of the SCSIT. These children ranged from 5.0 to 8.9 years of age (mean = 7.27, *SD* = .97). Of the 125 children in this sample, 71% were males, and 60% of the sample were in special education classes. The first factor to emerge from this analysis was designated a bilateral integration and sequencing (BIS) factor. Bilateral integration and sequencing are required for: (a) Sequencing Praxis (SPr), which loaded .78; (b) Bilateral

**Table 8**
**Factor Analysis of SIPT Scores of 1,750 Children of the Normative Sample**

| | Factor | | | |
|---|---|---|---|---|
| | Visuopraxis | 2 Somatopraxis | 3 Vestibular & Somatosensory | 4 Kinesthesia/ Motor Accuracy |
| **Test** | | | | |
| Space Visualization (SV) | .67 | –.07 | .05 | –.12 |
| Figure-Ground Perception (FG) | .52 | –.09 | .04 | .18 |
| Manual Form Perception (MFP) | .56 | –.02 | .08 | –.01 |
| Kinesthesia (KIN) | –.01 | .07 | .38 | .63 |
| Finger Identification (FI) | .06 | .38 | .44 | –.01 |
| Graphesthesia (GRA) | .00 | .54 | .25 | .09 |
| Localization of Tactile Stimuli (LTS) | .16 | .03 | .49 | –.08 |
| Praxis on Verbal Command (PrVC) | .43 | .27 | –.02 | –.15 |
| Design Copying (DC) | .63 | .12 | –.13 | .22 |
| Constructional Praxis (CPr) | .68 | .02 | .02 | –.01 |
| Postural Praxis (PPr) | .29 | .54 | .01 | –.37 |
| Oral Praxis (OPr) | –.18 | .80 | .04 | –.12 |
| Sequencing Praxis (SPr) | .28 | .53 | –.02 | .21 |
| Bilateral Motor Coordination (BMC) | .02 | .70 | –.17 | .18 |
| Standing and Walking Balance (SWB) | .11 | .39 | –.05 | .25 |
| Motor Accuracy (MAc) | .26 | .17 | –.27 | .57 |
| Postrotary Nystagmus (PRN) | –.03 | –.07 | .61 | .11 |
| **Correlation** | | | | |
| Factor 2 | .36 | | | |
| Factor 3 | .10 | .11 | | |
| Factor 4 | .16 | .12 | .00 | |

Motor Coordination (BMC), which loaded .69; (c) Standing and Walking Balance (SWB), which loaded .54; (d) Oral Praxis (OPr), which loaded .40; and (e) Manual Form Perception (MFP), which loaded .38. Although sequencing is generally considered central to praxis, the lack of positive correlations between Factor 1 and all other factors militates against relying on a common practic sequencing or a common somatopraxis relationship to explain the communality of the high loading tests on the BIS factor. Factor 2 has been labeled "praxis on verbal command." The Praxis on Verbal Command (PrVC) test has a high negative loading (–.59), and Postrotary Nystagmus (PRN) has a high positive loading (.76) on this factor. Figure-Ground Perception (FG) and Bilateral Motor Coordination have low negative loadings (–.36 and –.31, respectively), while Finger Identification (FI) has a low positive loading (.31). Factor 3, somatosensory processing with oral praxis, is characterized by a linkage among Localization of Tactile Stimuli (LTS), Kinesthesia (KIN), and Oral Praxis. Factor 4, visuopraxis, is identified by the high loadings for Motor Accuracy (MAc), Design Copying (DC), Space Visualization (SV), and Figure-Ground Perception. Factor 5, somatopraxis, is characterized by relatively high loadings of three praxis tests (Postural Praxis [PPr], .89; Constructional Praxis [CPr], .54; and Oral Praxis, .51), and moderate loadings of two tactile tests (Graphesthesia [GRA], .42; and Finger Identification, .30).

Comparing the factor analysis results from the learning-disabled, sensory integration disordered group with the results from the normative group, the visuopraxis factor and the somatopraxis factor are quite similar in both groups. A somatosensory processing factor also was found in both groups, but the composition of this factor differed substantially between the two groups. For the children in the normative sample, somatosensory processing was associated with vestibular function, as reflected in duration of postrotary nystagmus; for children with learning or sensory integrative deficits, however, postrotary nystagmus was not associated with the somatosensory processing factor, but loaded instead on a praxis on verbal command factor. The factor accounting for most of the variance among the children with learning or sensory integrative deficits was a bilateral integration and sequencing factor; no comparable factor was found for the normative sample.

Finally, a group of 176 children with learning or sensory integrative disorders was matched on age, grade, gender, geographic region, and parental education and occupation, with a group of 176 normal children from the standardization sample. In this matched sample, data were incomplete for 59 of the children with learning or sensory integrative disorders, resulting in a final sample of 293 children (117 children with learning or sensory integrative disorders, and 176 children from the standardization sample). This matched sample included 249 males and 49 females ranging from 5 years, 0 months to 8 years, 11 months (mean = 7 years, 4 months; *SD* = 1 year, 0 months). The sample included white (78.2%), black (13.0%), Hispanic (5.5%), and Asian (0.3%) children. Approximately 27% of these children were in special education classes. Approximately 87% were right-handed and approximately 13% were left-handed. Results of the principal-components analysis of SIPT scores for this matched sample are summarized in Table 10.

The first factor to emerge was a somatopraxis factor, best represented by Oral Praxis (OPr), with a .87 loading. Of the 17 SIPT measures, Oral Praxis is the most closely related to somatosensory processing. Graphesthesia (GRA) is also important in this factor, with a substantial factor loading of .72. Also noteworthy are the loadings of the other SIPT measures that constitute the bilateral

**Table 9**

**Factor Analysis of SIPT Scores of 125 Children With Learning or Sensory Integrative Deficits**

| | Factor | | | | |
|---|---|---|---|---|---|
| | **1** | **2** | **3** | **4** | **5** |
| | **Bilateral Integration & Sequencing** | **Praxis on Verbal Command** | **Somatosensory Processing & Oral Praxis** | **Visuopraxis** | **Somatopraxis** |
| **Test** | | | | | |
| Space Visualization (SV) | -.08 | -.11 | -.08 | .64 | .30 |
| Figure-Ground Perception (FG) | .20 | -.36 | .05 | .54 | -.02 |
| Manual Form Perception (MFP) | .38 | -.10 | .12 | .20 | .17 |
| Kinesthesia (KIN) | .24 | .13 | .74 | .02 | -.14 |
| Finger Identification (FI) | .24 | .31 | -.07 | .37 | .30 |
| Graphesthesia (GRA) | .57 | .09 | -.03 | -.04 | .42 |
| Localization of Tactile Stimuli (LTS) | -.27 | -.11 | .83 | .04 | .09 |
| Praxis on Verbal Command (PrVC) | .32 | -.59 | .14 | .06 | .14 |
| Design Copying (DC) | .18 | .00 | .06 | .67 | .06 |
| Constructional Praxis (CPr) | .07 | -.07 | .10 | .38 | .54 |
| Postural Praxis (PPr) | -.07 | -.03 | -.02 | .07 | .89 |
| Oral Praxis (OPr) | .40 | .00 | .37 | -.22 | .51 |
| Sequencing Praxis (SPr) | .78 | .04 | -.02 | .04 | .08 |
| Bilateral Motor Coordination (BMC) | .69 | -.31 | -.04 | .07 | -.10 |
| Standing and Walking Balance (SWB) | .54 | .15 | .16 | .26 | -.07 |
| Motor Accuracy (MAc) | -.03 | .20 | .09 | .78 | -.11 |
| Postrotary Nystagmus (PRN) | .06 | .76 | .04 | .07 | .01 |
| **Correlation** | | | | | |
| Factor 2 | -.08 | | | | |
| Factor 3 | -.26 | .04 | | | |
| Factor 4 | -.37 | .00 | .18 | | |
| Factor 5 | -.34 | .08 | .16 | .31 | |

integration and sequencing (BIS) constellation, namely Bilateral Motor Coordination [BMC] (.71), Sequencing Praxis [SPr] (.70), and Standing and Walking Balance [SWB] (.57). In addition, Praxis on Verbal Command (PrVC) loads .53, and Postural Praxis (PPr) loads .48. In this analysis, the BIS tests and the somatopraxis tests shared considerably more common variance than in the previous analysis of 125 dysfunctional children. The second factor to emerge was a visuopraxis factor, on which Space Visualization (SV) and Figure-Ground Perception (FG) had the highest loadings (.77 and .76, respectively), followed by Design Copying [DC] (.73), Motor Accuracy [MAc] (.65), and Constructional Praxis [CPr] (.53). The Postrotary Nystagmus (PRN) loading of .86 on the third factor shows a relationship between Postrotary Nystagmus and several other measures. On this factor, high Postrotary Nystagmus scores are associated with low scores on Praxis on Verbal Command (–.24), Figure-Ground Perception [FG] (–.24), Bilateral Motor Coordination (–.22), and Manual Form Perception [MFP] (–.22); low Postrotary Nystagmus scores are associated with low scores on Finger Identification [FI] (.32). A fourth factor was defined by Localization of Tactile Stimuli (LTS), with a loading of .91. A common somatosensory element is evident in this factor, as seen in the factor loadings for Kinesthesia [KIN] (.48), Finger Identification (.28), and Manual Form Perception (.27), although the ability to focus attention also may account for some of the communality.

**Cluster Analyses**

Of all the issues surrounding the empirical basis of a diagnostic inventory, the single most important one is whether or not the test can successfully identify clinically important groups of individuals. Thus, although questions of whether an individual scale measures a construct

**Table 10**
**Factor Analysis of SIPT Scores of 117 Children With Learning or Sensory Integrative Deficits and 176 Matched Children From the Normative Sample**

| | Factor | | | |
|---|---|---|---|---|
| | 1<br>Somatopraxis | 2<br>Visuopraxis | 3<br>Vestibular Functioning | 4<br>Somatosensory Processing |
| **Test** | | | | |
| Space Visualization (SV) | .04 | .77 | .10 | –.06 |
| Figure-Ground Perception (FG) | –.13 | .76 | –.24 | .17 |
| Manual Form Perception (MFP) | .20 | .34 | –.22 | .27 |
| Kinesthesia (KIN) | .29 | –.01 | .11 | .48 |
| Finger Identification (FI) | .17 | .28 | .32 | .28 |
| Graphesthesia (GRA) | .72 | .02 | .14 | –.01 |
| Localization of Tactile Stimuli (LTS) | –.10 | –.02 | .01 | .91 |
| Praxis on Verbal Command (PrVC) | .53 | .25 | –.24 | .08 |
| Design Copying (DC) | .17 | .73 | .11 | .01 |
| Constructional Praxis (CPr) | .24 | .53 | .09 | .13 |
| Postural Praxis (PPr) | .48 | .21 | .11 | .08 |
| Oral Praxis (OPr) | .87 | –.24 | .02 | .13 |
| Sequencing Praxis (SPr) | .70 | .25 | .01 | –.12 |
| Bilateral Motor Coordination (BMC) | .71 | .10 | –.22 | –.05 |
| Standing and Walking Balance (SWB) | .57 | .17 | .12 | .09 |
| Motor Accuracy (MAc) | .12 | .65 | .16 | –.11 |
| Postrotary Nystagmus (PRN) | –.04 | .07 | .86 | .06 |
| **Correlation** | | | | |
| Factor 2 | –.55 | | | |
| Factor 3 | .15 | –.05 | | |
| Factor 4 | .37 | –.28 | .06 | |

successfully are very important, the larger question is whether or not the SIPT battery, as a whole, can accurately identify groups of children in need of different kinds of remediation and services. The analyses of this section partially address that issue.

There are many different ways to study the clinical validation issue. One method is to take data from intact groups of individuals who have been identified as having one kind of disorder or another and then to compare and contrast their average score profiles with those of "normals." Such an approach does not seem appropriate with young children suffering from potential sensory integration disorders or learning disabilities, because while experts seem to agree that there are actually many different types of children given labels such as "learning disabled," the same experts generally argue that there is a lack of clinical and empirical consensus as to what such groups actually are. So, while a later part of this chapter *does* compare the mean profiles of groups of children given different global labels, those group profiles cannot be considered definitive typologies in an area such as sensory integration, where the introduction of a new test battery such as the SIPT will also serve to help delineate some major types of sensory integration disorder.

An alternate approach to the simple, contrasted groups strategy just discussed is to use empirical statistical methods to derive groups and then determine if these groups are clinically meaningful. This section discusses such an approach, which also provides the empirical justification for the WPS ChromaGraph™ groupings discussed in chapter 4. Specifically considered is the statistical method of cluster analysis that is used to determine which ways individual children can be empirically and meaningfully classified. Of course, such classifications are not meaningful unless they are clinically relevant. For this reason, a number of empirical and clinical studies were undertaken to evaluate the clinical significance of the statistically derived typology.

Cluster analysis refers to a generic set of statistical procedures that are used to determine typologies from observed data. Just as there are different kinds of factor and regression analysis, there are different kinds of cluster analysis methods used for determining typologies. Cluster analysis methods can be grouped into two major types of statistical procedures. The first kind are those methods called *agglomeration* or *joining techniques,* which first emphasize the idea that all individuals are unique, and then try to combine individuals into groups by selecting the individuals who seem most alike. A second kind of cluster analysis method is a *splitting technique,* which first posits that all individuals are from one large group, and then systematically splits up the group. Splitting techniques search for meaningful ways to divide large groups into approximately equal-sized pieces.

Of the two major kinds of clustering methods, the type usually preferred both for its theoretical assumptions and demonstrated results in known data sets is the agglomerative or joining cluster analysis. Of the agglomerative cluster analysis methods, the *Ward* method is usually shown to work best when it is assumed that the clusters overlap to some degree and are approximately equally sized. For instance, Lorr (1983) summarizes dozens of studies evaluating different cluster analysis techniques and concludes that for the kinds of issues and data similar to those of SIPT, the Ward method of cluster analysis is the most accurate. Ward's method of cluster analysis was employed with the SIPT data.

To cluster the respondents to the SIPT, the appropriate measures need to be selected for inclusion in the analyses. Because the clusters should not reflect performance on a single test but, rather, general tendencies in performance on a number of tests, it was considered appropriate to use a single summary measure from each test. The 17 variables employed were the major SIPT scores used in the factor analyses. To recap, the 17 variables were the time-adjusted accuracy score on Space Visualization (SV), the total accuracy score on Figure-Ground Perception (FG), the Manual Form Perception (MFP) total score, the Kinesthesia (KIN) total score, the Finger Identification (FI) total score, the Graphesthesia (GRA) total score, the Localization of Tactile Stimuli (LTS) total score, the total accuracy score on Praxis on Verbal Command (PrVC), the Design Copying (DC) total accuracy score, the total score on Constructional Praxis (CPr), the Postural Praxis (PPr) total score, the Oral Praxis (OPr) total score, the Sequencing Praxis (SPr) total score, the Bilateral Motor Coordination (BMC) total score, the total score on Standing and Walking Balance (SWB), the weighted Motor Accuracy (MAc) total score, and the total Postrotary Nystagmus (PRN) score. Because only a single measure was used from each test, it was felt that this strategy would be the most likely to uncover major clusters with true diagnostic implications.

The selection of a sample for the clustering is also appropriate, because the data should have at least several exemplars of a type in order to accurately recover the type in the analysis. It would be inappropriate to limit the sample to only normal children, because the cluster analysis then would be very unlikely to recover types of particular relevance to children with significant levels of impairment. Conversely, it would not be appropriate to limit the sample to children who all had a clinically identified disability, because then only types relevant to clinically defined impairment would likely be identified, rather than naturally occurring subgroups among children not identified for special treatment. Therefore, a heterogeneous group of children was used in this analysis, including both individuals identified as having clinically significant problems and those identified as not having such problems.

The same group of 293 children employed in the factor analysis was used for the cluster analyses. This group included 176 children identified as normal, or at least nonidentified for disorder, 28 children identified as having a sensory integration disorder, and 89 children identified as having a learning disability of one kind or another. Of the normal children, 139 were male and 37 were female. Of the impaired children, 106 were male and 11 were female. The mean age of the normal children was 7.21 years (*SD* = 1.07 years), while the mean age of the impaired children was 7.33 years (*SD* = 0.97).

Technically, the cluster analysis was conducted in the following way. First, the distances in the test score patterns among all pairs of the 293 children were calculated (squared Euclidean distances were used). Next, these distances were examined to determine how the children clustered using the Ward technique of maximizing the distance between clusters while minimizing the within-cluster variation. This is analogous, in a univariate sense, to combining the children into groups so as to make the analysis of variance between the groups as large as possible. Cluster solutions were generated to extract from 2 to 10 clusters; the 6-cluster solution was determined to be the most appropriate using statistical criteria and commonsense clinical criteria. Solutions with more than 6 clusters tend to split the children into very small groups, some of which have only a couple of members; solutions with fewer than 6 clusters tend to combine groups that can be delineated theoretically and clinically. (Note that the results of employing alternate clustering methods and measures of distance between individuals were also examined: the current solution of 6 clusters derived from the squared Euclidean distances and Ward's method was preferable to various combinations of Mahalanobis and city block distances with complete, average, and single-linkage clustering algorithms.)

After clustering empirically, using SPSSX programs, each child was assigned to a cluster. Then, to help interpret the clusters, the average *SD* scores on the 17 SIPT measures were calculated for each of the 6 subgroups. These means are presented in Table 11, along with the average *SD* score for all 6 subgroups.

Before interpreting the clusters, it should first be determined whether the groups differ from one another on the 17 SIPT measures. This issue can be assessed using a univariate analysis of variance on each of the 17 SIPT scores to see if they differentiate the 6 groups. Table 12 shows the univariate *F*-ratio associated with each of the 17 univariate analyses of variance. Note that technically, the significance levels associated with the tests are too liberal, because the groups were defined in the cluster analysis so as to make these *F* values as large as possible. Thus, to determine how much of the variance in the 17 tests is explained by the difference between the groups, coefficient eta squared was calculated from each *F* value. Eta squared is a correlation coefficient that tells how well each variable is related to the grouping variable. Eta squared can be interpreted as the percentage of the variance accounted for in the variable by the grouping. As can be seen in Table 12, much of the variance in each of the 17 tests is explainable as the difference between the six empirically derived groups. As a point of interest, it might be noted that the multivariate Wilks' lambda test applied in a multivariate analysis of variance was also significant ($p < .0001$).

Inspection of the means in Table 11 provides a good idea of how members of the different cluster groups formed on the SIPT. The empirical clustering results were combined with the clinical experience of the author in order to develop the following interpretations of the six SIPT types:

**Group 1: Low Average Bilateral Integration and Sequencing.** Group 1, which consists of approximately 19% (*n* = 55) of the children in the cluster analysis, had its lowest average *SD* scores on the tests identified with bilateral integration and sequencing (BIS) with approximately typical scores (mean *SD* scores around 0) on the remaining SIPT. The lowest scores for this group were on the BIS indicators of Standing and Walking Balance (SWB), Bilateral Motor Coordination (BMC), Oral Praxis (OPr), Sequencing Praxis (SPr), and Graphesthesia (GRA). Note, however, that the mean *SD* scores for these scales were at approximately one-half *SD* unit below the typical value expected for a child in the normal sample. As a result, the children fitting this group should not be considered to be profoundly dysfunctional but, rather, as showing a small degree of specific dysfunction. Some Group 1 children will, however, tend to have higher levels of general dysfunction than the average values shown in Table 11; in all cases it would be expected that these children would tend to have lower scores on the indicators of BIS than on the other SIPT.

**Group 2: Generalized Sensory Integrative Dysfunction.** Group 2, consisting of 11% (*n* = 34) of the sample, had the lowest *SD* scores on the SIPT, scoring far below average on almost all of the tests. Scores on the tests that require bilateral integration and sequencing (BIS) were consistently low, as were the scores on measures of visuo- and somatopraxis. This group's score on Praxis on Verbal Command (PrVC) was as low as that obtained by any

**Table 11**

**SIPT Means and Standard Deviations for the Six Cluster Groups[a]**

| Test | Group 1 | | Group 2 | | Group 3 | | Group 4 | | Group 5 | | Group 6 | | Total | |
|---|---|---|---|---|---|---|---|---|---|---|---|---|---|---|
| | *M* | *SD* | *M* | *SD* | *M* | *SD* | *M* | *SD* | *M* | *SD* | *M* | *SD* | *M* | *SD* |
| Space Visualization (SV) | –.03 | .67 | –1.36 | .79 | –.90 | .80 | –.32 | 1.03 | –.48 | .87 | .54 | .60 | –.27 | 1.00 |
| Figure-Ground Perception (FG) | .03 | 1.02 | –1.35 | 1.04 | –.60 | .70 | –.30 | 1.08 | –.81 | .76 | .60 | .89 | –.23 | 1.12 |
| Manual Form Perception (MFP) | –.13 | .87 | –1.60 | .97 | –.65 | 1.04 | –.09 | 1.17 | –.57 | .72 | .36 | .85 | –.28 | 1.12 |
| Kinesthesia (KIN) | –.34 | .96 | –1.60 | 1.50 | –1.20 | 1.21 | .14 | .74 | –.78 | 1.20 | .14 | .96 | –.40 | 1.22 |
| Finger Identification (FI) | .01 | .93 | –1.40 | .97 | –1.01 | 1.12 | –.51 | 1.01 | –.41 | .67 | .43 | .77 | –.34 | 1.09 |
| Graphesthesia (GRA) | –.81 | .93 | –2.13 | .88 | –1.01 | .91 | –.08 | .93 | –.72 | .80 | .42 | .86 | –.51 | 1.18 |
| Loc. of Tactile Stimuli (LTS) | –.24 | 1.04 | –.66 | 1.31 | –.41 | .92 | –.43 | 1.14 | –.12 | 1.09 | .61 | .81 | –.14 | 1.13 |
| Praxis on Verbal Command (PrVC) | .18 | .66 | –2.41 | .84 | –.15 | .68 | –.08 | .93 | –2.38 | .76 | .56 | .49 | –.39 | 1.31 |
| Design Copying (DC) | –.02 | .74 | –2.11 | .62 | –1.61 | .81 | –.43 | .96 | –1.07 | 1.00 | .74 | .74 | –.48 | 1.25 |
| Constructional Praxis (CPr) | .16 | .69 | –1.58 | 1.05 | –.88 | .71 | .07 | .62 | –.53 | .58 | .53 | .64 | –.17 | .97 |
| Postural Praxis (PPr) | –.52 | .95 | –2.14 | .93 | –.98 | 1.13 | –.55 | .93 | –.76 | .88 | .53 | .90 | –.55 | 1.22 |
| Oral Praxis (OPr) | –.94 | .99 | –2.20 | .72 | –.47 | .89 | .13 | .77 | –1.04 | .83 | .24 | .86 | –.50 | 1.16 |
| Sequencing Praxis (SPr) | –.66 | .90 | –2.06 | .74 | –1.21 | .84 | –.42 | .66 | –1.44 | .97 | .41 | 1.03 | –.66 | 1.16 |
| Bilateral Motor Coord. (BMC) | –.46 | .79 | –1.46 | .72 | –1.11 | .94 | –.24 | 1.07 | –1.56 | .59 | .47 | .99 | –.49 | 1.15 |
| Standing/Walking Balance (SWB) | –.77 | .79 | –2.28 | .63 | –1.28 | .94 | .12 | .77 | –1.17 | .85 | .49 | 1.05 | –.53 | 1.24 |
| Motor Accuracy (MAc) | –.17 | .65 | –1.03 | .48 | –.98 | .39 | –.19 | .73 | –.70 | .45 | .21 | .83 | –.34 | .79 |
| Postrotary Nystagmus (PRN) | –.36 | .88 | –.48 | .84 | –.63 | .78 | .11 | .78 | .47 | .84 | .09 | .71 | –.10 | .86 |

[a]Cluster Group Code:

Group 1 = Low Average Bilateral Integration and Sequencing
Group 2 = Generalized Sensory Integrative Dysfunction
Group 3 = Visuo- and Somatodyspraxia
Group 4 = Low Average Sensory Integration and Praxis
Group 5 = Dyspraxia on Verbal Command
Group 6 = High Average Sensory Integration and Praxis

**Table 12**

**Wilks' Lambda and Univariate *F*-Ratios Across the Six Cluster Groups**

| Test | Wilks' Lambda | $F^a$ | Significance |
|---|---|---|---|
| Space Visualization (SV) | .64 | 31.99 | $< .0001$ |
| Figure-Ground Perception (FG) | .71 | 23.98 | $< .0001$ |
| Manual Form Perception (MFP) | .73 | 20.73 | $< .0001$ |
| Kinesthesia (KIN) | .73 | 21.28 | $< .0001$ |
| Finger Identification (FI) | .71 | 23.95 | $< .0001$ |
| Graphesthesia (GRA) | .56 | 44.20 | $< .0001$ |
| Localization of Tactile Stimuli (LTS) | .85 | 10.44 | $< .0001$ |
| Praxis on Verbal Command (PrVC) | .31 | 126.70 | $< .0001$ |
| Design Copying (DC) | .43 | 76.86 | $< .0001$ |
| Constructional Praxis (CPr) | .52 | 53.12 | $< .0001$ |
| Postural Praxis (PPr) | .60 | 38.89 | $< .0001$ |
| Oral Praxis (OPr) | .53 | 51.04 | $< .0001$ |
| Sequencing Praxis (SPr) | .55 | 47.51 | $< .0001$ |
| Bilateral Motor Coordination (BMC) | .62 | 35.62 | $< .0001$ |
| Standing and Walking Balance (SWB) | .47 | 64.61 | $< .0001$ |
| Motor Accuracy (MAc) | .68 | 26.64 | $< .0001$ |
| Postrotary Nystagmus (PRN) | .84 | 10.59 | $< .0001$ |

[a]With 5, 287 degrees of freedom.

group. The children in this group scored relatively better on measures of tactile localization and postural praxis than on the other tests, although even these scores were considerably lower than average. The Oral Praxis (OPr) score of -2.2 indicates that oral dyspraxia is also part of this generalized sensory integrative dysfunction. In fact, deficits are present in all six practic areas tested by the SIPT, and they are accompanied by somatosensory deficits.

**Group 3: Visuo- and Somatodyspraxia.** This group, which included 12% ($n = 35$) of the sample, had scores that were below average on measures of tactile localization, visual praxis, and somatopraxis, with mean scores of -1.61 on Design Copying (DC), -1.20 on Kinesthesia (KIN), -1.01 on Finger Identification (FI), and -1.01 on Graphesthesia (GRA). Performance also was quite low on measures of bilateral integration and sequencing (BIS), as evident in mean scores of -1.28 on Standing and Walking Balance (SWB), -1.21 on Sequencing Praxis (SPr), and -1.11 on Bilateral Motor Coordination (BMC). This group's mean score of -.63 on Postrotary Nystagmus (PRN) was the lowest of the six cluster groups.

**Group 4: Low Average Sensory Integration and Praxis.** Group 4, consisting of 24% ($n = 71$) of the sample, achieved scores in the low average range on most of the tests. In contrast to most of the other cluster groups, children in Group 4 have a relatively flat profile of SIPT scores, with no outstanding strengths or weaknesses on any of the 17 measures.

**Group 5: Dyspraxia on Verbal Command.** Group 5, consisting of 10% ($n$ = 29) of the total sample, has a distinctive SIPT profile, with a considerable range of SIPT scores. Although most of the children's mean scores fell in the low average to mild dysfunction range, their Praxis on Verbal Command (PrVC) score of -2.38 was almost a full standard deviation lower than any of their other scores. Their scores also indicate a secondary problem with bilateral integration and sequencing (BIS). The children in this cluster group had average scores of at least one standard deviation below the mean on the following tests: -1.44 on Sequencing Praxis (SPr), -1.04 on Oral Praxis (OPr), -1.56 on Bilateral Motor Coordination (BMC), -1.17 on Standing and Walking Balance (SWB), and -1.07 on Design Copying (DC). The score profile for this group raises the question of whether their BIS problem is primarily one of bilateral integration, or whether it is largely a sequencing problem, which is generally considered to be a left cerebral hemisphere process.

**Group 6: High Average Sensory Integration and Praxis.** This cluster of children, consisting of 24% ($n = 69$) of the sample, had mean scores in the average to high average range on all of the tests. This group has a relatively flat profile, indicating no outstanding strengths or weaknesses.

**Summary.** Of the 293 children included in the cluster analyses, approximately 48% fell into groups identified as either low or high average in performance on the SIPT. (As shown in Table 13, more than 67% of the normal children in the sample fell into these two clusters, compared with only 18% of the learning-disabled/sensory integration deficient children.) A third group (19%) had a variable score pattern with low average bilateral integration and sequencing (BIS). The remaining cases grouped into three clusters, each representing a different pattern of sensory integrative dysfunction and praxis.

Group 1 has a low average, but clear and apparently pure, BIS score pattern. Group 2, with generalized sensory integrative dysfunction, scored low on virtually all of the SIPT measures. Group 3 was diagnosed as suffering from problems with visuo- and somatopraxis, with a secondary deficity in BIS; this group had the lowest Postrotary Nystagmus (PRN) score of any group. Group 4 generally scored in the low average range on the SIPT. Group 5 had severe difficulty with Praxis on Verbal Command (PrVC); this group had the highest Postrotary Nystagmus score of any group. It is noteworthy that all three of the groups with dysfunctional patterns and one of the low average groups scored low on the constellation of scores characterizing BIS. In contrast, group scores on

**Table 13**

**Representation of Normal, Learning-Disabled, and Sensory Integrative Deficient Children in the Six SIPT Cluster Groups**

| Children | Group 1 Low Average BIS | Group 2 Generalized SI Dysfunction | Group 3 Visuo- and Somatodyspraxia | Group 4 Low Average SI & Praxis | Group 5 Dyspraxia on Verbal Command | Group 6 High Average SI & Praxis | Total |
|---|---|---|---|---|---|---|---|
| Normal | 36 | 2 | 13 | 54 | 6 | 65 | 176 |
| Learning Disabled | 11 | 28 | 13 | 13 | 21 | 3 | 89 |
| SI Deficient | 8 | 4 | 9 | 4 | 2 | 1 | 28 |
| **Total** | 55 | 34 | 35 | 71 | 29 | 69 | 293 |

the Praxis on Verbal Command and on the visuo- and somatopraxis measures varied considerably. None of the groups did exceptionally poorly on the Localization of Tactile Stimuli (LTS) test, but the Postural Praxis (PPr) score, which is associated with tactile localization and with visuo- and somatopraxis, was depressed in all four of the dysfunctional groups.

## Criterion-Related Evidence: Concurrent

The SIPT is intended primarily for the detection, description, and explanation of current dysfunction, rather than for the prediction of some later criterion. Accordingly, most of the criterion-related validity evidence is concurrent in nature. The validity of most psychological tests is established through concurrent testing with a similar test, but such a process is inappropriate for the SIPT because there are no comparable sensory integration and praxis tests covering the same age range. Furthermore, such a procedure does not necessarily assure inferences about the meaning of a given test profile in the current life of a child. Such inferences are best collected by testing children with many known and different previously determined diagnoses, and through gradually accumulating evidence of the effects of treatment of sensory integrative dysfunction. In this respect, the prior use of the SCSIT with various populations contributes to the evolving ability to derive accurate inferences from SIPT scores. Co-occurrence, of course, is not to be confused with causation.

### Evidence by Diagnostic Groups

In the matched sample of 293 normal and dysfunctional children (described earlier in this chapter), the dysfunctional children differed significantly from the normal children on all 17 of the major SIPT measures (all $p < .01$). To determine how well the SIPT could actually differentiate between normal children and those children with learning disabilities or sensory integrative disorders, all 17 major SIPT scores were entered into a multiple discriminant analysis. Using the optimally weighted SIPT composite, approximately 88% of all children (91% of the normal children and 83% of the dysfunctional children) were classified correctly. Of course, classification from an unweighted composite, or from a smaller number of SIPT measures, will be considerably less accurate. However, using only the total weighted Motor Accuracy (MAc) score (the individual score that most effectively discriminated between the normal and dysfunctional groups), it was possible to correctly classify 81% of the normal children and 82% of the dysfunctional children.

There are, of course, many different varieties of dysfunction. Thus, a number of studies have examined and compared the performance of children in various different diagnostic groups. The results of these studies are summarized in Table 14.

**Learning disability and minimal brain dysfunction.** By definition, learning-disabled children have normal IQs. The high incidence of sensory integrative dysfunction among this group, as measured by the SIPT, supports the idea that intelligence and sensory integration are relatively distinct constructs. The mean SIPT scores of 195 learning-disabled children (137 boys, 58 girls; mean age = 7.31, *SD* = 1.0) all fall below the means for the normative sample (see Table 14). Most of the scores range between –0.5 and –1.5 *SD*. The wide standard deviations in this group indicate heterogeneity of scores. In general, the praxis test scores are a bit lower than the scores on form and space management and perception tests (Space Visualization [SV], Figure-Ground Perception [FG], Manual Form Perception [MFP], and Motor Accuracy [MAc]). A contrast among SIPT scores is characteristic of many individuals with learning disabilities or sensory integrative dysfunction, but computing mean scores for a group of dysfunctional children may often obscure those contrasts.

An early study (Ayres, 1965) using 35 measures, 14 of which were later standardized and eventually combined into the SCSIT, found statistically significant differences ($p < .05$) between a group of brain dysfunctional and learning-disabled children and a control group on 34 of the 35 measures. The tests, which later were incorporated into the SCSIT, all discriminated beyond the .01 level.

Later studies by Ayres (1969b, 1972b, 1977) and by Ayres, Mailloux, and Wendler (1987) all found generally low scores on the SCSIT by children determined by school personnel to be learning disabled and by children who were referred for evaluation because of known or suspected learning or behavior problems. McCarthy (1984) administered 14 of the 17 SCSIT to 71 learning-disabled children and found that all but one of the tests identified those with learning problems. The visual, kinesthetic, and tactile tests of the SCSIT were part of a large battery of tests used to determine the incidence of minimal brain dysfunction in Goteborg, Sweden (Gillberg, Rasmussen, Carlstrom, Svenson, & Waldenstrom, 1982). The diagnosis of minimal brain dysfunction was confirmed by extended neuropsychiatric examination. Kinnealey (1984) found that the tactile-kinesthetic tests of the SCSIT discriminated between 30 normal and 30 learning-disabled 8-year-olds with a $t$ of 7.13 ($p < .001$).

**Language disorders.** A group of 28 children (21 boys, 7 girls; mean age = 6.58, *SD* = 1.6) with an educational diagnosis of language disorder were given the SIPT. The mean scores of these children resemble those of SIPT Group 5: Dyspraxia on Verbal Command. The major

**Table 14**
**SIPT Means and Standard Deviations for Different Diagnostic Groups**

| | Autistic ($n$ = 7) | | Learning Disabled ($n$ = 195) | | Brain Injured ($n$ = 10) | | Mental Retardation ($n$ = 28) | | SI Dysfunction ($n$ = 36) | | Spina Bifida ($n$ = 21) | | Reading Disorder ($n$ = 60) | | Language Disorder ($n$ = 28) | | Cerebral Palsy ($n$ = 10) | | Normal ($n$ = 136) | |
|---|---|---|---|---|---|---|---|---|---|---|---|---|---|---|---|---|---|---|---|---|
| Test | *M* | *SD* | *M* | *SD* | *M* | *SD* | *M* | *SD* | *M* | *SD* | *M* | *SD* | *M* | *SD* | *M* | *SD* | *M* | *SD* | *M* | *SD* |
| SV | –.36 | .92 | –.71 | .85 | –1.03 | 1.01 | –1.51 | .97 | –.67 | 1.04 | –.74 | .63 | –.52 | .92 | –.75 | 1.15 | –.85 | .37 | .20 | .80 |
| FG | –.34 | 1.73 | –.75 | 1.07 | –1.31 | 1.29 | –1.73 | 1.68 | –.29 | 1.05 | –1.09 | .86 | –.92 | .79 | –.81 | 1.16 | –.68 | .88 | .32 | 1.06 |
| MFP | –1.08 | 1.27 | –1.02 | 1.23 | –1.90 | 1.30 | –2.79 | .32 | –.46 | .99 | –1.91 | 1.25 | –.99 | 1.10 | –1.17 | 1.14 | –.65 | .21 | .15 | .86 |
| KIN | –.67 | .87 | –1.09 | 1.36 | –1.69 | 1.59 | –2.73 | .55 | –.60 | 1.08 | –1.12 | 1.30 | –1.30 | 1.02 | –1.01 | 1.48 | –.60 | 1.54 | .16 | .83 |
| FI | –.24 | 1.07 | –1.02 | 1.03 | –.80 | 1.01 | –1.90 | .89 | –.73 | 1.05 | –.53 | 1.06 | –1.02 | 1.02 | –1.04 | 1.00 | –1.60 | 1.28 | –.10 | .97 |
| GRA | –.96 | .84 | –1.37 | 1.14 | –1.57 | 1.15 | –2.42 | .69 | –1.09 | 1.06 | –1.94 | .69 | –.63 | 1.18 | –1.17 | 1.01 | –1.28 | 1.47 | –.12 | 1.03 |
| LTS | –.54 | .31 | –.65 | 1.20 | –1.18 | 1.09 | –1.63 | 1.77 | –.61 | 1.20 | –1.38 | 1.12 | –.33 | 1.07 | –.86 | 1.04 | –1.80 | .94 | –.28 | 1.15 |
| PrVC | –2.09 | 1.02 | –1.40 | 1.36 | –1.58 | 1.50 | –3.00 | .00 | –.49 | 1.25 | –.99 | 1.24 | –1.01 | 1.32 | –1.74 | 1.38 | –.63 | 1.52 | .23 | .75 |
| DC | .06 | 1.66 | –1.60 | 1.12 | –1.43 | 1.35 | –3.00 | .00 | –.86 | 1.05 | –2.05 | 1.13 | –1.24 | 1.27 | –1.33 | 1.11 | –2.33 | .99 | .26 | .99 |
| CPr | –.72 | 1.11 | –.91 | .95 | –.83 | 1.02 | –2.17 | .53 | –.46 | .95 | –1.18 | 1.09 | –.60 | .88 | –.78 | .93 | –1.00 | .95 | .24 | .67 |
| PPr | –2.46 | .82 | –1.44 | 1.13 | –2.28 | 1.00 | –2.74 | .61 | –1.05 | 1.33 | –1.59 | .83 | –1.42 | 1.01 | –.92 | 1.08 | –1.73 | 1.07 | –.31 | .96 |
| OPr | –1.84 | .99 | –1.37 | 1.17 | –2.34 | .88 | –2.67 | .66 | –.77 | 1.23 | –2.05 | .79 | –.70 | 1.10 | –1.30 | .99 | –1.58 | 2.41 | –.14 | .98 |
| SPr | –2.38 | .93 | –1.48 | .98 | –1.56 | 1.11 | –2.36 | .74 | –1.17 | .87 | –1.13 | 1.00 | –.78 | .83 | –1.36 | .84 | –.93 | .76 | .21 | 1.01 |
| BMC | –2.18 | .61 | –1.15 | .99 | –1.68 | .91 | –1.85 | .49 | –.71 | 1.16 | –1.18 | .81 | –.58 | .92 | –1.47 | .54 | –1.23 | .86 | .09 | .95 |
| SWB | –1.57 | 1.35 | –1.58 | 1.11 | –2.17 | 1.17 | –2.87 | .31 | –1.46 | .98 | –2.98 | .11 | –.61 | 1.01 | –1.31 | 1.00 | –2.73 | .32 | .05 | .93 |
| MAc | –.37 | .81 | –1.04 | 1.02 | –1.97 | .97 | –2.44 | .83 | –.89 | 1.00 | –1.23 | 1.16 | –.47 | .86 | –.67 | 1.00 | –1.98 | .83 | .16 | .68 |
| PRN | –.70 | .90 | –.12 | 1.22 | 1.09 | 1.46 | –1.04 | 1.44 | –.84 | 1.00 | — | — | –.21 | .80 | –.05 | .77 | .19 | .31 | .05 | .95 |

likenesses to Group 5 are low Praxis on Verbal Command (PrVC), average Postrotary Nystagmus (PRN), and low scores on Oral Praxis (OPr), Sequencing Praxis (SPr), Bilateral Motor Coordination (BMC), and Standing and Walking Balance (SWB). Scores on most of the other SIPT measures are higher than these identifying scores. The somatosensory test scores are about one-half standard deviation lower than those of Group 5. Not all of these language-disordered children resemble those constituting Group 5. Fifteen (54%) of the language-disordered children scored –2.2 *SD* or lower on Praxis on Verbal Command, but 10 (36%) scored –1.1 *SD* or higher on the test. Seven (25%) of the group had Postrotary Nystagmus scores of –0.7 or lower.

**Sensory integrative dysfunction.** The mean SIPT scores of 36 children (34 boys, 2 girls; mean age = 6.87, *SD* = 1.10) with sensory integrative deficits, are shown in Table 14. The scores fall mainly between –0.5 and –1.0. The lowest score (–1.46) is Standing and Walking Balance (SWB). This score suggests the presence of dysfunctional children in the sample. The mean score of –.84 on Postrotary Nystagmus (PRN) probably reflects a selection process in accepting children for occupational therapy. Overall, the score profile for these children seems to reflect low average sensory integration and praxis, but it is unlikely that an individual with a similar profile would be considered dysfunctional. It should be noted that these group means obscure individual patterns of sensory integrative dysfunction (which have been presented and discussed earlier in this chapter).

**Reading disorder.** The mean SIPT scores of 60 children (40 boys, 20 girls; mean age = 7.08, *SD* = .90) with diagnosed reading disorders are shown in Table 14. In general, these children exhibit low average sensory integration and practic ability, with most scores falling no more than one standard deviation below the mean. Tests with scores of –1.0 or lower are Postural Praxis (PPr), Kinesthesia (KIN), Design Copying (DC), Praxis on Verbal Command (PrVC), Finger Identification (FI), and Manual Form Perception (MFP). Most areas evaluated by the SIPT, with the exception of bilateral integration and sequencing, are represented in this group of low scores. Note that the mean scores of these children do not differ appreciably from the mean scores of children with sensory integrative dysfunction. The reading-disordered group differs from the learning-disabled group in that the latter group generally scored more poorly on the praxis tests.

**Mental retardation.** The mean SIPT scores of 28 mentally retarded children (10 boys, 18 girls; mean age =

7.07, *SD* = 1.3) are shown in Table 14. Almost all of the mean scores for this group fall below –1.5. The Standing and Walking Balance (SWB) score is among the lowest. There is no clear pattern or contrast in the scores, although two of the most favorable scores were on the motor-free visual perception tests (Space Visualization [SV] and Figure-Ground Perception [FG]). The Postrotary Nystagmus (PRN) score of –1.04 indicates that many of these children have sensory processing problems which, using SIPT data alone, cannot be separated from low intellectual ability. All scores are low, and it appears that the sensory integrative and practic development of this group of children is very poor.

In an earlier study, McCracken (1975) found poor SCSIT tactile perception scores among 29 7- and 8-year-old (mean age = 8.5) educable mentally retarded children. The children's scores were significantly ($p < .05$) poorer than the SCSIT normative sample on Manual Form Perception, Finger Identification, Graphesthesia, and Perception of Double Tactile Stimuli, but not on Localization of Tactile Stimuli, where the mean score was –0.2 (*SD* = 1.1). Tactile performance was compared with IQ (mean = 59.3, *SD* = 7.8, range = 46–86), dividing the children into three IQ groups. The highest IQ group scored significantly higher on Graphesthesia than the lowest IQ group and scored significantly higher on Localization of Tactile Stimuli than did the middle group. Those children with higher IQs tended to have less tactilely defensive behavior, and the nondefensive children scored better ($p < .05$) on all tests than did those who were tactilely defensive. The study did not entirely support a trend for higher IQ children to have better tactile ability.

**Traumatic brain injury.** The mean SIPT scores of 10 children (6 boys, 4 girls; mean age = 7.57, *SD* = .76) who had suffered traumatic brain damage are shown in Table 14. The standard deviations of the SIPT scores for this group tend to be fairly high, suggesting fairly large differences among the children in this group. Although the mean scores may obscure important individual differences within this group, some general patterns are evident. Lowest scores are on Postural Praxis (PPr), Oral Praxis (OPr), Standing and Walking Balance (SWB), Motor Accuracy (MAc), and Manual Form Perception (MFP). Neuromotor incoordination may be depressing the Standing and Walking Balance and Motor Accuracy scores, and it is possible that the Design Copying (DC) score could be similarly affected. Apart from Postrotary Nystagmus (PRN), the rest of the test scores fall in the low to low average range. The mean Postrotary Nystagmus score of 1.09, an unfavorably high score, is consistent with the assumption that cerebral hemisphere damage is apt to release the inhibition of the nystagmatic reflex. The SIPT data show that the sensory integrative and practic status of these children at time of testing was, in general, poor. The most severe deficits were in postural and oral praxis.

**Spina bifida and meningomyelocele.** The mean SIPT scores of 21 children with the diagnosis of spina bifida (11 boys, 10 girls; mean age = 7.52, *SD* = .87) are shown in Table 14. The very low Standing and Walking Balance (SWB) score (–2.98) is considered a function of the neuromotor problems characteristic of spina bifida and is not considered a reflection of sensory integration. The Motor Accuracy (MAc) score could be similarly affected. Most of the other scores fall in the low range, indicating some general difficulty in sensory integration and praxis. Excluding scores on Postrotary Nystagmus (PRN) and Standing and Walking Balance from interpretation, the lowest scores (Oral Praxis [OPr], Design Copying [DC], Graphesthesia [GRA], Manual Form Perception [MFP], Postural Praxis [PPr], and Localization of Tactile Stimuli [LTS]) suggest visuo- and somatodyspraxia. Visual perception tasks making little practic demand (Space Visualization [SV] and Figure-Ground Perception [FG]) were easier for this group than those tests requiring practic ability (Design Copying and Constructional Praxis [CPr]). Bilateral integration and sequencing appears to be poor, as reflected in the scores on Oral Praxis, Bilateral Motor Coordination (BMC), Sequencing Praxis (SPr), and Graphesthesia. On practic ability tests, the group scored best (–.99) on Praxis on Verbal Command (PrVC). In view of the low Postural Praxis score (–1.59), it is likely that poor postural praxis was a major contributor to the low Praxis on Verbal Command score. The linguistic interpretation component of Praxis on Verbal Command may be satisfactory. Scores on the form and space, visuopraxis, and somatosensory tests show some contrast, with the complexity of the task apparently contributing to score depression. Manual Form Perception, Graphesthesia, and Design Copying performance involves mental operations that are more complex than those required in Kinesthesia (KIN), Finger Identification (FI), Figure-Ground Perception, or Space Visualization. In summary, these children with spina bifida show deficits in complex somatosensory processing and in visuo- and somatopractic development. Visual space perception and figure-ground perception not requiring a motor response are easier than visual space tasks involving praxis. Finger identification is adequate.

Several previous studies have examined perceptual and sensory processing status in children with spina bifida and meningomyelocele disorders. Brunt (1980) administered 13 of the SCSIT to 26 boys and 15 girls (ages 4 years to 9 years, 10 months) with meningomyelocele. The standing balance and all tactile tests except Double Tactile Stimuli were omitted. On only one test (Right-Left Discrimination) did the children score above 0.0 *SD*. The poorest performance was on Motor Accuracy, with scores of –2.69 *SD* and –2.82 *SD* for the preferred and nonpreferred hands, respectively. Other test scores below –1.0 were Space Visualization, Design Copying, and Manual Form Perception. A factor analysis of the scores was interpreted as describing three factors: the first represented constructional and gestural apraxia with loadings by Position in Space, Design Copying, Manual Form Perception, Imitation of Postures, and Bilateral Motor Coordination; the second factor reflected bilateral coordination, and the third, ataxic-like movements.

Grimm (1976) studied hand-motor function and tactile perception in 17 myelomeningocele children, ages 6 years to 8 years, 11 months. Of these children, 14 showed impaired hand function, and 8 showed tactile perception deficit as measured by the SCSIT. Scores of 82% of the children fell more than two standard deviations below the mean on Finger Identification, Graphesthesia, Localization of Tactile Stimuli, and Double Tactile Stimuli. In general, there was no significant association between hand-motor function and tactile perception scores, suggesting they were unrelated factors. In contrast to the Grimm study of tactile and hand function, Gressang (1974) found age-adequate performance on Space Visualization and Figure-Ground Perception of the SCSIT by 21 girls and 8 boys, 4–9 years of age, with diagnoses of myelomeningocele and hydrocephalus.

**Cerebral palsy.** The mean SIPT scores of 10 children (6 boys, 4 girls; mean age = 6.10, *SD* = 1.4) diagnosed as having cerebral palsy are shown in Table 14. The majority of the scores are low. The scores on Standing and Walking Balance (SWB), Motor Accuracy (MAc), and possibly Design Copying (DC) must be considered depressed by the neuromotor incoordination typical of cerebral palsy. This group as a whole has trouble with both visuopraxis and somatopraxis. Poor tactile perception is associated with the dyspraxia. The strengths, as indicated by the group's mean scores, lie in form and space perception not involving motor execution and in following verbal directions to assume various postures.

**Emotional disturbance.** Sensory integrative dysfunction has been found in children with behavior disorders or emotional disturbance. In a factor analytic study, the SCSIT was administered to 87 children (66 boys, 21 girls; ages 5.0–9.3 years, mean = 7.6) referred to a child guidance clinic because of behavior problems. The children did not have severe emotional disturbances, but academic problems were common complaints. All SCSIT scores were below the normative scores (Silberzahn, 1975).

In Rider's (1973) study, 14 male and 6 female children (ages 6.5–12.5 years) in public school classes for the emotionally disturbed were matched for sex and age with 15 male and 8 female nonemotionally disturbed children, and administered the SCSIT. All scores of the emotionally disturbed children were lower than those of the control children (all $p < .05$).

Doyle and Higginson (1984) studied 52 learning-disabled students to evaluate the relationships among self-concept and: (a) school achievement, (b) maternal self-esteem, and (c) sensory integration abilities. Of these variables, only perceptual-motor abilities, as measured by the SCSIT, contributed to reported self-concept of learning-disabled students. Ayres and Heskett (1972) reported average SCSIT scores more than two standard deviations below the mean on a 7-year-old schizophrenic girl.

**Orofacial cleft.** Scores on the SCPNT and SCSIT of 43 boys and 27 girls (ages 4–9 years) with orofacial cleft were significantly lower than the normative scores (Chapparo, Yerxa, Nelson, & Wilson, 1981). Exceptions were scores on Figure-Ground Perception, Position in Space, Right-Left Discrimination, and Imitation of Postures. The mean SCPNT score was –2.67 *SD* ($p < .05$). The children exhibited vestibular and tactile processing and bilateral integration problems. The authors linked these deficits to the symptom complex *vestibular and bilateral integration.*

**Exposure to Agent Orange.** Becker (1982) administered the SCSIT to six children, ages 4–8 years (mean = 5.6, *SD* = 1.2), whose fathers were exposed to Agent Orange in Vietnam and to six matched control children, ages 4–8.7 years (mean = 5.9, *SD* = 1.4). She compared the two groups on general level of sensory integrative functioning, perceptual-motor functioning, eye-hand coordination, visual-motor integration, and somatosensory functioning. The Agent Orange group scored lower on all of the parameters, with statistically significant differences ($p < .05$) between scores obtained on the composite score of sensory integrative functioning in general, as well as on the perceptual-motor group (Imitation of Postures, Bilateral Motor Coordination, Right-Left Discrimination, Crossing the Midline, Standing Balance: Open, and Standing Balance: Closed).

### Evidence From Comparison With Alternate Tests

Another approach to the evaluation of test validity is to correlate test scores with scores on alternate measures, some of which are presumed to assess similar abilities and others of which are presumed to assess different abilities. The pattern of correlations is then examined to determine whether or not the obtained results are consistent with these theoretical expectations.

Both the SIPT and the Kaufman Assessment Battery for Children (K-ABC) (Kaufman & Kaufman, 1983) were administered to a combined sample of normal ($n = 47$), learning-disabled ($n = 35$), and sensory integration disordered ($n = 9$) children. Correlations of the subscales on the two tests are shown in Tables 15, 16, and 17. It is reassuring to note that the SIPT measures of sequential processing, especially Sequencing Praxis (SPr) and Bilateral Motor Coordination (BMC), have higher correlations with the K-ABC Sequential Processing scale than with the Simultaneous Processing scale and, in fact, have the highest correlations with the Sequential Processing scale of any SIPT tests. In addition, the SIPT tests that should, in theory, require little or no sequential processing (i.e., Finger Identification [FI], Localization of Tactile Stimuli [LTS], and Postrotary Nystagmus [PRN]) generally show the lowest correlations with the K-ABC Sequential Processing scale.

As expected, the simple tactile tests in the SIPT (i.e., Finger Identification and Localization of Tactile Stimuli) generally show the lowest correlations with the K-ABC scales, while the complex praxis tests generally show the highest correlations with the K-ABC. The processes common to the SIPT praxis tests and the K-ABC are probably of a complex cognitive nature. The Oral Praxis (OPr) and Sequencing Praxis correlations with the K-ABC Mental Processing Composite suggest that a CNS function basic to somatopraxis may also underlie the mental processing and problem-solving ability evaluated by the K-ABC.

Parts of the Luria-Nebraska Neuropsychological Battery, Children's Revision (Golden, Hemmeke, & Purisch, 1980a) assess parameters that are similar to those assessed by the SCSIT. Both tests were designed to evaluate neurological dysfunction. Kinnealey (1984) administered the tactile-kinesthetic sections of both tests to 30 8-year-old normal and 30 8-year-old learning-disabled children and obtained a correlation of .73 ($p < .001$) between total scores. In a comparable study of the motor tests of the Luria and the SCSIT, Su and Yerxa (1984) obtained a .83 correlation between scores of the two tests administered to 30 dysfunctional children, 8 years of age, who had been referred to three private occupational therapy clinics for evaluation or treatment of sensory integrative dysfunction.

Although the Bruininks-Oseretsky Test of Motor Proficiency (BOTMP) (Bruininks, 1978) is a test of motor skill rather than of sensory integration, it does contain a number of tests requiring practic ability. Ziviani, Poulsen, and O'Brien (1982) administered both the SCSIT and the BOTMP to 32 boys and 17 girls with learning disabilities (ages 4 years, 10 months to 12 years, 2

**Table 15**

**Pearson Product-Moment Correlations Between SIPT Scores and Standardized K-ABC Scores for a Combined Group of Normal ($n = 47$), Learning-Disabled ($n = 35$), and Sensory Integrative Disordered ($n = 9$) Children**

| | K-ABC Scale | | | | | | | |
|---|---|---|---|---|---|---|---|---|
| SIPT | Arithmetic | Riddles | Decoding | Understanding | Sequential Processing | Simultaneous Processing | Mental Proc Composite | Achievement |
| SV | .47 | .41 | .43 | .12 | .43 | .43 | .50 | .19 |
| FG | .54 | .61 | .50 | .30 | .55 | .50 | .53 | .26 |
| MFP | .56 | .49 | .37 | .15 | .54 | .61 | .59 | .15 |
| KIN | .67 | .54 | .54 | .36 | .59 | .55 | .58 | .36 |
| FI | .33 | .27 | .28 | .02 | .22 | .32 | .24 | .12 |
| GRA | .55 | .36 | .50 | .28 | .51 | .44 | .43 | .31 |
| LTS | .39 | .27 | .32 | .10 | .30 | .29 | .27 | .21 |
| PrVC | .68 | .68 | .53 | .41 | .63 | .60 | .65 | .24 |
| DC | .73 | .60 | .64 | .46 | .57 | .66 | .66 | .34 |
| CPr | .70 | .56 | .53 | .30 | .57 | .66 | .68 | .24 |
| PPr | .57 | .46 | .44 | .30 | .55 | .50 | .52 | .28 |
| OPr | .60 | .57 | .51 | .39 | .59 | .60 | .61 | .24 |
| SPr | .66 | .46 | .65 | .45 | .67 | .59 | .63 | .33 |
| BMC | .56 | .45 | .59 | .54 | .56 | .48 | .51 | .26 |
| SWB | .68 | .59 | .48 | .42 | .58 | .53 | .54 | .44 |
| MAc | .62 | .59 | .61 | .49 | .52 | .58 | .51 | .29 |
| PRN | .04 | .05 | .08 | .16 | .12 | .03 | .08 | .14 |

*Note.* All correlations greater than .18 are significant ($p < .05$).

**Table 16**

**Pearson Product-Moment Correlations Between SIPT Scores and Standardized K-ABC Scores for a Group of Normal ($N$ = 47) Children**

| | K-ABC Scale | | | | | | | |
|---|---|---|---|---|---|---|---|---|
| SIPT | Arithmetic | Riddles | Decoding | Understanding | Sequential Processing | Simultaneous Processing | Mental Proc Composite | Achievement |
| SV | .24 | .41 | .21 | -.20 | .17 | .20 | .23 | -.02 |
| FG | .12 | .46 | .26 | .19 | .28 | .15 | .25 | .05 |
| MFP | .04 | -.01 | -.07 | -.23 | .30 | .15 | .26 | -.13 |
| KIN | .20 | .28 | .36 | .27 | .22 | .03 | .19 | .27 |
| FI | .09 | .07 | -.09 | -.35 | -.16 | .19 | .01 | -.06 |
| GRA | .01 | -.09 | -.02 | .00 | -.06 | -.14 | -.15 | .16 |
| LTS | .11 | .03 | .10 | .04 | .03 | .08 | .15 | .11 |
| PrVC | .14 | .22 | .10 | -.03 | .21 | .18 | .28 | .24 |
| DC | .21 | .29 | .43 | .24 | .20 | .15 | .25 | .17 |
| CPr | .11 | .14 | .15 | -.06 | .22 | .29 | .36 | -.10 |
| PPr | .16 | .15 | .14 | .12 | .28 | .02 | .19 | .14 |
| OPr | .00 | .33 | -.04 | .14 | .06 | .13 | .19 | -.05 |
| SPr | .23 | .16 | .45 | .10 | .46 | .38 | .46 | .17 |
| BMC | .34 | .20 | .45 | .53 | .31 | .20 | .32 | .06 |
| SWB | .21 | .16 | .18 | .23 | .15 | -.01 | .07 | .38 |
| MAc | -.16 | .23 | .21 | .28 | .05 | -.03 | .12 | .02 |
| PRN | -.08 | -.21 | -.02 | .41 | .04 | -.26 | -.19 | .16 |

*Note.* All correlations greater than .24 are significant ($p < .05$).

**Table 17**

**Pearson Product-Moment Correlations Between SIPT Scores and Standardized K-ABC Scores for a Group of Learning-Disabled ($N$ = 35) Children**

| | K-ABC Scale | | | | | | | |
|---|---|---|---|---|---|---|---|---|
| SIPT | Arithmetic | Riddles | Decoding | Understanding | Sequential Processing | Simultaneous Processing | Mental Proc Composite | Achievement |
| SV | .13 | -.21 | .15 | -.06 | .27 | .28 | .33 | .02 |
| FG | .41 | .40 | .33 | .12 | .30 | .45 | .48 | .47 |
| MFP | .30 | .22 | -.10 | -.23 | .15 | .57 | .51 | .05 |
| KIN | .45 | .07 | .13 | .02 | .36 | .41 | .47 | .22 |
| FI | -.30 | -.41 | -.12 | .03 | -.17 | -.12 | -.21 | -.32 |
| GRA | .21 | -.21 | .18 | .00 | .32 | .27 | .36 | .11 |
| LTS | .07 | .17 | .02 | .27 | .09 | .09 | .03 | .06 |
| PrVC | .41 | .55 | .11 | .26 | .44 | .31 | .47 | .41 |
| DC | .51 | .14 | .27 | .22 | .17 | .49 | .45 | .37 |
| CPr | .59 | .10 | .11 | .03 | .26 | .44 | .44 | .24 |
| PPr | .41 | .07 | .08 | .00 | .25 | .46 | .47 | .13 |
| OPr | .40 | .14 | .23 | .18 | .45 | .55 | .63 | .29 |
| SPr | .49 | -.08 | .47 | .48 | .47 | .15 | .37 | .40 |
| BMC | .24 | -.00 | .38 | .32 | .29 | .14 | .24 | .32 |
| SWB | .50 | .34 | -.03 | .09 | .36 | .34 | .44 | .21 |
| MAc | .28 | .07 | .22 | .35 | -.01 | .20 | .14 | .24 |
| PRN | .04 | -.14 | -.15 | -.16 | -.12 | .19 | .08 | -.13 |

*Note.* All correlations greater than .28 are significant ($p < .05$).

months; *SD* = 1.2 years), and obtained significant ($p <$ .01) correlations between the BOTMP Fine Motor scale and 13 of the SCSIT. Those SCSIT correlating .372 ($p <$ .01) or higher with the BOTMP Fine Motor scale were Space Visualization, Position in Space, Design Copying, Manual Form Perception, Finger Identification, Graphesthesia, Double Tactile Stimuli, Imitation of Postures, Bilateral Motor Coordination, Standing Balance: Open, Standing Balance: Closed, Crossing the Midline, and Motor Accuracy. There were fewer significant correlations between the SCSIT and the BOTMP Gross Motor scale. The tests correlating significantly ($p <$ .01) with the Gross Motor scale were Kinesthesia, Finger Identification, Localization of Tactile Stimuli, Imitation of Postures, Bilateral Motor Coordination, Standing Balance: Open, and Standing Balance: Closed. Those correlating significantly ($p <$ .01) with the BOTMP Composite score were Space Visualization, Design Copying, Kinesthesia, Manual Form Perception, Imitation of Postures, Bilateral Motor Coordination, Standing Balance: Open, Standing Balance: Closed, and Motor Accuracy. It is interesting to note that the SCSIT with the highest correlations with any scale were Imitation of Postures (.75), Standing Balance: Closed (.69), Standing Balance: Open (.69), Graphesthesia (.68), Bilateral Motor Coordination (.61), and Design Copying (.60), suggesting a common tactile-practic-postural domain tapped by both tests.

The Bender-Gestalt Test (Bender, 1938) has been used frequently to detect neuropsychological deficits in children. Kimball (1977) administered the Bender-Gestalt and the SCSIT (excluding the standing balance tests) to 26 children, ages 60 to 108 months (mean = 84.81, *SD* = 12.84), with suspected sensory integrative dysfunction. Scores on the Bender predicted the combined score for all SCSIT space perception measures ($r$ = .65; $p <$ .01) and the combined score for all SCSIT tactile-praxis tests ($r$ = .61; $p <$ .01). The SCSIT Motor Accuracy and Design Copying scores were associated most closely with the Bender, with correlations of .71 and .72 (both $p <$ .01), respectively. Position in Space had a correlation of .50 ($p <$ .01). The Bender did not correlate significantly with tests of postural and bilateral integration.

The somatopractic domain was determined (Ayres, 1969a) to be a common area of assessment in the Gesell Developmental Schedules (Gesell & Amatruda, 1974) and a group of perceptual-motor tests consisting of three of the subtests of the Marianne Frostig Developmental Test of Visual Perception (Frostig, Lefever, & Whittlesey, 1961) and early versions of most of the SCSIT. The tests were administered to 92 children (mean age = 70 months, *SD* = 14 months) who had been given the Gesell—by physicians—at a mean age of 57 weeks. The mean scores on the 16 perceptual-motor tests of children who earlier had above average Developmental Quotients (DQs) were significantly higher ($p <$ .01) than those of similar children matched for age and who, at Gesell testing, had below average DQs. The one exception to this level of statistical significance was the Double Tactile Stimuli test. A stepwise multiple regression resulted in an $R$ of .75 between the full perceptual-motor battery and the Gesell total DQ. Most of the association could be predicted by Kinesthesia, Graphesthesia, Motor Accuracy, Frostig Spatial Relations subtest, and the Ayres Space Test. The Frostig Spatial Relations subtest, like Design Copying of the SCSIT, required the child to copy designs on a dot grid. At the time of the study, Design Copying of the SCSIT had not yet been standardized.

Most of the perceptual-motor tests had higher correlations with the Gesell Adaptive scale score than with the Motor or Total scales. One of the strongest associations held between the Gesell Adaptive score and eight test (later SCSIT) scores related to the somatopractic domain. Correlations on these tests ranged from .50 to .60, whereas only one (Frostig Spatial Relations) of five visual perception tests held a correlation (.57) in the .50s. (The Spatial Relations subtest taps practic skill.) Because praxis is the ability to conceptualize, plan, and execute adaptive interaction with the physical world, it is logical that tests in this domain would share the greatest variance with the Adaptive scale score on the Gesell.

## Evidence Supporting Individual Test Validity

In this section, both the SIPT and earlier versions of each test are discussed. In discriminant analyses, each of the 17 tests in the SIPT showed significant ($p <$ .01) ability to discriminate between normal ($n$ = 176) and dysfunctional ($n$ = 117) children.

### Space Visualization (SV)

The SIPT Space Visualization test is similar to the Space Visualization task in the SCSIT, but the scoring and normative data differ. In a sample of 1,750 normative children (see Table 18), scores on Space Visualization correlated significantly with Design Copying (DC), Constructional Praxis (CPr), Sequencing Praxis (SPr), Praxis on Verbal Command (PrVC), and Figure-Ground Perception (FG). In a group of 125 children with learning or sensory integrative deficits (also see Table 18), tests correlating most highly with Space Visualization were Design Copying, Constructional Praxis, Motor Accuracy (MAc), Postural Praxis (PPr), and Figure-Ground Perception. Finally, in a matched sample of dysfunctional and normal children (see Table 18), Space Visualization correlated most highly with Design Copying, Constructional Praxis, Motor Accuracy, and Postural Praxis. In a combined sample of normal, learning-disabled, and sensory integration disordered children ($N = 91$), scores on the Space Visualization were significantly correlated with scores on the K-ABC Mental Processing Composite ($r = .50$) and with the K-ABC Sequential Processing ($r = .43$), Simultaneous Processing ($r = .43$), Arithmetic ($r = .47$), and Riddles ($r = .41$) scores. However, although the Space Visualization is correlated with performance on a wide range of cognitive tasks on the K-ABC, there is some indication that the correlations may depend more upon the visuospatial perception skills that underlie many of the K-ABC scales rather than a common higher-level cognitive component. In particular, it is interesting to note that the Space Visualization score is significantly correlated with the K-ABC Reading/Decoding scale ($r = .43$), but not with scores on the K-ABC Reading/Understanding scale ($r = .12$).

In factor analyses, Space Visualization had moderate to high loadings on the visuopraxis factors. Other tests with high loadings on those factors were Design Copying, Constructional Praxis, and Motor Accuracy. These results suggest that there is a practic as well as a visual perception aspect to Space Visualization. That aspect appears to be involved in ideation and motor planning.

In summary, Space Visualization correlates most highly with the SIPT tests of visuopraxis (Motor Accuracy and Figure-Ground Perception) and helps to define the visuopraxis factors. It also correlates substantially with most of the K-ABC scales. It shares sufficient variance with several of the praxis tests to suggest that there may be a practic ideation and planning component to the Space Visualization test.

Space Visualization was first published as the Ayres Space Test (Ayres, 1962a) in a form that had 60 items. When scores on these 60 items were factor analyzed, the major categories of content extracted were perceptual speed, configuration type, directionality, and visualized rotation. The visualized rotation factor appeared in the performance of the 6- to 10-year-old children. Because the test was terminated after a given number of errors, that older age group was most likely to reach the items requiring a mental rotation of a block (Ayres, 1962a).

Sleeper (1962) found that a group of 38 cerebral palsied individuals, ages 7 to 44 years, scored widely on the Ayres Space Test. The scores correlated .47 ($p < .01$) with a test of perception of verticality. A negligible correlation of .11 between scores on the Ayres Space Test and body balance suggested that the balance

**Table 18**

**Pearson Product-Moment Correlation Coefficients Between SIPT Scores of 1,750 Children in the Normative Sample and 125 Children With Learning or Sensory Integrative Deficits**

| | SV | FG | MFP | KIN | FI | GRA | LTS | PrVC | DC | CPr | PPr | OPr | SPr | BMC | SWB | MAc | PRN |
|---|---|---|---|---|---|---|---|---|---|---|---|---|---|---|---|---|---|
| SV | 1.00 | .21 | .20 | .09 | .13 | .15 | .10 | .22 | .30 | .28 | .19 | .08 | .27 | .14 | .16 | .16 | .02 |
| FG | .34 | 1.00 | .22 | .12 | .13 | .14 | .07 | .15 | .28 | .22 | .13 | .09 | .27 | .15 | .12 | .18 | .02 |
| MFP | .30 | .38 | 1.00 | .11 | .16 | .16 | .10 | .21 | .28 | .24 | .18 | .11 | .22 | .17 | .17 | .17 | .02 |
| KIN | .07 | .16 | .26 | 1.00 | .16 | .18 | .10 | .11 | .16 | .16 | .08 | .12 | .22 | .13 | .15 | .17 | .06 |
| FI | .29 | .31 | .29 | .17 | 1.00 | .26 | .16 | .19 | .18 | .15 | .22 | .23 | .26 | .22 | .13 | .13 | .08 |
| GRA | .31 | .32 | .37 | .21 | .28 | 1.00 | .10 | .19 | .23 | .17 | .25 | .28 | .37 | .28 | .19 | .14 | .07 |
| LTS | .09 | .16 | .11 | .30 | .07 | .11 | 1.00 | .08 | .09 | .11 | .11 | .10 | .10 | .09 | .07 | .07 | .01 |
| PrVC | .27 | .41 | .37 | .20 | .04 | .28 | .13 | 1.00 | .28 | .26 | .25 | .18 | .31 | .24 | .20 | .18 | .03 |
| DC | .44 | .43 | .40 | .26 | .41 | .39 | .11 | .30 | 1.00 | .44 | .27 | .16 | .41 | .30 | .23 | .38 | .03 |
| CPr | .44 | .46 | .42 | .29 | .49 | .46 | .14 | .35 | .52 | 1.00 | .30 | .12 | .30 | .22 | .20 | .21 | .06 |
| PPr | .39 | .24 | .32 | .06 | .35 | .41 | .12 | .31 | .30 | .54 | 1.00 | .34 | .34 | .27 | .20 | .13 | .02 |
| OPr | .28 | .26 | .37 | .35 | .29 | .50 | .27 | .37 | .29 | .45 | .44 | 1.00 | .28 | .36 | .25 | .12 | .03 |
| SPr | .26 | .30 | .38 | .28 | .34 | .54 | .08 | .35 | .41 | .44 | .30 | .43 | 1.00 | .51 | .25 | .31 | .06 |
| BMC | .32 | .36 | .29 | .14 | .28 | .39 | .17 | .34 | .25 | .26 | .18 | .35 | .52 | 1.00 | .26 | .28 | .00 |
| SWB | .26 | .35 | .39 | .32 | .37 | .34 | .08 | .30 | .42 | .32 | .25 | .43 | .38 | .30 | 1.00 | .23 | .06 |
| MAc | .40 | .25 | .20 | .20 | .34 | .23 | .13 | .05 | .43 | .33 | .20 | .13 | .31 | .17 | .33 | 1.00 | –.03 |
| PRN | .07 | –.06 | .03 | .04 | .12 | .04 | –.01 | –.22 | .06 | –.04 | –.01 | –.02 | .04 | –.15 | .12 | .10 | 1.00 |

*Note.* Normative sample correlation coefficients are in the upper-right section of the table; correlation coefficients for the learning or sensory integrative deficits sample are in the lower-left section of the table.

**Table 19**

**Pearson Product-Moment Correlation Coefficients Between SIPT Scores in a Group Composed of 117 Children With Learning or Sensory Integrative Deficits and 176 Matched Children From the Normative Sample**

| | SV | FG | MFP | KIN | FI | GRA | LTS | PrVC | DC | CPr | PPr | OPr | SPr | BMC | SWB | MAc | PRN |
|---|---|---|---|---|---|---|---|---|---|---|---|---|---|---|---|---|---|
| SV | | | | | | | | | | | | | | | | | |
| FG | .40 | | | | | | | | | | | | | | | | |
| MFP | .32 | .38 | | | | | | | | | | | | | | | |
| KIN | .21 | .18 | .25 | | | | | | | | | | | | | | |
| FI | .35 | .27 | .25 | .27 | | | | | | | | | | | | | |
| GRA | .38 | .26 | .35 | .30 | .38 | | | | | | | | | | | | |
| LTS | .17 | .18 | .19 | .27 | .26 | .20 | | | | | | | | | | | |
| PrVC | .39 | .38 | .40 | .31 | .29 | .40 | .20 | | | | | | | | | | |
| DC | .56 | .48 | .41 | .35 | .41 | .46 | .20 | .51 | | | | | | | | | |
| CPr | .51 | .40 | .37 | .41 | .42 | .46 | .20 | .48 | .62 | | | | | | | | |
| PPr | .45 | .27 | .31 | .26 | .38 | .48 | .24 | .43 | .46 | .49 | | | | | | | |
| OPr | .31 | .23 | .30 | .35 | .35 | .50 | .27 | .42 | .36 | .39 | .46 | | | | | | |
| SPr | .45 | .37 | .37 | .37 | .40 | .55 | .17 | .49 | .60 | .48 | .46 | .50 | | | | | |
| BMC | .38 | .34 | .33 | .25 | .34 | .41 | .23 | .44 | .42 | .38 | .35 | .48 | .63 | | | | |
| SWB | .37 | .34 | .35 | .39 | .37 | .50 | .21 | .51 | .54 | .52 | .43 | .48 | .51 | .42 | | | |
| MAc | .49 | .33 | .22 | .27 | .30 | .33 | .18 | .33 | .58 | .41 | .35 | .49 | .50 | .41 | .39 | | |
| PRN | .11 | .00 | .07 | .12 | .20 | .18 | .08 | .01 | .17 | .11 | .12 | .15 | .14 | .01 | .20 | .13 | |

difficulties of these cerebral palsied persons were more apt to be a function of neuromotor disorder than of sensory processing.

Studies using the Ayres Space Test have shown the capacity of the test to discriminate between dysfunctional children and those selected without regard to perceptual skill. In a sample of 100 children with and 50 children without suspected dysfunction (Ayres, 1965), the difference in scores yielded a $t$-ratio of 6.32, where a 2.58 ratio was required at $p < .01$. A $t$-ratio of 4.23 was obtained between the mean Ayres Space Test scores of these children who, earlier in their lives, scored above versus (scored) below 100 on the Gesell Developmental Schedules (Ayres, 1969a).

The 60 items of the Ayres Space Test were reduced to 30 items when the test was restandardized and incorporated into the SCSIT as the Space Visualization test. In factor analyses, both the original and revised forms of the Space Visualization have consistently loaded on a form and space perception factor serving more than the visual sensory modality (Ayres, 1965, 1966b, 1966d, 1972b, 1976a; Silberzahn, 1975), but they also have shown associative bonds and shared factor loadings with tactile and praxis tests (Ayres, Mailloux, & Wendler, 1987).

The Ayres Space Test identified poor space perception in two thirds of a group of 15 right-handed adults with left hemiparesis and 16 right-handed adults with right hemiparesis (Ayres, 1962b). The distribution of scores of the remaining one third of the hemiparetics was similar to a matched control group. There was no significant difference between scores of the left and right hemiplegics. In this sample a rho reliability-stability coefficient of .95 was obtained when odd and even items of the test were given one week apart.

Comparing performance of 18 language-disabled children, ages 7.5–9.5 years, with matched control children on spatial representation tasks (including the Ayres Space Test) involving anticipatory imagery, Savich (1984) found the language disabled were less accurate. The results suggested to the author that difficulty with dynamic cognitive representation could be the cause of the deficits demonstrated by language-disabled children.

That children perform better on Space Visualization when it is the first of the SCSIT to be administered was determined on 50 children, 5–7 years old (Ward, 1973).

Space Visualization provides an opportunity to observe hand usage in an unobtrusive manner. This opportunity gave rise to the development of two procedures yielding scores derived from hand usage during, but independent of, the Space Visualization score—namely the Space Visualization Contralateral Use (SVCU) score (Ayres, 1976b) and the Preferred Hand Use (PHU) score. Cermak, Quintero, and Cohen (1980), in a study of 30 children between the ages of 4 years, 0 months and 8 years, 11 months, found developmental age trends in SVCU, but the variability at each age for PHU was so great that there were no significant differences among age groups on PHU.

Atwood and Cermak (1986) found that the greater the placement distance between the Space Visualization test blocks, the lower the SVCU score. They concluded that a three-inch distance between the blocks was the best distance to elicit mild midline crossing deficits and yet be adequate for visual regard. They also found that the SVCU score did not change over three trials if the distance between the blocks remained the same.

Studying 23 6- to 8-year-old learning-disabled children and 23 normal children of the same age, Stilwell (1981) demonstrated that SVCU scores discriminate ($p <$ .05) between learning-disabled and normal children. Similarly, Cermak and Ayres (1984) found lower ($p < .05$) SVCU scores among 5-, 6-, and 8-year-old learning-disabled children than among normal children. (The difference at age 7 years did not reach significance.) The learning-disabled sample, as diagnosed by schools or through psychological testing, consisted of 179 children; the normal population consisted of 30 children, half boys and half girls, in each of four age groups, 5 to 8 years. Eighty to 90% in each group preferred the right hand. A total group-by-age analysis revealed a significant group effect ($p < .001$), with the learning-disabled sample scoring lower. A significant age effect with younger children scoring lower ($p < .001$) was also demonstrated.

A number of factor analyses were made of correlations of SIPT scores of both the normative and dysfunctional children in order to explore the meaning of the SVCU and PHU scores in these children. Untabled factor analysis of scores of 1,750 normative children showed that Bilateral Motor Coordination (BMC) and SVCU consistently loaded on the same factor, thus reinforcing their bilateral integration communality. Low scores on SVCU were associated with many right-to-left atypical approaches on the Design Copying task. Other analyses of the normative children's scores found that children with low SVCU scores also made more right-to-left errors on the Constructional Praxis test. In the normative children, high frequency right-hand usage on Space Visualization was not only associated with a high SVCU score but also with low Design Copying right-to-left scores and with good Oral Praxis and Standing and Walking Balance performance. These associations suggest that the SVCU correlations are not entirely due to the presence of normal left-handers in the group. On several untabled factor analyses of scores of dysfunctional children, low SVCU scores showed strongest factorial linkage with poor performance on the somatosensory tests and on Praxis on Verbal Command, Bilateral Motor Coordination (BMC), Sequencing Praxis, Manual Form Perception (MFP), Oral Praxis (OPr), Standing and Walking Balance (SWB), and with low scores on Postrotary Nystagmus (PRN). In these analyses, low SVCU scores were also associated with a high incidence of right-left errors in Constructional Praxis, and with reversals, inversions, and right-to-left errors in Design Copying.

In a number of different analyses, SVCU was as apt to be associated with high as with low PRN scores. The SVCU mean of 81 dysfunctional children with PRN scores below the mean was –0.29 *SD* (*SD* = 1.0), and the SVCU mean of 81 dysfunctional children with PRN scores above the mean was –0.36 *SD* (*SD* = .98). In high and low PRN scoring groups, low SVCU scores correlated positively with short PRN (loading = –.36; *N* = 81) and negatively with long PRN (loading = –.29; *N* = 42). Stated otherwise, low SVCU scores were associated with both too low and too high PRN scores. SVCU loaded .29 and .48 on factors identified as disorder of praxis on verbal command in analyses of 42 children with PRN scores above the mean. The Praxis on Verbal Command accuracy score loaded .71 and .88 on those factors. Similarly, SVCU had a loading of –.46 on a right cerebral hemisphere factor generated from scores of 37 dysfunctional children; both the counterclockwise and clockwise values for PRN also loaded on this factor (.54 and .50, respectively). Additional factor loadings indicated poor performance on Constructional Praxis, Design Copying, and on left-hand Finger Identification (FI), Graphesthesia (GRA), and Kinesthesia (KIN). These analyses showing possible left and right cerebral hemisphere dysfunction indicate that the SVCU score may be vulnerable to cerebral disorder. SVCU also had a very low (.21) loading on the vestibular-bilateral integration factor (*N* = 125 dysfunctional children) where counterclockwise PRN had a loading of .60 and clockwise PRN had a loading of .58, indicating depression of the nystagmatic response in both directions. SVCU also had moderate to substantial loadings on factors on which PRN held negligible loadings. SVCU correlations with PRN are not inevitable. SVCU scores may be negatively affected by the same conditions that contribute to either exceptionally low or high Postrotary Nystagmus scores. Low SVCU scores may also be associated with cerebral dysfunction. A low SVCU score is most apt to reflect poor functional integration of the two sides of the body. It also may be associated with a tendency toward drawing lines right to left, reversing a block in Constructional Praxis, or reversing a design. It appears that many different conditions may interfere with the normal development of preferred-hand use or with the use of each hand in extracorporeal space.

The Preferred Hand Use (PHU) score is based on the proportion of preferred- versus nonpreferred-hand use during Space Visualization performance. The mean PHU score of 1,750 normative children was .70 (*SD* = .22);

among 125 dysfunctional children, the mean PHU score was .64 ($SD$ = .21). In both normative and dysfunctional groups, girls had higher PHU scores than did boys. Approximately 11% of the normative and 18% of the dysfunctional children were designated left-handed by SIPT examiners. For both samples, PHU correlated positively with right-hand preference, but only for the normative group was the correlation significant ($r$ = .15; $p <$ .001). The correlation suggests that left-handers were slightly less lateralized in hand preference than were right-handers during Space Visualization performance. Correlations between PHU and SVCU ranged from .47 ($p <$ .001) for the total normative sample and .48 ($p <$ .001) for right-handers only to .52 ($p <$ .001) for right-handed dysfunctionals.

In the normative sample, left-handedness (not PHU) had a correlation of .43 ($p <$ .001) with right-to-left approaches on Design Copying; in the dysfunctional group, this correlation coefficient was .50 ($p <$ .001).

In summary, data from analysis of scores of the normative children indicate that strength of hand dominance as judged by the PHU score obtained from Space Visualization hand usage is related to: (a) the BIS score constellation, (b) right-left orientation, (c) preferred-hand Motor Accuracy, (d) Oral Praxis, and (e) Postrotary Nystagmus. The data support the idea that limitations in development of unilateral hand preference may be associated with poor functional integration of the two sides of the body, with right-left confusion, diminished preferred-hand visuomotor coordination, and with depressed duration postrotary nystagmus. These associations are not necessarily strong and may easily be obscured by other linkages in the dysfunctional child.

**Figure-Ground Perception (FG)**

The SIPT Figure-Ground Perception test is the same as the SCSIT Figure-Ground Perception test, but administration method is different. In a sample of 1,750 normative children (see Table 18), Figure-Ground Perception accuracy scores correlated most strongly with Design Copying [DC] (.28), Sequencing Praxis [SPr] (.27), Manual Form Perception [MFP] (.22), Constructional Praxis [CPr] (.22), and Space Visualization [SV] (.21). In a group of 125 dysfunctional children, the highest correlations with Figure-Ground Perception were Constructional Praxis (.46), Design Copying (.43), Praxis on Verbal Command [PrVC] (.41), and Manual Form Perception (.38). In the matched sample of 293 normative and dysfunctional children (see Table 19), Figure-Ground Perception correlated strongly with Design Copying (.48) and Constructional Praxis (.40), but the relationship with Praxis on Verbal Command was weaker (.38). Figure-Ground Perception accuracy also correlated significantly with the major K-ABC scales (see Table 15).

Figure-Ground Perception accuracy had higher correlations with the other SIPT scores than did Figure-Ground Perception time. The correlation between Figure-Ground Perception accuracy and Figure-Ground Perception time was .61 in the sample of 1,750 normal children; in the group of 125 dysfunctional children, this correlation was only .39. These correlations show that in the normative group there was a closer relationship between the number of correct Figure-Ground Perception items and the total time taken to perform the test than there was in the dysfunctional group. Of the two Figure-Ground Perception parameters, accuracy is the more meaningful.

In factor analyses, Figure-Ground Perception accuracy consistently loaded on the visuopraxis factors (see Tables 8, 9, and 10). In most of the analyses, Figure-Ground Perception had a lower loading on visuopraxis than did Design Copying, suggesting that Figure-Ground Perception is primarily visual in its assessment, with little or no practic component to the task.

These correlations and loadings reflect the fact that Figure-Ground Perception is the most purely visual of the SIPT. Suggestions of its sensitivity to cerebral integrity are also seen in the negative correlations between Figure-Ground Perception and Postrotary Nystagmus (PRN). In the sample of 125 children with learning or sensory integrative deficits (see Table 9), Figure-Ground Perception had a factor loading of –.36 on the praxis on verbal command factor, on which Praxis on Verbal Command loaded –.59 and Postrotary Nystagmus loaded .76. That is, this factor is characterized by long-duration Postrotary Nystagmus, by poor performance on Praxis on Verbal Command and poor Figure-Ground Perception discrimination.

All data considered, Figure-Ground Perception is one of the SIPT that is least related to somatosensory processing and praxis. The test does appear to reflect higher level CNS activity.

Much of the validity data on the Figure-Ground Perception test of the SCSIT is germane to the SIPT Figure-Ground Perception test. An early form of this test, referred to as *superimposed figures*, discriminated (critical ratio = 5.19; $p <$ .01) between 100 children with and 50 children without suspected perceptual deficits (Ayres, 1965). In a more refined form, the test items containing pictures of common objects only were administered to children who earlier in life scored above versus below 100 on the Gesell Developmental Schedules. The difference in scores was significant ($p <$ .01). In this sample, the test had a correlation of .43 with the total Gesell score. It was associated most strongly ($r$ = .47) with the Adaptive subscale and least strongly ($r$ = .32) with the Gross Motor subscale (Ayres, 1969b).

In a factor analysis including an early version of this test, perceiving a figure against a background showed relative independence from tests of visual form and space perception (Ayres, 1965). Later versions of the test tended to share factor loadings with other tests of praxis, including the tactile tests (Ayres, 1966a). The first published form of the Southern California Figure-Ground Visual Perception Test (Ayres, 1966c) loaded most strongly with other tests of visual perception and, to a lesser extent, with tactile tests (Ayres, 1972b, 1976a; Silberzahn, 1975).

Studying 118 educable mentally handicapped children (IQs = 45–85), Brooks and Clair (1971) found that figure-ground perception was related to IQ ($p < .05$) and to word recognition ($p < .05$). The authors felt the results supported the idea that visual figure-ground perception is related to reading skills at the readiness and first reader levels.

Because the SCSIT version of Figure-Ground Perception discriminated well at older ages, it often has been used with adults. Petersen and Wikoff (1983) administered the Figure-Ground Perception to 100 men, ages 19–57 years (mean = 30.97), chosen to cover a fairly broad range of education, occupation, and socioeconomic status. The scores for this group formed a normal distribution with no restriction in range, and were not related to age, education, or socioeconomic status. Scores were considerably higher than the mean scores for the oldest children (10-year-olds) in the normative sample. A correlation of –.67 between Figure-Ground Perception scores and scores on the Embedded Figures Test (EFT) (Witkin, 1950) led to the conclusion that Figure-Ground Perception was a satisfactory alternative to the well-known EFT and had advantages for assessing figure-ground perception in persons with motor impairment or who were under psychotropic medication. (Low scores on the EFT and high scores on Figure-Ground Perception are favorable.)

SCSIT Figure-Ground Perception scores almost identical to those obtained by Petersen and Wikoff (1983) were obtained from 124 women from 18 to 81 years of age (Petersen, Goar, & Van Deusen, 1985). Moderate correlations were found between Figure-Ground Perception scores and age and educational level. A split-half reliability of .87 and a test-retest (spanning 4 to 6 weeks) reliability of .71 were reported for this sample.

#### Manual Form Perception (MFP)

An early version of the Manual Form Perception test demonstrated the capacity of a test of manual perception to differentiate between normal and dysfunctional children (critical ratio = 7.91), and showed a close factorial association with visual perception skills and, to a lesser extent, with kinesthesia and the variables that identify dyspraxia (Ayres, 1965). Later factor analyses, including the Manual Form Perception test of the SCSIT, found similar associations (Ayres, 1966b, 1972b, 1976a). The factor analytic studies supported the idea of a form and space perception ability that incorporated and integrated tactile, kinesthetic, and visual sensory information. Kinnealey (1984) administered the SCSIT version of Manual Form Perception to 30 normal and 30 learning-disabled children. A comparison of the scores of the two groups yielded a $t$ of 3.21 ($p < .01$).

The SIPT Manual Form Perception test is more complex than the Manual Form Perception Test in the SCSIT. Performance on Part II of the SIPT version appears to be especially dependent upon functional integration of the two sides of the body. In a combined group of normal ($n = 47$), learning-disabled ($n = 35$), and sensory integration disordered ($n = 9$) children, the total score on Manual Form Perception was significantly correlated with most of the major K-ABC scales. It had a correlation of .61 with the K-ABC Simultaneous Processing score, .54 with the K-ABC Sequential Processing score, and .59 with the K-ABC Mental Processing Composite. In the SIPT normative sample, Manual Form Perception correlated most strongly with Design Copying [DC] (.28), Constructional Praxis [CPr] (.24), Figure-Ground Perception [FG] (.22), Sequencing Praxis [SPr] (.22), and Praxis on Verbal Command [PrVC] (.21), as shown in Table 18. Among a dysfunctional sample of 125, the tests showing the highest correlations with Manual Form Perception were Constructional Praxis (.42), Design Copying (.40), Standing and Walking Balance [SWB] (.39), Sequencing Praxis (.38), Graphesthesia [GRA] (.37), Praxis on Verbal Command (.37), and Oral Praxis [OPr] (.37). Part-score correlations of both the normative and dysfunctional groups fail to identify Manual Form Perception with either the right or left cerebral hemisphere.

Manual Form Perception loaded primarily on visuopraxis factors (see Tables 8, 9, and 10), indicating a strong visualization component in the test. The factor matrix derived from scores of a matched sample of dysfunctional and normative children (see Table 10) shows Manual Form Perception loading .20 on the somatopraxis factor and .27 on the localization of tactile stimuli factor. The somatosensory basis of Manual Form Perception is not as well demonstrated as the visualization basis. One of the highest Manual Form Perception factor loadings—.38 on the BIS factor (see Table 9)—emphasizes its bilateral integration component. (However, Manual Form Perception did not load appreciably on BIS factors that were generated using different samples of dysfunctional children or different factor analytic techniques, suggesting that its role in identifying functional integration of the two sides of the body may be somewhat limited.)

The evidence indicates that Manual Form Perception is primarily a test of haptic form and space perception and visualization. In certain situations it may reflect the functional integration of the two sides of the body. Performance on this test also may be affected by more general problem-solving competency, as well as by form and space perception.

**Motor Accuracy (MAc)**

The SIPT version of the Motor Accuracy test continues that test's long history as a contributor to the understanding of sensory integrative dysfunction. The Motor Accuracy task has remained essentially unchanged throughout that history, but scoring methods have changed. Motor Accuracy is primarily a visuomotor test, but the data suggest that it also depends upon vestibular and somatic sensory processing as well as visual perception. Both right- and left-hand performance requires visual space management, for which the right cerebral hemisphere is generally considered more efficient than the left hemisphere. Both the preferred- and nonpreferred-hand performance appears to tap a fundamental sensorimotor process related to most of the other SIPT. The SIPT data analyzed and reported here, unless otherwise indicated, are based on the Motor Accuracy weighted total score (i.e., the sum of right-hand and left-hand weighted scores, compared against the appropriate time-adjusted norms).

In a combined group of normal ($n$ = 47), learning-disabled ($n$ = 35) and sensory integration disordered ($n$ = 9) children, Motor Accuracy was significantly correlated with all of the major K-ABC scales (see Table 15). In a group of 1,750 children from the SIPT normative sample (see Table 18), Motor Accuracy scores correlated most highly with Design Copying [DC] (.38), Sequencing Praxis [SPr] (.31), Bilateral Motor Coordination [BMC] (.28), and Constructional Praxis [CPr] (.21). In a group of 125 dysfunctional children, highest Motor Accuracy correlations were with Design Copying (.43), Space Visualization [SV] (.40), Finger Identification [FI] (.34), Constructional Praxis (.33), and Standing and Walking Balance [SWB] (.33). Similar relationships (see Table 19) were found in a group of dysfunctional children and in the matched sample of dysfunctional and normal children (although a stronger relationship with Sequencing Praxis is evidenced in correlations of .49 and .50 in the latter groups). The elements that these tests share with Motor Accuracy are: (a) visual manipulation of space, (b) a basic somatopractic ability, (c) vestibulosomatosensory processing, and (d) functional integration of the two sides of the body and extracorporeal space.

In a group of 125 children with learning or sensory integrative deficits, the correlation between Motor Accuracy scores for the preferred and the nonpreferred hands was .49. Comparison of preferred- and nonpreferred-hand correlations with other SIPT scores shows interesting relationships. Some of the higher correlations (with correlations for the alternate hand shown in parentheses) are: MAc nonpreferred hand with Design Copying Part I Accuracy, .45 (.39); MAc preferred hand with Design Copying Part II Should Have, .41 (.37); MAc preferred hand with Constructional Praxis Part II Accuracy, .25 (.26); MAc preferred hand with Postural Praxis (PPr), .33 (.22); MAc preferred hand with Oral Praxis (OPr), .32 (.15); and MAc nonpreferred hand with Sequencing Praxis, .30 (.25). The largest and most important difference between preferred- and nonpreferred-hand correlations was with Oral Praxis, which is consistent with the positive correlation between frequency of right-hand usage and Oral Praxis scores in the normative sample.

The fundamental somatomotor nature of eye-hand coordination as measured by Motor Accuracy is also seen in the part scores of 1,750 children from the normative sample. In this group, both the preferred- and nonpreferred-hand scores correlated significantly ($p <$ .05) with almost all of the part scores of the other SIPT. The highest correlations of the Motor Accuracy preferred-hand score were with Design Copying Part II Should Have (.30), Design Copying Part I Accuracy (.28), Sequencing Praxis Hand score (.25), and Bilateral Motor Coordination Arms score (.23). The highest Motor Accuracy nonpreferred-hand correlations were with Design Copying Part II Should Have (.31), Design Copying Part I Accuracy (.30), Sequencing Praxis Hand score (.28), Sequencing Praxis Fingers score (.25), and Bilateral Motor Coordination Arms score (.25). It appears that it is the basic integrity of the sensorimotor system, rather than eye-hand practic ability, that accounts for the correlations linking Motor Accuracy not only to another eye-hand test (Design Copying), but also to two tests (Sequencing Praxis and Bilateral Motor Accuracy) that tap the functional integration of the two sides of the body.

In factor analysis of SIPT scores from 125 children with sensory integrative or learning disorders (see Table 9), the total Motor Accuracy score had a .78 loading on the visuopraxis factor, emphasizing the visual component of this task; for the matched sample (see Table 10), Motor Accuracy had a loading of .65 on the visuopraxis factor. The kinesthetic component of the Motor Accuracy task is also evident in the factor loadings: For the 1,750 children from the normative sample (shown in Table 8), Motor Accuracy had a .57 loading on the fourth factor, which was essentially a motor accuracy-kinesthesia doublet.

Although the method of scoring Motor Accuracy has been changed for the SIPT, the task itself is the same as in earlier versions of the Motor Accuracy test. The 1964 version was shown to differentiate between dysfunctional and control groups with a *t*-ratio of 5.19 ($p < .01$) (Ayres, 1965) and between those who earlier in life scored above versus below 100 on the Gesell Developmental Schedules with a *t*-ratio of 4.13 ($p < .01$). In the Gesell study, in which almost all of the tests (later combined into the SCSIT) were administered to 92 children, Motor Accuracy had the highest correlation (.59) with the total Gesell scores. The correlation with the Adaptive subscale was .60, with the Fine Motor subscale .62, and with the Gross Motor subscale .49 (Ayres, 1969a).

The mean Motor Accuracy score of a group of 87 children in a child guidance clinic was more than one standard deviation below the mean of the 1,964 normative group (Silberzahn, 1975). In a group of 30 children referred to a private occupational therapy clinic because of suspected sensory integrative dysfunction (Su & Yerxa, 1984), the mean right-hand score was -1.45 *SD*, and the mean left-hand score was -1.96 *SD* (using the standard deviations derived from the 1980 normative data).

In factor analytic studies, Motor Accuracy and MAc-R (1964 and 1980 versions) showed natural associations with other tests identifying dyspraxia (Ayres, 1965; Ayres, Mailloux, & Wendler, 1987); with visual-motor skills (Ayres, 1966a, 1966b); with ocular pursuits (Ayres, Mailloux, & Wendler, 1987); and as a doublet when both right and left hands were entered (Ayres, 1972b, 1977). These studies showed that Motor Accuracy provided an adequate measure of eye-hand coordination and that such coordination may be affected by practic skill and visuo-ocular ability.

Su and Yerxa (1984) reported correlations of .83 for the right-hand and .77 for the left-hand Motor Accuracy-R score with the total score of the seven subtests of motor skills of the Luria-Nebraska Neuropsychological Battery: Children's Revision (Golden, Hammeke, Purisch, 1987). They also obtained correlations of .91 for the right-hand and .85 for the left-hand MAc-R score with a total score composed of summed scores on all SCSIT in which a motor or practic function was under test. These correlations support the idea that MAc provides a fair sampling of aspects of visuomotor and practic ability.

Smith (1983) studied the difference between right- and left-hand MAc-R scores of 120 right-handed children and 68 left-handed children. There were no less than 8, nor more than 20, children at each whole-year age group. Right- and left-hand difference scores were obtained by subtracting the nonpreferred-hand adjusted score from the preferred-hand adjusted score. The mean difference between hands for all children was 6.7 points ($SD = 4.8$), suggesting the normative range of difference between right- and left-hand skill performance on the MAc-R was 2–11 points. There were no significant differences in sex or handedness in the adjusted or difference scores.

In studying 63 normal boys (5, 7, and 9 years of age), Mandell, Nelson, and Cermak (1984) found no difference in degree of lateralization of right- and left-handed children on the MAc-R, although 5- and 7-year-old left-handers were less lateralized ($p < .05$) on the Displacing subtest of the Minnesota Rate of Manipulation Tests (1969). As a group, the left-handed children were less well lateralized on the Placing subtest of the Purdue Pegboard (Purdue Research Foundation, 1948). The authors suggested the left-handers showed as much lateralization as the right-handers on the MAc-R because of practice effects in use of a pencil. This finding suggests that the score of the hand the child has used for writing, regardless of whether it should be his or her tool-using hand, will be optimized through practice.

**Kinesthesia (KIN)**

An early version of the Kinesthesia test was able to discriminate ($p < .01$) between dysfunctional and control groups (Ayres, 1965), and the SCSIT version of the Kinesthesia test discriminated equally well between children who earlier scored above versus below 100 on the Gesell Developmental Schedules (Ayres, 1969a). In the Gesell study discussed earlier, scores on the Kinesthesia test correlated .47 with the total score, .50 with the Adaptive subscale, and .31 with the Gross Motor subscale. Comparing SCSIT Kinesthesia scores of 30 normal and 30 learning-disabled children, Kinnealey (1984) found a significant difference ($p < .0001$) between the mean scores of the two groups. In factor analyses of the SCSIT, the Kinesthesia test loaded with tests of visual and haptic perception (Ayres, 1965; Silberzahn, 1975), and with motor planning and tactile tests (Ayres, 1965, 1966a, 1966b, 1976a; Ayres, Mailloux, & Wendler, 1987).

Elfant (1977) found that a group of blinded adults scored less well on the SCSIT Kinesthesia test than did a group of sighted adults. She suggested the difference might be in visualizing ability. Lindley (1973) found that 20 early blind adults were significantly more accurate in reproducing length of arm movement, but did not differ from sighted adults in reproducing arm position during administration of an adapted form of this test. Hsu and Nelson (1981) reported that adults scored significantly better than the SCSIT standardization sample of 8-year-olds. This finding suggests that the ceiling on that test is not reached by 8-year-old children.

The SIPT Kinesthesia test is similar in design and administration to the Kinesthesia test of the SCSIT, although the scoring method differs. In a group of normal ($n = 47$), learning-disabled ($n = 35$), and sensory integration disordered ($n = 9$) children, Kinesthesia was significantly correlated with all of the major K-ABC scales (see

Table 15). The Kinesthesia test correlated most strongly in a group of 1,750 normative children (see Table 18) with Sequencing Praxis [SPr] (.22), Graphesthesia [GRA] (.18), Motor Accuracy [MAc] (.17), Finger Identification [FI] (.16), Design Copying [DC] (.16), and Constructional Praxis [CPr] (.16). In the group of 125 children with learning or sensory integrative deficits (also see Table 18), the highest correlations of Kinesthesia were with Oral Praxis [OPr] (.35), Standing and Walking Balance [SWB] (.32), Localization of Tactile Stimuli [LTS] (.30), Constructional Praxis (.29), Sequencing Praxis (.28), Manual Form Perception [MFP] (.26), and Design Copying (.26). Similar associations were found in the matched sample of 293 dysfunctional and normative children (see Table 19). Highest Kinesthesia correlations in this group were with Constructional Praxis (.41), Standing and Walking Balance (.39), Sequencing Praxis (.37), Oral Praxis (.35), and Design Copying (.35). These correlations probably reflect the contribution of the kinesthetic sense to body balance, to eye-hand coordination, and to sequencing, oral, and visual praxis.

In factor analyses, Kinesthesia carried highest loadings on, and helped to identify, somatosensory processing factors. The results presented in Table 9 show a .74 Kinesthesia loading on the somatosensory processing and oral praxis factor in a group of 125 dysfunctional children. In an analysis of 1,750 normal children (Table 8), Kinesthesia had a loading of .63 on a sensorimotor factor, and a loading of .38 on a factor reflecting a combination of vestibular and somatic processing.

**Finger Identification (FI)**

Disorders of finger identification have been recognized and studied since Gerstmann (1940) identified a syndrome consisting of finger agnosia, disorientation for right and left, agraphia, and acalculia. The SIPT Finger Identification test was correlated with Sequencing Praxis [SPr] (.26), Graphesthesia [GRA] (.26), Oral Praxis [OPr] (.23), Postural Praxis [PPr] (.22), and Bilateral Motor Coordination [BMC] (.22) in a group of 1,750 normal children (see Table 18). In a group of 125 children with learning or sensory integrative deficits (also see Table 18), Finger Identification correlations were highest with Constructional Praxis [CPr] (.49), Design Copying [DC] (.41), Graphesthesia (.38), Standing and Walking Balance [SWB] (.37), and Postural Praxis (.35). The two lowest correlations were with Localization of Tactile Stimuli [LTS] (.07) and Praxis on Verbal Command [PrVC] (.04). Major correlations with the other SIPT in the matched sample of 293 dysfunctional and normal children (see Table 19) were consistent with these values. The strongest Finger Identification affiliations were with Graphesthesia (the most complex of the tactile tests) and with both visuo- and somatopraxis tests. The contrast of the .04 correlation with Praxis on Verbal Command and the .49 correlation with Constructional Praxis in the group of 125 dysfunctional children suggests that the right cerebral hemisphere may be more efficient in Finger Identification performance than is the left cerebral hemisphere. There were no interpretable pertinent differences between right- and left-hand Finger Identification scores in either the normative or dysfunctional groups. Correlations between right- and left-hand scores were .42 and .47, respectively, in these two groups.

In factor analyses, Finger Identification loaded most strongly on somatopraxis or somatosensory processing factors (which it helped to define), on visuopraxis factors, and on factors reflecting positive correlations between vestibular (Postrotary Nystagmus) and somatosensory processing (see Tables 8, 9, and 10). These data show a depressing effect common to Postrotary Nystagmus and Finger Identification. In addition, Finger Identification loads positively on factors where Postrotary Nystagmus has a high positive loading and Praxis on Verbal Command a negative loading.

The SCSIT version of the Finger Identification test showed a strong capacity to differentiate: (a) between 100 dysfunctional and 50 nondysfunctional children (critical ratio = 7.91) (Ayres, 1965); (b) between children who earlier in life scored above versus below 100 on the Gesell Developmental Schedules ($t$- ratio = 4.66) (Ayres, 1969a); and (c) between 30 learning-disabled and 30 normal children ($t = 4.70; p < .0001$) (Kinnealey, 1984). In a sample of 92 children (mean age = 69.2 months, $SD$ = 13.6 months), the SCSIT Finger Identification score correlated .56 with the total score on the Gesell Developmental Schedules (Ayres, 1969a). A correlation of .58 was found between SCSIT Finger Identification and the Adaptive subscale, .35 with the Gross Motor subscale, and .56 with the Fine Motor subscale.

In factor analyses, an early form of the Finger Identification test had a substantial loading (.58) on a tactile-motor planning factor identified as developmental apraxia (Ayres, 1965). The SCSIT version of Finger Identification loaded with other tactile tests given normal children (Ayres, 1966b) and with variables associated with praxis (loading = .62) and visual perception (loading = .49) in a mixed group of normals and non-normals (Ayres, 1966d). In a factor analysis of scores on a battery of tests including the SCSIT, the Illinois Test of Psycholinguistic Abilities (Kirk, McCarthy, & Kirk, 1968), and academic tests administered to 148 learning-disabled children, Finger Identification carried a loading of .51 on a factor identified as form and space perception. Other tactile tests also loaded on this factor, thus showing an affinity between tactile analysis in the SCSIT scores and ocular and postural responses of 128 learning-disabled children. Finger Identification shared loadings on a common factor with the SCSIT Graphesthesia and Localization of Tactile Stimuli. The loadings were .50, .61, and .54, respectively (Ayres, 1976a). In a child guidance population, the SCSIT Finger Identification test dominated (loading .69) a factor that was also represented by the SCSIT Imitation of Postures with a loading of .35 (Silberzahn, 1975).

Hsu and Nelson (1981) reported that 51 normal adults scored significantly better on Finger Identification than the oldest group (8 years of age) in the SCSIT standardization sample, with many adults achieving perfect scores.

To summarize, the ability to identify fingers shows closest ties with other tactile tests, next closest with tests of praxis, and then with tests of visual perception. It is an ability vulnerable to dysfunction. The Finger Identification test helps define visuo- and somatopraxis factors. This linkage reaffirms the close relationship between tactile processing and praxis. Conditions that depress the Postrotary Nystagmus score are also apt to depress the Finger Identification score. Some of data from the dysfunctional sample studied suggest that the right cerebral hemisphere may be more efficient in finger identification than the left cerebral hemisphere. However, the close relationship between finger identification and visuopraxis may reflect more than just a common site of dysfunction; it may be that tactile sensory processing is fundamental to certain behavioral expressions of praxis.

**Graphesthesia (GRA)**

The SIPT Graphesthesia test presents a more powerful test than the previous SCSIT Graphesthesia test. In a combined sample of normal ($n$ = 47), learning-disabled ($n$ = 35), and sensory integration disordered ($n$ = 9) children, SIPT Graphesthesia scores were significantly correlated with scores on all of the major K-ABC subscales (see Table 9). In a sample of 1,750 normal children (see Table 18), SIPT scores that correlated most highly with the Graphesthesia scores were Sequencing Praxis [SPr] (.37), Oral Praxis [OPr] (.28), Bilateral Motor Coordination [BMC] (.28), and Finger Identification [FI] (.26). Similar relationships were found among the sample of 125 children (Table 18) with learning or sensory integrative disorders. In that group, the SIPT scores that correlated most highly with Graphesthesia were Sequencing Praxis (.54), Oral Praxis (.50), Constructional Praxis [CPr] (.46), and Postural Praxis [PPr] (.41). Similarly, in a matched group of dysfunctional and normative children (see Table 19), the highest correlations were with Sequencing Praxis (.55), Oral Praxis (.50), and Standing and Walking Balance [SWB] (.50). The consistency and strength of the correlations between Graphesthesia and Sequencing Praxis are notable.

There were no recognizably pertinent right-hand versus left-hand Graphesthesia correlation differences in the group of 1,750, but there were several correlation differences of interest in the group of 125 dysfunctional children. The frequency of left-hand usage on Space Visualization (SV) correlated .27 with both right- and left-hand accuracy on Graphesthesia, compared to correlations of .08 and .04 with Space Visualization right-hand usage. One possible interpretation of this difference is that the Graphesthesia task is performed more effectively, but not exclusively, by the right cerebral hemisphere. A unilateral right cerebral hemisphere problem would lower both right and left scores on Graphesthesia and reduce the tendency to use the left hand. Such an interpretation also would be consistent with the fact that the Graphesthesia left-hand accuracy was more highly correlated with both preferred- and nonpreferred-hand Motor Accuracy (MAc) scores (.24 and .14, respectively) than was Graphesthesia right-hand accuracy (.09 and .06, respectively).

On factor analyses, Graphesthesia loaded substantially on and helped to define the somatopraxis factors. It also loaded on vestibulosomatosensory and somatosensory processing factors. The relationship between Graphesthesia and bilateral integration and sequencing (BIS) is seen most clearly in Table 9. These extensive loadings indicate the sensitivity of Graphesthesia to deficits in complex tactile processing and the role of interpreting complex tactile stimuli into planned action.

An early form of the Graphesthesia test differentiated between 100 dysfunctional children and 50 control children with a critical ratio of 4.72 (Ayres, 1965). An improved form of the test differentiated with a $t$-ratio of 4.20 between a group of 88 children (mean age = 69.2 weeks, $SD$ = 13.6) whose total score on the Gesell Developmental Schedules fell above versus below 100 at the mean age of 57.7 weeks ($SD$ = 41.0). In this same study, the correlation of Graphesthesia with the earlier total Gesell score was .56, with the Adaptive subscale .54, with Fine Motor subscale .51, and with the Gross Motor subscale .40. All values cited were significant beyond the .01 level (Ayres, 1969a). The SCSIT version of Graphesthesia differentiated between 30 normal and 30 dysfunctional children with a $t$-ratio of 5.79 ($p < .0001$) (Kinnealey, 1984).

In summary, Graphesthesia is primarily a test of complex tactile processing, but there are additional components as well. One of those components is praxis and another is related to BIS. Some data suggest that the right cerebral hemisphere may be more efficient at performing the Graphesthesia test than the left cerebral hemisphere. If this is the case, the use of Graphesthesia right-left hand differences in interpreting unilateral dysfunction must be done with caution.

**Localization of Tactile Stimuli (LTS)**

The SIPT Localization of Tactile Stimuli test provides information on the most pure and simple tactile processing task. Virtually no praxis is required in performance of the test. In a group of 1,750 children from the SIPT normative sample (see Table 18), Localization of Tactile Stimuli correlated most highly with Finger Identification [FI] (.16), Constructional Praxis [CPr] (.11), and Postural Praxis [PPr] (.11). In a group of 125 dysfunctional children (also Table 18), highest LTS correlations were with Kinesthesia [KIN] (.30), Oral Praxis [OPr] (.27), Bilateral Motor Coordination [BMC] (.17) and Figure-Ground Perception [FG] (.16). Correlations within the matched sample of dysfunctional and normal children (see Table 19) also show Localization of Tactile Stimuli sharing variance with the somatosensory tests, with visuopraxis, and with Postural Praxis and Oral Praxis.

In factor analyses, Localization of Tactile Stimuli loaded strongly on and helped define somatosensory processing factors (see Tables 8, 9, and 10). The considerable contribution of tactile perception to Oral Praxis is suggested by the third factor in Table 9. The tendency for Localization of Tactile Stimuli to share variance with Kinesthesia is seen in the third factor of Tables 8 and 9, and the fourth factor of Table 10.

The design of the SIPT Localization of Tactile Stimuli test is essentially unchanged from the SCSIT Localization of Tactile Stimuli, although the scoring method is different. Therefore, the validity data on the SCSIT version is germane to the SIPT version. An early form of Localization of Tactile Stimuli differentiated between 100 dysfunctional and 50 nondysfunctional children with a critical ratio of 3.13 (Ayres, 1965), and the SCSIT version of Localization of Tactile Stimuli discriminated between children who earlier scored above versus below 100 on the Gesell Developmental Schedules with a *t*-ratio of 3.80 (Ayres, 1969a). Kinnealey (1984) obtained a *t*-ratio of 3.42 ($p < .001$) when comparing 8-year-old dysfunctional and normal children.

In factor analytic studies, SCSIT Localization of Tactile Stimuli showed the closest bonds with other tactile and with praxis test scores (Ayres, 1965, 1966b, 1966d, 1969b, 1972b, 1976a; Silberzahn, 1975). It is this relative purity of tactile, as opposed to a possible form visualization, skill that made it an attractive and desirable inclusion in tests of tactile function. The correlation of .51 between the Localization of Tactile Stimuli score and the total Gesell score in the study cited earlier (Ayres, 1969) suggests that Localization of Tactile Stimuli has relevance beyond the tactile system. Its correlation with the Adaptive subscale of the Gesell was .48 and with the Fine Motor subscale .52.

In summary, Localization of Tactile Stimuli, the simplest and most exclusively tactile test in the SIPT, shows the closest positive praxis association with Postural Praxis and Oral Praxis, but it also can be associated with visuopraxis. Localization of Tactile Stimuli is often positively correlated with Finger Identification and Kinesthesia. Taken alone, Localization of Tactile Stimuli is not particularly helpful in identifying possible unilateral cerebral dysfunction. Overall, the data reaffirm the close relationship between tactile processing and certain practic abilities.

**Postrotary Nystagmus (PRN)**

The Southern California Postrotary Nystagmus Test (SCPNT) (Ayres, 1975) served well as it was designed, and it was incorporated unchanged into the SIPT as the Postrotary Nystagmus test. Only duration of elicited vestibular nystagmus is observed in Postrotary Nystagmus. Both high and low scores are unfavorable on this test.

There are a number of ways in which the integrity of central nervous system (CNS) processing of vestibular sensory input is manifested: (a) protective, righting, and postural responses; (b) equilibrium; (c) muscle tone; (d) muscle cocontraction or coinnervation; and (e) ocular responses. These manifestations indirectly reflect integration of vestibular as well as somatic and visual information. Postural reactions are particularly dependent upon input from the gravity receptors. Ocular responses are closely associated with stimulation of the semicircular canals. Maintaining a stable visual field, an important function of the vestibulo-ocular reflexes, is critical to operating within the environment. Several CNS pathways and many mechanisms are involved in maintaining vestibulo-ocular integrity.

Elicited vestibular postrotary nystagmus is one of many responses to stimulation of the semicircular canals. When electronystagmographic readings of nystagmus are made, many aspects of the beats are considered, such as velocity of the slow phase, frequency, amplitude, and rhythmicity. These parameters are not easily observed during or following manual rotation, but duration following rotation can be measured. It has been established by a number of studies that duration of postrotary nystagmus is a meaningful index of some aspect of sensory integration, although the exact nature of that meaning will need investigation for many years.

Using electronystagmography, Fisher, Mixon, and Herman (1986) sought to determine just which measurement of visuo-vestibular function would best reflect the type of inadequate central nervous system processing of vestibular inputs often seen in individuals with learning disabilities and motor incoordination. They compared reflex eye responses of 10 normal adults (mean age = 30 years) with 7 adults (mean age = 31 years) with a history of learning disabilities and showing suspected vestibular dysfunction as judged by current or past clinical observation. The investigators looked at a number of different vestibulo-ocular responses to three different types of stimuli. All data were gathered with the subjects in the dark. Discriminant analysis determined that the vestibulo-ocular reflex elicited by trapezoidal rotation most accurately classified the subjects into their respective groups and that, of the types of measurements obtained from trapezoidal stimuli, the duration of postrotary nystagmus was the most discriminating. The classification rate of duration was 94.1%. (Postrotary Nystagmus of the SIPT provides trapezoidal rotation.) The overall results showed that adults with suspected vestibular dysfunction associated with learning disability and incoordination demonstrated normal cortical optokinetic nystagmus and phasic (transient) or peripheral vestibular responses, but demonstrated impaired tonic (prolonged) processing of vestibular inputs within indirect brain stem pathways. The authors concluded that a diagnosis of vestibular dysfunction based on a meaningful cluster of symptoms including duration of postrotary nystagmus and clinical observation or tests of postural reactions is valid. Their data also support the premise that vestibular dysfunction at a brain stem level contributes to learning disability or deficits in gross motor control.

Keating (1979) compared the duration of postrotary nystagmus as measured by SCPNT and by electronystagmography of 20 normal women, 25 to 30 years of age, and obtained a correlation of .899 ($p < .01$). Both measurements were made in the light. In contrast, Royeen, Lesinski, Ciani, and Schneider (1981) found no agreement between SCPNT and otolaryngological evaluations of a 7-year-old boy with a SCPNT score of −1.8 *SD*, nor between short (6 seconds) nystagmus duration in a 10-year-old boy. Both boys scored poorly in two areas of auditory processing.

Originally, locally normed (Ayres, 1975) on 111 boys and 115 girls (ages 5 through 9 years) enrolled in public schools in Los Angeles County, SCPNT normative data were expanded and compared with several other sample populations. Punwar (1982) tested 372 children, ages 3 through 10 years, and found no significant differences (except for 5-year-old boys whose mean score was longer) between her scores and those obtained in southern California. She found no significant differences in responsivity by age, sex, or interaction. Similarly, no significant difference was found between nystagmus duration of 41 4-year-old Seattle children and the Los Angeles normative sample. However, in the Seattle sample, boys had significantly longer duration than did girls (Crowe, Deitz, & Siegner, 1984). Longer ($p < .01$) duration nystagmus was obtained in 222 Syracuse, New York, children, ages 5 to 9 years, compared to the Los Angeles County children (Kimball, 1981). Although the duration of postrotary nystagmus from 3 to 10 years of age remains relatively constant (Punwar, 1982), the duration is longer in adults (Keating, 1979; Shuer, Clark, & Azen, 1980). Limited SCPNT normative scores on 4-year-olds were reported by Clayton, Cochran, Edwards, and Reep (1985, 1986).

A number of postural and muscle parameters believed to be reflective of central vestibular system integrity have been associated with low scores on the SCPNT. These parameters include reduced oculomotor control skills (Ottenbacher, Watson, Short, & Biderman, 1979), walking with eyes closed and open, standing balance with eyes closed and open, muscle tone, and prone extension posture (Clyse & Short, 1983; Ottenbacher, 1978). Short, Watson, Ottenbacher, and Rogers (1983) found that regression analysis of 145 4-year-old children on measures of muscle tone, muscle cocontraction, standing balance, prone extension and supine flexion postures, and asymmetric tonic neck reflex indicated that these combined variables accounted for only 13.5% of the variance of postrotary nystagmus duration. However, for children exhibiting nystagmus duration lower than −1.0 *SD*, these variables accounted for 50% of the variance on the SCPNT. The prone extension posture, standing balance with eyes closed, and muscle tone accounted for 37% of the variance.

In contrast, Bundy and Fisher (1981) did not find SCPNT scores to predict the prone extension posture in a study of 343 normal children in the first, second, and third grades of three public schools. The authors felt their study offered support for the division of static from dynamic vestibular function.

The lack of high correlation between SCPNT scores and muscle and postural measures of vestibular sensory processing suggests that postrotary nystagmus and postural responses reflect different CNS functions which may or may not be closely associated. Dutton's (1985) review of the SCPNT research led her to conclude that the test could be used with confidence if its interpretation was conservative.

SCPNT has been used with a number of different diagnostic groups to help define the sensory integrative status. Interpreted as a central nervous system underreactivity to vestibular sensory input and not as a peripheral disorder, short nystagmus duration has been found in a number of studies of learning-disabled children. Ayres (1978) found 50% of 102 boys and 26 girls—ages 6 to 10 years (mean = 97.6 months, *SD* = 14.3 months) with learning disabilities—had SCPNT scores of -1.1 *SD* or lower. The expected number of scores that low in a normal sample is only 14%. In a group of 73 boys and 19 girls (ages 53 to 120 months) who appeared to be of normal intelligence, but with diagnoses of learning disability, minimal brain dysfunction, or perceptual-motor disorder, 46% were found to have SCPNT scores less than -1.0 *SD* (Ottenbacher, 1978). In another sample of 109 (71 boys, 38 girls) learning-disabled children, ages 58–116 months, Ottenbacher (1980) found 45% had SCPNT scores of less than -1.0 *SD*. On the other hand, 14% of the scores were above 1.0 *SD*, indicating prolonged nystagmus. The children with excessive nystagmus duration showed greater neuropsychological impairment as indicated by combined SCSIT scores below -1.0 *SD*. The SCSIT scores accounted for less than 10% of shared variance with normal postrotary nystagmus duration, but for 23% of shared variance in a group exhibiting short nystagmus. The SCSIT scores that most closely differentiated the long and short nystagmus groups were Motor Accuracy, Imitation of Postures, Bilateral Motor Coordination, Space Visualization, Position in Space, and Design Copying. The tactile-kinesthetic group of tests did not differentiate between the long and short nystagmus groups. Ottenbacher (1980) interpreted these results as evidence that prolonged nystagmus is associated with deficits in higher-level integrated cortical functioning.

Morrison, Hinshaw, and Carte (1985) found low nystagmus duration in a sample of 37 learning-disabled children. The nystagmus scores did not correlate significantly with equilibrium reactions. There was a marginally significant association with primitive reflexes ($r = .32; p < .10$), with greater duration of nystagmus tending to associate with lower reflex scores.

Among 52 children—66 to 136 months of age, medically referred because of perceptual-motor difficulties, educational difficulties, or both—those with SCPNT scores below -1.0 *SD* showed poorer human figure drawing skill than those with average scores (Watson, Ottenbacher, Short, Kittrell, & Workman, 1981). Similarly, postrotary nystagmus duration shared a significant amount of variance with human figure drawing scores of 40 learning-disabled children, ages 59 to 146 months (Ottenbacher, Abbott, Haley, & Watson, 1984). The authors interpreted their findings to support the assertion that some learning-disabled children have deficits in vestibular processing that may affect performance on cognitive-perceptual tasks.

Low SCPNT scores have also been associated with behavior disorders. In a group of 31 boys and 14 girls (ages 51–114 months) with learning disabilities, boys judged by their teachers to have the most socially inappropriate behavior scored significantly lower ($p < .05$) on the SCPNT than those with more socially appropriate behavior. The same was not true of the girls, even though their nystagmus durations were essentially the same as those of the boys (Ottenbacher, Watson, & Short, 1979).

Among 17 boys (mean age = 106.8 months, *SD* = 15.4), previously judged to be good or poor responders to Ritalin, Kimball (1986) predicted response to Ritalin with SCPNT scores. The poor responders showed prolonged nystagmus and an adverse reaction to test procedure; children with decreased postrotary nystagmus duration and nonaversive reactions to SCPNT administration tended to be good responders to Ritalin.

In a sample of 23 emotionally disturbed children (14 boys, 9 girls), 5.9 to 12.4 years of age, the majority had a SCPNT score of -1.0 *SD* or lower (Watson, Ottenbacher, Workman, Short, & Dickman, 1982). The authors thought the finding implicated "an important vestibular contribution to the characteristics of these subjects" (p. 71). Of 43 children (41 boys, 2 girls, ages 6–13 years) enrolled in a public school center for behaviorally disordered children, 32.5% were found to have hyporeactive nystagmus as measured by the SCPNT (Royeen, 1984).

Underreactivity of the CNS to vestibular input, as indicated by short nystagmus duration, was found: (a) in a group of 16 mildly retarded adults, ages 18–33 years (Shuer, Clark, & Azen, 1980); (b) in 18 female subjects with adolescent idiopathic scoliosis (Jensen & Wilson, 1979); (c) in 35 children, ages 5–9 years with Down's syndrome (Zee-Chen & Hardman, 1983); (d) in 15 children, ages 3.5 to 5.5 years, with a history of otitis media (Schaaf, 1985); and (e) in 111 children (ages 3 years, 9 months to 15 years, 4 months) with either articulation, speech and language disorders, or language disorder (Stilwell, Crowe, & McCallum, 1978). In contrast, 8 boys and 4 girls, ages 4.0 to 6.3 years, with nonparalytic strabismus demonstrated abnormally prolonged nystagmus ($p < .001$) (Slavik, 1982). The children with strabismus also showed delays in fine motor adaptive and gross motor areas of the Denver Developmental Screening Test.

Depressed nystagmus has been found in children who were considered high risk at birth. Deitz and Crowe (1985) looked at the SCPNT scores of 95 4½-year-old children (46 boys, 49 girls) who had been high-risk infants. Eight boys and three girls (11.6%) scored 0 (zero) seconds; 29 girls and 20 boys (52%) scored within the normal range between ± 1.0 *SD*. The zero nystagmus group performed more poorly than the normal nystagmus group (all $p < .05$) on the Peabody Developmental Gross Motor Scale (Folia & Dubose, 1974), scissor cutting, the WPPSI-Verbal scale, standing balance, and the Frostig Developmental Test of Visual Perception (Frostig, Lefever, & Whittlesey, 1961). Group differences in scores on the WPPSI-Performance Scale and on the prone extension posture did not reach significance.

DeGangi (1982) found that a significant proportion of 55 high-risk infants (3 to 24 months of age) who had abnormal or suspect neurological function and delayed motor skills exhibited nystagmus of significantly lesser or greater than normal duration when compared with a control group of 20 normal full-term infants. Nystagmus was observed using a modified SCPNT board. Chi-square analyses found a significant ($p < .05$) relationship between abnormal duration (either long or short) and the Motor Scale of the Bayley Scales of Infant Development (Bayley, 1969), but not the Mental Scale of the Bayley.

Potter and Silverman (1984) found that 58.8% of a group of 34 deaf children, ages 5–9 years, scored −1.0 *SD* or lower on the SCPNT ($p < .001$), and nearly half of the group showed no observable nystagmatic response. To correct for the variable of head and trunk control while testing nystagmus in young children, Kennedy (1983) developed a revised procedure which correlated .88 ($p < .05$) with the standard procedure employed with 5- and 6-year-olds.

As part of the SIPT, the Postrotary Nystagmus administration procedure is the same as that of the SCPNT, but the normative data are different. Of 1,750 normative children, 86 (4.9%) scored 0 seconds on PRN. There were no normative cases scoring 0 seconds on the SCPNT. Only 20 (1.1%) of the SIPT normative sample demonstrated nystagmus longer than approximately 2.0 *SD* by SCPNT norms. The mean PRN score of 125 dysfunctional children was −.375 *SD*. The range was −2.22 to 1.88 *SD*. Their PRN clockwise scores correlated .39 with their counterclockwise scores. The data indicate a greater prevalence of low than high PRN scores among the dysfunctional sample studied.

Because both too high and too low PRN scores are unfavorable, correlations between total PRN scores and other variables are often clouded in samples studied. In a sample of 1,750 normative children, PRN was not significantly correlated with scores on any of the other SIPT. However, in the sample of 125 children with learning or sensory integrative deficits (see Table 18), PRN had a significant negative correlation with Praxis on Verbal Command [PrVC] (−.22) and with Bilateral Motor Coordination [BMC] (−.15). In the matched sample of dysfunctional and normal children (see Table 19), PRN had significant positive correlations with Design Copying [DC] (.17), suggesting a relationship between short nystagmus and poor visual space management. In that sample, PRN also correlated with Finger Identification [FI] (.20), Graphesthesia [GRA] (.18), Oral Praxis [OPr] (.15), and Sequencing Praxis [SPr] (.14).

Positive association of PRN scores with somatosensory test scores is seen in the third factor of the principal-components analysis of the SIPT scores of 1,750 normative children (see Table 8). The PRN loading on that factor is .61. Factor analyses also demonstrated the negative relationship of PRN with Praxis on Verbal Command among the matched sample of normal and dysfunctional children (see Table 10). This negative relationship is consistent with the significant negative correlation between PRN and Praxis on Verbal Command ($r = -.22$; $p < .05$) for the dysfunctional group, but is in contrast to a positive (nonsignificant) correlation of .03 for the normative group (see Table 18). It is consistent with an earlier analysis of scores of learning-disabled children where PRN loaded −.37 on an auditory-language factor (Ayres, 1977). The negative relationship between PRN and Praxis on Verbal Command among dysfunctional groups is interpreted as reflecting insufficient higher level CNS inhibition on the oculomotor reflex generated at lower CNS levels. This interpretation is consistent with the mean PRN score of 1.09 *SD* (*SD* = 1.46) in a sample of 10 brain-injured children (see Table 8).

In summary, the data support the contention that both exceptionally high and exceptionally low PRN scores are suggestive of dysfunction. Either indication of dysfunction may result in similar patterns of low scores, and those patterns vary a great deal. When a PRN score fell within normal limits, it had little relevance for other SIPT scores, but when it was higher than the average range, it was most apt to be associated with low Praxis on Verbal Command. When the PRN score was well below average, scores on the complex somatosensory tests were most apt to be depressed, but scores on those tests could also be depressed by the same conditions that contributed to high PRN. Scores on the somatopraxis tests, the visuopraxis tests, and SVCU scores were also apt to be negatively influenced by the same conditions that contribute to depressed nystagmus. PRN appears to have a direct relationship to management of visuospace and somatosensory processing. PRN associations with several SIPT are also indirect in that conditions that contribute to high PRN scores also contribute to some low SIPT scores.

**Standing and Walking Balance (SWB)**

Scores on the SIPT Standing and Walking Balance correlated significantly with scores on a number of other SIPT. In a sample of 1,750 children from the normative sample (see Table 18), highest correlations were with Bilateral Motor Coordination [BMC] (.26), Sequencing Praxis [SPr] (.25), Oral Praxis [OPr] (.25), Motor Accuracy [MAc] (.23), Design Copying [DC] (.23), Praxis on Verbal Command [PrVC] (.20), Postural Praxis [PPr] (.20), and Constructional Praxis [CPr] (.20). Among the 125 dysfunctional children (also Table 18), the highest

Standing and Walking Balance correlations were with Oral Praxis (.43), Design Copying (.42), Manual Form Perception [MFP] (.39), Sequencing Praxis [SPr] (.38), Finger Identification [FI] (.37), Figure-Ground Perception [FG] (.35), Graphesthesia [GRA] (.34), Motor Accuracy [MAc] (.33), Constructional Praxis [CPr] (.32), and Kinesthesia [KIN] (.32). In this dysfunctional sample, Standing and Walking Balance: Open correlated .79 with Standing and Walking Balance: Closed; in the normative group that correlation was .62. There were no meaningful differences between the Open and Closed correlations with the rest of the SIPT.

In a combined group of normal ($n$ = 47), learning-disabled ($n$ = 35), and sensory integration disordered ($n$ = 9) children, Standing and Walking Balance scores were significantly correlated with all of the major K-ABC scales (see Table 15). In fact, Standing and Walking Balance had the highest correlation of any of the SIPT (.44) with the K-ABC Achievement Scale.

These extensive Standing and Walking Balance correlations suggest that some process needed for body balance may contribute to performance on many of the other SIPT, and may also be implicated in the more cognitive tasks that are included in the K-ABC. Three major relationships are indicated by the pattern of correlations between Standing and Walking Balance and the other SIPT: (a) an association with the tests assessing bilateral integration and sequencing (BIS) (Bilateral Motor Coordination, Sequencing Praxis, Oral Praxis, and Graphesthesia); (b) correlations with other tests tapping proprioception (Oral Praxis, Kinesthesia, Bilateral Motor Coordination, Sequencing Praxis, Motor Accuracy, Postural Praxis, and Manual Form Perception); and (c) an association with tests of visuoconstruction (Constructional Praxis and Design Copying). Because children with definite neuromotor problems were not included in either the normative or the dysfunctional sample, the most likely basis for these relationships is integration of sensation from the vestibular (especially macular) and proprioceptive systems. The correlations could also be interpreted as suggesting that Standing and Walking Balance and all tests correlating significantly with it have a common vulnerability to dysfunction.

These relationships are reflected in the factor analyses of scores of both normative and dysfunctional children. In these analyses, Standing and Walking Balance generally had its highest loadings on the visuo- and somatopraxis and BIS factors. This relationship between Standing and Walking Balance and BIS reaffirms a previously established relationship between postural responses and bilateral integration (Ayres, 1969b). It should be noted, however, that in other analyses, Standing and Walking Balance sometimes had quite low loadings on the BIS factors, indicating that it is not an invariable concomitant of the BIS function. In separate analyses of scores of dysfunctional children whose Postrotary Nystagmus (PRN) scores were either above or below the test's normative mean, low SWB scores were more closely associated with PRN scores above the mean than with PRN scores below the mean. On a number of untabled factor analyses of scores of children with low PRN, SWB seldom carried a high loading where PRN also had a high loading. However, on some analyses of scores of children with PRN scores above the mean, SWB (either with eyes open or eyes closed) loaded moderately with a sign opposite to that of the PRN loadings. These factor analyses suggest that in the sample studied, SWB was more likely to be negatively influenced by the conditions associated with prolonged postrotary nystagmus than with short postrotary nystagmus. However, the correlations of .12 and .06 between SWB and PRN (see Table 18) indicate that such a relationship is not invariable.

The SIPT groups varied considerably in their scores on Standing and Walking Balance. The three dysfunctional groups (Groups 2, 3, and 5) all scored below −1.0 *SD*, indicating a vulnerability of body balance to dysfunction regardless of type of sensory integrative dysfunction. That vulnerability is also inferred from the fact that one of the lowest means for dysfunctional children on the SIPT was on Standing and Walking Balance. The data strengthen the notion that a process important in body balance also is important in academic learning.

Earlier tests of standing balance discriminated between dysfunctional and control children ($p < .01$) (Ayres, 1965) and between children who earlier in life scored above versus below 100 on the Gesell Developmental Schedules ($p < .01$) (Ayres, 1969a). Su and Yerxa (1984) reported a wide range of scores (mean *SD* score = 1.47) on Standing Balance: Eyes Open of the SCSIT and a mean *SD* score of −1.21 on Standing Balance: Eyes Closed for 30 8-year-old children with sensory integrative dysfunction. These authors also reported that scores on these tests held the lowest correlation (.40 and .53, respectively) of any of the SCSIT sensorimotor tests with the total motor skills items of the Luria-Nebraska Neuropsychological Battery (Golden, Hemmeke, & Purisch, 1980). The latter tests upper extremity function only.

In an early factor analysis, scores on balance tests with eyes open and eyes closed formed a doublet (Ayres, 1965), but on subsequent analyses, the variables have loaded with measures of somatomotor and practic ability (Ayres, 1966d, 1972b; Ayres, Mailloux, & Wendler, 1987; Silberzahn, 1975).

When a preliminary form of the SCSIT test of standing balance was administered to 92 children with a mean age of 69.2 months ($SD$ = 13.6 months), the correlation of scores between standing balance and the scores on the total Gesell Developmental Schedules given at 56.8 weeks ($SD$ = 40.5 weeks) was .38 ($p < .01$). Standing balance scores correlated significantly with the Adaptive subscale score (.40), with the Fine Motor score (.38), and with the Gross Motor score (.34) (Ayres, 1969a). These findings suggest a common underlying sensory integrative function subserving both body balance and early adaptive ability.

In a sample of 50 learning-disabled children, 25 of whom had vestibular disorder and 25 of whom did not, Standing Balance: Closed correlated ($p < .005$) with the Bruininks Oseretsky Balance subtest (Bruininks, 1978) and with ($p < .005$) one of five measures of tilt reactions and equilibrium. Standing Balance: Open correlated ($p <$ .05) with one of the tilt reactions. The fact that there were so few significant correlations led the authors to suggest that tilt reactions test something different from Standing Balance: Open and Standing Balance: Closed (Bundy, Fisher, Freeman, Liebert, and Israelovitz, 1987). Potter and Silverman (1984) reported that among 34 severely or profoundly deaf children with IQs ranging from 85 to 151, 44.1% of the scores on Standing Balance: Open were below -1.0 $SD$ and 35.3% of the Standing Balance: Closed scores were below -1.0 $SD$. There was no significant association between the Standing Balance: Open or Standing Balance: Closed scores with SCPNT scores. Standing Balance: Open was one of the major predictors of a child's tendency to make reversals or draw right to left in executing the figures of Design Copying of the SIPT (McAtee, 1987).

In summary, Standing and Walking Balance correlated with many of the other SIPT. The data show relationships to bilateral integration and sequencing, visuopraxis, somatopraxis, and to other tests tapping proprioception. SWB performance appears to be vulnerable to the same conditions that contribute to prolonged nystagmus and to the other patterns of sensory integrative dysfunction identified by the SIPT. Because SWB is apt to be affected by any type of sensory integrative dysfunction, it serves better to identify dysfunction than it does to differentiate between sensory integrative patterns of dysfunction.

**Praxis on Verbal Command (PrVC)**

In a sample of 1,750 children from the SIPT normative group (see Table 18), correlations between Praxis on Verbal Command and the other SIPT scores particularly reflected the somatopractic aspect of this test. Highest correlations were with Sequencing Praxis [SPr] (.31), Design Copying [DC] (.28), Constructional Praxis [CPr] (.26), Postural Praxis [PPr] (.25), and Bilateral Motor Coordination [BMC] (.24). Similar patterns of correlations were found for the matched sample of dysfunctional and normal children (see Table 19), and for the sample of 125 children with learning or sensory integrative disorders (Table 18). In this latter group of dysfunctional children, the SIPT showing the highest correlations with Praxis on Verbal Command were Figure-Ground Perception [FG] (.41), Manual Form Perception [MFP] (.37), Oral Praxis [OPr] (.37), Sequencing Praxis (.35), Constructional Praxis (.35), and Bilateral Motor Coordination (.34). In addition, Praxis on Verbal Command had a correlation of -.22 with Postrotary Nystagmus (PRN) in this group (in contrast to the .03 correlation with PRN in the normative group). In a separate group consisting of normal ($n$ = 47), learning-disabled ($n$ = 35), and sensory integration disordered ($n$ = 9) children, scores on Praxis on Verbal Command were significantly correlated with all of the major K-ABC scales (Table 15).

In factor analyses, Praxis on Verbal Command loaded most strongly on the visuopraxis factors, but it also loaded on the somatopraxis factors. When five factors were extracted for the group of 125 dysfunctional children (see Table 9), Praxis on Verbal Command loaded -.59 and Postrotary Nystagmus loaded .76 on a factor identified as a praxis on verbal command factor. This relationship between high Postrotary Nystagmus and low PrVC also appeared in several other analyses and supports the hypothesis that a low PrVC score may reflect a problem with left cerebral hemisphere integrity. The praxis on verbal command factor did not correlate appreciably with the other factors. PrVC also loaded .32 on the bilateral integration and sequencing (BIS) factor.

In summary, Praxis on Verbal Command shows some linkage with both visuopraxis and somatopraxis, but is distinguished from the other praxis tests by its negative correlation with Postrotary Nystagmus. This negative correlation, along with some of the other data, suggests that PrVC may be impaired by left cerebral hemisphere inefficiency. A BIS deficit is apt to accompany low PrVC, but that kind of BIS dysfunction does not necessarily stem from the same conditions that are associated with the BIS dysfunction seen in other groups.

**Design Copying (DC)**

Of all the SIPT, Design Copying has one of the highest saturations of a common praxis element. It especially taps practic management of two-dimensional visual space. Design Copying and Constructional Praxis (CPr) both require constructional praxis. Numerous data analyses suggest that practic construction is an especially vulnerable human ability. While Design Copying is one of the best single indicators of dysfunction, it should be noted that some dysfunctional children do perform adequately on this task.

In a combined group of normal, learning-disordered, and sensory integration disordered children (see Table 15), Design Copying was significantly correlated with all of the major K-ABC scales. It had the highest correlation (.73) with K-ABC Arithmetic scores of any of the SIPT. In addition, Design Copying and Constructional Praxis had the highest correlations with the K-ABC Simultaneous Processing Scale (both .66), and with the K-ABC Mental Processing Composite (.66 and .68, respectively).

In a group of 1,750 children from the normative sample (see Table 18), Design Copying was significantly correlated with many of the other SIPT, especially with Constructional Praxis (.44), Sequencing Praxis [SPr] (.41), and Motor Accuracy [MAc] (.38). Lowest correlations were with Localization of Tactile Stimuli [LTS] (.09) and Postrotary Nystagmus [PRN] (.03). Among 125 dysfunctional children (also Table 18), correlations with other SIPT were similar but higher—the highest being with Constructional Praxis (.52), Space Visualization [SV] (.44), Figure-Ground Perception [FG] (.43), and Motor Accuracy (.43). Lowest correlations in the dysfunctional sample were with Localization of Tactile Stimuli (.11) and Postrotary Nystagmus (.06). Design Copying correlations in the group of 293 matched dysfunctional and normal children (see Table 19) showed essentially the same relationships, but with a notably higher correlation (.51) with Praxis on Verbal Command (PrVC). This latter correlation would appear to reflect a common cognitive inefficiency.

Design Copying Part I accuracy correlated .52 with the Part II Should Have score in the sample of 1,750 normative children and .66 in the sample of 125 dysfunctional children. In the normative sample, all Part II approach parameters correlated negatively ($p < .05$) with the Part I accuracy and with the Part II Should Have scores (i.e., the higher the accuracy and the better the Should Have score, the fewer the approach errors). In the normative group, segmentation errors and additions correlated highest with the Part II Should Have score (–.34 and –.30, respectively; both $p < .01$). The pattern of correlations between approach parameters, Part I Accuracy and Part II Should Have parameters were similar to those for the normative sample. However, in the dysfunctional sample, the correlations between Part II Should Have parameters and right-left errors, inversions, and jogs were close to zero, and were not statistically significant. Similarly, the correlations between Part I accuracy and right-to-left errors and jogs were trivial and not statistically significant.

On all of the tabled factor analyses (Tables 8, 9, and 10) Design Copying had either the highest or one of the highest loadings on the visuopraxis factors, and had one of the lowest loadings on the somatopraxis factors. The factor loadings show that Design Copying is more purely visual than is Constructional Praxis—the latter being more closely related to total body praxis. On some factors, Design Copying correlated positively with the other praxis tests, with the exception of Oral Praxis (OPr), with which it often correlated negatively.

Studying the scores on the SCSIT and on the atypical approach parameters to the SIPT Design Copying test in a small sample of 6- to 8-year-old dysfunctional children, McAtee (1987) found that the atypical approach parameters assessed various aspects of practic management of space and a vestibular-bilateral component reflected in directional confusion. Scores on additions, segmentation, inversion, and right-to-left associated most closely with visual perception tests of the SCSIT and were judged to particularly reflect aspects of visual space management. Scores on additions, segmentation, inversion, and jogs were influenced by deficits in praxis, as seen by their relationship to tests requiring motor planning. The vestibular-bilateral or directional component in reversal scores was assumed from: (a) the prediction of those scores by Standing Balance: Open and Crossing the Midline; (b) the prediction of segmentation scores by Bilateral Motor Coordination (BMC) and Standing Balance: Open (in addition to Space Visualization and Finger Identification); and (c) the correlation ($r = -.50; p < .05$) between right-to-left score with Standing Balance: Open. McAtee suggested that boundary errors may not be indicative of a discrete disorder, for these errors correlated with a number of other scores, yet were predicted by only Double Tactile Stimuli. McAtee concluded that the approach parameters appeared to measure related aspects of sensory integration, but that the correlations among the parameters were not high enough to suggest that any two parameters were measuring the same trait.

The SCSIT version of Design Copying—simpler than the SIPT version—has shown linkage to a number of patterns of dysfunction. In a Q-factor analysis (Ayres, 1969b,), low SCSIT Design Copying scores were associated with children who exhibited deficits in: (a) postural and bilateral integration; (b) auditory language, sequencing, and reading problems; and (c) those with poorer coordination on the left than the right side of the body along with deficiencies in postural and bilateral integration. In other factor analyses, SCSIT Design Copying loaded equally on a form and space perception factor and on a praxis factor (Ayres, 1972a) and, in another study (Silberzahn, 1975), had moderate loadings on both a form and space perception factor and an eye-hand coordination with visualization factor.

**Constructional Praxis (CPr)**

The Constructional Praxis test taps a practic skill

unique to the SIPT (and unique within the SIPT), yet the data indicate that it also tests an aspect of praxis that is common to most of the SIPT and is related to many complex cognitive tasks. In a combined sample of normal, learning-disabled, and sensory integration disordered children (Table 15), Constructional Praxis had a higher correlation (.68) with the K-ABC Mental Processing Composite than did any of the other SIPT, and also had the highest correlation (.66) with the K-ABC Simultaneous Processing Scale.

In a sample of 1,750 normative children (see Table 18), Constructional Praxis correlated significantly with all of the other SIPT except Postrotary Nystagmus (PRN). Highest correlations were with Design Copying [DC] (.44), Postural Praxis [PPr] (.30), and Sequencing Praxis [SPr] (.30). Among the 125 dysfunctional children (also Table 18), Constructional Praxis correlated highest with Postural Praxis (.54), Design Copying (.52), Finger Identification [FI] (.49), Figure-Ground Perception [FG] (.46), and Graphesthesia [GRA] (.46). Major correlations in the matched sample of 293 (see Table 19) are similar to those in the normal and dysfunctional samples, the highest correlation being with Design Copying (.62). Overall, these correlations suggest that three-dimensional construction involves more than visual space perception. They also indicate that Constructional Praxis assesses a basic visuosomatopraxis function.

In factor analyses, Constructional Praxis loaded most strongly on the visuopraxis factors, but it also loaded substantially on a somatopraxis factor where Postural Praxis had the highest loading (.89) (Table 9). Although Constructional Praxis right-for-left errors did not correlate significantly with Design Copying right-to-left or with Design Copying reversals, an untabled factor analysis linked CPr: right-for-left with DC: right-to-left and with Space Visualization Contralateral Use (SVCU) in the normative sample. The better the SVCU score, the fewer the CPr: right-for-left errors. Similar relationships were seen among the dysfunctional children. CPr: right-for-left helped identify a leftward orientation factor. Although CPr: right-for-left may reflect right-left confusion in some cases, on the whole the data indicate that unless errors on the right-for-left parameter are supported by other scores suggesting right-left confusion, they are best interpreted as simply poor praxis.

It should be noted that dysfunctional children's scores may vary considerably on Constructional Praxis, so it serves better for differential diagnosis than for detection of a problem. SIPT Group 2, Generalized Dysfunction, scored most poorly (–1.6) on Constructional Praxis, and SIPT Group 3, Visuo- and Somatodyspraxia, scored at the borderline (–0.9) level.

**Postural Praxis (PPr)**

Postural Praxis of the SIPT correlated with most of the other SIPT both in the normative and the dysfunctional samples, indicating a high saturation of a common praxis element. In a sample of 1,750 normative children (see Table 18), the tests that correlated most strongly with Postural Praxis were Oral Praxis [OPr] (.34), Sequencing Praxis [SPr] (.34), Constructional Praxis [CPr] (.30), Design Copying [DC] (.27), Bilateral Motor Coordination [BMC] (.27), Praxis on Verbal Command [PrVC] (.25), and Finger Identification [FI] (.22). In a sample of 125 dysfunctional children (also Table 18), the tests that correlated most strongly with Postural Praxis were Constructional Praxis (.54), Oral Praxis (.44), Graphesthesia [GRA] (.41), Space Visualization [SV] (.39), Finger Identification (.35), Manual Form Perception [MFP] (.32), Praxis on Verbal Command (.31), and Design Copying (.30). Similar patterns of correlations were found for the matched sample of 293 normal and dysfunctional children (see Table 19). These correlations point up the central practic ability common to many praxis tests, but especially to Postural Praxis, Oral Praxis, Sequencing Praxis, and the tests of visual construction. The postural praxis component of Praxis on Verbal Command performance probably underlies the strength of the correlation between that test and Postural Praxis. The tactile-praxis association established in prior research is also seen in these correlations. Part-score correlations link Postural Praxis scores as strongly with left as with right somatosensory test scores. Further evidence for the centrality of the practic ability assessed by Postural Praxis can be seen in its correlations with the K-ABC scales. In a mixed group of normal, learning-disabled, and sensory integration disordered children, Postural Praxis was significantly correlated with all of the major K-ABC scales.

In factor analyses, Postural Praxis loaded substantially on the somatopraxis factors. The association between postural practic ability and visually directed manipulation of objects (Constructional Praxis) is evident in the fifth factor of Table 9. The tactile-praxis linkage is evident across factors in all three analyses.

Postural Praxis of the SIPT has evolved from earlier tests assessing the ability to imitate body postures. The Imitation of Postures test, first published as one of the Southern California Perceptual Motor Tests (Ayres, 1968), was later incorporated into the SCSIT. An early form of the Imitation of Postures test discriminated between a dysfunctional and a control group with a critical ratio of 8.39 ($p < .01$) (Ayres, 1965). Imitation of Postures was consistently the best identifying variable (with factor loadings of .58, .77, .50, and .74) in factors designated as somatomotor or developmental dyspraxia in a number of factor analyses (Ayres, 1965, 1966d,

1972b, 1977). When preliminary versions of the SIPT praxis tests were entered into a factor analysis along with scores on the SCSIT, Imitation of Postures carried loadings of .55 and .43 on the praxis factors and correlated most strongly with Motor Accuracy-R, left hand (.49), a constructed block building test (.48), and a group of visual tests (.46) (Ayres, Mailloux, & Wendler, 1987).

Administered to a group with a mean age of 5.5 years, Imitation of Postures was a fair predictor of the total Gesell Developmental Schedules score given earlier at an average age of 56.8 weeks (*SD* = 40.5 weeks). The correlation of .58 between Imitation of Postures and the total Gesell score was exceeded only by Motor Accuracy, with a correlation of .59. Imitation of Postures correlated highest (.59) with the Adaptive subscale score, next (.57) with the Fine Motor score, and least (.48) with the Gross Motor score (Ayres, 1969a). The strong associations between both the tactile and praxis tests with the Adaptive subscale score indicate that the tactile-praxis component of development is related to successful early adaptive ability. It is less closely related to early motor development.

Because a group of 15 children with articulation disorders performed significantly more poorly than a normal control group on tests of motor impairment, but not on Imitation of Postures, Cermak, Ward, and Wood (1986) suggested that, as a group, children with articulation problems are not dyspraxic.

In summary, Postural Praxis evaluates a practic ability that is common to many of the SIPT, but especially to Oral Praxis and to visual construction. The tactile-postural praxis linkage is reaffirmed. Postural Praxis is one of the SIPT most vulnerable to the conditions that depress Postrotary Nystagmus (PRN) scores. Postural Praxis had a major role in differentiating among the different SIPT cluster groups, and it assumes an important role in differential diagnosis.

**Oral Praxis (OPr)**

The Oral Praxis test is one of the few tests in the SIPT that requires no vision for execution of the task. That uniqueness is reflected in its correlations and factor loadings. In a sample of 1,750 normative children (see Table 18), Oral Praxis correlated highest with Bilateral Motor Coordination [BMC] (.36), Postural Praxis [PPr] (.34), Graphesthesia [GRA] (.28), Sequencing Praxis [SPr] (.28), and Standing and Walking Balance [SWB] (.25). In a sample of 125 children with learning and sensory integrative deficits (also Table 18), the tests that correlated most highly with Oral Praxis were Graphesthesia (.50), Constructional Praxis [CPr] (.45), Postural Praxis (.44), Sequencing Praxis (.43), Praxis on Verbal Command [PrVC] (.37), and Manual Form Perception [MFP] (.37). In a matched sample of dysfunctional and normal children (see Table 19), Oral Praxis correlated most strongly with Graphesthesia (.50), Sequencing Praxis (.50), Motor Accuracy [MAc] (.49), Standing and Walking Balance (.48), and Postural Praxis (.46). Oral Praxis also had substantial and significant correlations with all of the major K-ABC scales. Part-score correlations in both the normative and dysfunctional groups link Oral Praxis as strongly with left as with right somatosensory test scores. All data considered, these correlations point up three major components of Oral Praxis: (a) somatosensory association, (b) bilateral integration and sequencing, and (c) motor planning subserving the entire body. There also may be a lesser association with eye-hand coordination.

Factor analyses also demonstrated these Oral Praxis components. In an analysis of scores of the normative sample of 1,750 children, Oral Praxis had the highest loading (.80) on the somatopraxis factor (see Table 8). It also had high loadings (.51 and .87) on the somatopraxis factors for the dysfunctional sample and for the matched sample of dysfunctional and normal children (see Tables 9 and 10). Oral Praxis also loaded moderately on the somatosensory processing and oral praxis factor (.37) for the dysfunctional group, and on the bilateral integration and sequencing factor (.40) (see Table 9).

SIPT groups varied considerably in their Oral Praxis scores, indicating that Oral Praxis can contribute much to differential diagnosis. Group 2, Generalized Dysfunction, scored especially poorly (–2.2 *SD*); Group 5, Dyspraxia on Verbal Command, scored –1.0 *SD;* and Group 1, Low Average BIS, scored –0.9 *SD* on Oral Praxis (see Table 11).

**Sequencing Praxis (SPr)**

In a sample of 1,750 children from the SIPT normative sample (see Table 18), Sequencing Praxis correlated substantially with all other tests in the SIPT, except Localization of Tactile Stimuli (LTS) and Postrotary Nystagmus (PRN). Highest correlations were with Bilateral Motor Coordination [BMC] (.51), Design Copying [DC] (.41), Graphesthesia [GRA] (.37), Postural Praxis [PPr] (.34), Praxis on Verbal Command [PrVC] (.31), and Motor Accuracy [MAc] (.31). In a group of 125 dysfunctional children (also Table 18), Sequencing Praxis was significantly correlated with most other tests in the SIPT. Highest correlations in this dysfunctional sample were with Graphesthesia (.54), Bilateral Motor Coordination (.52), Constructional Praxis [CPr] (.44), Oral Praxis [OPr] (.43), Design Copying (.41), and Standing and Walking Balance [SWB] (.38). Strong correlations also were found in the group of 293 matched dysfunctional and normative children (see Table 19). Highest Sequencing Praxis correlations in the group of 293 were with Bilateral Motor Coordination (.63), Design Copying (.60), Graphesthesia (.55), Standing and Walking Balance (.51), Oral Praxis (.50) and Motor Accuracy (.50). In a combined sample of normal, learning-disabled, and sensory integration disordered children (see Table 15), Sequencing Praxis was significantly correlated with all of the major K-ABC scales, and had the highest correlation (.67) with the K-ABC Sequential Processing scale of any tests in the SIPT. For the dysfunctional sample (Table 17), the correlation between Sequencing Praxis and the K-ABC Sequential Processing scale (.47) was significant, but the correlation with the K-ABC Simultaneous Processing scale (.15) was not.

In factor analyses, Sequencing Praxis loaded primarily on the somatopraxis factors for the normal (see Table 8) and matched (see Table 10) samples. In the sample of 125 dysfunctional children, however, its primary loading was on the bilateral integration and sequencing (BIS) factor (Table 9). Although sequencing has been considered a left cerebral hemisphere function, there is no evidence from these analyses to suggest that Sequencing Praxis can be used to detect unilateral dysfunction. In an effort to understand just what Sequencing Praxis assesses other than BIS, an additional factor analysis examined scores from the normative group omitting Bilateral Motor Coordination and Oral Praxis, the two tests with the most obvious task similarity to Sequencing Praxis. When two factors were generated, Sequencing Praxis carried a modest loading (.44) on a somatopraxis factor and slightly lower loading on a visuopraxis factor. When three to six factors were generated, the somatic foundation of Sequencing Praxis was emphasized. These data indicate that Sequencing Praxis has, at least in part, a somatic base and evaluates a central practic ability that subserves most aspects of praxis evaluated by the SIPT, and assesses BIS of movement.

In the cluster analyses, Sequencing Praxis was one of five tests which, as a group, identified the BIS function. The six SIPT cluster groups (see Table 11) showed marked differences in their Sequencing Praxis scores, with the dysfunctional groups exhibiting consistently low scores (–2.1 to –1.2 *SD*) on this test, as did the different diagnostic groups shown in Table 14.

In summary, Sequencing Praxis is particularly sensitive to dysfunction, makes few high-level cognitive demands, primarily measures bilateral integration and sequencing, and secondarily measures praxis as a fundamental skill.

**Bilateral Motor Coordination (BMC)**

Bilateral Motor Coordination assumes a more influential role as part of the SIPT than it did in the earlier SCSIT. In a group of 1,750 children from the SIPT normative sample (see Table 18), Bilateral Motor Coordination correlated most strongly with Sequencing Praxis [SPr] (.51), Oral Praxis [OPr] (.36), Design Copying [DC] (.30), Motor Accuracy [MAc] (.28), and Graphesthesia [GRA] (.28). In a sample of 125 dysfunctional children (also Table 18), Bilateral Motor Coordination correlated most strongly with Sequencing Praxis (.52), Graphesthesia (.39), Figure-Ground Perception [FG] (.36), Oral Praxis (.35), and Praxis on Verbal Command [PrVC] (.34). In this sample, it had a notable negative correlation with Postrotary Nystagmus [PRN] (–.15). Similar relationships were found in the matched sample of dysfunctional and normal children (see Table 19), but in this sample there was no significant correlation between Bilateral Motor Coordination and Postrotary Nystagmus.

In the group of 1,750 children from the normative sample, the Bilateral Motor Coordination arms score had a higher correlation with the Space Visualization Contralateral Use (SVCU) score than did any other non-Space Visualization part score ($r = .13; p < .01$). Because the SVCU score is not dependent upon the sequencing of action, this association strengthens the bilateral integration interpretation of Bilateral Motor Coordination. BMC: Arms correlated with BMC: Feet, .54 and .56 in the normative and dysfunctional groups, respectively.

In factor analyses, Bilateral Motor Coordination loaded strongly on the somatopraxis factors for the normative and matched samples (see Tables 8 and 10), and helped to identify the bilateral integration and sequencing (BIS) factor for the dysfunctional sample (see Table 9). These factor loadings show that Bilateral Motor Coordination shares a practic quality in common with Oral Praxis and Postural Praxis. Postural Praxis requires very little bilateral integration, thus suggesting that Bilateral Motor Coordination performance is partly dependent upon a general somatopractic ability. On the vestibular functioning (see Tables 8 and 10) and praxis on verbal command (Table 9) factors, Bilateral Motor Coordination tends to have substantial negative loadings, whereas Postrotary Nystagmus has high positive loadings. This inverse relationship, also evident in the simple correlations discussed previously, suggests that the same conditions that contribute to prolonged nystagmus may cause impairments in bilateral motor coordination. It should be noted, however, that low scores on Bilateral Motor Coordination were not *invariably* associated with long duration of Postrotary Nystagmus; in fact, in the group of 1,750 children from the normative sample (see Table 18) and in the matched sample of dysfunctional and normal children (see Table 19), there was virtually no correlation between Bilateral Motor Coordination and Postrotary Nystagmus. It seems likely that Bilateral Motor Coordination performance may be depressed by unilateral cerebral inefficiency of either side of the brain. There is no indication that Bilateral Motor Coordination taps a normally lateralized function.

An early version of the Bilateral Motor Coordination test differentiated between a dysfunctional and a control group with a critical ratio of 5.04 ($p < .01$) (Ayres, 1965). The test also discriminated with a $t$-ratio of 3.94 ($p < .01$) between children who earlier in life scored above versus below 100 on the Gesell Developmental Schedules. In the Gesell study the test showed a correlation of .57 with the total Gesell score, .54 with the Adaptive subscale score, .46 with the Fine Motor subscale score, and .39 with the Gross Motor score (Ayres, 1969a).

Through factor analysis (Ayres, 1965, 1969a), Bilateral Motor Coordination helped identify the pattern of dysfunction initially referred to as deficit in integration of function of the two sides of the body. Tapping the same bilateral integration function was probably responsible for Bilateral Motor Coordination loading on the same factor as preliminary versions of Sequencing Praxis and Oral Praxis (Ayres, Mailloux, & Wendler, 1987). Oral Praxis and Sequencing Praxis both have a number of items requiring the ability to coordinate the two sides of the body. Performance on an early version of Bilateral Motor Coordination also tapped praxis, as reflected in several factor analyses (Ayres, 1965, 1966d, 1972b; Silberzahn, 1975).

In summary, the data show five aspects to Bilateral Motor Coordination: (a) Bilateral Motor Coordination performance requires somatopractic ability; (b) Bilateral Motor Coordination primarily evaluates functional integration of the two sides of the body, and possibly sequencing; (c) Bilateral Motor Coordination is a major contributor to the diagnosis of BIS function; (d) Bilateral Motor Coordination is sensitive to the conditions that produce prolonged postrotary nystagmus, and probably is also sensitive to conditions that depress postrotary nystagmus; and (e) Bilateral Motor Coordination variance is not appreciably associated with test parameters that reflect directionality or right-left orientation.

## Criterion-Related Evidence: Predictive

The SIPT, as was the SCSIT, is designed to be diagnostic and prescriptive rather than predictive. The SIPT does not predict academic achievement, behavior problems, or any diagnostic category. The SIPT may help the clinician understand the functioning of the nervous system subserving learning and behavior. It provides insight into behavior so that an intervention course can be planned. This test objective contributes to limited data on the predictive validity of the SIPT.

Some studies have used retrospective analysis to gain insight into which tests might predict a child's response to occupational therapy for sensory integrative dysfunction. Studying the academic effect of sensory integrative procedures with 46 experimental and 46 control learning-disabled children (mean ages = 98.3 and 100.3 months, respectively), stepwise discriminant analysis determined that the pre-program score of −1.1 *SD* or lower on SCPNT was the best predictor of academic change. This was a good predictor of academic success for both experimental and control children: It predicted lack of achievement for control children receiving special education but no occupational therapy, and favorable ($p < .05$) achievement for children receiving both therapy and special education. The mean ages of the children with low SCPNT scores were 97.13 (*SD* = 13.25) and 98.43 (*SD* = 12.21) for the experimental and control groups, respectively (Ayres, 1978).

Carte, Morrison, Sublett, Vemura, and Setrakian (1984) failed to obtain a similar prediction of short nystagmus duration facilitating academic success through occupational therapy in a sample of 87 learning-disabled children, ages 6–11 years (mean = 102.5 months, *SD* = 16.7 months). It should be noted that most persons past 8 years of age with sensory integrative disorder are only fair to poor candidates for occupational therapy for sensory processing disorders. In reviewing studies on the effectiveness of sensory integrative procedures, Ottenbacher (1982) found that therapy had its greatest effect on children diagnosed as aphasic or "at risk," but he noted that these children represented the youngest age groups treated.

Scores on tests of Standing Balance: Open and Standing Balance: Closed of the SCSIT, when combined with measures of prone extension and muscle tone, predicted which learning-disabled children with short nystagmus duration would respond to sensory integrative procedures with an increase in nystagmus duration (Ottenbacher, Short, & Watson, 1980). Those authors also found (1979) that they could meaningfully categorize learning-disabled children according to their nystagmus duration responses; those with short durations responded to therapy with increased duration while others displayed decreases.

## Validity Generalizations

Given the similarities between the SCSIT and SCPNT and the SIPT, certain expectations can be made concerning SIPT application in specific populations. Appropriate diagnostic conditions for SIPT use and in which dysfunction can be expected to be identified in some group members are: (a) learning disabilities, (b) behavior and emotional problems, (c) language disorder, (d) developmental delay, (e) motor and perceptual-motor incoordination, (f) mental retardation, (g) high risk in infancy, (h) autism, and (i) spina bifida or meningomyelocele. The presence of sensory integrative dysfunction in some children in high-candidacy groups in no way implies dysfunction in all children in that category. Symptom complexes that have been identified through analyses of the performance of learning-disabled children, specifically those with developmental dyspraxia and vestibular-bilateral integration deficit, cannot be generalized to other diagnostic categories without further study.

There are insufficient data to enable use of specific SIPT score constellations as predictors of response to therapy. The exact ways in which identified sensory integrative and practic dysfunction may contribute to a presenting problem are not yet clear and remain an enticing area for future investigation.

Most validity data are based on use of the entire SIPT or SCSIT, or on the study of postrotary nystagmus duration only. Whereas patterns of scores across a number of SIPT can support meaningful inferences, inferences based on only a single test are subject to error. When Postrotary Nystagmus (PRN) is used independently, concomitant clinical observations of other vestibular functions should be made before interpreting the score.

Some of the information generated by the SCSIT and SCPNT can be generalized to use of the SIPT in similar situations; other information must be used with caution. A few of the tests in the SIPT have changed very little from previously published versions. The Postrotary Nystagmus test of the SIPT is the same as the previous SCPNT, but with the scores standardized against new normative data. Motor Accuracy (MAc), Space Visualization (SV), and Figure-Ground Perception (FG) have changed only in that their scoring methods have been refined and scores standardized against the new norms.

Only minor design changes were made in Finger Identification (FI), Kinesthesia (KIN), and Localization of Tactile Stimuli (LTS). Greater changes were made in Graphesthesia (GRA), Manual Form Perception (MFP), Bilateral Motor Coordination (BMC), Design Copying (DC), Standing and Walking Balance (SWB) (formerly Standing Balance: Open and Standing Balance: Closed), and Postural Praxis (PPr) (formerly Imitation of Postures). Previous findings from use of these tests can be expected to hold some generalizability to use of the SIPT, based on accumulation of evidence. The four entirely new praxis tests—Praxis on Verbal Command (PrVC), Sequencing Praxis (SPr), Oral Praxis (OPr), and Constructional Praxis (CPr)—have only a brief history and statistical analysis of data from a few diagnostic groups. With these tests, validity generalization is limited.

## Differential Prediction

The SIPT scores of children from different ethnic groups in the SIPT normative sample were compared using multivariate analysis of variance (MANOVA). Ethnic identification was available for 1,719 of the 1,750 children who provided complete data on the SIPT. In this sample, 1,339 of the children were white, 213 were black, 104 were of Hispanic origin, 36 were of Asian origin, and 27 belonged to other ethnic groups (e.g., American Indian, Aleut). The mean scores for each ethnic group are shown in Table 20.

There were no significant ethnic differences in scores on Finger Identification (FI), Oral Praxis (OPr), Bilateral Motor Coordination (BMC), Standing and Walking Balance (SWB), and Postrotary Nystagmus (PRN). Significant differences among ethnic groups were found for the remaining 12 tests in the SIPT. Children of Asian origin scored higher than other ethnic groups on Space Visualization (SV), Figure-Ground Perception (FG), Manual Form Perception (MFP), Kinesthesia (KIN), Graphesthesia (GRA), Design Copying (DC), Constructional Praxis (CPr), Postural Praxis (PPr), Sequencing Praxis (SPr), and Motor Accuracy (MAc). Black children scored significantly lower than white or Asian children on Space Visualization, Figure-Ground Perception, Manual Form Perception, Kinesthesia, Graphesthesia, Localization of Tactile Stimuli, Design Copying, Constructional Praxis, Postural Praxis, and Sequencing Praxis. On most of the tests, white and Hispanic children fell somewhere between these two extremes. However, on Praxis on Verbal Command (PrVC), children of Hispanic origin scored significantly lower than most other groups, probably reflecting the fact that English was not the primary language for many of these children.

Racial similarities and differences in SCSIT performance also have been reported. For example, Fox (1968) administered the Illinois Test of Psycholinguistic Abilities, the Ayres Space Test, the SCSIT Motor Accuracy test and the Southern California Figure-Ground Visual Perception test to 194 Wisconsin Headstart enrollees, with American Indian, black, and white children represented. Multivariate analysis of variance yielded significant main effects for race and for sex. All of the significant racial differences were between black and American Indian groups, with Indian children outperforming black children on Motor Accuracy and on the Ayres Space Test.

Knorn (1985) found differences in SCSIT scores

**Table 20**

**SIPT Means and Standard Deviations for Different Ethnic Groups From the Normative Sample**

| Test | Asian (n = 36) | | Black (n = 213) | | Hispanic (n = 104) | | White (n = 1,339) | | Other (n = 27) | | Total (n = 1,719) | |
|---|---|---|---|---|---|---|---|---|---|---|---|---|
| | *M* | *SD* | *M* | *SD* | *M* | *SD* | *M* | *SD* | *M* | *SD* | *M* | *SD* |
| Space Visualization (SV) | .39 | .89 | –.42 | .98 | –.18 | .92 | .07 | .96 | –.07 | .88 | .00 | .98 |
| Figure–Ground Perception (FG) | .19 | 1.05 | –.23 | .95 | –.14 | .94 | .05 | .98 | –.19 | .89 | .00 | .98 |
| Manual Form Perception (MFP) | .41 | .89 | –.24 | 1.08 | –.02 | 1.00 | .03 | .96 | .19 | .92 | .00 | .98 |
| Kinesthesia (KIN) | .29 | .98 | –.15 | 1.02 | –.05 | 1.04 | .05 | .93 | .11 | .81 | .03 | .95 |
| Finger Identification (FI) | .17 | .79 | –.11 | 1.11 | .00 | 1.04 | .01 | .97 | .22 | 1.19 | .00 | .99 |
| Graphesthesia (GRA) | .40 | .87 | –.25 | 1.04 | .10 | 1.05 | .02 | .96 | .02 | .96 | .00 | .98 |
| Loc. of Tactile Stimuli (LTS) | .06 | 1.06 | –.20 | 1.02 | .11 | .96 | .04 | .97 | –.13 | 1.00 | .00 | .98 |
| Praxis on Verbal Command (PrVC) | –.10 | .86 | –.40 | 1.10 | –.70 | 1.21 | .15 | .81 | –.75 | 1.29 | .01 | .93 |
| Design Copying (DC) | .49 | .93 | –.47 | 1.04 | –.15 | 1.04 | .10 | .91 | .06 | .79 | .02 | .95 |
| Constructional Praxis (CPr) | .45 | .83 | –.44 | .89 | –.07 | .84 | .11 | .83 | .00 | .81 | .03 | .93 |
| Postural Praxis (PPr) | .34 | .88 | –.19 | 1.05 | –.03 | 1.03 | .03 | .98 | –.14 | 1.03 | .00 | 1.00 |
| Oral Praxis (OPr) | .15 | .78 | .01 | 1.02 | .09 | 1.01 | –.01 | .98 | .22 | .89 | .00 | .98 |
| Sequencing Praxis (SPr) | .37 | 1.01 | –.28 | 1.04 | –.23 | .96 | .05 | .98 | .16 | 1.06 | .00 | .99 |
| Bilateral Motor Coord. (BMC) | –.03 | 1.02 | –.06 | .96 | –.13 | .96 | .02 | .99 | .04 | 1.15 | .00 | .99 |
| Standing/Walking Balance (SWB) | .19 | .98 | –.13 | 1.06 | –.02 | 1.01 | .01 | .95 | .02 | .94 | .00 | .97 |
| Motor Accuracy (MAc) | .21 | .80 | –.18 | .89 | –.23 | .96 | .05 | .95 | –.10 | .90 | .00 | .96 |
| Postrotary Nystagmus (PRN) | –.03 | .99 | –.07 | .84 | .10 | .93 | .02 | .91 | –.19 | .97 | .01 | .93 |

between a group of 20 male and 16 female German children (5 to 5½ years of age) and the Los Angeles County norms for children of that age. The German children had significantly higher scores than the Los Angeles norms on Imitation of Postures, Bilateral Motor Coordination, and Right-Left Discrimination, and significantly lower scores than the Los Angeles normative group on Motor Accuracy, Standing Balance: Open, Standing Balance: Closed, Double Tactile Stimuli, Crossing Midline of the Body, and Graphesthesia.

Saeki, Clark, and Azen (1985) compared the scores of 82 Japanese children born in the U.S. and 98 Japanese children born in Japan with the U.S. normative data for the revised version of Motor Accuracy and the SCSIT Design Copying test. Both Japanese groups performed significantly better than the original normative group on Design Copying and on Motor Accuracy right-hand performance; there was no significant difference, however, in Motor Accuracy left-hand scores. For Design Copying, there was also a significant cultural effect, along with a significant interaction between age and culture, with the greatest cultural differences appearing at the oldest ages (7 to 10 years of age).

## Reliability

### Test-Retest Reliability

Test-retest reliability measures the extent to which test scores for an individual are consistent across different testings over time. Insofar as the constructs assessed by the SIPT are assumed to be fairly stable over time, a good measure of these constructs should have fairly high test-retest reliability.

The test-retest reliability of the SIPT was evaluated in a sample of 41 dysfunctional children (24 boys, 17 girls; mean age = 6.5 years, *SD* = 1.3 years) and 10 normal children (4 boys, 6 girls; mean age = 6.8 years, *SD* = 1.4 years). Each child was tested twice with the SIPT, with an interval of one to two weeks between the first and second testing. The test-retest reliability coefficients are shown in Table 21.

As a group, the praxis tests had the highest test-retest reliability, but reliabilities for most of the other tests were acceptable. However, four of the 17 tests showed low test-retest reliability. These included Postrotary Nystagmus (PRN), two of the somatosensory tests (Kinesthesia [KIN] and Localization of Tactile Stimuli [LTS]), and

**Table 21**
**SIPT Test-Retest Reliabilities**

| | Combined Sample | | | | | | Learning-Disabled Sample | | | | | |
|---|---|---|---|---|---|---|---|---|---|---|---|---|
| | Test | | Retest | | | | Test | | Retest | | | |
| Test/Score | *M* | *SD* | *M* | *SD* | *r* | (*n*) | *M* | *SD* | *M* | *SD* | *r* | (*n*) |
| **Space Visualization (SV)** | | | | | | | | | | | | |
| *Time-adjusted accuracy* | –.42 | .93 | –.25 | 1.17 | .69 | (49) | –.58 | .87 | –.48 | 1.11 | .62 | (39) |
| Accuracy | –.51 | 1.20 | –.31 | 1.38 | .70 | | –.80 | 1.08 | –.64 | 1.31 | .63 | |
| Time | –.03 | 1.08 | .05 | .98 | .70 | | –.23 | .96 | –.08 | .90 | .68 | |
| Contralateral use | –.31 | .97 | –.04 | 1.01 | .61* | | –.08 | 1.10 | .30 | 1.13 | .49* | |
| Preferred hand use | –.05 | 1.05 | –.04 | 1.03 | .64 | | .15 | 1.12 | –.03 | 1.12 | .66 | |
| **Figure-Ground Perception (FG)** | | | | | | | | | | | | |
| *Accuracy* | –.67 | 1.24 | –.28 | 1.02 | .56* | (47) | –.90 | 1.23 | –.41 | 1.02 | .54* | (38) |
| Time | –.22 | 1.22 | –.56 | .93 | .34 | | –.34 | 1.10 | –.54 | .98 | .47 | |
| **Manual Form Perception (MFP)** | | | | | | | | | | | | |
| *Total accuracy* | –.99 | 1.10 | –.34 | 1.24 | .70* | (31) | –1.07 | 1.14 | –.43 | 1.25 | .69* | (26) |
| Total time | –.00 | .92 | –.06 | .73 | .52 | | –.02 | .97 | –.00 | .74 | .54 | |
| Part I accuracy | –.73 | .89 | –.22 | 1.01 | .48* | | –.81 | .88 | –.38 | 1.02 | .46* | |
| Part I right accuracy | –.56 | 1.15 | –.05 | 1.08 | .33* | | –.65 | 1.18 | –.17 | 1.06 | .31* | |
| Part I left accuracy | –.95 | 1.08 | –.49 | 1.24 | .54* | | –.98 | 1.05 | –.60 | 1.25 | .51* | |
| Part I time | .06 | .98 | .20 | .87 | .49 | | .12 | 1.04 | .35 | .85 | .45 | |
| Part I right time | .00 | .98 | .16 | .95 | .31 | | .05 | 1.04 | .30 | .97 | .27 | |
| Part I left time | .11 | 1.09 | .24 | 1.05 | .58 | | .20 | 1.17 | .40 | 1.05 | .55 | |
| Part II accuracy | –.71 | .98 | –.32 | 1.02 | .71* | | –.80 | .93 | –.44 | 1.03 | .70* | |
| Part II right accuracy | –.60 | 1.12 | –.21 | 1.22 | .68* | | –.66 | 1.10 | –.26 | 1.21 | .80* | |
| Part II left accuracy | –.82 | 1.04 | –.44 | 1.06 | .50* | | –.94 | .96 | –.62 | 1.08 | .40 | |
| Part II time | –.04 | 1.08 | –.43 | .88 | .63* | | –.13 | 1.05 | –.54 | .83 | .71* | |
| Part II right time | .04 | 1.15 | –.64 | .92 | .37* | | –.02 | 1.12 | –.74 | .84 | .49* | |
| Part II left time | –.13 | 1.21 | –.24 | 1.03 | .67 | | –.23 | 1.17 | –.34 | 1.02 | .73 | |
| **Kinesthesia (KIN)** | | | | | | | | | | | | |
| *Total accuracy* | –.94 | 1.31 | –.55 | 1.19 | .50* | (46) | –1.29 | 1.19 | –.76 | 1.15 | .33* | (37) |
| Right hand accuracy | –.81 | 1.31 | –.34 | 1.24 | .44* | | –1.22 | 1.15 | –.58 | 1.22 | .29* | |
| Left hand accuracy | –.78 | 1.30 | –.64 | 1.11 | .45 | | –1.01 | 1.30 | –.81 | 1.07 | .34 | |
| **Finger Identification (FI)** | | | | | | | | | | | | |
| *Total accuracy* | –.52 | 1.33 | –.62 | 1.28 | .74 | (46) | –.67 | 1.33 | –.76 | 1.27 | .75 | (38) |
| Right hand accuracy | –.41 | 1.26 | –.51 | 1.23 | .59 | | –.56 | 1.27 | –.61 | 1.25 | .64 | |
| Left hand accuracy | –.38 | 1.26 | –.52 | 1.21 | .65 | | –.50 | 1.30 | –.68 | 1.19 | .65 | |
| **Graphesthesia (GRA)** | | | | | | | | | | | | |
| *Total accuracy* | –.28 | 1.37 | –.21 | 1.37 | .74 | (42) | –.60 | 1.37 | –.55 | 1.31 | .72 | (32) |
| Right hand accuracy | –.33 | 1.34 | –.17 | 1.27 | .68 | | –.62 | 1.32 | –.51 | 1.21 | .64 | |
| Left hand accuracy | –.24 | 1.29 | –.17 | 1.33 | .59 | | –.51 | 1.31 | –.39 | 1.31 | .53 | |

*Note.* Major score for each test is designated by italics.

*Indicates significant practice effects on retest ($p < .05$).

*table continued on next page . . .*

**Table 21 (Continued)**
**SIPT Test-Retest Reliabilities**

| | Combined Sample | | | | | | Learning-Disabled Sample | | | | | |
|---|---|---|---|---|---|---|---|---|---|---|---|---|
| | Test | | Retest | | | | Test | | Retest | | | |
| **Test/Score** | *M* | *SD* | *M* | *SD* | *r* | (*n*) | *M* | *SD* | *M* | *SD* | *r* | (*n*) |
| **Localization of Tactile Stimuli (LTS)** | | | | | | | | | | | | |
| *Total accuracy* | -.42 | 1.08 | -.29 | 1.24 | .53 | (47) | -.65 | 1.02 | -.54 | 1.16 | .54 | (37) |
| Right hand accuracy | -.51 | 1.11 | -.23 | 1.08 | .50 | | -.74 | 1.07 | -.42 | 1.04 | .50 | |
| Left hand accuracy | -.09 | .88 | -.18 | 1.12 | .44 | | -.21 | .88 | -.40 | 1.08 | .44 | |
| **Praxis on Verbal Command (PrVC)** | | | | | | | | | | | | |
| *Total accuracy* | -.87 | 1.34 | -.61 | 1.43 | .86* | (48) | -1.17 | 1.31 | -.91 | 1.44 | .88* | (38) |
| Total time | -.80 | 1.29 | -.44 | 1.34 | .88* | | -1.11 | 1.23 | -.75 | 1.30 | .87* | |
| **Design Copying (DC)** | | | | | | | | | | | | |
| *Total accuracy* | -.47 | 1.45 | -.08 | 1.60 | .93* | (36) | -.68 | 1.50 | -.36 | 1.64 | .94* | (27) |
| Adjusted accuracy | -.49 | 1.39 | -.26 | 1.60 | .96 | | -.67 | 1.42 | -.45 | 1.64 | .96 | |
| Part I accuracy | -.85 | 1.27 | -.27 | 1.51 | .90* | | -1.04 | 1.26 | -.57 | 1.39 | .91* | |
| Part II accuracy | -.70 | 1.59 | -.75 | 1.59 | .91 | | -1.04 | 1.62 | -1.05 | 1.70 | .91 | |
| Atypical approach parameters: | | | | | | | | | | | | |
| Boundaries | .28 | 1.20 | .50 | 1.32 | .56 | | .43 | 1.27 | .61 | 1.30 | .51 | |
| Additions | .19 | 1.12 | .19 | 1.20 | .65 | | .38 | 1.21 | .43 | 1.30 | .62 | |
| Segmentations | .09 | 1.34 | -.03 | 1.27 | .85 | | .08 | 1.36 | -.03 | 1.32 | .84 | |
| Reversals | .25 | 1.30 | .21 | 1.10 | .25 | | .40 | 1.41 | .13 | .92 | .25 | |
| Right/left errors | -.21 | 1.19 | -.07 | 1.04 | .57 | | -.27 | 1.20 | -.06 | .99 | .53 | |
| Inversions | -.08 | .83 | .36 | 1.31 | .55* | | -.16 | .67 | .26 | 1.24 | .48 | |
| Jogs | -.28 | .83 | -.26 | .88 | .62 | | -.29 | .87 | -.32 | .92 | .59 | |
| **Constructional Praxis (CPr)** | | | | | | | | | | | | |
| *Total accuracy* | -.39 | 1.12 | -.42 | 1.10 | .70 | (51) | -.54 | 1.17 | -.56 | 1.09 | .67 | (41) |
| Part I accuracy | -.44 | 1.29 | -.35 | 1.24 | .72 | | -.52 | 1.39 | -.39 | 1.35 | .74 | |
| Part II accuracy | -.38 | 1.24 | -.46 | 1.41 | .63 | | -.62 | 1.21 | -.73 | 1.22 | .67 | |
| Part II errors: | | | | | | | | | | | | |
| Displacement 1–2.5 cm | -.51 | .86 | -.15 | 1.03 | .21 | | -.51 | .87 | -.05 | .99 | .27 | |
| Displacement > 2.5 cm | -.14 | .91 | -.32 | .77 | .25 | | -.19 | .90 | -.36 | .65 | .02 | |
| Rotation > 15 degrees | -.37 | .75 | -.20 | 1.00 | .52 | | -.44 | .70 | -.31 | 1.02 | .64 | |
| Reversals | -.34 | 1.05 | .01 | 1.29 | .67* | | -.43 | 1.06 | .05 | 1.33 | .70* | |
| Incorrect but logical | .37 | 1.20 | -.11 | 1.04 | .23* | | .43 | 1.24 | .06 | 1.10 | .24 | |
| Gross mislocation | .55 | 1.34 | .59 | 1.35 | .73 | | .78 | 1.41 | .82 | 1.42 | .69 | |
| Omission | .03 | .89 | -.01 | .75 | .78 | | .10 | .99 | .05 | .84 | .77 | |
| **Postural Praxis (PPr)** | | | | | | | | | | | | |
| *Total accuracy* | -.52 | 1.30 | -.12 | 1.46 | .86* | (49) | -.65 | 1.37 | -.30 | 1.50 | .88* | (39) |
| **Oral Praxis (OPr)** | | | | | | | | | | | | |
| *Total accuracy* | -.47 | 1.53 | -.25 | 1.43 | .90* | (49) | -.76 | 1.50 | -.52 | 1.41 | .89* | (39) |

*Note.* Major score for each test is designated by italics.

*Indicates significant practice effects on retest ($p < .05$).

*table continued on next page . . .*

**Table 21 (Continued)**
**SIPT Test-Retest Reliabilities**

| | Combined Sample | | | | | | Learning-Disabled Sample | | | | | |
|---|---|---|---|---|---|---|---|---|---|---|---|---|
| | Test | | Retest | | | | Test | | Retest | | | |
| Test/Score | *M* | *SD* | *M* | *SD* | *r* | (*n*) | *M* | *SD* | *M* | *SD* | *r* | (*n*) |
| **Sequencing Praxis (SPr)** | | | | | | | | | | | | |
| *Total accuracy* | -.77 | 1.16 | -.55 | 1.22 | .84* | (47) | -1.03 | 1.10 | -.74 | 1.23 | .84* | (38) |
| Hand accuracy | -.75 | 1.29 | -.52 | 1.29 | .75 | | -1.05 | 1.24 | -.75 | 1.28 | .71 | |
| Finger accuracy | -.49 | .92 | -.29 | .98 | .85* | | -.66 | .87 | -.42 | 1.00 | .87* | |
| **Bilateral Motor Coordination (BMC)** | | | | | | | | | | | | |
| *Total accuracy* | -.74 | 1.07 | -.52 | 1.14 | .82* | (45) | -1.08 | .82 | -.76 | 1.08 | .77* | (36) |
| Arms accuracy | -.64 | 1.14 | -.49 | 1.22 | .82 | | -.98 | .95 | -.77 | 1.15 | .77 | |
| Feet accuracy | -.70 | .79 | -.55 | .90 | .38 | | -.90 | .65 | -.70 | .76 | .70 | |
| **Standing and Walking Balance (SWB)** | | | | | | | | | | | | |
| *Total score* | -1.50 | 1.42 | -1.68 | 1.33 | .86 | (48) | -1.88 | 1.32 | -2.06 | 1.19 | .80 | (38) |
| Eyes open | -1.57 | 1.32 | -1.63 | 1.34 | .85 | | -1.97 | 1.18 | -1.99 | 1.25 | .81 | |
| Eyes closed | -1.24 | 1.48 | -1.53 | 1.27 | .80* | | -1.56 | 1.43 | -1.90 | 1.11 | .78 | |
| Right foot | -1.43 | 1.23 | -1.38 | 1.11 | .71 | | -1.40 | 1.29 | -1.65 | 1.11 | .65 | |
| Left foot | -1.23 | 1.32 | -1.45 | 1.11 | .70 | | -1.50 | 1.36 | -1.68 | 1.08 | .66 | |
| **Motor Accuracy (MAc)** | | | | | | | | | | | | |
| *Weighted total accuracy* | -.44 | 1.20 | -.43 | 1.20 | .84 | (45) | -.59 | 1.27 | -.53 | 1.29 | .84 | (35) |
| Unweighted total accuracy | -.59 | 1.25 | -.52 | 1.24 | .85 | | -.77 | 1.31 | -.65 | 1.32 | .85 | |
| Preferred hand weighted accuracy | -.27 | 1.27 | -.33 | 1.28 | .72 | | -.40 | 1.35 | -.43 | 1.39 | .76 | |
| Preferred hand unweighted accuracy | -.43 | 1.38 | -.43 | 1.33 | .74 | | -.61 | 1.44 | -.54 | 1.43 | .78 | |
| Nonpreferred hand weighted accuracy | -.56 | 1.29 | -.55 | 1.28 | .80 | | -.74 | 1.35 | -.68 | 1.39 | .82 | |
| Nonpreferred hand unweighted accuracy | -.70 | 1.30 | -.65 | 1.32 | .82 | | -.89 | 1.36 | -.82 | 1.40 | .82 | |
| **Postrotary Nystagmus (PRN)** | | | | | | | | | | | | |
| *Average nystagmus* | .03 | .77 | .11 | .67 | .48 | (39) | -.03 | .87 | .12 | .72 | .47 | (29) |
| Average clockwise | .09 | .94 | .10 | .96 | .40 | | .05 | 1.05 | .11 | 1.04 | .37 | |
| Average counterclockwise | -.05 | .90 | .11 | .85 | .32 | | -.08 | 1.03 | .11 | .95 | .31 | |

*Note.* Major score for each test is designated by italics.

*Indicates significant practice effects on retest ($p < .05$).

one visual test (Figure-Ground Perception [FG]). The SIPT test-retest reliabilities for Kinesthesia and Localization of Tactile Stimuli are quite similar to those reported for the SCSIT versions of these tests (Ayres, 1980b). The reliability for the SIPT Figure-Ground Perception is slightly higher than the reported reliability of the earlier SCSIT Figure-Ground Perception test.

It should be noted that the test-retest reliability coefficient of .49 for Postrotary Nystagmus is considerably lower than the test-retest coefficients obtained earlier with this test. Although it is conceivable that the test-retest reliability of Postrotary Nystagmus is negatively affected when the test is administered after other tests in the SIPT, the coefficient of .49 is probably an underestimate of the actual reliability of this test. For example, the SCPNT Manual (Ayres, 1975) reported a two-week test-retest reliability coefficient of .83 in a sample of 42 children presumed normal. Kimball (1981) obtained a test-retest correlation of .80 in a sample of 63 normal children, ages 5 through 9 years, with a test-retest interval of approximately 2½ years. Punwar (1982) reported a .82 test-retest correlation between scores of 56 normal children from 3 through 10 years of age, tested two weeks apart. In 1985, Dutton reviewed the published SCPNT reliability data and found that the test-retest reliability coefficients ranged from .79 to .81 for normal 4- to 11-year-old children.

**Interrater Reliability**

Interrater reliability indicates the extent to which a child's test scores remain the same when his or her performance is evaluated, recorded, and scored by different examiners. Most tests have some margin for human error. For example, examiners may differ in the accuracy and precision with which they measure the time the child takes to perform a task, or in the leniency with which they evaluate the accuracy of the child's performance. A high interrater reliability coefficient is an indication that the child's score will be very similar when his or her performance is evaluated by different examiners.

To evaluate interrater reliability, the SIPT was administered to 63 children from 5 years, 0 months to 8 years, 11 months of age (50 boys, 13 girls; mean age = 7.26 years, *SD* = 1.04). This sample included 19 children with diagnosed reading disorders, 41 children with other learning disabilities, and 3 children with spina bifida. Eight examiners participated in the interrater reliability study, and each child's performance on the SIPT was evaluated, rated, and scored by two different examiners. All of the interrater reliability coefficients for the major SIPT scores were very high, ranging from .94 to .99. These coefficients are shown in Table 22.

Interrater reliabilities also were computed from this sample for all of the SIPT part scores. Most of the part scores had interrater reliabilities greater than .90, and all had reliabilities of .85 or above, with exception of the Figure-Ground Perception (FG) Time score (.77) and some of the Design Copying (DC) atypical approach parameters. The following reliability coefficients were obtained for the Design Copying atypical approach parameters: (a) boundary errors, .97; (b) additions, .51; (c) segmentations, .91; (d) reversals, .37; (e) right-to-left errors, .81; (f) inversions, .61; (g) jogs, .31; and (f) distortions, .67.

Overall, these correlations indicate that different examiners, trained in the use of the SIPT, will obtain similar results from the instrument. It should be noted, however, that all of the examiners in the interrater reliability study had completed a comprehensive SIPT administration course. Interrater reliabilities probably would be considerably lower among untrained examiners.

**What is the "best" subset of SIPT variables for discriminating among various groups?** Throughout the Manual, the emphasis has been upon administering the entire SIPT in order to make the best possible classification of the strengths and weaknesses of an individual child. In general, it should be firmly emphasized that the maximum utility of the SIPT is achieved only when all 17 measures are administered. However, in real clinical settings, there are factors such as limited testing time, child inattention, parental consent to specific procedures, and other considerations that may make it unfeasible to administer the entire SIPT. This section of the Manual makes specific recommendations on which of the 17 tests to administer depending upon the specific purposes of the assessment.

The recommendations made here about appropriate subsets of the SIPT for different decisions are entirely derived from statistical (discriminant) analyses. Statistical criteria have been employed to determine the best subsets of SIPT that yield information necessary to make certain clinical decisions. For instance, if we have to decide whether a child fits into one of the six SIPT cluster analysis groups as opposed to all other groups, we might need to use a different "best" subset of SIPT procedures than if we are merely trying to differentiate whether the child is best categorized as a member of SIPT Group 5 (Dyspraxia on Verbal Command) or SIPT Group 2 (Generalized Sensory Integrative Dysfunction). That is, *specific clinical decisions will require different best subsets of SIPT measures* in order to make the decision in the most accurate manner.

Technically, the best subsets of SIPT measures were determined by utilizing stepwise discriminant analysis on different combinations of problems. Stepwise discriminant analysis seeks to determine, using *statistical* criteria, which subset of variables best discriminates among two or more groups. When this type of statistical analysis is used, a premium is placed upon selecting tests that give unique information, or information that is not redundant with that provided by another test. So, in developing the subsets, usually two tests that are highly correlated with one another will not both be included in the subset, unless each contributes *uniquely* to an understanding of why the groups are different.

It should be emphasized that there are several well-known limitations to the methods of stepwise analysis which relate to this generic type of procedure, and not specifically to the SIPT. First, stepwise methods do tend to yield results that are optimized for the specific sample of clients tested. That is, new samples of children might yield slightly different best subsets of tests for making discriminations; in other words, stepwise procedures tend to be the multivariate statistical methods whose results are most in need of crossvalidation in new samples. Second, the overall validity of the selection of the best subsets is dependent upon the validity of the separation of

**Table 22**
**SIPT Interrater Reliabilities**

| Test/Score | *r* | (*n*) |
|---|---|---|
| **Space Visualization (SV)** | | |
| *Time-adjusted accuracy* | .99 | (63) |
| Accuracy | .99 | |
| Time | .93 | |
| Contralateral use | .94 | |
| Preferred hand use | | |
| **Figure-Ground Perception (FG)** | | |
| *Accuracy* | .99 | (58) |
| Time | .77 | |
| **Manual Form Perception (MFP)** | | |
| *Total accuracy* | .99 | (47) |
| Total time | .98 | |
| Part I accuracy | .97 | |
| Part I right accuracy | .93 | |
| Part I left accuracy | .94 | |
| Part I time | .90 | |
| Part I right time | .88 | |
| Part I left time | .86 | |
| Part II accuracy | .99 | |
| Part II right accuracy | .95 | |
| Part II left accuracy | .99 | |
| Part II time | .99 | |
| Part II right time | .99 | |
| Part II left time | .97 | |
| **Kinesthesia (KIN)** | | |
| *Total accuracy* | .99 | (60) |
| Right hand accuracy | .89 | |
| Left hand accuracy | .90 | |
| **Finger Identification (FI)** | | |
| *Total accuracy* | .95 | (62) |
| Right hand accuracy | .92 | |
| Left hand accuracy | .95 | |
| **Graphesthesia (GRA)** | | |
| *Total accuracy* | .96 | (54) |
| Right hand accuracy | .93 | |
| Left hand accuracy | .94 | |
| **Localization of Tactile Stimuli (LTS)** | | |
| *Total accuracy* | .99 | (59) |
| Right hand accuracy | .99 | |
| Left hand accuracy | .99 | |
| **Praxis on Verbal Command (PrVC)** | | |
| *Total accuracy* | .98 | (62) |
| Total time | .99 | |
| **Design Copying (DC)** | | |
| *Total accuracy* | .97 | (58) |
| Adjusted accuracy | .96 | |
| Part I accuracy | .98 | |
| Part II accuracy | .91 | |
| Atypical approach parameters: | | |
| Boundaries | .97 | |
| Additions | .51 | |
| Segmentations | .91 | |
| Reversals | .37 | |
| Right/left errors | .81 | |
| Inversions | .61 | |
| Jogs | .31 | |
| **Constructional Praxis (CPr)** | | |
| *Total accuracy* | .98 | (63) |
| Part I accuracy | .93 | |
| Part II accuracy | .97 | |
| Part II errors: | | |
| Displacement 1–2.5 cm | .93 | |
| Displacement > 2.5 cm | .80 | |
| Rotation > 15 degrees | .87 | |
| Reversals | .97 | |
| Incorrect but logical | .95 | |
| Gross mislocation | .99 | |
| Omission | .99 | |
| **Postural Praxis (PPr)** | | |
| *Total accuracy* | .96 | (62) |
| **Oral Praxis (OPr)** | | |
| *Total accuracy* | .94 | (63) |
| **Sequencing Praxis (SPr)** | | |
| *Total accuracy* | .99 | (51) |
| Hand accuracy | .99 | |
| Finger accuracy | .99 | |
| **Bilateral Motor Coordination (BMC)** | | |
| *Total accuracy* | .96 | (48) |
| Arms accuracy | .96 | |
| Feet accuracy | .88 | |
| **Standing and Walking Balance (SWB)** | | |
| *Total score* | .99 | (60) |
| Eyes open | .99 | |
| Eyes closed | .98 | |
| Right foot | .94 | |
| Left foot | .98 | |
| **Motor Accuracy (MAc)** | | |
| *Weighted total accuracy* | .99 | (62) |
| Unweighted total accuracy | .99 | |
| Preferred hand weighted accuracy | .98 | |
| Preferred hand unweighted accuracy | .98 | |
| Nonpreferred hand weighted accuracy | .98 | |
| Nonpreferred hand unweighted accuracy | .98 | |
| **Postrotary Nystagmus (PRN)** | | |
| *Average nystagmus* | .98 | (56) |
| Average clockwise | .94 | |
| Average counterclockwise | .96 | |

*Note.* Major score for each test is designated by italics.

the children tested into meaningful and real clinical groups. To the extent that good classifications of the children are used to generate subsets of tests, subsets which are valid for use in making appropriate clinical decisions will be obtained.

Overall, a number of different subsets of SIPT variables were generated, with each subset generated by a stepwise discriminant analysis. In each case, the intent of the stepwise discriminant analysis was to differentiate between one or more of the six SIPT groups identified through cluster analysis and discussed in detail earlier in this Manual.

Three major kinds of stepwise discriminant analyses were conducted. In the first analysis, all six SIPT groups were examined simultaneously to determine the best subset of variables for differentiating between all six groups at once. This subset of variables will be the single best set of variables for general purpose assessments when no specific hypotheses are identified in advance and when it is impossible to use the full battery.

The second major kind of stepwise discriminant analyses were run to determine the best subset of SIPT variables to differentiate each of the six SIPT groups from another SIPT group. There are 15 different analyses (and subsets of variables) because there are 15 unique ways of pairing the SIPT groups with one another. Users should find the relevant comparison in which they are interested in the second section of Table 23. These subsets of variables will be most useful when it is necessary to discriminate individuals in one group from individuals in another group and when it is not possible to administer the entire battery.

The third major kind of stepwise analysis was designed to differentiate members of one group from members of all other groups combined. For instance, the analysis of the best subset for differentiating Group 1 individuals compared Group 1 individuals to a combined sample comprised of all Group 2, Group 3, Group 4, Group 5, and Group 6 children. There is one analysis of this kind for each SIPT group.

Each of the stepwise discriminant analyses was conducted using the SPSSX discriminant analysis program. The stepwise method chosen was the minimization of Wilks' lambda, a method which is usually considered the best for such analyses. For interpretation, note that a Wilks' lambda value of 1.0 indicates perfect differentiation among groups, whereas a value of 0.0 indicates no differentiation. The variable addition and deletion was accomplished using a forward stepping algorithm with the *F*-to-enter and *F*-to-remove values set at the critical value of *F* equivalent to the .01 level of significance for the relevant degrees of freedom for the problem. The results from these analyses are presented in Table 23.

Examining the first analysis, which compares all groups, 13 SIPT were selected as the best set of variables for differentiating among all six groups at once. These 13 variables are listed in Table 23 in order of their ability to uniquely discriminate the six SIPT groups from one another. The first such indicator presented is Praxis on Verbal Command (PrVC), which provided the most discrimination among the various SIPT groups (Wilks' lambda equal to .31). The next best variable for differentiating among all SIPT groups was determined by controlling the ability of Praxis on Verbal Command to discriminate among the groups; the next best indicator was Design Copying [DC] (Wilks' lambda equal to .18). The third best indicator, controlling for the effects of Praxis on Verbal Command and Design Copying to differentiate among the groups, was Oral Praxis [OPr] (Wilks' lambda equal to .12). The remaining indicators were determined in the same way, controlling for the effects of the variables that had been previously identified.

There are a few caveats that should be noted when using the results of the stepwise discriminant analyses for clinical interpretation. Variables were selected for their ability to differentiate among groups by a statistical criterion; that is, those variables that differentiated among groups at the .01 level of significance were included in the best set of SIPT for that particular combination of groups. However, these statistical decisions may not necessarily reflect clinically meaningful differences. Thus, the results presented in Table 23 should be carefully considered in making informed clinical decisions about using subsets of the SIPT.

**Table 23**
**Discrimination Among SIPT Groups**

| Comparison | Wilks' Lambda |
|---|---|
| **Section I** | |
| **All Groups** | |
| Praxis on Verbal Command (PrVC) | .31 |
| Design Copying (DC) | .18 |
| Oral Praxis (OPr) | .12 |
| Standing & Walking Balance (SWB) | .10 |
| Postural Praxis (PPr) | .08 |
| Bilateral Motor Coordination (BMC) | .07 |
| Finger Identification (FI) | .06 |
| Postrotary Nystagmus (PRN) | .06 |
| Kinesthesia (KIN) | .05 |
| Localization of Tactile Stimuli (LTS) | .05 |
| Graphesthesia (GRA) | .04 |
| Figure-Ground Perception (FG) | .04 |
| Constructional Praxis (CPr) | .04 |
| **Section II** | |
| **Low Average BIS (1) vs. General SI Dysfunction (2)** | |
| Praxis on Verbal Command (PrVC) | .25 |
| Design Copying (DC) | .16 |
| Oral Praxis (OPr) | .13 |
| Postural Praxis (PPr) | .11 |
| Figure-Ground Perception (FG) | .10 |
| **Low Average BIS (1) vs. Visuo- & Somatodyspraxia (3)** | |
| Design Copying (DC) | .49 |
| Motor Accuracy (MAc) | .43 |
| Constructional Praxis (CPr) | .40 |
| Figure-Ground Perception (FG) | .37 |
| **Low Average BIS (1) vs. Low Average SI & Praxis (4)** | |
| Oral Praxis (OPr) | .72 |
| Standing & Walking Balance (SWB) | .58 |
| Finger Identification (FI) | .51 |
| Graphesthesia (GRA) | .46 |
| Postural Praxis (PPr) | .44 |
| **Low Average BIS (1) vs. Dyspraxia on Verbal Command (5)** | |
| Praxis on Verbal Command (PrVC) | .24 |
| Bilateral Motor Coordination (BMC) | .21 |
| **Low Average BIS (1) vs. High Average SI & Praxis (6)** | |
| Graphesthesia (GRA) | .68 |
| Postural Praxis (PPr) | .54 |
| Standing & Walking Balance (SWB) | .43 |
| Space Visualization (SV) | .38 |
| Oral Praxis (OPr) | .34 |
| Praxis on Verbal Command (PrVC) | .31 |
| Figure-Ground Perception (FG) | .28 |
| Localization of Tactile Stimuli (LTS) | .26 |
| Postrotary Nystagmus (PRN) | .24 |
| **General SI Dysfunction (2) vs. Visuo- & Somatodyspraxia (3)** | |
| Praxis on Verbal Command (PrVC) | .31 |
| Oral Praxis (OPr) | .22 |
| Manual Form Perception (MFP) | .20 |
| Postrotary Nystagmus (PRN) | .18 |
| **General SI Dysfunction (2) vs. Low Average SI & Praxis (4)** | |
| Standing & Walking Balance (SWB) | .29 |
| Oral Praxis (OPr) | .19 |
| Praxis on Verbal Command (PrVC) | .14 |
| Motor Accuracy (MAc) | .13 |
| Kinesthesia (KIN) | .18 |
| **General SI Dysfunction (2) vs. Dyspraxia on Verbal Command (5)** | |
| Graphesthesia (GRA) | .59 |
| Standing & Walking Balance (SWB) | .41 |
| Manual Form Perception (MFP) | .33 |
| Postural Praxis (PPr) | .27 |
| Postrotary Nystagmus (PRN) | .23 |
| **General SI Dysfunction (2) vs. High Average SI & Praxis (6)** | |
| Praxis on Verbal Command (PrVC) | .16 |
| Design Copying (DC) | .10 |
| Oral Praxis (OPr) | .08 |
| Postural Praxis (PPr) | .06 |
| Graphesthesia (GRA) | .06 |
| Localization of Tactile Stimuli (LTS) | .05 |
| Space Visualization (SV) | .05 |
| **Visuo- & Somatodyspraxia (3) vs. Low Average SI & Praxis (4)** | |
| Standing & Walking Balance (SWB) | .61 |
| Kinesthesia (KIN) | .45 |
| Constructional Praxis (CPr) | .37 |
| Motor Accuracy (MAc) | .32 |
| Postrotary Nystagmus (PRN) | .30 |
| **Visuo- & Somatodyspraxia (3) vs. Dyspraxia on Verbal Command (5)** | |
| Praxis on Verbal Command (PrVC) | .29 |
| Postrotary Nystagmus (PRN) | .25 |
| Localization of Tactile Stimuli (LTS) | .22 |
| **Visuo- & Somatodyspraxia (3) vs. High Average SI & Praxis (6)** | |
| Design Copying (DC) | .32 |
| Kinesthesia (KIN) | .26 |
| Bilateral Motor Coordination (BMC) | .21 |
| Postural Praxis (PPr) | .18 |
| Graphesthesia (GRA) | .16 |

*table continued on next page . . .*

**Table 23 (Continued)**
**Discrimination Among SIPT Groups**

| Comparison | Wilks' Lambda |
|---|---|
| **Low Average SI & Praxis (4) vs. Dyspraxia on Verbal Command (5)** | |
| Praxis on Verbal Command (PrVC) | .41 |
| Oral Praxis (OPr) | .34 |
| Standing & Walking Balance (SWB) | .31 |
| Localization of Tactile Stimuli (LTS) | .26 |
| Kinesthesia (KIN) | .22 |
| Figure-Ground Perception (FG) | .21 |
| **Low Average SI & Praxis (4) vs. High Average SI & Praxis (6)** | |
| Localization of Tactile Stimuli (LTS) | .54 |
| Finger Identification (FI) | .48 |
| Postural Praxis (PPr) | .44 |
| Praxis on Verbal Command (PrVC) | .41 |
| Figure-Ground Perception (FG) | .38 |
| Space Visualization (SV) | .36 |
| **Dyspraxia on Verbal Command (5) vs. High Average SI & Praxis (6)** | |
| Praxis on Verbal Command (PrVC) | .16 |
| Bilateral Motor Coordination (BMC) | .13 |
| Space Visualization (SV) | .12 |
| Design Copying (DC) | .11 |
| Postural Praxis (PPr) | .10 |
| Graphesthesia (GRA) | .09 |
| **Section III** | |
| **Low Average BIS (1) vs. all others** | |
| Praxis on Verbal Command (PrVC) | .96 |
| Oral Praxis (OPr) | .87 |
| Finger Identification (FI) | .83 |
| Standing & Walking Balance (SWB) | .80 |
| Constructional Praxis (CPr) | .78 |
| Graphesthesia (GRA) | .76 |
| **General SI Dysfunction (2) vs. all others** | |
| Praxis on Verbal Command (PrVC) | .68 |
| Oral Praxis (OPr) | .58 |
| Constructional Praxis (CPr) | .53 |
| Graphesthesia (GRA) | .51 |
| Manual Form Perception (MFP) | .50 |
| Bilateral Motor Coordination (BMC) | .49 |
| **Visuo- & Somatodyspraxia (3) vs. all others** | |
| Design Copying (DC) | .89 |
| Praxis on Verbal Command (PrVC) | .81 |
| Kinesthesia (KIN) | .78 |
| Postrotary Nystagmus (PRN) | .76 |
| Bilateral Motor Coordination (BMC) | .74 |
| Oral Praxis (OPr) | .71 |
| **Low Average SI & Praxis (4) vs. all others** | |
| Oral Praxis (OPr) | .90 |
| Localization of Tactile Stimuli (LTS) | .85 |
| Kinesthesia (KIN) | .81 |
| Finger Identification (FI) | .77 |
| Standing & Walking Balance (SWB) | .73 |
| Postural Praxis (PPr) | .71 |
| **Dyspraxia on Verbal Command (5) vs. all others** | |
| Praxis on Verbal Command (PrVC) | .74 |
| Postrotary Nystagmus (PRN) | .69 |
| Postural Praxis (PPr) | .67 |
| Bilateral Motor Coordination (BMC) | .65 |
| **High Average SI & Praxis (6) vs. all others** | |
| Design Copying (DC) | .71 |
| Postural Praxis (PPr) | .63 |
| Localization of Tactile Stimuli (LTS) | .58 |
| Bilateral Motor Coordination (BMC) | .54 |

# REFERENCES


Aram, D.M., & Horwitz, S.J. (1983). Sequential and nonspeech practic abilities in developmental verbal apraxia. *Developmental Medicine and Child Neurology, 25,* 197–206.

Atwood, R.M., & Cermak, S.A. (1986). Crossing the midline as a function of distance from midline. *American Journal of Occupational Therapy, 40,* 685–690.

Ayres, A.J. (1962a). *The Ayres Space Test.* Beverly Hills, CA: Western Psychological Services.

Ayres, A.J. (1962b). Perception of space of adult hemiplegic patients. *Physical Medicine and Rehabilitation, 43,* 552–555.

Ayres, A.J. (1964). *Southern California Motor Accuracy Test.* Los Angeles: Western Psychological Services.

Ayres, A.J. (1965). Patterns of perceptual-motor dysfunction in children: A factor analytic study. *Perceptual and Motor Skills, 20,* 335–368.

Ayres, A.J. (1966a). *Southern California Kinesthesia and Tactile Perception Tests Manual.* Los Angeles: Western Psychological Services.

Ayres, A.J. (1966b). Interrelation among perceptual-motor abilities in a group of normal children. *American Journal of Occupational Therapy, 22,* 288–292.

Ayres, A.J. (1966c). *Southern California Figure-Ground Visual Perception Test.* Beverly Hills, CA: Western Psychological Services.

Ayres, A.J. (1966d). Interrelation among perceptual-motor functions in children. *American Journal of Occupational Therapy, 20,* 68–71.

Ayres, A.J. (1968). *Southern California Perceptual-Motor Tests.* Los Angeles: Western Psychological Services.

Ayres, A.J. (1969a). Relation between Gesell Developmental Quotients and latter perceptual-motor performance. *American Journal of Occupational Therapy, 23,* 11–17.

Ayres, A.J. (1969b). Deficits in sensory integration in educationally handicapped children. *Journal of Learning Disabilities, 2,* 160–168.

Ayres, A.J. (1972a). *Southern California Sensory Integration Tests.* Los Angeles: Western Psychological Services.

Ayres, A.J. (1972b). Types of sensory integrative dysfunction among disabled learners. *American Journal of Occupational Therapy, 26,* 13–18.

Ayres, A.J. (1975). *Southern California Postrotary Nystagmus Test Manual.* Los Angeles: Western Psychological Services.

Ayres, A.J. (1976a). *The effect of sensory integrative therapy on learning disabled children.* (A final report of a research project). Los Angeles: University of Southern California.

Ayres, A.J. (1976b). *Interpreting the Southern California Sensory Integration Tests.* Los Angeles: Western Psychological Services.

Ayres, A.J. (1977). Cluster analyses of measures of sensory integration. *American Journal of Occupational Therapy, 31,* 362–366.

Ayres, A.J. (1978). Learning disabilities and the vestibular system. *Journal of Learning Disabilities, 11,* 18–29.

Ayres, A.J. (1980a). *Southern California Motor Accuracy Test: Revised 1980.* Los Angeles: Western Psychological Services.

Ayres, A.J. (1980b). *Southern California Sensory Integration Tests.* Los Angeles: Western Psychological Services.

Ayres, A.J., & Heskett, W.M. (1972). Sensory integrative dysfunction in a young schizophrenic girl. *Journal of Autism and Childhood Schizophrenia, 2,* 174–181.

Ayres, A.J., Mailloux, Z., & Wendler, C.L. (1987). Developmental dyspraxia: Is it a unitary function? *Occupational Therapy Journal of Research, 7,* 93–110.

Ayres, A.J., & Reid, W. (1966). The self-drawing as an expression of perceptual-motor dysfunction. *Cortex, 2,* 254–265.

Bayley, N. (1969). *The Bayley Scales of Infant Development.* New York: The Psychological Corporation.

Becker, M.S. (1982). Level of sensory integrative functioning in children of Vietnam veterans exposed to Agent Orange. *Occupational Therapy Journal of Research, 2,* 234–244.

Beery, K.E. (1982). *Revised Administration, scoring, and teaching Manual for the Developmental Test of Visual-Motor Integration.* Cleveland: Modern University Press.

Bender, L. (1938). *A visual-motor gestalt test and its clinical use.* (Research Monograph, No. 3). New York: American Orthopsychiatric Association.

Bender, L. (1956). *Psychopathology of Children with organic brain disorders.* Springfield, IL: Charles C. Thomas.

Benson, D. F., & Barton, M.I. (1970). Disturbances of constructional ability. *Cortex, 6,* 19–46.

Benton, A.L. (1955). Development of finger-localization capacity in school children. *Child Development, 26,* 225–230.

Benton, A.L. (1959). *Right-left Discrimination and Finger Localization.* New York: Paul B. Haeber.

Benton, A.L., & Fogel, M.I. (1970). Disturbances of constructional ability. *Cortex, 6,* 19–46.

Berges, J., & Lezine, I. (1975). The imitation of gestures. *Clinics in Developmental Medicine, No. 18.*

Brooks, C.R., & Clair, T.N. (1971). Relationships among visual figure-ground perception, word recognition, IQ, and chronological age. *Perceptual and Motor Skills, 33,* 59–62.

Bruininks, R.H. (1978). *Bruininks-Oseretsky Test of Motor Proficiency.* Circle Pines, MN: American Guidance Services.

Brunt, D. (1980). Characteristics of upper limb movements in a sample of meningomyelocele children. *Perceptual and Motor Skills, 51,* 431–437.

Bundy, A.C., & Fisher, A.G. (1981). The relationship of prone extension to other vesticular function. *American Journal of Occupational Therapy, 35,* 782–787.

Bundy, A.C., Fisher, A.G., Freeman, M., Lieberg, G., & Israelovitz, T.E. (1987). Concurrent validity of equilibrium tests in boys with learning disabilities with and without vestibular dysfunction. *American Journal of Occupational Therapy, 41,* 28–34.

Carte, E., Morrison, D., Sublett, J., Vermura, A., & Setrakian, W. (1984). Sensory integration therapy: a trial of a specific neurodevelopmental therapy for the remediation of learning disabilities. *Developmental and Behavioral Pediatrics, 5,* 189–194.

Cermak, S.A., & Ayres, A.J. (1984). Crossing the body midline in learning-disabled and normal children. *American Journal of Occupational Therapy, 38,* 35–39.

Cermak, S.A., Quintero, E.J., & Cohen, P.M. (1980). Developmental age trends in crossing the body midline in normal children. *American Journal of Occupational Therapy, 34* (5), 313–319.

Cermak, S.A., Ward, E.A., & Wood, L.M. (1986). The relationship between articulation disorders and motor coordination in children. *The American Journal of Occupational Therapy, 40,* 546–550.

Chapparo, C.J., Yerxa, E.J., Nelson, J.G., & Wilson, L. (1981). Incidence of sensory integrative dysfunction among children with orofacial cleft. *American Journal of Occupational Therapy, 35,* 96–100.

Clayton, K., Cochran, C., Edwards, N., & Reep, B. (1985). The effects of rotary movement on nystagmus in normal 4-year-olds. *Occupational Therapy Journal of Research, 5,* 197–198.

Clayton, K., Cochran, C., Edwards, N., & Reep, B. (1986). The effects of rotary movement on nystagmus in normal 4-year-olds. *Occupational Therapy Journal of Research, 6,* 47–48.

Clyse, S.J., & Short, M.A. (1983). The relationship between dynamic balance and postrotary nystagmus in learning disabled children. *Physical and Occupational Therapy in Pediatrics, 3*(3), 25–32.

Critchley, M. (1953). *The Parietal Lobes.* London: Edward Arnold.

Crowe, T.K., Deitz, J.C., & Siegner, C.B. (1984). Postrotary nystagmus response of normal four-year-old children. *Physical and Occupational Therapy in Pediatrics, 4*(2), 19–28.

Cruickshank, W.M., & Bice, H.V. (1955). Personality characteristics. In W.M. Cruickshank & G.M. Raus (Eds.), *Cerebral Palsy. Its Individual and Community Problems.* Syracuse, NY: Syracuse University Press.

Cruickshank, W.M., Podosek, E., & Thomas, E.R. (1965). *Perception and cerebral palsy; a study in figure-background relationship.* (Special Education and Rehabilitation Monographs, series 2). Syracuse, NY: Syracuse University Press.

Dee, H.L., & Benton, A.L. (1970). A cross-model investigation of spatial performance in patients with unilateral cerebral disease. *Cortex, 6,* 261–272.

DeGangi, G.A. (1982). The relationship of vestibular responses and developmental functions in high risk infants. *Physical and Occupational Therapy in Pediatrics, 2* (2/3), 35–49.

Deitz, J.C., & Crowe, T.K. (1985). Developmental status of children exhibiting postrotary nystagmus durations of zero seconds. *Physical & Occupational therapy in pediatrics, 5*(2/3), 69–79.

Denny-Brown, A. (1958). The nature of apraxia. *Journal of Nervous and Mental Disease, 126,* 9–32.

deQuiros, J.B. (1976). Diagnosis of vestibular disorders in the learning disabled. *Journal of Learning Disabilities, 9,* 39–47.

deQuiros, J.B., & Schrager, O.L. (1978). *Neuropsychological Fundamentals in Learning Disabilities.* San Rafael, CA: Academic Therapy Publications.

DeRenzi, E. (1985). Methods of limb apraxia examination and their bearing on the interpretation of the disorder. In E.A. Roy (Ed.), *Neuropsychological studies of apraxia and related disorders,* (pp. 45–64). Amsterdam, Netherlands: Elsevier (North Holland).

DeRenzi, E., Faglioni, P., & Sorgato, P. (1982). Modality-specific and supramodal mechanisms of apraxia. *Brain, 105* (2), 301–312.

DeRenzi, E., Motti, F., & Michelli, P. (1980). Imitating gestures. A Quantitative approach to ideomotor apraxia. *Archives of Neurology, 37,* 6–10.

Doyle, B.A., & Higginson, D.C. (1984). Relationships among self-concept and school achievement, maternal self-esteem and sensory integration abilities for learning disabled children, ages 7 to 12 years. *Perceptual and Motor Skills, 58,* 177–178.

Dutton, R.E. (1985). Reliability and clinical significance of the Southern California Postrotary Nystagmus Test. *Physical & Occupational Therapy in Pediatrics, 5*(2/3), 57–67.

Elfant, I.L. (1977). Correlation between kinesthetic discrimination and manual dexterity. *American Journal of Occupational Therapy, 31,* 23–28.

Ettlinger, G. (1969). Apraxia considered as a disorder of movements that are language-dependent: evidence from cases of brain bi-section. *Cortex, 5,* 285–289.

Fisher, A.G., Mixon, J., & Herman, R. (1986). The validity of the clinical diagnosis of vestibular dysfunction. *The Occupational Therapy Journal of Research, 6,* 3–20.

Folio, R., & Dubose, R.F. (1974). Peabody Developmental Motor Scales (Revised experimental edition). *IMRID Behavioral Science Monograph* (No. 25).

Fox, F.H. (1968). *A description of language and perceptual function of culturally deprived children.* Unpublished doctoral dissertation, University of Wisconsin, Madison.

Frank, J., & Levinson, H.N. (1973). Dysmetric dyslexia and dyspraxia: Hypothesis and study. *Journal of American Academy of Child Psychiatry, 12,* 690–701.

Frank, J., & Levinson, H.N. (1975-76). Dysmetric dyslexia and dyspraxia: Synapsis of continuing research project. *Academic Therapy, 11,* 133–143.

Frostig, M., Lefever, D.W., & Whittlesey, J.R.B. (1961). *Marianne Frostig Developmental Test of Visual Perception* (3rd. ed.). Palo Alto, CA: Consulting Psychologists Press.

Fukuda, T. (1975). Postural behavior and motion sickness. *Acta Otolaryngologica Supplement, 330,* 9–14.

Gainotti, G., Miceli, G., & Caltagirone, C. (1977). Constructional apraxia in left brain-damaged patients: a planning disorder? *Cortex, 13,* 109–118.

Gerstmann, J. (1940). Syndrome of finger agnosia, disorientation for right and left, agraphia and acalculia. *Archives of Neurology and Psychiatry, 44,* 398–408.

Gesell, A., & Amatruda, C.S. (1974). *Developmental Diagnosis.* New York: Paul B. Hoeber.

Gillberg, C., Rasmussen, P., Carlstrom, G., Svenson, B., & Waldenstrom, E. (1982). Perceptual, motor and attentional deficits in six-year-old children. Epidemiological aspects. *Journal of Child Psychology and Psychiatry, 23,* 131–144.

Golden, C.J., Hemmeke, T.A., & Purisch, A.D. (1980). *The Luria-Nebraska Neuropsychological Battery.* Los Angeles: Western Psychological Services.

Golden, C.J., Hammeke, T., & Purisch, A.D. (1987). *Luria-Nebraska Neuropsychological Battery: Children's Revision.* Los Angeles: Western Psychological Services.

Gooddy, W., & Reinhold, N. (1952). Some aspects of human orientation in space. *Bruin, 75,* 472–509.

Gressang, J.D. (1974). Perceptual processes of children with myelomeningocele and hydrocephalus. *American Journal of Occupational Therapy, 28,* 226–230.

Grimm, R.A. (1976). Hand function and tactile perception in a sample of children with myelomeningocele. *American Journal of Occupational Therapy, 30,* 235–240.

Gubbay, S.S., Ellis, E., Walton, J.N., & Court, S.D.M. (1965). Clumsy Children: a study of apraxic and agnosic defects in 21 children. *Brain, 88,* 295–312.

Halstead, W.C. (1947). *Brain and Intelligence: A Quantitative Study of the Frontal Lobes.* Chicago: University of Chicago Press.

Haworth, M.R. (1970). *The Primary Visual Motor Test: Test Manual and Scoring Instructions.* New York: Grune & Stratton.

Head, H. (1920). *Studies in neurology* (Vol. II). London: Oxford University Press.

Hecaen, H., & Albert, M.L. (1978). *Human neuropsychology.* New York: Wiley Interscience.

Hecaen, H., Penfield, W., Bertrand, C., & Malmo, R. (1956). The syndrome of apractognosia due to lesions of the minor hemisphere. *A.M.A. Archives of Neurology and Psychiatry, 75,* 400–434.

Hecaen, H., & Rondot, P. (1985). Apraxia as a disorder of a septem of signs. In E.A. Roy (Ed.), *Neuropsychological studies of apraxia and related disorders* (pp. 75–97). Amsterdam, Netherlands: Elsevier (North Holland).

Hsu, Y., & Nelson, D.L. (1981). Adult performance on the Southern California Kinesthesia and Tactile Perception Tests. *American Journal of Occupational Therapy, 35,* 788–791.

Jensen, G.M., & Wilson, K.B. (1979). Horizontal postrotary nystagmus response in female subjects with adolescent ideopathic scoliosis. *Journal of American Physical Therapy Association, 59,* 1226–1233.

Kaufman, A.S., & Kaufman, N.L. (1983). *Kaufman Assessment Battery for Children.* Circle Pines, MN: American Guidance Service.

Keating, N. (1979). A comparison of duration of nystagmus as measure by the Southern California Postrotary Nystagmus Test and electronystagmography. *American Journal of Occupational Therapy, 33,* 92–97.

Kennedy, K.S. (1983). The Southern California Postrotary Nystagmus Test: Development of a revised procedure for use with preschool children. *Occupational Therapy Journal of Research, 3,* 93–103.

Kephart, N.C. (1971). *The Slow Learner in the Classroom.* Columbus, OH: Charles E. Merrill.

Kertesz, A., & Hooper, P. (1982). Praxis and language: the extent and variety of apraxia in aphasia. *Neuropsychologia, 20,* 275–286.

Kimball, J.G. (1977). The Southern California Sensory Integration Tests (Ayres) and the Bender Gestalt: A creative study. *American Journal of Occupational Therapy, 31,* 294–299.

Kimball, J.G. (1981). Normative comparison of the Southern California Postrotary Nystagmus Test: Los Angeles vs. Syracuse data. *American Journal of Occupational Therapy, 35,* 21–25.

Kimball, J.G. (1986). Prediction of methylphenidate (Ritalin) responsiveness through sensory integrative testing. *American Journal of Occupational Therapy, 40,* 241–248.

Kimura, D. (1977). Acquisition of a motor skill after left-hemisphere damage. *Brain, 100,* 527–542.

Kinnealey, M. (1984). *Reliability and validity of two tests of tactile function.* Unpublished dissertation, Temple University, Philadelphia.

Kirk, S.A., McCarthy, J.J., & Kirk, W.D. (1968). *Illinois Test of Psycholinguistic Abilities.* Urbana: University of Illinois Press.

Knorn, P. (1985). *Translation of the Southern California Sensory Integration Tests and empirical investigation of the interrater reliability and the sensory integration of 5- to 5½-year-old German children.* Unpublished graduate thesis, Department of Psychology, University of Bielefeld, Federal Republic of Germany.

Lesny, I.A. (1980). Developmental dyspraxia-dysgnosia or a cause of congenital children's clumsiness. *Brain and Development, 2,* 69–71.

Lindley, S. (1973). Kinesthetic perception in early blind adults. *American Foundation for the Blind, Research Bulletin No. 25,* 75–191.

Llorens, L.A. (1968). Identification of the Ayres' syndromes in emotionally disturbed children: An exploratory study. *American Journal of Occupational Therapy, 22,* 286–288.

Lorr, M. (1983). Cluster analysis for social scientists. San Francisco: Jossey-Bass.

Luria, A.R. (1966). *Higher Cortical Function in Man*. New York: Basic Books.

Mandell, R.J., Nelson, D.L., & Cermak, S.A. (1984). Differential laterality of hand function in right-handed and left-handed boys. *American Journal of Occupational Therapy, 38,* 114–120.

McAtee, S.M. (1987). A correlational study of the Design Copying atypical approach parameters and the *Southern California Sensory Integration Tests*: A pilot study. Unpublished master's thesis, University of Southern California, Los Angeles.

McCarthy, P.J. (1984). A study of the SCSIT as a task for identifying neurophysiological dysfunctions in learning disabled children. *Dissertation Abstracts International, 44,* (12A), 3660.

McCracken, A. (1975). Tactile function of educable mentally retarded children. *American Journal of Occupational Therapy, 29,* 397–402.

*The Minnesota Rate of Manipulation Tests.* (1969). Circle Pines, MN: American Guidance Service, Inc.

Morrison, D.C., Hinshaw, S.P., & Carte, E.T. (1985). Signs of behavioral dysfunction in a sample of learning disabled children: stability and concurrent validity. *Perceptual and Motor Skills, 61,* 863–872.

Nielsen, J.M. (1938). Disturbance of the body scheme. *Bulletin of Los Angeles Neurological Society, 3,* 127–135.

Nielsen, J.M. (1946). *Agnosia, apraxia, aphasia: Their value in cerebral localization*. New York: Paul B. Haeber.

Nielsen, J.M. (1955). Agnosias, apraxias, speech, and phasia. In A.B. Baker (Ed.), *Clinical neurology, Vol. I* (pp. 352–378). New York: Paul B. Haeber.

Ottenbacher, K. (1978). Identifying vestibular processing dysfunction in learning-disabled children. *American Journal of Occupational Therapy, 32,* 217-221.

Ottenbacher, K. (1980). Excessive postrotary nystagmus duration in learning-disabled children. *American Journal of Occupational Therapy, 34,* 40–44.

Ottenbacher, K. (1982). Sensory integration therapy: affect or effect. *American Journal of Occupational Therapy, 36,* 571–578.

Ottenbacher, K., Abbott, C., Haley, D., & Watson, P.J. (1984). Human figure drawing ability and vestibular processing dysfunction in learning-disabled children. *Journal of Clinical Psychology, 40,* 1084–1089.

Ottenbacher, K., Short, M.A., & Watson, P.J. (1979). Nystagmus duration changes of learning-disabled children during sensory integrative therapy. *Perceptual and Motor Skills, 48,* 1159–1164.

Ottenbacher, K., Short, M.A., & Watson, P.J. (1980). The use of selected clinical observations to predict postrotary nystagmus change in learning-disabled children. *Physical and Occupational Therapy in Pediatrics, 1,* 31–37.

Ottenbacher, K., Watson, P.J., & Short, M.A. (1979). Association between nystagmus hyporesponsivity and behavioral problems in learning-disabled children. *American Journal of Occupational Therapy, 33,* 317–322.

Ottenbacher, K., Watson, P.J., Short, M.A., & Biderman, M.D. (1979). Nystagmus and ocular fixation difficulties in learning-disabled children. *American Journal of Occupational Therapy, 33,* 717–721.

Petersen, P., Goar, D., & Van Deusen, J. (1985). Performance of female adults on the Southern California Visual Figure-Ground Perception Test. *American Journal of Occupational Therapy, 39,* 525–530.

Petersen, P., & Wikoff, R.L. (1983). The performance of adult males on the Southern California Figure-Ground Visual Perception Test. *American Journal of Occupational Therapy, 37,* 554–560.

Poeck, K. (1982). The two types of motor apraxia. *Archives Italiennes de Biologie, 120,* 361–369.

Poeck, K. (1985). Clues to the nature of disruptions of limb praxis. In E.A. Roy (Ed.), *Neuropsychological studies of apraxia and related disorders* (pp. 99–109). Amsterdam, Netherlands: Elsevier (North Holland).

Poeck, K., & Orgass, B. (1971). The concept of the body schema: a critical review and some experimental results. *Cortex, 7,* 254–277.

Potter, C.N., & Silverman, L.N. (1984). Characteristics of Vestibular function and static balance skills in deaf children. *Physical Therapy, 64,* 1071–1075.

Punwar, A. (1982). Expanded normative data: Southern California Postrotary Nystagmus Test. *American Journal of Occupational Therapy, 36,* 183–187.

Purdue Research Foundation. (1948). *Purdue Pegboard Test.* Chicago: Science Research Associates, Inc.

Rider, B.A. (1973). Perceptual-motor dysfunction in emotionally disturbed children. *American Journal of Occupational Therapy, 26,* 316–320.

Roy, E.A., & Square, P.A. (1985). Common considerations in the study of limb, verbal, and oral apraxia. In E.A. Roy (Ed.), *Neuropsychological studies of apraxia and related disorders* (pp. 111–161). Amsterdam, Netherlands: Elsevier (North Holland).

Royeen, C.B. (1984). Incidence of hypoactive nystagmus among behaviorally disordered children. *Occupational Therapy Journal of Research, 4,* 237–240.

Royeen, C.B. (1986). The development of a touch scale for measuring defensiveness in children. *American Journal of Occupational Therapy, 40,* 414–419.

Royeen, C.B., Lesinski, G., Ciani, S., & Schneider, D. (1981). Relationship of the Southern California Sensory Integration Tests, the Southern California Postrotary Nystagmus Test, and clinical observations accompanying them to evaluation in otolaryngology, ophthalmology, and audiology: Two descriptive studies. *American Journal of Occupational Therapy, 35,* 443–450.

Saeki, K., Clark, F.A., & Azen, S.P. (1985). Performance of Japanese and Japanese-American children on the Motor Accuracy-Revised and Design Copying Tests of the Southern California Sensory Integration Tests. *American Journal of Occupational Therapy, 39,* 103–109.

Savich, P.A. (1984). Anticipatory imagery ability in normal and language-disabled children. *Journal of Speech and Hearing research, 27,* 494–501.

Schaaf, R. (1985). The frequency of vestibular disorder in developmentally delayed preschooler with otitis media. *American Journal of Occupational Therapy, 39,* 247–252.

Schilder, P. (1942). *Mind: Perception and Thought in Their Constructional Aspects.* New York: International Universities Press.

Schilder, P. (1950). *The Image and Appearance of the Human Body.* New York: International Universities Press.

Short, M.A., Watson, P.J., Ottenbacher, K., & Rogers, C. (1983). Vestibular-proprioceptive function in 4-year-olds: Normative and regression analyses. *American Journal of Occupational Therapy, 37*(2), 102–109.

Shuer, J., Clark, F., & Azen, S.P. (1980). Vestibular function in mildly mentally retarded adults. *American Journal of Occupational Therapy, 34,* 664–670.

Silberzahn, M. (1975). Sensory integrative function in a child guidance population. *American Journal of Occupational Therapy, 29,* 28–34.

Slavik, B.A. (1982). Vestibular function in children with non-paralytic strabismus. *Occupational Therapy Journal of Research, 2,* 220–233.

Sleeper, M.L. (1962). *Correlation of body balance and space perception in cerebral palsied individuals.* Unpublished master's thesis, University of Southern California, Los Angeles.

Smith, S. (1983). Performance difference between hands in children on the Motor Accuracy Test-Revised. *American Journal of Occupational Therapy, 37,* 96–101.

Spreen, O., & Benton, A.L. (1963). *Sentence Repetition Test: Administration, scoring, and preliminary norms.* Unpublished manuscript, University of Iowa.

Steinberg, M., & Rendle-Short, J. (1977). Vestibular dysfunction in young children with minor neurological impairment. *Developmental Medicine and Child Neurology, 19,* 639–651.

Stanley, J.C. (1971). Reliability. In R.L. Thorndike (Ed.), *Educational Measurement* (pp. 356–442). Washington, DC: American Council on Education.

Stilwell, J.M. (1981). Relationship between development of the body-righting reaction and manual midline crossing behavior in the learning disabled. *American Journal of Occupational Therapy, 35,* 391–398.

Stilwell, J.M., Crowe, T.K., & McCallum, L.W. (1978). Postrotary nystagmus duration as a function of communication disorder. *American Journal of Occupational Therapy, 32,* 222–228.

Strauss, A.A., & Lehtinen, L.E. (1947). *Psychopathology and Education of the Brain Injured Child, Vol. I.* New York: Grune & Stratton.

Su, R.V., & Yerxa, E.J. (1984). Comparison of the motor tests of the SCSIT and the LNNB-C. *Occupational Therapy Journal of Research, 4,* 96–107.

Tibling, L. (1969). The rotary response in children. *Acta Oto-Laryngologica, 68,* 459–469.

Tuohimaa, P. (1978). Vestibular disturbances after acute mild head injury. *Acta Oto-Laryngologica Supplement. 357,* 3–67.

Ward, S.E. (1973). Order effect on selected perceptual motor tests. *American Journal of Occupational Therapy, 27,* 321–325.

Watson, P.J., Ottenbacher, K., Short, M.A., Kittrell, J., & Workman, E.A. (1981). Human figure drawings of learning disabled children with hyporesponsive postrotary nystagmus. *Physical and Occupational Therapy in Pediatrics, 1*(4), 21–25.

Watson, P.J., Ottenbacher, K., Workman, E.A., Short, M.A., & Dickman, D.A. (1982). Visual motor difficulties in emotionally disturbed children with hyporesponsive nystagmus. *Physical and Occupational Therapy in Pediatrics, 2*(2–3), 67–72.

Witkin, H.A. (1950). Individual differences in ease of perception of embedded figures. *Journal of Personality, 19,* 1–15.

Zee-Chen, E.L.F., & Hardman, M.L. (1983). Postrotary nystagmus response in children with Down's syndrome. *American Journal of Occupational Therapy, 37,* 260–265.

Ziviani, J., Poulsen, A., & O'Brien, A. (1982). Correlation of the Bruininks-Oseretsky Test of Motor Proficiency with the Southern California Sensory Integration Tests. *American Journal of Occupational Therapy, 36,* 519–523.

**IMPORTANT:** You must enter this Transmittal Number on all protocol sheets for this child.

PLEASE DO NOT MARK INSIDE THIS BOX

# wps TEST REPORT™

Western Psychological Services • 12031 Wilshire Blvd. • Los Angeles, CA 90025-1251

SENSORY INTEGRATION AND PRAXIS TESTS (SIPT)

# TRANSMITTAL SHEET

A. Jean Ayres, Ph.D.

## MARKING INSTRUCTIONS

1. Use only a soft (No. 2), black-leaded pencil. Do **not** use ballpoint pen, felt-tipped pen, or mechanical pencil. (Use ink only for the shipping label below.)
2. Completely fill in the circles.
3. To change a response, erase the first mark completely and fill in the correct circle.
4. When entering a number on this sheet or the protocol sheets (e.g., Transmittal Number, time score), write the number in the boxes below the heading and then fill in the corresponding circles below the boxes. Be sure that your entry is the same number of digits as requested; for example, if the Transmittal Number is only six digits rather than eight, enter zeros in the first two columns (e.g., 00929546).
5. **Make no stray marks on the protocol sheets.** Write only in designated spaces, response circles, and boxes labeled "Observations." **Do not write in any margins or in any spaces not clearly indicated as acceptable for writing.**

## TEST DIRECTIONS

1. First fill in the demographic information on the other side of this sheet. (After testing, remember to fill in the sections regarding your observations.) This sheet must be completed for each child and must accompany any protocol sheets submitted for scoring.
2. Follow the administration directions in the Manual. Cues to the directions appear on the protocol sheets.
3. **It is very important that you enter the Transmittal Number (printed at the top of this sheet) on all protocol sheets for a child.** Please double-check before submitting for scoring.
4. After administration, check all protocol sheets for incomplete identifying information, stray marks, light marks, improper writing materials, and damages or tears.
5. In the spaces below, enter your name and complete address. Also, indicate all protocol sheets being submitted for this child.
6. Send all protocol sheets with this Transmittal Sheet to WPS TEST REPORT. The scannable sheets are returned to you with the reports. To identify the protocol sheets as belonging to a particular child, you should enter the child's name in the appropriate box on each protocol sheet, and note the Transmittal Number for the child in your records before sending in these sheets for processing.

This Computerized Test Product
AUTHORIZED
By The Test Publisher

4 5 6 7 8 9
76410-321
Printed in U.S.A.

For WPS Use Only

① ② ③ ④ ⑤ ⑥ ⑦ ⑧ ⑨ ⑩ ⑪ ⑫ ⑬ ⑭ ⑮ ⑯ ⑰ ⑱ ⑲ ⑳ ㉑

**IMPORTANT: The information below must be completed for processing.**

| WHICH PROTOCOL SHEETS ARE ENCLOSED FOR THIS CHILD? | |
|---|---|
| ① SV | ⑩ MAc |
| ② FG | ⑪ SPr |
| ③ SWB | ⑫ OPr |
| ④ DC | ⑬ MFP |
| ⑤ PPr | ⑭ KIN |
| ⑥ BMC | ⑮ FI |
| ⑦ PrVC | ⑯ GRA |
| ⑧ CPr | ⑰ LTS |
| ⑨ PRN | ○ ALL 17 |

| TESTING DATE | | | | | |
|---|---|---|---|---|---|
| Month | | Day | | Year | |
| ⓪ | ⓪ | ⓪ | ⓪ | ⓪ | ⓪ |
| ① | ① | ① | ① | ① | ① |
| | ② | ② | ② | ② | ② |
| | ③ | ③ | ③ | ③ | ③ |
| | ④ | | ④ | ④ | ④ |
| | ⑤ | | ⑤ | ⑤ | ⑤ |
| | ⑥ | | ⑥ | ⑥ | ⑥ |
| | ⑦ | | ⑦ | ⑦ | ⑦ |
| | ⑧ | | ⑧ | ⑧ | ⑧ |
| | ⑨ | | ⑨ | ⑨ | ⑨ |

**Print clearly (in ink) your name and complete address, including zip code. This will be used as the shipping label to mail your reports.** →

From: **WPS TEST REPORT**
12031 Wilshire Blvd.
Los Angeles, CA 90025-1251

To:

W-260C

**Complete the background information below for the child.**
**Age, sex, and writing hand must be completed. All others are optional.**

**CHILD'S NAME**
**(Enter first name, middle initial, last initial)**

| First Name | MI | Last |
|---|---|---|
| A–Z | A–Z | A–Z |

**CHILD'S AGE**

| Yrs. | Mos. |
|---|---|
| 4, 5, 6, 7, 8, 9, 10, 11, 12+ | 0–9, 0–1 |

**CHILD'S SEX**

- ◯ Male
- ◯ Female

**CHILD'S WRITING HAND**

- ◯ Right
- ◯ Left
- ◯ Both

**CHILD'S GRADE**

Ⓟ Ⓚ ① ② ③ ④ ⑤ ⑥ ⑦+

**IS THE CHILD IN A SPECIAL EDUCATION CLASS?**

- ◯ No
- ◯ Yes (If yes, mark *all* descriptors below that apply)
  - ◯ Special school
  - ◯ Special class
  - ◯ Resource room at least 40% of time
  - ◯ Regular classroom most of time
  - ◯ Behaviorally/emotionally disordered
  - ◯ Hyperactive (ADD)
  - ◯ Learning disabled
  - ◯ Mentally retarded
  - ◯ Physically handicapped
  - ◯ Sensory impaired (deaf, blind)
  - ◯ Speech impaired

**CHILD'S ETHNIC BACKGROUND**

- ◯ Asian
- ◯ White
- ◯ Black
- ◯ Other
- ◯ Hispanic

**CHILD'S IQ**

| IQ Score | IQ Test Administered |
|---|---|
| 0–1, 0–9, 0–9 | ◯ WISC-R<br>◯ WPPSI<br>◯ Stanford-Binet<br>◯ K-ABC<br>◯ McCarthy<br>◯ Other: ________ |

**PARENTS' OCCUPATIONS**

| Occupational Level | Father | Mother |
|---|---|---|
| Executive, advanced professional | ◯ | ◯ |
| Business manager, lower professional, teacher | ◯ | ◯ |
| Administrative personnel, small business owner | ◯ | ◯ |
| Clerical, sales, technical | ◯ | ◯ |
| Skilled manual | ◯ | ◯ |
| Semi-skilled, machine operator | ◯ | ◯ |
| Unskilled | ◯ | ◯ |
| Not employed outside the home | ◯ | ◯ |

**CHILD'S HOME ZIP CODE**

0–9 0–9 0–9 0–9 0–9

**DOES THE CHILD HAVE A KNOWN NEUROLOGICAL IMPAIRMENT?**

◯ Yes ◯ No

AFTER THE TESTS ARE COMPLETED, FILL IN THIS INFORMATION BASED ON YOUR OBSERVATIONS.

**TACTILE DEFENSIVENESS**

- ◯ No defensive responses
- ◯ Mild defensiveness
- ◯ Moderate defensiveness
- ◯ Severe defensiveness

**CLINICAL OBSERVATIONS**

| Observation | Poor | Slightly Deficient | Adequate |
|---|---|---|---|
| Prone extension | ◯ | ◯ | ◯ |
| Supine flexion | ◯ | ◯ | ◯ |
| Ocular pursuits | ◯ | ◯ | ◯ |

**DO NOT WRITE HERE**

# wps TEST REPORT™

Western Psychological Services • 12031 Wilshire Blvd. • Los Angeles, CA 90025-1251

SENSORY INTEGRATION AND PRAXIS TESTS (SIPT)

## 1. SPACE VISUALIZATION (SV)

Protocol Sheet

A. Jean Ayres, Ph.D.

**DIRECTIONS:** If you are right-handed, seat the child to your left. For Trial I: **"Which of these blocks fits this big black hole?"** If child doesn't put it in hole: **"Put the block in the hole."** If child chooses wrong block: **"Try the other one."** For Trial II: **"Here's another. Look first; then show me which one fits. . . Put it in the hole. . . That's right. Think about it first; then pick up the block that fits the hole."** After Item 3 or 4: **"Look at BOTH blocks. Choose carefully. The first one you move counts as your choice. You want to be right the first time."**

For test items, record accuracy (R = Right, W = Wrong), hand (L = Left, B = Both, R = Right), and number of seconds (maximum = 25 seconds). Keep track of number of errors (marked "W" for accuracy) and discontinue after fifth error, not necessarily consecutive.

After the test, fill in the circles corresponding to the number of seconds. Enter time as two digits (e.g., 6 seconds as 06).

- Maximum time per item: **25** seconds
- Discontinue after **5** errors, not necessarily consecutive

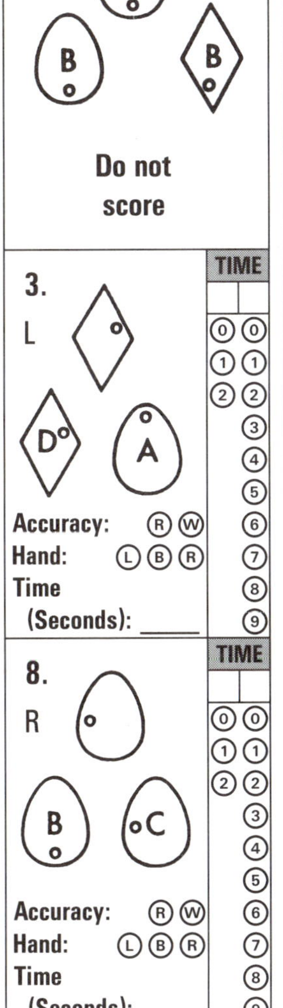

Trial I
L
Do not score
3.
L
Accuracy: R W
Hand: L B R
Time (Seconds): ____
8.
R


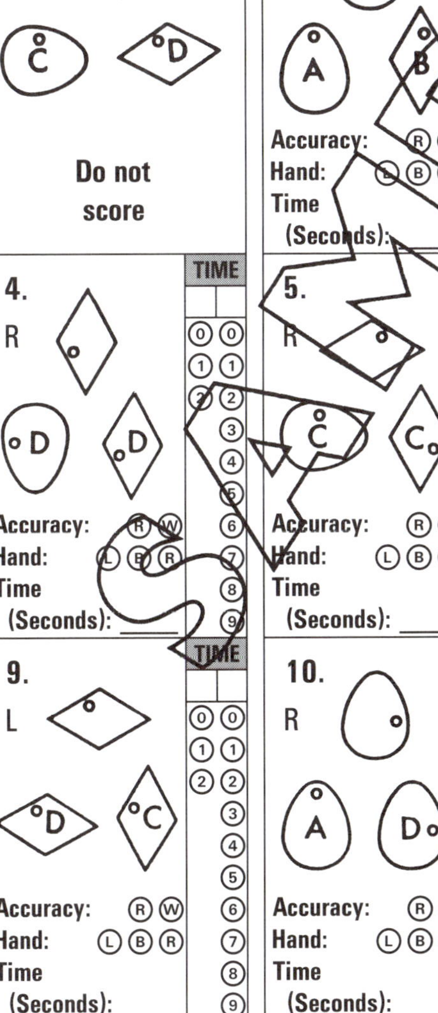


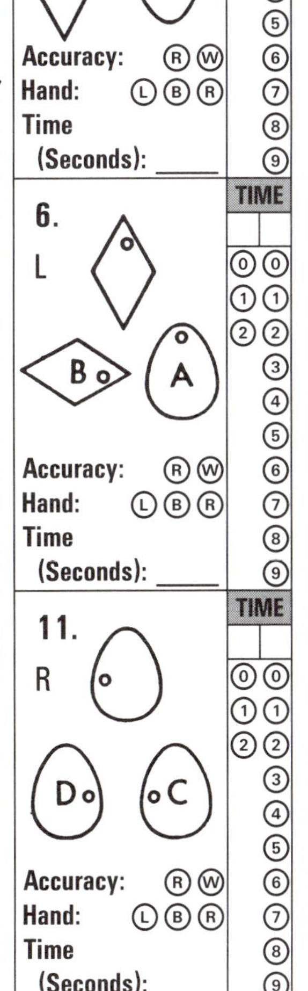


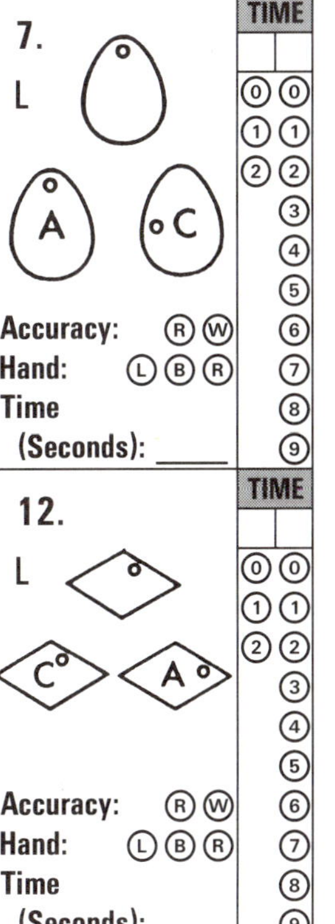


Number of Errors (Items 1-12):

CONTINUE ON BACK

4 5 6 7 8 9
76412-321
Printed in U.S.A.
W-260B(1)

For WPS Use Only

- Maximum time per item: **25** seconds
- Discontinue after **5** errors, not necessarily consecutive

**DO NOT WRITE OUTSIDE THE BOXES**

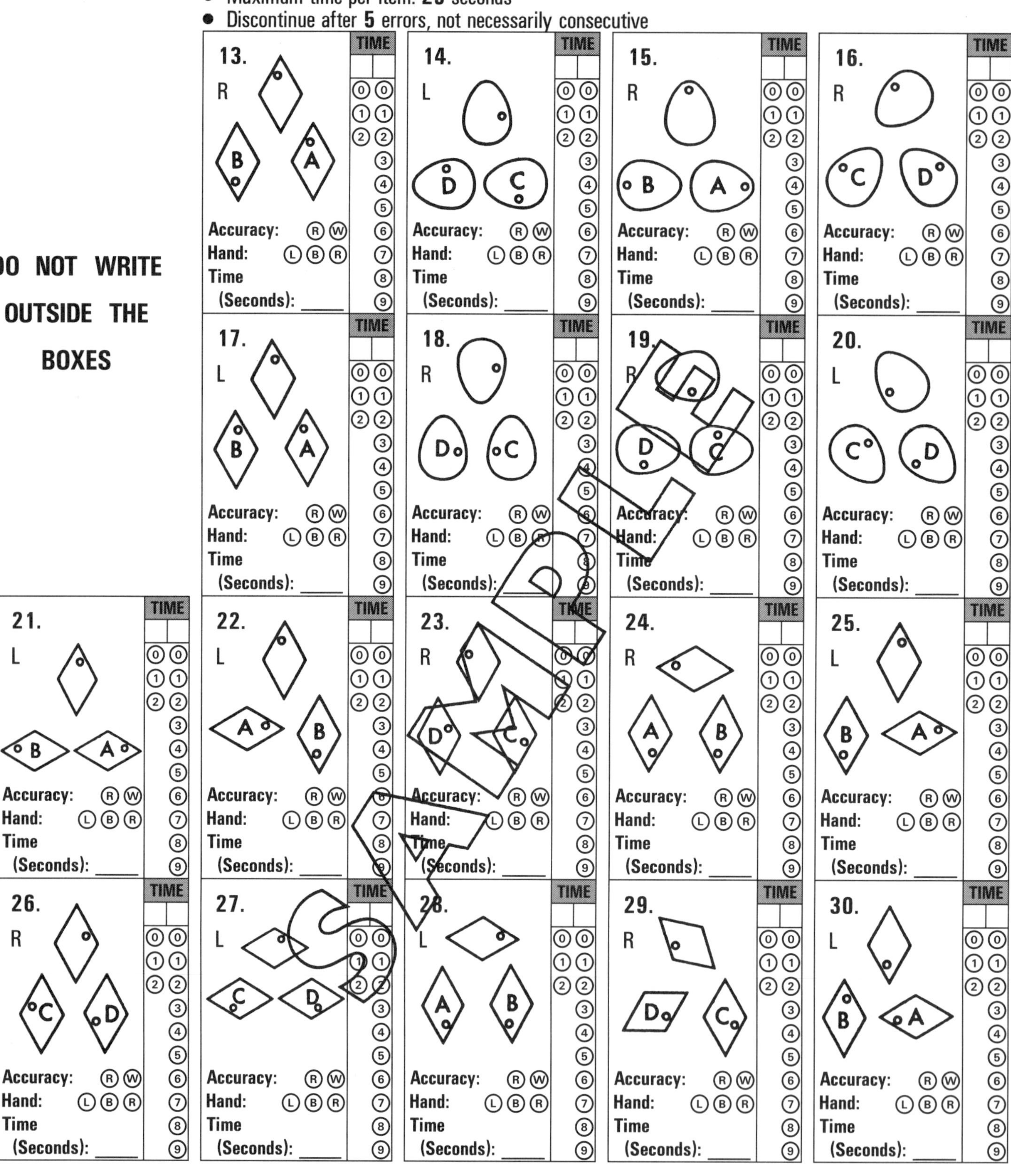


Number of Errors (Continued from front side):

**DO NOT WRITE OUTSIDE THE BOXES**

# wps TEST REPORT™

Western Psychological Services • 12031 Wilshire Blvd. • Los Angeles, CA 90025-1251

SENSORY INTEGRATION AND PRAXIS TESTS (SIPT)

# 2. FIGURE-GROUND PERCEPTION (FG)

Protocol Sheet

A. Jean Ayres, Ph.D.

**DIRECTIONS:** Seat the child to your left. For Trial I: **"Three of these pictures are up here. Which three are they?. . . These three are not up here, are they? That is the way it will be each time I turn the page. Find three pictures down here which are up here. Look carefully because it can be tricky."**

For Trial II: **"Now you will look at designs instead of pictures of things. Three of these designs are part of this one. They are hidden in this upper design, just as some of these pictures** (turn back to Plates 8A and 8B) **were part of this upper figure. . .** (Point to Design 1 of Plate IIB) **This design is a cross, but not like the crossed lines up here, so it is not part of the upper design. This one** (Design 2) **is part of the design up here. Can you see it? Can you see this one** (Design 3) **up here? Some of the lines in this upper picture are not always in the choices down here. Which of these** (Designs 4, 5, and 6) **is hidden in this design?. . . The rest of the designs will be something like this one. Find the three designs here that are up here."** If the child has not made three choices on each item by 50 seconds, ask the child to choose. Fill in the circles for the child's *three* selections. Correct choices are shaded. Reinstruct if two errors are made on Item 1. Record the number of errors and time taken to make three selections (maximum = 60 seconds). Stop timing at last error. Discontinue after seven errors, four on last three plates attempted. If not on last three plates, continue until four errors on three plates.

After the test, fill in the circles corresponding to the number of seconds. Enter time as two digits (e.g., 6 seconds as 06).

REQUIRED TRANSMITTAL NUMBER

| | | | | | | |
|---|---|---|---|---|---|---|
| 0 | 0 | 0 | 0 | 0 | 0 | 0 |
| 1 | 1 | 1 | 1 | 1 | 1 | 1 |
| 2 | 2 | 2 | 2 | 2 | 2 | 2 |
| 3 | 3 | 3 | 3 | 3 | 3 | 3 |
| 4 | 4 | 4 | 4 | 4 | 4 | 4 |
| 5 | 5 | 5 | 5 | 5 | 5 | 5 |
| 6 | 6 | 6 | 6 | 6 | 6 | 6 |
| 7 | 7 | 7 | 7 | 7 | 7 | 7 |
| 8 | 8 | 8 | 8 | 8 | 8 | 8 |
| 9 | 9 | 9 | 9 | 9 | 9 | 9 |

CHILD'S NAME:

**Trial I**

| | | |
|---|---|---|
| ① | ② | ③ |
| ④ | ⑤ | ⑥ |

- Maximum time per plate: **60** seconds
- Discontinue after **7** errors, **4** of which are on last **3** plates attempted

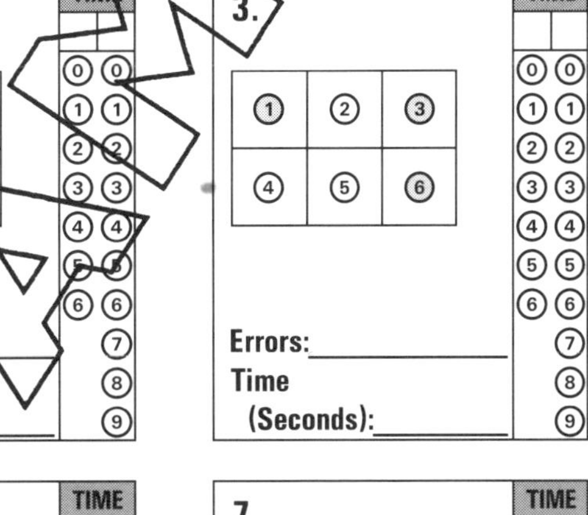

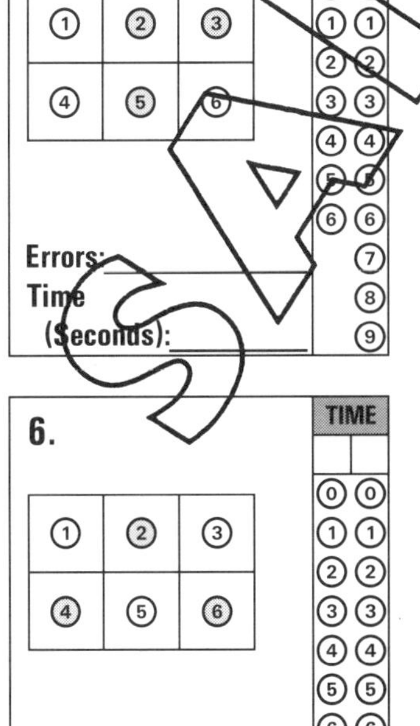

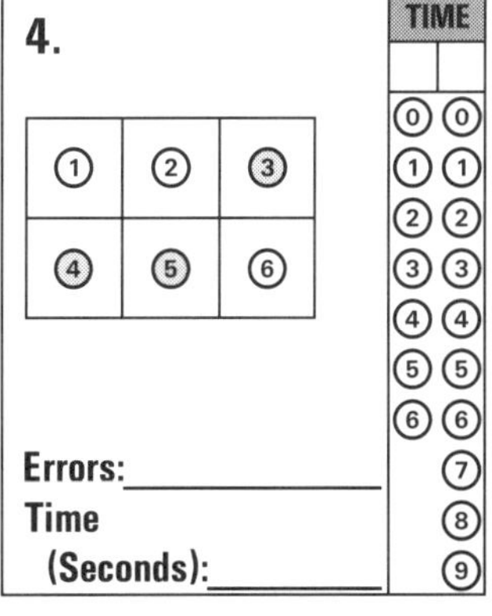

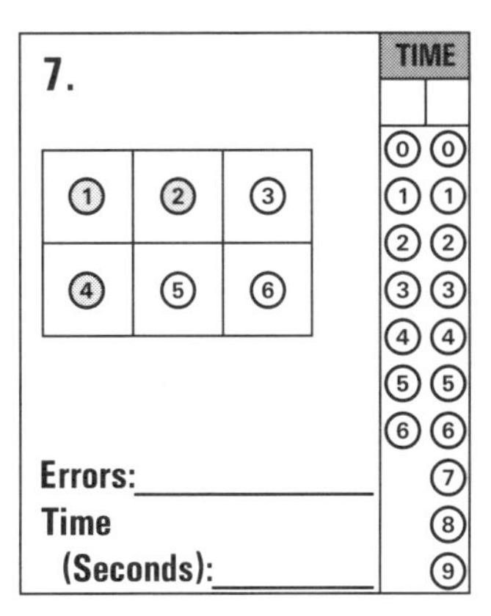

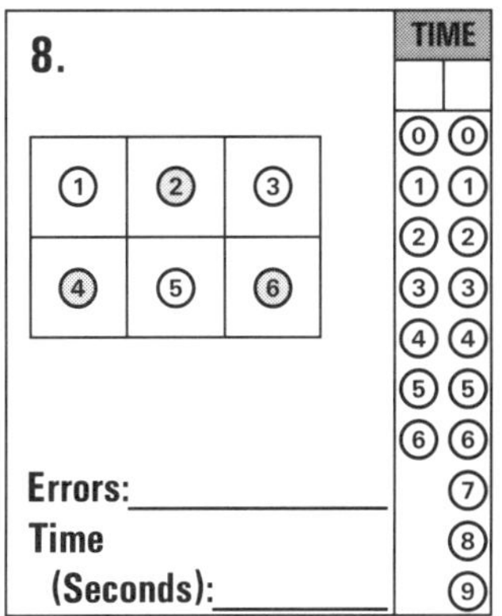

**1.**

| | | |
|---|---|---|
| ① | ② | ③ |
| ④ | ⑤ | ⑥ |

Errors:______

Time (Seconds):______

TIME: 0 0 / 1 1 / 2 2 / 3 3 / 4 4 / 5 5 / 6 6 / 7 / 8 / 9

**2.**

| | | |
|---|---|---|
| ① | ② | ③ |
| ④ | ⑤ | ⑥ |

Errors:______

Time (Seconds):______

TIME: 0 0 / 1 1 / 2 2 / 3 3 / 4 4 / 5 5 / 6 6 / 7 / 8 / 9

**3.**

| | | |
|---|---|---|
| ① | ② | ③ |
| ④ | ⑤ | ⑥ |

Errors:______

Time (Seconds):______

TIME: 0 0 / 1 1 / 2 2 / 3 3 / 4 4 / 5 5 / 6 6 / 7 / 8 / 9

**4.**

| | | |
|---|---|---|
| ① | ② | ③ |
| ④ | ⑤ | ⑥ |

Errors:______

Time (Seconds):______

TIME: 0 0 / 1 1 / 2 2 / 3 3 / 4 4 / 5 5 / 6 6 / 7 / 8 / 9

**5.**

| | | |
|---|---|---|
| ① | ② | ③ |
| ④ | ⑤ | ⑥ |

Errors:______

Time (Seconds):______

TIME: 0 0 / 1 1 / 2 2 / 3 3 / 4 4 / 5 5 / 6 6 / 7 / 8 / 9

**6.**

| | | |
|---|---|---|
| ① | ② | ③ |
| ④ | ⑤ | ⑥ |

Errors:______

Time (Seconds):______

TIME: 0 0 / 1 1 / 2 2 / 3 3 / 4 4 / 5 5 / 6 6 / 7 / 8 / 9

**7.**

| | | |
|---|---|---|
| ① | ② | ③ |
| ④ | ⑤ | ⑥ |

Errors:______

Time (Seconds):______

TIME: 0 0 / 1 1 / 2 2 / 3 3 / 4 4 / 5 5 / 6 6 / 7 / 8 / 9

**8.**

| | | |
|---|---|---|
| ① | ② | ③ |
| ④ | ⑤ | ⑥ |

Errors:______

Time (Seconds):______

TIME: 0 0 / 1 1 / 2 2 / 3 3 / 4 4 / 5 5 / 6 6 / 7 / 8 / 9

Number of Errors (Items 1-8):

**CONTINUE ON BACK**



4 5 6 7 8 9

76415-321

Printed in U.S.A.

W-260B(2)

For WPS Use Only

| | | | | | | |
|---|---|---|---|---|---|---|
| 1 | 2 | 3 | 4 | 5 | 6 | 7 |
| 8 | 9 | 10 | 11 | 12 | 13 | 14 |
| 15 | 16 | 17 | 18 | 19 | 20 | 21 |

**DO NOT WRITE OUTSIDE THE BOXES**

**OBSERVATIONS:**

**Trial II**

| ① | ② | ③ |
|---|---|---|
| ④ | ⑤ | ⑥ |

- Maximum time per plate: **60** seconds
- Discontinue after **7** errors, **4** of which are on last **3** plates attempted

**9.** ① ② ③ / ④ ⑤ ⑥ — TIME — Errors:____ Time (Seconds):____

**10.** ① ② ③ / ④ ⑤ ⑥ — TIME — Errors:____ Time (Seconds):____

**11.** ① ② ③ / ④ ⑤ ⑥ — TIME — Errors:____ Time (Seconds):____

**12.** ① ② ③ / ④ ⑤ ⑥ — TIME — Errors:____ Time (Seconds):____

**13.** ① ② ③ / ④ ⑤ ⑥ — TIME — Errors:____ Time (Seconds):____

**14.** ① ② ③ / ④ ⑤ ⑥ — TIME — Errors:____ Time (Seconds):____

**15.** ① ② ③ / ④ ⑤ ⑥ — TIME — Errors:____ Time (Seconds):____

**16.** ① ② ③ / ④ ⑤ ⑥ — TIME — Errors:____ Time (Seconds):____

Number of Errors (Continued from front side):

SENSORY INTEGRATION AND PRAXIS TESTS (SIPT)

# 3. STANDING AND WALKING BALANCE (SWB)

Protocol Sheet

A. Jean Ayres, Ph.D.

**WPS TEST REPORT**™

Western Psychological Services • 12031 Wilshire Blvd. • Los Angeles, CA 90025-1251

**REQUIRED TRANSMITTAL NUMBER**

**DIRECTIONS:** If you think the child has a severe balance problem, begin on Item 1. However, for most children, begin on Item 4.

Demonstrate each item and give verbal directions. Record the number of seconds (or number of steps on Items 9-10). If the child does not receive the maximum score on the first attempt, allow a second attempt and, if necessary (on Items 5-16), a third attempt. Before each repeated attempt: **"Good. Try again. Try to stand longer (walk further)."** Discontinue each item when the child meets the maximum score or when the required number of attempts has been made.

If the child does not receive the maximum score on Item 4 (10 seconds), administer Items 3, 2, and 1 until the child receives the maximum score on two items. Then continue with Item 5.

Discontinue the test when all attempts on three consecutive items are scored as 0 or 1. After the test, fill in the circles corresponding to the best response (highest score) on each item.

**CHILD'S NAME:**

- Begin on Item 4 for most children; if maximum not reached, administer Items 3, 2, 1
- Discontinue when **3** consecutive items are scored as 0 or 1

| Item | Attempts | | | Best Response |
|---|---|---|---|---|
| 1. **OPEN** Stand feet together, toes even (Discontinue 5 seconds) | 1 | 2 | | 0 1 2 3 4 5 |
| 2. **CLOSED** Stand feet together, toes even (Discontinue 5 seconds) | 1 | 2 | | 0 1 2 3 4 5 |
| 3. **OPEN** Stand heel to toe (Discontinue 10 seconds) | 1 | 2 | | 0 1 2 3 4 5 6 7 8 9 10 |
| Start here 4. **CLOSED** Stand heel to toe (Discontinue 10 seconds) | 1 | 2 | | 0 1 2 3 4 5 6 7 8 9 10 |
| 5. **OPEN** Stand, balance R foot (Discontinue 10 seconds) | 1 | 2 | 3 | 0 1 2 3 4 5 6 7 8 9 10 |
| 6. **OPEN** Stand, balance L foot (Discontinue 10 seconds) | 1 | 2 | 3 | 0 1 2 3 4 5 6 7 8 9 10 |
| 7. **CLOSED** Stand, balance R foot (Discontinue 10 seconds) | 1 | 2 | 3 | 0 1 2 3 4 5 6 7 8 9 10 |
| 8. **CLOSED** Stand, balance L foot (Discontinue 10 seconds) | 1 | 2 | 3 | 0 1 2 3 4 5 6 7 8 9 10 |
| 9. **OPEN** Walk heel/toe; count correct steps (Discontinue 15 steps) | 1 | 2 | 3 | 0 1 2 3 4 5 6 7 8 9 10 11 12 13 14 15 |
| 10. **CLOSED** Walk heel/toe; count correct steps (Discontinue 15 steps) | 1 | 2 | 3 | 0 1 2 3 4 5 6 7 8 9 10 11 12 13 14 15 |
| 11. **OPEN** Balance R foot on wood dowel (Discontinue 15 seconds) | 1 | 2 | 3 | 0 1 2 3 4 5 6 7 8 9 10 11 12 13 14 15 |
| 12. **OPEN** Balance L foot on wood dowel (Discontinue 15 seconds) | 1 | 2 | 3 | 0 1 2 3 4 5 6 7 8 9 10 11 12 13 14 15 |
| 13. **CLOSED** Balance R foot on wood dowel (Discontinue 10 seconds) | 1 | 2 | 3 | 0 1 2 3 4 5 6 7 8 9 10 |
| 14. **CLOSED** Balance L foot on wood dowel (Discontinue 10 seconds) | 1 | 2 | 3 | 0 1 2 3 4 5 6 7 8 9 10 |
| 15. **OPEN** Balance both feet on wood dowel (Discontinue 15 seconds) | 1 | 2 | 3 | 0 1 2 3 4 5 6 7 8 9 10 11 12 13 14 15 |
| 16. **CLOSED** Balance both feet on wood dowel (Discontinue 15 seconds) | 1 | 2 | 3 | 0 1 2 3 4 5 6 7 8 9 10 11 12 13 14 15 |

This Computerized Test Product
AUTHORIZED
By The Test Publisher

**For WPS Use Only**

4 5 6 7 8 9
76418-321
Printed in U.S.A.
W-260B(3)

DO NOT WRITE

ON THIS SIDE

OF SHEET

SENSORY INTEGRATION AND PRAXIS TESTS (SIPT)

# 4. DESIGN COPYING (DC)

Protocol Sheet

A. Jean Ayres, Ph.D.

## PART I

**DIRECTIONS FOR PART I: "Here are some pictures for you to copy. The pictures are up here. You draw them down here."** Trial item: **"Here is a line. It begins on this dot** (top left) **and stops on this dot. You draw it down here. Begin on this dot and draw the line to this dot. . . . That's right. This line is between this dot and this dot. There is no line here or here** (lower dots). **Your picture looks like this one."** Part I: **"Now draw these pictures. Draw them down here. Begin on a dot and stop on a dot. Draw carefully. We can't erase. Start with this one. Draw a line down here."** After Item 1: **"I am going to watch you and I will draw the same thing you draw. You watch your pictures."**

Draw the child's approach on Items 2-9, unless the child is age 4. Discontinue Part I after two consecutive items are scored 0 on accuracy (0 = incorrect, 1 = partially correct, 2 = correct) and go to Part II.

Refer to Manual to score accuracy and approach on the "Should Not Have" parameters. When more than one attempt is made, score **least** accurate attempt.

REQUIRED TRANSMITTAL NUMBER

| | | | | | | |
|---|---|---|---|---|---|---|
| 0 | 0 | 0 | 0 | 0 | 0 | 0 |
| 1 | 1 | 1 | 1 | 1 | 1 | 1 |
| 2 | 2 | 2 | 2 | 2 | 2 | 2 |
| 3 | 3 | 3 | 3 | 3 | 3 | 3 |
| 4 | 4 | 4 | 4 | 4 | 4 | 4 |
| 5 | 5 | 5 | 5 | 5 | 5 | 5 |
| 6 | 6 | 6 | 6 | 6 | 6 | 6 |
| 7 | 7 | 7 | 7 | 7 | 7 | 7 |
| 8 | 8 | 8 | 8 | 8 | 8 | 8 |
| 9 | 9 | 9 | 9 | 9 | 9 | 9 |

CHILD'S NAME:

• Discontinue Part I when **2** consecutive items are scored 0 on accuracy

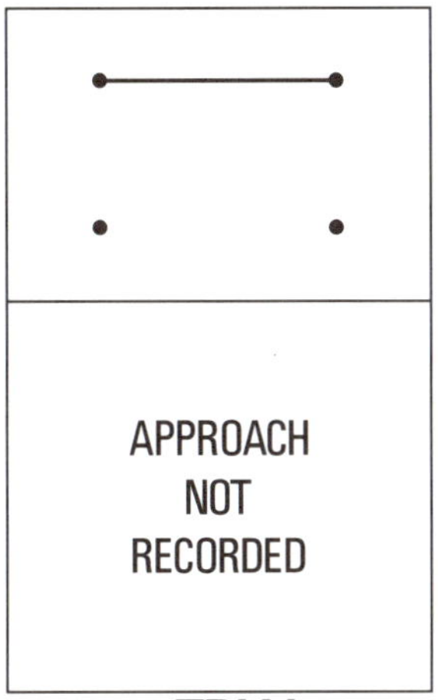

APPROACH NOT RECORDED

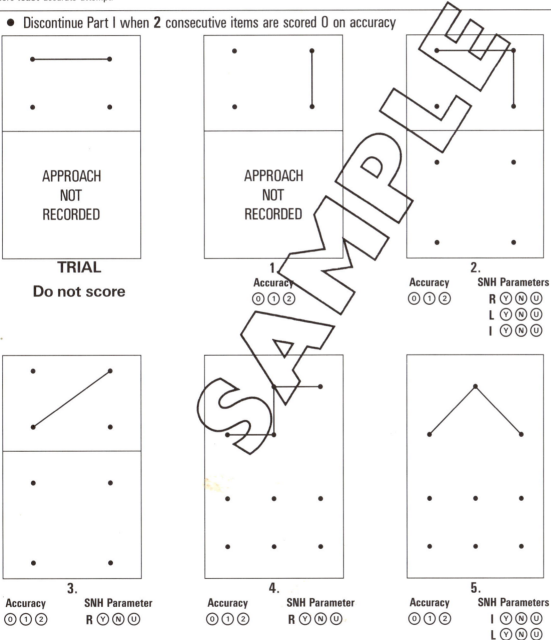

**TRIAL**

**Do not score**

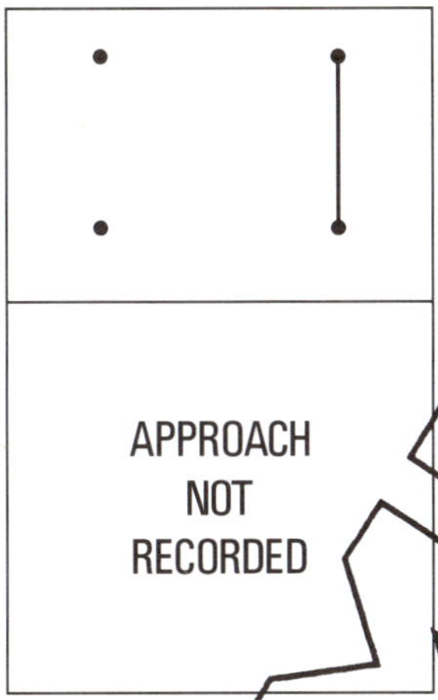

APPROACH NOT RECORDED

**1.**
Accuracy ⓪ ① ②

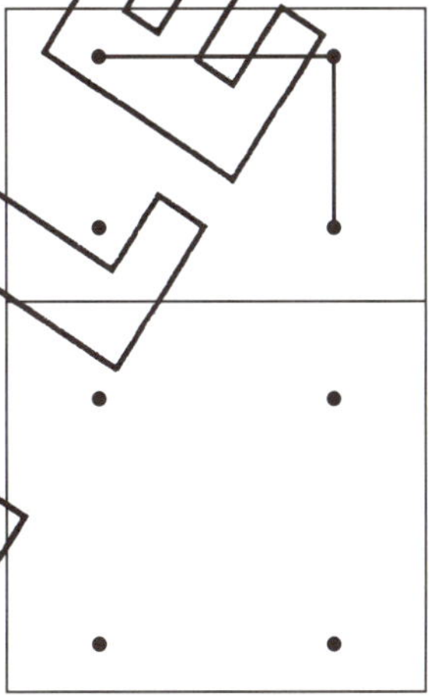

**2.**
Accuracy ⓪ ① ②
SNH Parameters
R Ⓨ Ⓝ Ⓤ
L Ⓨ Ⓝ Ⓤ
I Ⓨ Ⓝ Ⓤ

**DO NOT SEPARATE THE PAGES OF THIS BOOKLET**

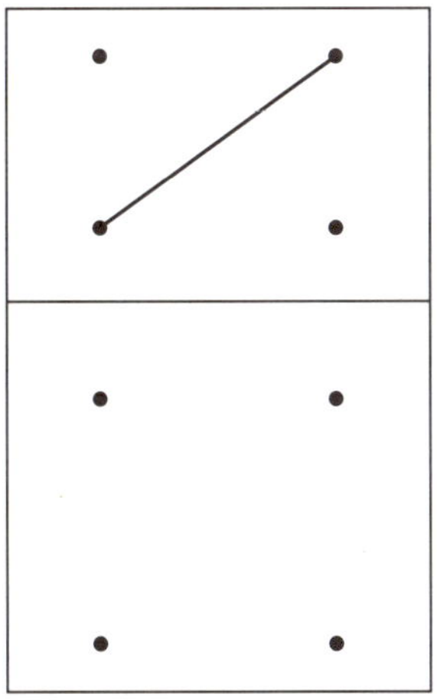

**3.**
Accuracy ⓪ ① ②
SNH Parameter
R Ⓨ Ⓝ Ⓤ

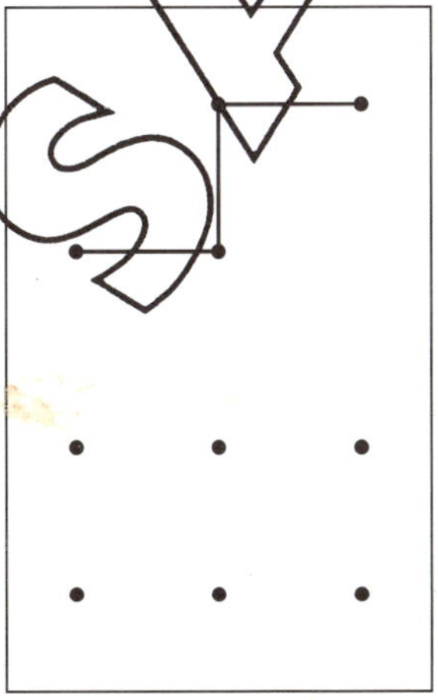

**4.**
Accuracy ⓪ ① ②
SNH Parameter
R Ⓨ Ⓝ Ⓤ

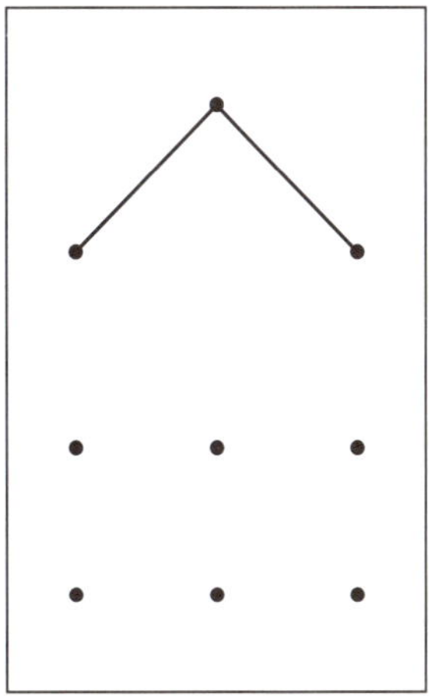

**5.**
Accuracy ⓪ ① ②
SNH Parameters
I Ⓨ Ⓝ Ⓤ
L Ⓨ Ⓝ Ⓤ

This Computerized Test Product
AUTHORIZED
By The Test Publisher

4 5 6 7 8 9
Printed in U.S.A.
76420-321
W-260B(4)

For WPS Use Only

| | | | | | | |
|---|---|---|---|---|---|---|
| 1 | 2 | 3 | 4 | 5 | 6 | 7 |
| 8 | 9 | 10 | 11 | 12 | 13 | 14 |
| 15 | 16 | 17 | 18 | 19 | 20 | 21 |

DO NOT

WRITE

HERE

OBSERVATIONS:

## PART I CONTINUED

- Discontinue Part I when **2** consecutive items are scored 0 on accuracy

6.
Accuracy
⓪ ① ②
SNH Parameters
R Ⓨ Ⓝ Ⓤ
I Ⓨ Ⓝ Ⓤ

7.
Accuracy
⓪ ① ②
SNH Parameters
R Ⓨ Ⓝ Ⓤ
L Ⓨ Ⓝ Ⓤ
I Ⓨ Ⓝ Ⓤ

8.
Accuracy
⓪ ① ②
SNH Parameters
R Ⓨ Ⓝ Ⓤ
I Ⓨ Ⓝ Ⓤ

9.
Accuracy
⓪ ① ②
SNH Parameters
R Ⓨ Ⓝ Ⓤ
I Ⓨ Ⓝ Ⓤ

APPROACH NOT RECORDED

10.
Accuracy
⓪ ① ②

APPROACH NOT RECORDED

11.
Accuracy
⓪ ① ②

APPROACH NOT RECORDED

12.
Accuracy
⓪ ① ②

APPROACH NOT RECORDED

13.
Accuracy
⓪ ① ②

## PART II

**DIRECTIONS FOR PART II: "Now draw these designs. Draw them in the empty spaces."** Draw the child's approach to all Part II items. Discontinue Part II after three consecutive items are scored "N" on gross approximation (SH Parameter 1 on all items). When a correction is made on Part II, score more accurate line(s) and disregard less accurate line(s). Refer to Manual to score "Should Have" and "Should Not Have" parameters. Do not score SNH parameters for 4-year-olds.

- Discontinue Part II when **3** consecutive items are scored "N" on SH Parameter 1 (gross approximation)

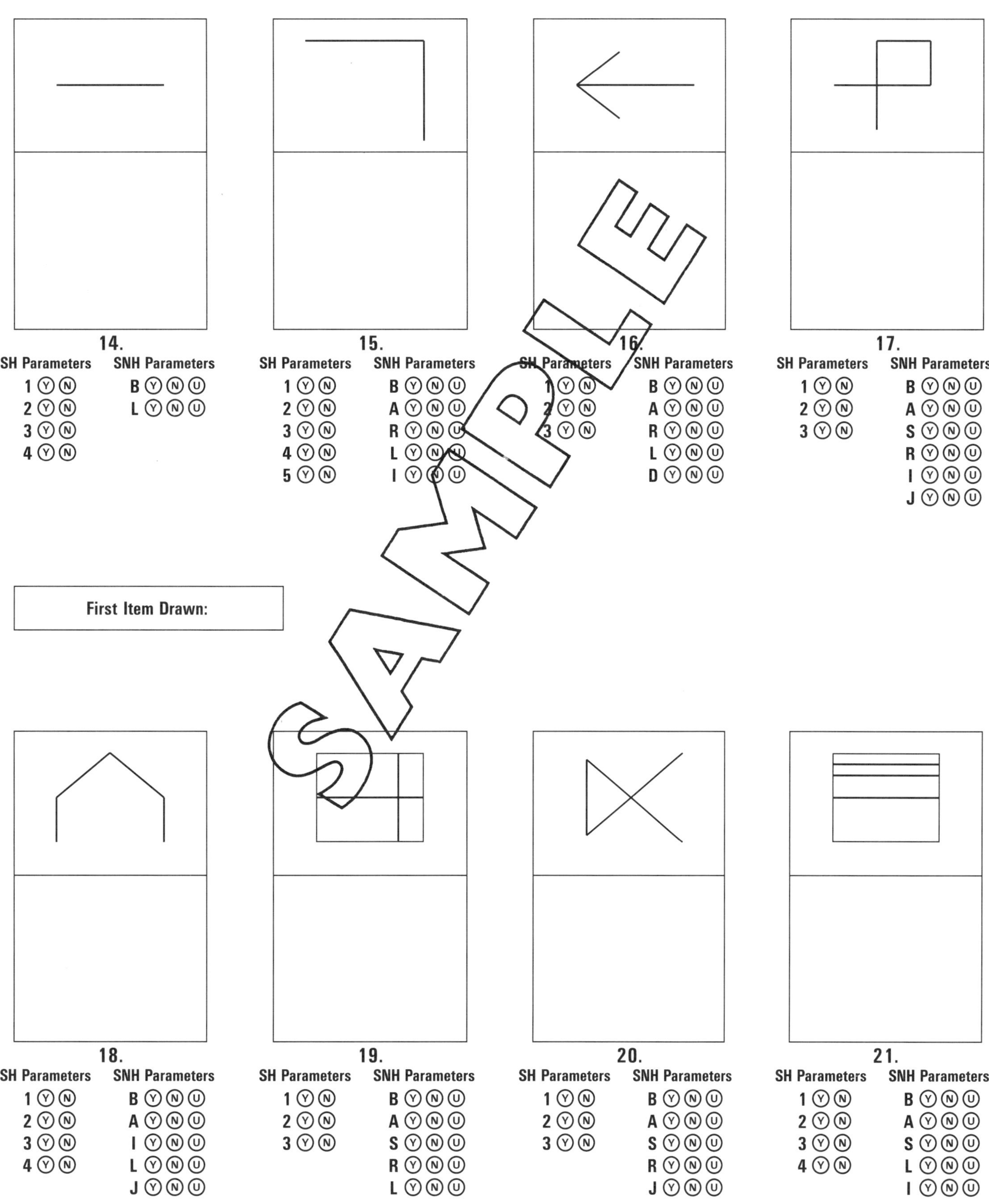

**14.**

| SH Parameters | SNH Parameters |
|---|---|
| 1 (Y) (N) | B (Y) (N) (U) |
| 2 (Y) (N) | L (Y) (N) (U) |
| 3 (Y) (N) | |
| 4 (Y) (N) | |

**15.**

| SH Parameters | SNH Parameters |
|---|---|
| 1 (Y) (N) | B (Y) (N) (U) |
| 2 (Y) (N) | A (Y) (N) (U) |
| 3 (Y) (N) | R (Y) (N) (U) |
| 4 (Y) (N) | L (Y) (N) (U) |
| 5 (Y) (N) | I (Y) (N) (U) |

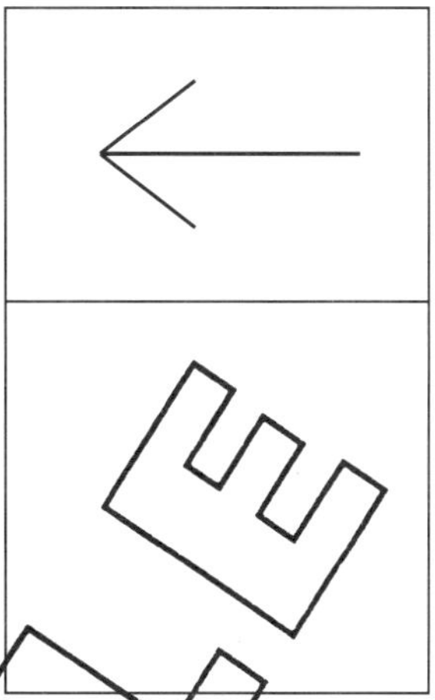

**16.**

| SH Parameters | SNH Parameters |
|---|---|
| 1 (Y) (N) | B (Y) (N) (U) |
| 2 (Y) (N) | A (Y) (N) (U) |
| 3 (Y) (N) | R (Y) (N) (U) |
| | L (Y) (N) (U) |
| | D (Y) (N) (U) |

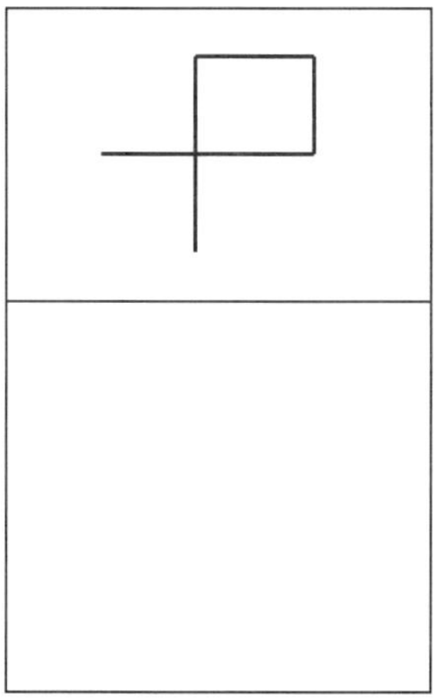

**17.**

| SH Parameters | SNH Parameters |
|---|---|
| 1 (Y) (N) | B (Y) (N) (U) |
| 2 (Y) (N) | A (Y) (N) (U) |
| 3 (Y) (N) | S (Y) (N) (U) |
| | R (Y) (N) (U) |
| | I (Y) (N) (U) |
| | J (Y) (N) (U) |

First Item Drawn:

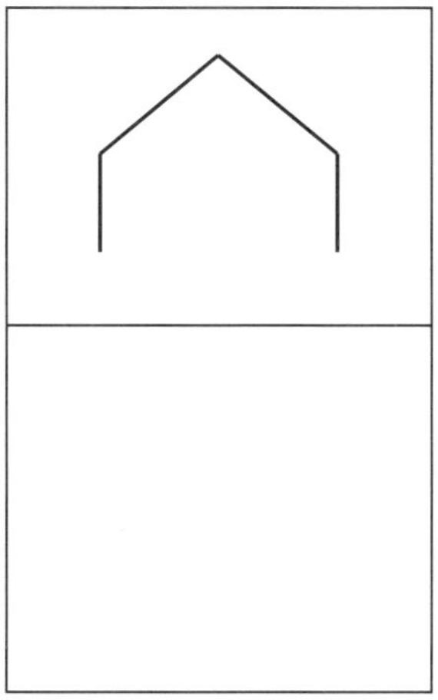

**18.**

| SH Parameters | SNH Parameters |
|---|---|
| 1 (Y) (N) | B (Y) (N) (U) |
| 2 (Y) (N) | A (Y) (N) (U) |
| 3 (Y) (N) | I (Y) (N) (U) |
| 4 (Y) (N) | L (Y) (N) (U) |
| | J (Y) (N) (U) |

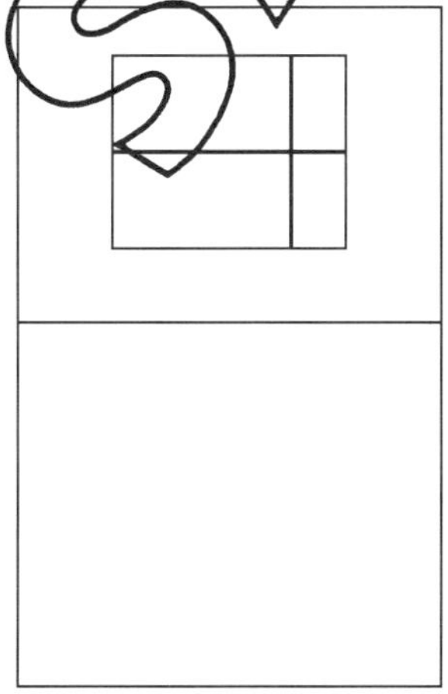

**19.**

| SH Parameters | SNH Parameters |
|---|---|
| 1 (Y) (N) | B (Y) (N) (U) |
| 2 (Y) (N) | A (Y) (N) (U) |
| 3 (Y) (N) | S (Y) (N) (U) |
| | R (Y) (N) (U) |
| | L (Y) (N) (U) |

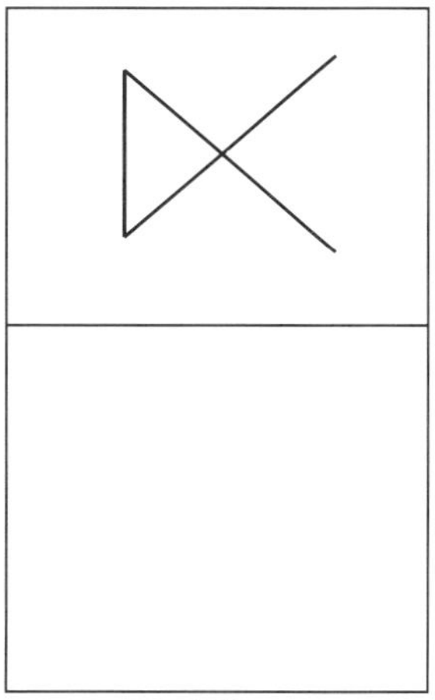

**20.**

| SH Parameters | SNH Parameters |
|---|---|
| 1 (Y) (N) | B (Y) (N) (U) |
| 2 (Y) (N) | A (Y) (N) (U) |
| 3 (Y) (N) | S (Y) (N) (U) |
| | R (Y) (N) (U) |
| | J (Y) (N) (U) |

**21.**

| SH Parameters | SNH Parameters |
|---|---|
| 1 (Y) (N) | B (Y) (N) (U) |
| 2 (Y) (N) | A (Y) (N) (U) |
| 3 (Y) (N) | S (Y) (N) (U) |
| 4 (Y) (N) | L (Y) (N) (U) |
| | I (Y) (N) (U) |

**OBSERVATIONS:**

| REQUIRED TRANSMITTAL NUMBER | | | | | | |
|---|---|---|---|---|---|---|
| | | | | | | |
| 0 | 0 | 0 | 0 | 0 | 0 | 0 |
| 1 | 1 | 1 | 1 | 1 | 1 | 1 |
| 2 | 2 | 2 | 2 | 2 | 2 | 2 |
| 3 | 3 | 3 | 3 | 3 | 3 | 3 |
| 4 | 4 | 4 | 4 | 4 | 4 | 4 |
| 5 | 5 | 5 | 5 | 5 | 5 | 5 |
| 6 | 6 | 6 | 6 | 6 | 6 | 6 |
| 7 | 7 | 7 | 7 | 7 | 7 | 7 |
| 8 | 8 | 8 | 8 | 8 | 8 | 8 |
| 9 | 9 | 9 | 9 | 9 | 9 | 9 |

**CHILD'S NAME:**



## PART II CONTINUED

- Discontinue Part II when **3** consecutive items are scored "N" on SH Parameter 1 (gross approximation)

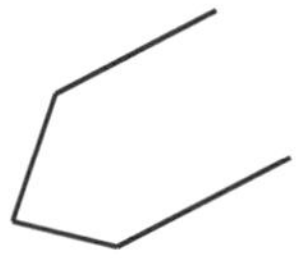

**22.**

SH Parameters: 1 (Y) (N); 2 (Y) (N); 3 (Y) (N)

SNH Parameters: B (Y) (N) (U); A (Y) (N) (U); R (Y) (N) (U); L (Y) (N) (U); I (Y) (N) (U); J (Y) (N) (U)

**23.**

SH Parameters: 1 (Y) (N); 2 (Y) (N); 3 (Y) (N)

SNH Parameters: B (Y) (N) (U); A (Y) (N) (U); S (Y) (N) (U)

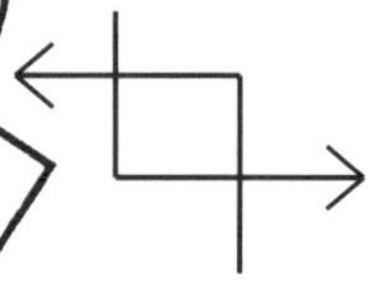

**24.**

SH Parameters: 1 (Y) (N); 2 (Y) (N); 3 (Y) (N); 4 (Y) (N); 5 (Y) (N)

SNH Parameters: B (Y) (N) (U); A (Y) (N) (U); S (Y) (N) (U); R (Y) (N) (U); D (Y) (N) (U)

**25.**

SH Parameters: 1 (Y) (N); 2 (Y) (N); 3 (Y) (N); 4 (Y) (N); 5 (Y) (N)

SNH Paramete[rs]: B (Y) (N) (U); A (Y) (N) (U); S (Y) (N) (U); R (Y) (N) (U)

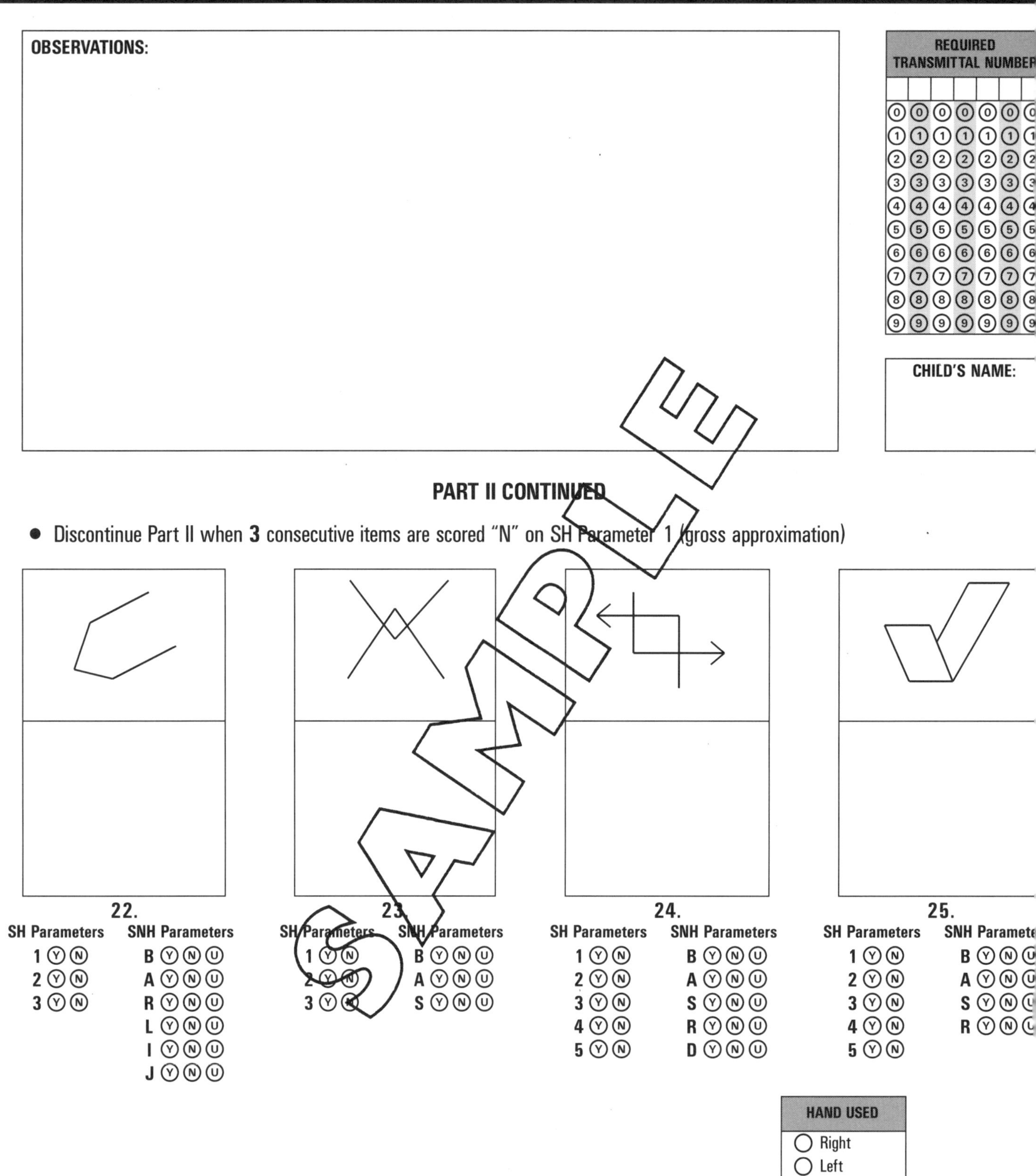

**HAND USED**
- ◯ Right
- ◯ Left
- ◯ Both

**DO NOT SEPARATE THE PAGES OF THIS BOOKLET**

**DO NOT WRITE HERE**

# wps TEST REPORT™

Western Psychological Services • 12031 Wilshire Blvd. • Los Angeles, CA 90025-1251

SENSORY INTEGRATION AND PRAXIS TESTS (SIPT)

## 5. POSTURAL PRAXIS (PPr)

Protocol Sheet

A. Jean Ayres, Ph.D.

**DIRECTIONS:** Sit opposite the child. Trial item: **"You make your arms and hands do the same thing that mine do. See how fast you can do it."** Assume mirror image of illustration. After child imitates: **"This hand is over my ear, this hand is on top of it, and I'm leaning to the side. It's as though you were looking in a mirror."** Assume neutral position. Item 1: **"Now do this one. Do it quickly."**

Hold each position until the child correctly imitates it, or for 7 seconds. In all cases, score 0 if the posture was incorrect or completed after 7 seconds. Score 1 if the correct posture was assumed in 4-7 seconds, or the posture deviated as described in the scoring criteria. Score 2 if the correct posture was assumed within 3 seconds. If the position changed frequently, the accurate position must be maintained for 2 of the first 3 seconds to receive a score of 2; otherwise, score the position at 7 seconds.

Do not penalize for nonmirrored response. However, the first time this occurs, encourage the child to assume a mirrored position.

If unsure of the child's score, you may draw any deviations over the illustrations on this sheet for later scoring. Also note the response time in these cases. However, all such marks must be erased.

**REQUIRED TRANSMITTAL NUMBER**

| | | | | | | |
|---|---|---|---|---|---|---|
| 0 | 0 | 0 | 0 | 0 | 0 | 0 |
| 1 | 1 | 1 | 1 | 1 | 1 | 1 |
| 2 | 2 | 2 | 2 | 2 | 2 | 2 |
| 3 | 3 | 3 | 3 | 3 | 3 | 3 |
| 4 | 4 | 4 | 4 | 4 | 4 | 4 |
| 5 | 5 | 5 | 5 | 5 | 5 | 5 |
| 6 | 6 | 6 | 6 | 6 | 6 | 6 |
| 7 | 7 | 7 | 7 | 7 | 7 | 7 |
| 8 | 8 | 8 | 8 | 8 | 8 | 8 |
| 9 | 9 | 9 | 9 | 9 | 9 | 9 |

**CHILD'S NAME:**

- Administer all items
- Hold each posture until child correctly imitates or for **7** seconds

**TRIAL**

**Do not score**

**1.**

**Arm abducted more than 45°, not forward 45° or more, 1 or more fingers touch shoulder**

⓪ Incorrect or exceeded 7 seconds
① Correct in 4-7 seconds, or arm abducted 45° or less or forward 45° or more
② Correct in 3 seconds

**2.**

**Fingers curve over opposite shoulder, no fingernails visible except thumb**

⓪ Incorrect or exceeded 7 seconds
① Correct in 4-7 seconds, or 1 or more fingernails visible but all 5 above sternal end of clavicle
② Correct in 3 seconds

**3.**

**Arms crossed, fingers and part of palms on sides of head**

⓪ Incorrect or exceeded 7 seconds
① Correct in 4-7 seconds, or arms crossed, fingers only either hand on sides of head
② Correct in 3 seconds

**4.**

**Tips of thumb and finger(s) of each hand form opening, can see pupils**

⓪ Incorrect or exceeded 7 seconds
① Correct in 4-7 seconds, or 2 openings but cannot see 1 or 2 pupils
② Correct in 3 seconds

**5.**

**Arms crossed, hands cover center of opposite patella**

⓪ Incorrect or exceeded 7 seconds
① Correct in 4-7 seconds, or entire hands on distal half of thighs
② Correct in 3 seconds

**CONTINUE ON BACK**

This Computerized Test Product
AUTHORIZED
By The Test Publisher

4 5 6 7 8 9
76421-321
Printed in U.S.A.

W-260B(5)

**For WPS Use Only**

1 2 3 4 5 6 7
8 9 10 11 12 13 14
15 16 17 18 19 20 21

- Administer all items
- Hold each posture until child correctly imitates or for **7** seconds

**6.**

**Elbows cupped by opposite hands**

(0) Incorrect or exceeded 7 seconds
(1) Correct in 4-7 seconds, or only 1 elbow cupped
(2) Correct in 3 seconds

**7.**

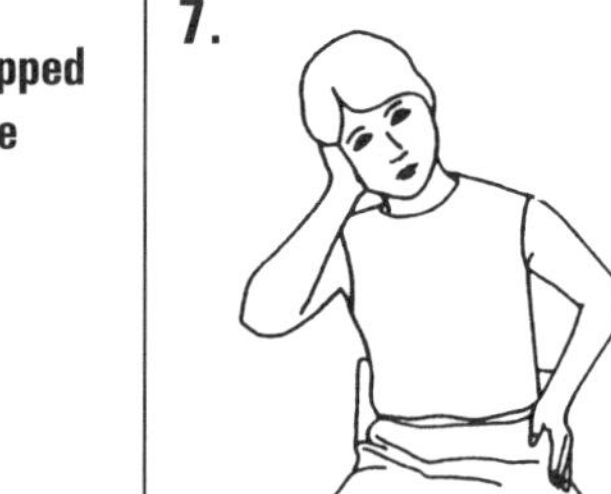

**One hand on side of head, other hand on hip, head and trunk leaning**

(0) Incorrect or exceeded 7 seconds
(1) Correct in 4-7 seconds, or head or trunk alone leaning or hand on thigh, buttock, or above waist.
(2) Correct in 3 seconds

**8.**

**Thumbs and index fingers form linked shapes**

(0) Incorrect or exceeded 7 seconds
(1) Correct in 4-7 seconds, or other or additional fingers and thumbs form linked shapes
(2) Correct in 3 seconds

**9.**

**Wrists crossed, palmar surface of 2 fingertips of each hand touch**

(0) Incorrect or exceeded 7 seconds
(1) Correct in 4-7 seconds, or only 1 finger either hand touches fingers of other hand
(2) Correct in 3 seconds

**10.**

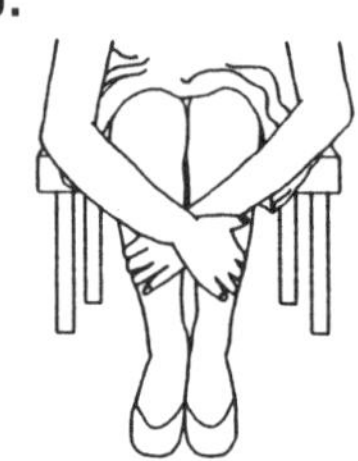

**Wrists/arms crossed, all 5 fingers each hand touch contralateral leg between knee and ankle**

(0) Incorrect or exceeded 7 seconds
(1) Correct in 4-7 seconds, or some but not all 5 fingers each hand touch
(2) Correct in 3 seconds

**11.**

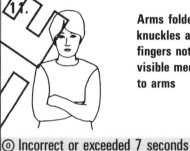

**Arms folded, knuckles and fingers not visible medial to arms**

(0) Incorrect or exceeded 7 seconds
(1) Correct in 4-7 seconds, or knuckles or fingers of either hand visible medial to arms
(2) Correct in 3 seconds

**12.**

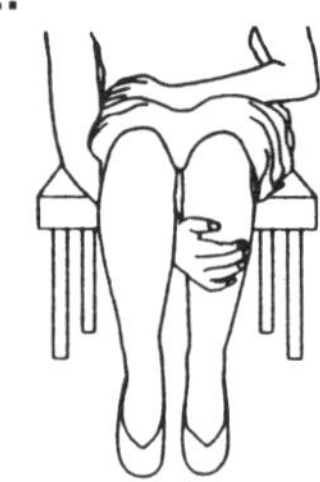

**Arm behind leg, grasps other leg between knee and ankle**

(0) Incorrect or exceeded 7 seconds
(1) Correct in 4-7 seconds, or hand touches but does not grasp other leg
(2) Correct in 3 seconds

**13.**

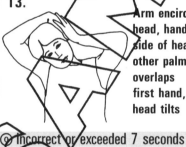

**Arm encircles head, hand flat on side of head, other palm overlaps first hand, head tilts**

(0) Incorrect or exceeded 7 seconds
(1) Correct in 4-7 seconds, or head not tilted or only fingers overlap first hand
(2) Correct in 3 seconds

**14.**

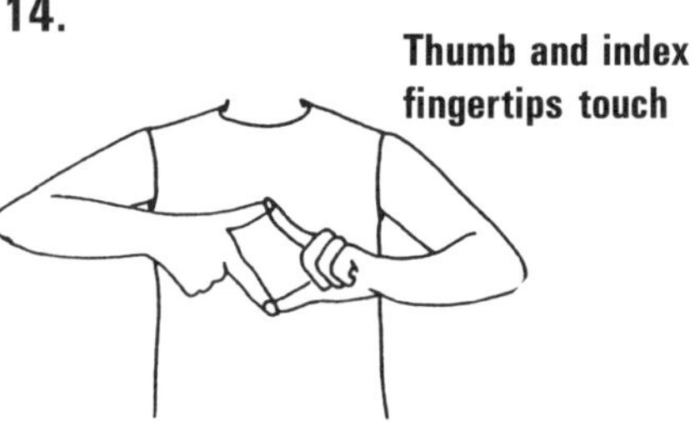

**Thumb and index fingertips touch**

(0) Incorrect or exceeded 7 seconds
(1) Correct in 4-7 seconds, or middle finger(s) substituted for index finger(s)
(2) Correct in 3 seconds

**15.**

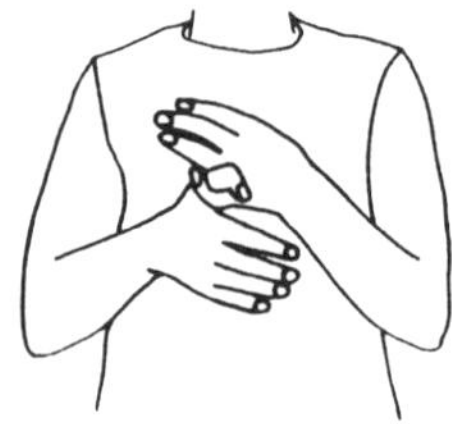

**Thumb and little finger hooked, not all joints need flex, palmar approach**

(0) Incorrect or exceeded 7 seconds
(1) Correct in 4-7 seconds, or dorsal approach
(2) Correct in 3 seconds

**16.**

**Hands grasp opposite leg between ankle and knee, flexion of 2 fingers each hand, 10 fingers visible**

(0) Incorrect or exceeded 7 seconds
(1) Correct in 4-7 seconds, or fewer than 10 fingers visible or touch not grasp
(2) Correct in 3 seconds

**17.**

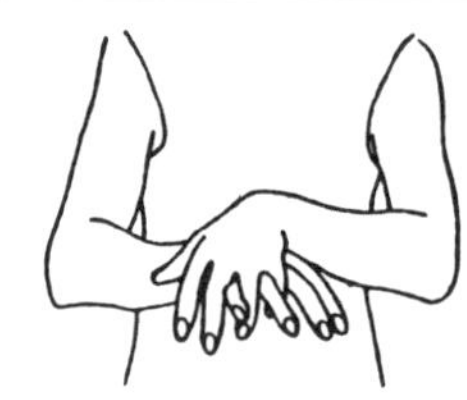

**Palm of 1 hand on dorsum of other, palms down, ring and little fingers interposed**

(0) Incorrect or exceeded 7 seconds
(1) Correct in 4-7 seconds, or 1 or 3 fingers or ulnar side interposed
(2) Correct in 3 seconds



**DO NOT WRITE HERE**

SENSORY INTEGRATION AND PRAXIS TESTS (SIPT)

# wps TEST REPORT™

Western Psychological Services • 12031 Wilshire Blvd. • Los Angeles, CA 90025-1251

# 6. BILATERAL MOTOR COORDINATION (BMC)

Protocol Sheet

A. Jean Ayres, Ph.D.

**DIRECTIONS FOR ARMS ITEMS:** Sit facing the child. Trial item: **"Watch my hands move. When they are through moving, you do the same thing.** (Demonstrate.) **(That's correct.) Be sure to move smoothly like this** (move child's hands). **When I begin with this hand** (left), **you begin with this hand** (child's right). **When I begin with this hand** (right), **you begin with this hand** (child's left)." If insufficient number of movements: **"Do it as many times as I did it."** Item 1: **"Watch me do another one.** (Demonstrate.) **Now you do it."**

Do not penalize the child for starting with the wrong hand, but remind the child on first (but only on first) wrong hand use. Fill in the circle corresponding to the child's item score (0 = incorrect, nonreciprocal, dysrhythmic, segmented, incomplete; 1 = approximately correct or slightly irregular; 2 = correct and well coordinated). Discontinue when four consecutive items are scored as 0. Then give feet items.

**DIRECTIONS FOR FEET ITEMS:** Do not give feet items to 4-year-olds. Trial item: **"Now we'll do the same thing with our feet. Watch me.** (Demonstrate.) **You do it."** Score each item as 0, 1, or 2 as for the arms items. Discontinue when two consecutive items are scored as 0.

**REQUIRED TRANSMITTAL NUMBER**

| | | | | | | |
|---|---|---|---|---|---|---|
| 0 | 0 | 0 | 0 | 0 | 0 | 0 |
| 1 | 1 | 1 | 1 | 1 | 1 | 1 |
| 2 | 2 | 2 | 2 | 2 | 2 | 2 |
| 3 | 3 | 3 | 3 | 3 | 3 | 3 |
| 4 | 4 | 4 | 4 | 4 | 4 | 4 |
| 5 | 5 | 5 | 5 | 5 | 5 | 5 |
| 6 | 6 | 6 | 6 | 6 | 6 | 6 |
| 7 | 7 | 7 | 7 | 7 | 7 | 7 |
| 8 | 8 | 8 | 8 | 8 | 8 | 8 |
| 9 | 9 | 9 | 9 | 9 | 9 | 9 |

**CHILD'S NAME:**

## ARMS

- Discontinue when **4** consecutive items are scored 0 and go on to feet items

| | Incorrect | Approximately Correct | Correct |
|---|---|---|---|
| **TRIAL I:** L, R, L, R (4 seconds) | | | |
| **1.** R, L, R, L (4 seconds) | 0 | 1 | 2 |
| **2.** L, R, L, R (4 seconds) | 0 | 1 | 2 |
| **3.** Both, clap clap, both, clap clap (4 seconds) | 0 | 1 | 2 |
| **4.** LR, LR, LR (3 seconds) | 0 | 1 | 2 |
| **5.** Both, both both, both, both both (4 seconds) | 0 | 1 | 2 |
| **6.** Crossed RL, RL, RL (3 seconds) | 0 | 1 | 2 |
| **7.** LL, RR, LL, RR (4 seconds) | 0 | 1 | 2 |
| **8.** RR, LL, RR, LL (4 seconds) | 0 | 1 | 2 |
| **9.** L, RR, L, RR (4 seconds) | 0 | 1 | 2 |
| **10.** R, LL, R, LL (4 seconds) | 0 | 1 | 2 |

## FEET

- Do not administer to 4-year-olds
- Discontinue when **2** consecutive items are scored as 0

| | Incorrect | Approximately Correct | Correct |
|---|---|---|---|
| **TRIAL II:** L, R, L, R (2 seconds) | | | |
| **11.** R, L, R, L (2 seconds) | 0 | 1 | 2 |
| **12.** LL, RR, LL, RR (3 seconds) | 0 | 1 | 2 |
| **13.** R, LL, R, LL (2 seconds) | 0 | 1 | 2 |
| **14.** L, RR, L, RR (2 seconds) | 0 | 1 | 2 |

**OBSERVATIONS:**







4 5 6 7 8 9
76422-321
Printed in U.S.A.

W-260B(6)

**For WPS Use Only**

| | | | | | | |
|---|---|---|---|---|---|---|
| 1 | 2 | 3 | 4 | 5 | 6 | 7 |
| 8 | 9 | 10 | 11 | 12 | 13 | 14 |
| 15 | 16 | 17 | 18 | 19 | 20 | 21 |

SENSORY INTEGRATION AND PRAXIS TESTS (SIPT)

# WPS TEST REPORT

Western Psychological Services • 12031 Wilshire Blvd. • Los Angeles, CA 90025-1251

# 7. PRAXIS ON VERBAL COMMAND (PrVC)

Protocol Sheet

A. Jean Ayres, Ph.D. and Zoe K. Mailloux, M.A.

**DIRECTIONS:** Face the child and read the trial command aloud. If correct: **"That's correct. Now we'll do another."** If incorrect: **"Put your hand on your nose like this** (place child's hand on his or her nose). **Now we'll do another."** Administer all items. The child sits for Items 1-10 and stands for Items 11-24. For children ages 4 and 5, repeat command at 5 seconds if the action is not yet completed accurately. Do not demonstrate, assist, or give cues. You and child assume neutral position between items.

Record accuracy for each item (0 = incorrect, or correct but not within time limit; 1 = correct within time limit). Also record the time when the child correctly completes the action or, if incorrect, the maximum time allowed (15 seconds for ages 4-5, 10 seconds for ages 6-8). After the test is completed, fill in the circles corresponding to the time scores.

**REQUIRED TRANSMITTAL NUMBER**

| | | | | | | |
|---|---|---|---|---|---|---|
| 0 | 0 | 0 | 0 | 0 | 0 | 0 |
| 1 | 1 | 1 | 1 | 1 | 1 | 1 |
| 2 | 2 | 2 | 2 | 2 | 2 | 2 |
| 3 | 3 | 3 | 3 | 3 | 3 | 3 |
| 4 | 4 | 4 | 4 | 4 | 4 | 4 |
| 5 | 5 | 5 | 5 | 5 | 5 | 5 |
| 6 | 6 | 6 | 6 | 6 | 6 | 6 |
| 7 | 7 | 7 | 7 | 7 | 7 | 7 |
| 8 | 8 | 8 | 8 | 8 | 8 | 8 |
| 9 | 9 | 9 | 9 | 9 | 9 | 9 |

**CHILD'S NAME:**

- Administer all items
- Time limit is **15** seconds for ages 4-5, **10** seconds for ages 6-8

**TRIAL: "Put one hand on your nose."**

| ITEM | ACCURACY Incorrect | ACCURACY Correct | TIME Seconds | TIME |
|---|---|---|---|---|
| 1. "Put one hand on your nose and one hand on your stomach." | 0 | 1 | ____ | 0 1 2 3 4 5 6 7 8 9 10 11 12 13 14 15 |
| 2. "Put one foot on the other foot." | 0 | 1 | ____ | 0 1 2 3 4 5 6 7 8 9 10 11 12 13 14 15 |
| 3. "Put both arms out to the side." | 0 | 1 | ____ | 0 1 2 3 4 5 6 7 8 9 10 11 12 13 14 15 |
| 4. "Put one hand on your foot and one hand on your head." | 0 | 1 | ____ | 0 1 2 3 4 5 6 7 8 9 10 11 12 13 14 15 |
| 5. "Put one foot on your other knee." | 0 | 1 | ____ | 0 1 2 3 4 5 6 7 8 9 10 11 12 13 14 15 |
| 6. "Put your elbows together." | 0 | 1 | ____ | 0 1 2 3 4 5 6 7 8 9 10 11 12 13 14 15 |
| 7. "Cross your legs and bend to the front." | 0 | 1 | ____ | 0 1 2 3 4 5 6 7 8 9 10 11 12 13 14 15 |
| 8. "Put the backs of your hands together." | 0 | 1 | ____ | 0 1 2 3 4 5 6 7 8 9 10 11 12 13 14 15 |
| 9. "Put one elbow on the back of your hand." | 0 | 1 | ____ | 0 1 2 3 4 5 6 7 8 9 10 11 12 13 14 15 |
| 10. "Put the bottoms of your feet together." | 0 | 1 | ____ | 0 1 2 3 4 5 6 7 8 9 10 11 12 13 14 15 |

**CONTINUE ON BACK**

This Computerized Test Product
AUTHORIZED
By The Test Publisher

**For WPS Use Only**

| | | | | | | |
|---|---|---|---|---|---|---|
| 1 | 2 | 3 | 4 | 5 | 6 | 7 |
| 8 | 9 | 10 | 11 | 12 | 13 | 14 |
| 15 | 16 | 17 | 18 | 19 | 20 | 21 |

4 5 6 7 8 9
76423-321
Printed in U.S.A.
W-260B(7)

**DO NOT WRITE OUTSIDE THE BOX**

| OBSERVATIONS: |
|---|
| |

- **Ask the child to stand for the remaining items**
- Time limit is **15** seconds for ages 4-5, **10** seconds for ages 6-8

| ITEM | ACCURACY Incorrect | Correct | Seconds | TIME |
|---|---|---|---|---|
| **11.** "Put one foot on the other foot." | 0 | 1 | ____ | 0 1 2 3 4 5 6 7 8 9 10 11 12 13 14 15 |
| **12.** "Put your hands together and your feet apart." | 0 | 1 | ____ | 0 1 2 3 4 5 6 7 8 9 10 11 12 13 14 15 |
| **13.** "Put your toes together and the back of your feet apart." | 0 | 1 | ____ | 0 1 2 3 4 5 6 7 8 9 10 11 12 13 14 15 |
| **14.** "Put your feet together and your hands apart." | 0 | 1 | ____ | 0 1 2 3 4 5 6 7 8 9 10 11 12 13 14 15 |
| **15.** "Put your knees together and your feet apart." | 0 | 1 | ____ | 0 1 2 3 4 5 6 7 8 9 10 11 12 13 14 15 |
| **16.** "Put the back part of your feet together and your toes apart." | 0 | 1 | ____ | 0 1 2 3 4 5 6 7 8 9 10 11 12 13 14 15 |
| **17.** "Put one hand on your stomach and one hand on your back." | 0 | 1 | ____ | 0 1 2 3 4 5 6 7 8 9 10 11 12 13 14 15 |
| **18.** "Put one arm to the front and one arm back." | 0 | 1 | ____ | 0 1 2 3 4 5 6 7 8 9 10 11 12 13 14 15 |
| **19.** "Put both hands on your head and bend your knees." | 0 | 1 | ____ | 0 1 2 3 4 5 6 7 8 9 10 11 12 13 14 15 |
| **20.** "Put both hands behind your back and bend to the front." | 0 | 1 | ____ | 0 1 2 3 4 5 6 7 8 9 10 11 12 13 14 15 |
| **21.** "Put one hand on the chair and one knee on the chair." | 0 | 1 | ____ | 0 1 2 3 4 5 6 7 8 9 10 11 12 13 14 15 |
| **22.** "Put one arm to the front and one leg to the front." | 0 | 1 | ____ | 0 1 2 3 4 5 6 7 8 9 10 11 12 13 14 15 |
| **23.** "Put your knees and feet together and bend your knees." | 0 | 1 | ____ | 0 1 2 3 4 5 6 7 8 9 10 11 12 13 14 15 |
| **24.** "Put both hands and one foot on the chair." | 0 | 1 | ____ | 0 1 2 3 4 5 6 7 8 9 10 11 12 13 14 15 |

wps TEST REPORT™
Western Psychological Services • 12031 Wilshire Blvd. • Los Angeles, CA 90025-1251

SENSORY INTEGRATION AND PRAXIS TESTS (SIPT)

# 8. CONSTRUCTIONAL PRAXIS (CPr)

Protocol Sheet

A. Jean Ayres, Ph.D.

## STRUCTURE I

**DIRECTIONS FOR STRUCTURE I:** Sit to right of child with two feet of space to left of child's midline. Remove Blocks 1, 2, 3, and 4 from box one at a time. **"I'm going to build a house** (place Block 1). **Here is the floor** (place Block 2). **Here are the walls** (place Blocks 3 and 4). **We line up the blocks VERY carefully** (finger corners of blocks). **Now you build a house like mine here** (point). **I'll put the first block here."** Place Block 1 at child's midline 10 cm (4in.) from table edge. Place Blocks 3 and 4 20 cm (8in.) distal to Block 1 and Block 2 behind them. Use same orientation as in box. If child begins house at wrong place, place Block 1 in original place. **"Keep this one here. Now you build the rest."** If child places Block 2 proximal to Block 1, return Block 2 to previous position. **"Build your building here."** If child disassembles, ask child to rebuild. **"We are going to build some more."** Allow 3 minutes.

**"Here is the roof** (place Blocks 5 and 6 on your structure) **and here is the chimney** (add Block 7). (Place child's Blocks 5, 6, and 7 in that order behind child's structure.) **Now build the rest of your house with these."** Allow 3 minutes or 1 minute of nonproductivity.

If the criterion for a block is met, score as 1; if the criterion is *not* met, score as 0. After scoring: **"Now take your building down. Put the blocks there** (upper left of table) **and get ready to build a different building."**

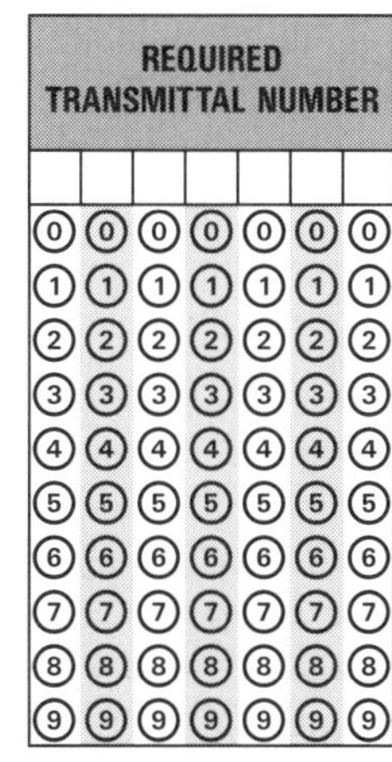


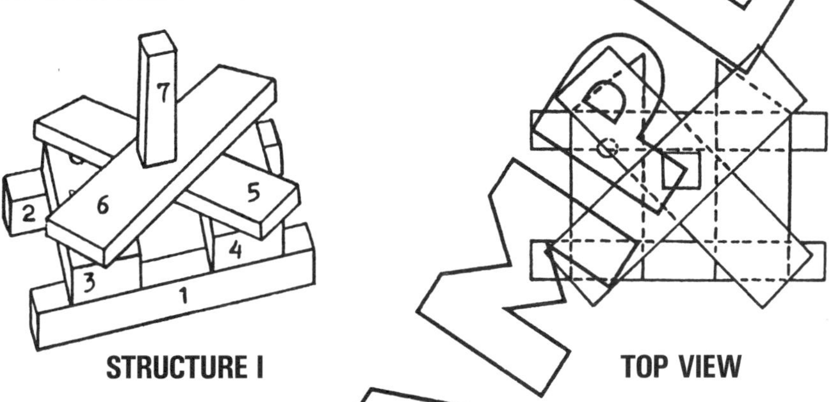




| BLOCK | CRITERION | SCORE No | SCORE Yes |
|---|---|---|---|
| 1 | Parallel to edge of table + or - 15° | 0 | 1 |
| 2 | 1. **1** and **2** not touching; can support **3** and **4** | 0 | 1 |
| | 2. Ends line up within 2½ cm of ends of **1** | 0 | 1 |
| 3 | 1. **3** to the L of **4** (If to the R, score **3** as **4** and **4** as **3** hereafter) | 0 | 1 |
| | 2. Rests on both **1** and **2** | 0 | 1 |
| | 3. On L half of **1** and **2**; **1** and **2** extend to L of **3** | 0 | 1 |
| | 4. Angled end distal; correct side up | 0 | 1 |
| 4 | 1. Rests on both **1** and **2** | 0 | 1 |
| | 2. On R half of **1** and **2; 1** and **2** extend to R of **4** | 0 | 1 |
| | 3. Angled end distal; correct side up | 0 | 1 |
| 5 | 1. One end on **3** only, other on **4** only | 0 | 1 |
| | 2. L intersection upper half of **3,** R intersection lower half of **4** | 0 | 1 |
| 6 | 1. One end over **3** only, other over **4** only | 0 | 1 |
| | 2. R intersection upper half of **4,** L intersection lower half of **3** (If **5** and **6** rotated 90°, read "upper" as "lower" and vice versa) | 0 | 1 |
| 7 | 1. Entirely on **6;** any orientation | 0 | 1 |
| | 2. Upright, entirely on **6** and entirely over **5** | 0 | 1 |

CONTINUE ON BACK

4 5 6 7 8 9 Printed in U.S.A.
76424-321
W-260B(8)

For WPS Use Only
1 2 3 4 5 6 7
8 9 10 11 12 13 14
15 16 17 18 19 20 21

## STRUCTURE II

**DIRECTIONS FOR STRUCTURE II:** Do not administer Structure II to 4-year-olds.

Add additional blocks, keeping same orientation. Place model 7.5 cm (3 in.) to right of child's right shoulder and 24 cm (10 in.) from front of table. **"Now build one like this. Here are the blocks** (point). **Make them look just like this** (point). **I'll start the building. I'll put this one here.** (Place Block 1 so proximal end is 10 cm [4 in.] from table front edge and 10 cm [4 in.] to left of child's midline.) **Now you build the rest here** (gesture)." If child moves Block 1, return block to original position. "**Keep this one here. Now you build the rest.**" If child builds to left of Block 1, return block used to block area. "**Build your building HERE** (gesture)." Tell child he or she may stand up and look at structure but may not move model.

Allow 15 minutes or 2 minutes of nonproductivity. Score each block only on single most erroneous parameter applicable.

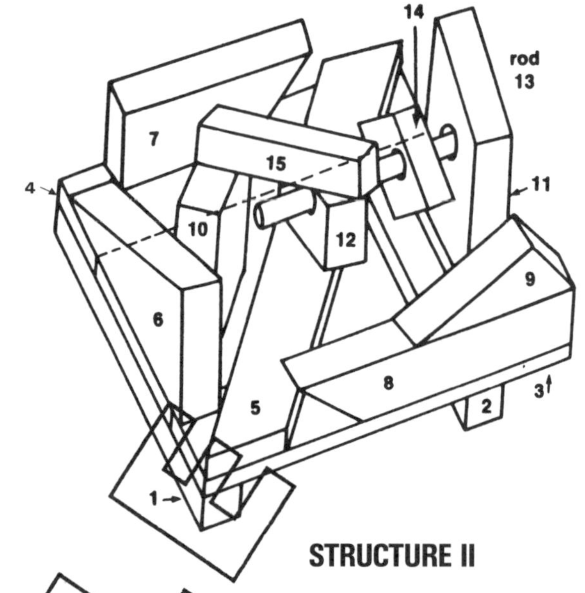


STRUCTURE II



| PARAMETER | BLOCK NUMBER | | | | | | | | | | | | | | |
|---|---|---|---|---|---|---|---|---|---|---|---|---|---|---|---|
| | 1 | 2 | 3 | 4 | 5 | 6 | 7 | 8 | 9 | 10 | 11 | 12 | 13 | 14 | 15 |
| 1. Displacement 1 to 2½ cm | ■ | ○ | ○ | ○ | ○ | ○ | ○ | ○ | ○ | ○ | ○ | ○ | ■ | ■ | ○ |
| 2. Displacement greater than 2½ cm | ■ | ○ | ○ | ○ | ○ | ○ | ○ | ○ | ○ | ○ | ○ | ○ | ■ | ■ | ○ |
| 3. Rotation greater than 15 degrees | ■ | ○ | ○ | ○ | ○ | ○ | ○ | ○ | ○ | ○ | ○ | ■ | ■ | ■ | ○ |
| 4. Upside down, R/L, end for end | ■ | ○ | ○ | ○ | ○ | ○ | ○ | ○ | ○ | ○ | ○ | ○ | ■ | ■ | ○ |
| 5. Incorrect but logical | ■ | ○ | ○ | ○ | ○ | ○ | ○ | ○ | ○ | ○ | ○ | ○ | ○ | ○ | ○ |
| 6. Gross mislocation | ■ | ○ | ○ | ○ | ○ | ○ | ○ | ○ | ○ | ○ | ○ | ○ | ○ | ○ | ○ |
| 7. Omission | ○ | ○ | ○ | ○ | ○ | ○ | ○ | ○ | ○ | ○ | ○ | ○ | ○ | ○ | ○ |
| 8. O.K. | ○ | ○ | ○ | ○ | ○ | ○ | ○ | ○ | ○ | ○ | ○ | ○ | ○ | ○ | ○ |

**DO NOT WRITE HERE**

**OBSERVATIONS:**

**wps TEST REPORT**™

Western Psychological Services • 12031 Wilshire Blvd. • Los Angeles, CA 90025-1251

SENSORY INTEGRATION AND PRAXIS TESTS (SIPT)

# 9. POSTROTARY NYSTAGMUS (PRN)

Protocol Sheet

A. Jean Ayres, Ph.D.

**REQUIRED TRANSMITTAL NUMBER**

| | | | | | | |
|---|---|---|---|---|---|---|
| 0 | 0 | 0 | 0 | 0 | 0 | 0 |
| 1 | 1 | 1 | 1 | 1 | 1 | 1 |
| 2 | 2 | 2 | 2 | 2 | 2 | 2 |
| 3 | 3 | 3 | 3 | 3 | 3 | 3 |
| 4 | 4 | 4 | 4 | 4 | 4 | 4 |
| 5 | 5 | 5 | 5 | 5 | 5 | 5 |
| 6 | 6 | 6 | 6 | 6 | 6 | 6 |
| 7 | 7 | 7 | 7 | 7 | 7 | 7 |
| 8 | 8 | 8 | 8 | 8 | 8 | 8 |
| 9 | 9 | 9 | 9 | 9 | 9 | 9 |

**CHILD'S NAME:**

**DIRECTIONS:** The child sits cross-legged in center of nystagmus board and holds on to front edge. Head is tilted forward 30°. **"I am going to turn you around 10 times. While I'm turning you, hold your head like this. Don't move your head while you're turning. When you stop, look up and look at the wall. I will look at you but don't you look at me; you look at the wall."** Turn child to his/her left (counterclockwise), pushing the child's left knee each time it comes around.

After completing 10 rotations in 20 seconds, stop the child so he/she is facing the wall. **"Look at the wall. Don't look at me.** (After a few seconds:) **Keep looking at the wall."** Record duration of nystagmus to nearest whole second. Be sure the time is entered as two digits (e.g., enter 6 seconds as 06).

After a 30-second rest: **"Now we'll go the other way. Keep your head this way** (position head if necessary) **and when you stop, look at the wall."** Record duration to nearest whole second. Avoid including random eye movements or secondary nystagmus in time.

After the SIPT is completed, administer this test again and record duration to the left and right.

## FIRST TEST

**COUNTERCLOCKWISE (TO THE LEFT)**

Number of Seconds:______

**CLOCKWISE (TO THE RIGHT)**

Number of Seconds:______

## RETEST

**COUNTERCLOCKWISE (TO THE LEFT)**

Number of Seconds:______

**CLOCKWISE (TO THE RIGHT)**

Number of Seconds:______

| No. of Seconds | | No. of Seconds | | No. of Seconds | | No. of Seconds | |
|---|---|---|---|---|---|---|---|
| 0 | 0 | 0 | 0 | 0 | 0 | 0 | 0 |
| 1 | 1 | 1 | 1 | 1 | 1 | 1 | 1 |
| 2 | 2 | 2 | 2 | 2 | 2 | 2 | 2 |
| 3 | 3 | 3 | 3 | 3 | 3 | 3 | 3 |
| 4 | 4 | 4 | 4 | 4 | 4 | 4 | 4 |
| 5 | 5 | 5 | 5 | 5 | 5 | 5 | 5 |
| 6 | 6 | 6 | 6 | 6 | 6 | 6 | 6 |
| 7 | 7 | 7 | 7 | 7 | 7 | 7 | 7 |
| 8 | 8 | 8 | 8 | 8 | 8 | 8 | 8 |
| 9 | 9 | 9 | 9 | 9 | 9 | 9 | 9 |

**OBSERVATIONS:**

This Computerized Test Product
AUTHORIZED
By The Test Publisher

W-260B(9)
4 5 6 7 8 9
76425-321
Printed in U.S.A.

**For WPS Use Only**

| | | | | | | |
|---|---|---|---|---|---|---|
| 1 | 2 | 3 | 4 | 5 | 6 | 7 |
| 8 | 9 | 10 | 11 | 12 | 13 | 14 |
| 15 | 16 | 17 | 18 | 19 | 20 | 21 |

DO NOT WRITE

ON THIS SIDE

OF SHEET

**wps TEST REPORT**™

Western Psychological Services • 12031 Wilshire Blvd. • Los Angeles, CA 90025-1251

SENSORY INTEGRATION AND PRAXIS TESTS (SIPT)

# 10. MOTOR ACCURACY (MAc)

Protocol Sheet

A. Jean Ayres, Ph.D.

**DIRECTIONS:** Sit opposite the child. Tape the test sheet flat in front of the child. Give pen to child's previously determined writing hand. Demonstration: **"Watch me. I'm drawing a line on top of this black line.... Now you draw a line on top of this black line."** If necessary, advise child to go slower or faster.

Test with preferred hand: **"Now draw a line on top of this black line beginning here and going around to here. Draw carefully."** Begin timing at horizontal short-dashed line one inch above starting point. Encourage child to take neither less than 30 nor more than 120 seconds. Discourage child's retracing line. Stop timing at horizontal short-dashed line. Child fills in skipped portions.

Turn over test sheet to test nonpreferred hand. **"Now you will draw a line with the other hand. You will draw on this black line beginning here and go around to here."** Encourage child to take neither less than 30 nor more than 90 seconds.

Record number of whole seconds as three digits (e.g. 52 as 052). Measure distance to nearest half inch that child's line is off solid line or outside areas bounded by short, medium, and long broken lines (e.g. 6½ as 06.5). If child's line is not outside an area, record 00.0. Each hand's distance measurements must decrease in value, from the solid through the long broken line.

**REQUIRED TRANSMITTAL NUMBER**

| | | | | | | |
|---|---|---|---|---|---|---|
| 0 | 0 | 0 | 0 | 0 | 0 | 0 |
| 1 | 1 | 1 | 1 | 1 | 1 | 1 |
| 2 | 2 | 2 | 2 | 2 | 2 | 2 |
| 3 | 3 | 3 | 3 | 3 | 3 | 3 |
| 4 | 4 | 4 | 4 | 4 | 4 | 4 |
| 5 | 5 | 5 | 5 | 5 | 5 | 5 |
| 6 | 6 | 6 | 6 | 6 | 6 | 6 |
| 7 | 7 | 7 | 7 | 7 | 7 | 7 |
| 8 | 8 | 8 | 8 | 8 | 8 | 8 |
| 9 | 9 | 9 | 9 | 9 | 9 | 9 |

**CHILD'S NAME:**

## RIGHT HAND

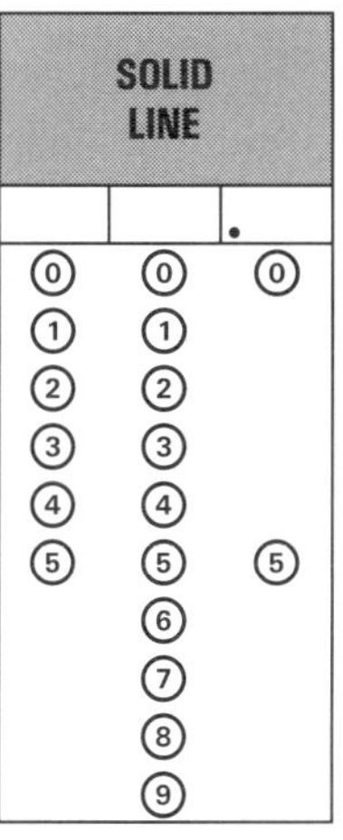

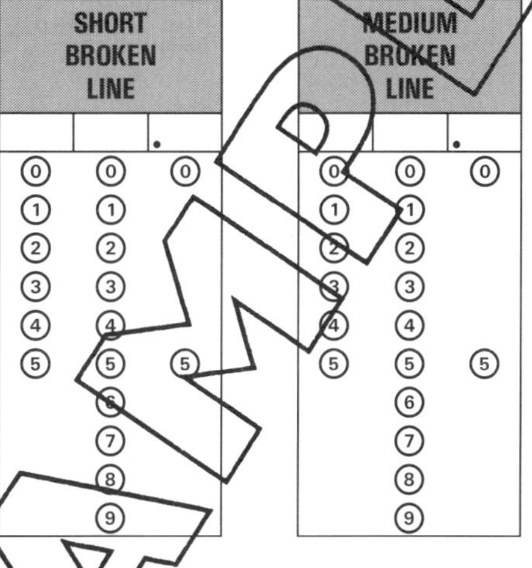

| TIME (Seconds) | | | SOLID LINE | | . | SHORT BROKEN LINE | | . | MEDIUM BROKEN LINE | | . | LONG BROKEN LINE | | . |
|---|---|---|---|---|---|---|---|---|---|---|---|---|---|---|
| 0 | 0 | 0 | 0 | 0 | 0 | 0 | 0 | 0 | 0 | 0 | 0 | 0 | 0 | 0 |
| 1 | 1 | 1 | 1 | 1 | | 1 | 1 | | 1 | 1 | | 1 | 1 | |
| 2 | 2 | 2 | 2 | 2 | | 2 | 2 | | 2 | 2 | | 2 | 2 | |
| 3 | 3 | 3 | 3 | 3 | | 3 | 3 | | 3 | 3 | | 3 | 3 | |
| | 4 | 4 | 4 | 4 | | 4 | 4 | | 4 | 4 | | 4 | 4 | |
| | 5 | 5 | 5 | 5 | 5 | 5 | 5 | 5 | 5 | 5 | 5 | 5 | 5 | 5 |
| | 6 | 6 | | 6 | | | 6 | | | 6 | | | 6 | |
| | 7 | 7 | | 7 | | | 7 | | | 7 | | | 7 | |
| | 8 | 8 | | 8 | | | 8 | | | 8 | | | 8 | |
| | 9 | 9 | | 9 | | | 9 | | | 9 | | | 9 | |

**PREFERRED HAND (First Hand Tested)**

○ Right ○ Left

## LEFT HAND

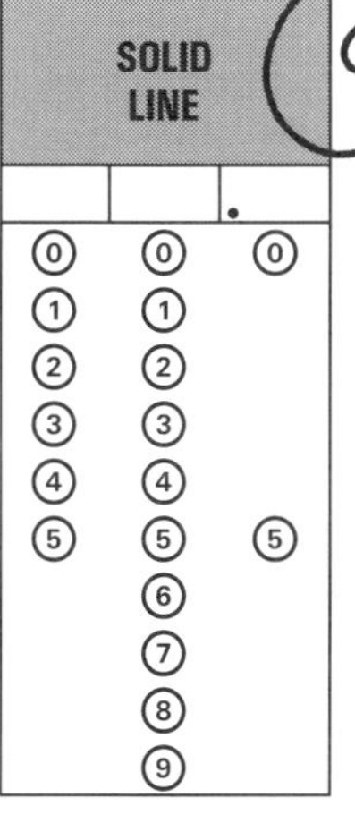

| TIME (Seconds) | | | SOLID LINE | | . | SHORT BROKEN LINE | | . | MEDIUM BROKEN LINE | | . | LONG BROKEN LINE | | . |
|---|---|---|---|---|---|---|---|---|---|---|---|---|---|---|
| 0 | 0 | 0 | 0 | 0 | 0 | 0 | 0 | 0 | 0 | 0 | 0 | 0 | 0 | 0 |
| 1 | 1 | 1 | 1 | 1 | | 1 | 1 | | 1 | 1 | | 1 | 1 | |
| 2 | 2 | 2 | 2 | 2 | | 2 | 2 | | 2 | 2 | | 2 | 2 | |
| 3 | 3 | 3 | 3 | 3 | | 3 | 3 | | 3 | 3 | | 3 | 3 | |
| | 4 | 4 | 4 | 4 | | 4 | 4 | | 4 | 4 | | 4 | 4 | |
| | 5 | 5 | 5 | 5 | 5 | 5 | 5 | 5 | 5 | 5 | 5 | 5 | 5 | 5 |
| | 6 | 6 | | 6 | | | 6 | | | 6 | | | 6 | |
| | 7 | 7 | | 7 | | | 7 | | | 7 | | | 7 | |
| | 8 | 8 | | 8 | | | 8 | | | 8 | | | 8 | |
| | 9 | 9 | | 9 | | | 9 | | | 9 | | | 9 | |

**For WPS Use Only**

| | | | | | | |
|---|---|---|---|---|---|---|
| 1 | 2 | 3 | 4 | 5 | 6 | 7 |
| 8 | 9 | 10 | 11 | 12 | 13 | 14 |
| 15 | 16 | 17 | 18 | 19 | 20 | 21 |



4 5 6 7 8 9
76426-321
Printed in U.S.A.
W-260B(10)

DO NOT WRITE

ON THIS SIDE

OF SHEET

**wps TEST REPORT**™

Western Psychological Services • 12031 Wilshire Blvd. • Los Angeles, CA 90025-1251

SENSORY INTEGRATION AND PRAXIS TESTS (SIPT)

# 11. SEQUENCING PRAXIS (SPr)

Protocol Sheet

A. Jean Ayres, Ph.D. and Zoe K. Mailloux, M.A.

**REQUIRED TRANSMITTAL NUMBER**

0 1 2 3 4 5 6 7 8 9

**CHILD'S NAME:**

**DIRECTIONS:** Sit opposite child. Trial: **"I am going to move my hands. When I stop moving, you do the same thing. If I use this hand** (hold up left), **you use this hand** (touch child's right). **If I use this hand** (hold up right), **you use this one** (touch child's left). **If I use both hands** (hold up both), **you use both hands** (touch both of child's hands). **Now do this."** If incorrect position or number of taps on Trial I, say **"Watch me again,"** and repeat trial.

Demonstrate test items only once. If child begins any item before you finish: **"Wait until I finish; then you do it."** Repeat demonstration.

Fill in the 0, 1, or 2 circle corresponding to the child's score on each subitem:

0 = Executed with wrong hand position or movement, or too few or too many motions in sequence
1 = Started with wrong hand position or movement but started over and completed correctly
2 = Completed with correct hand positions in correct sequence and correct number of motions and/or taps

Use of incorrect hand does not affect score.

Discontinue each item when two consecutive subitems are incorrect (scored as 0) and go on to next item.

Discontinue test when subitems a and b of two consecutive items are scored as 0.

Do not administer finger items (Items 7-9) to 4-year-olds.

## HAND ITEMS

- Discontinue each item when **2** consecutive subitems are scored 0 and go on to next item
- Discontinue the test when subitems a and b of **2** consecutive items are scored as 0

**TRIALS:** I. R, R  II. L, L  III. R, L, R

### Item 1

| | Sequence | Score |
|---|---|---|
| a. | Clap, Clap | 0 1 2 |
| b. | Clap, Clap, B | 0 1 2 |
| c. | Clap, Clap, B, B | 0 1 2 |
| d. | Clap, Clap, B, B, Clap | 0 1 2 |
| e. | Clap, Clap, B, B, Clap, Clap | 0 1 2 |

### Item 2

| | Sequence | Score |
|---|---|---|
| a. | L, L | 0 1 2 |
| b. | R, R | 0 1 2 |
| c. | B, B | 0 1 2 |
| d. | B, B, Clap | 0 1 2 |
| e. | B, B, Clap, Clap | 0 1 2 |
| f. | B, B, Clap, Clap, Clap, B | 0 1 2 |

**CONTINUE ON BACK**

This Computerized Test Product
AUTHORIZED
By The Test Publisher

4 5 6 7 8 9
76428-321
Printed in U.S.A.
W-260B(11)

**For WPS Use Only**

1 2 3 4 5 6 7 8 9 10 11 12 13 14 15 16 17 18 19 20 21

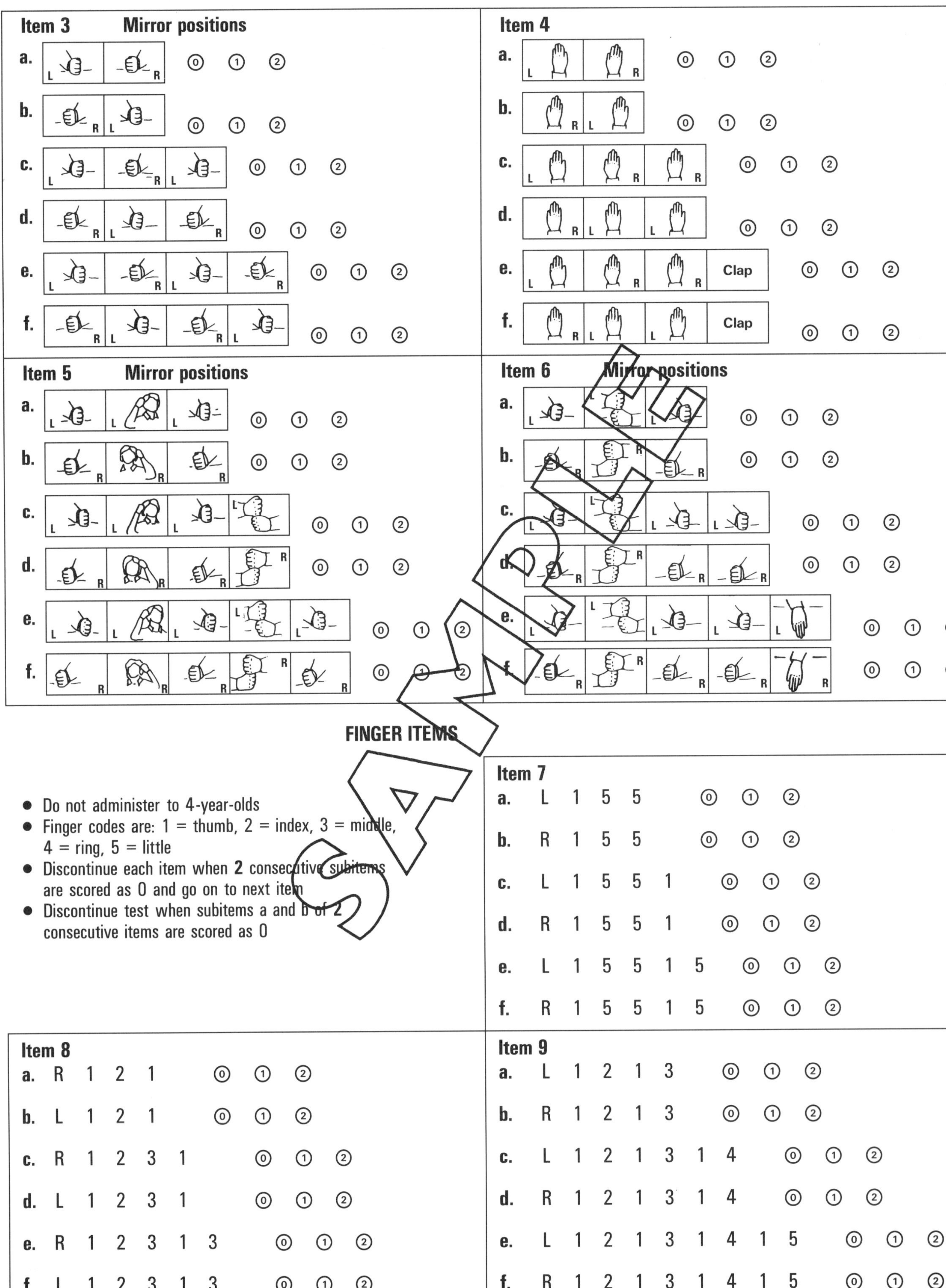

**Item 3** Mirror positions

a. L R ⓪ ① ②

b. R L ⓪ ① ②

c. L R L ⓪ ① ②

d. R L R ⓪ ① ②

e. L R L R ⓪ ① ②

f. R L R L ⓪ ① ②

**Item 4**

a. L R ⓪ ① ②

b. R L ⓪ ① ②

c. L R R ⓪ ① ②

d. R L L ⓪ ① ②

e. L R R Clap ⓪ ① ②

f. R L L Clap ⓪ ① ②

**Item 5** Mirror positions

a. L L L ⓪ ① ②

b. R R R ⓪ ① ②

c. L L L L ⓪ ① ②

d. R R R R ⓪ ① ②

e. L L L L L ⓪ ① ②

f. R R R R R ⓪ ① ②

**Item 6** Mirror positions

a. L L L ⓪ ① ②

b. R R R ⓪ ① ②

c. L L L L ⓪ ① ②

d. R R R R ⓪ ① ②

e. L L L L L ⓪ ① ②

f. R R R R R ⓪ ① ②

## FINGER ITEMS

- Do not administer to 4-year-olds
- Finger codes are: 1 = thumb, 2 = index, 3 = middle, 4 = ring, 5 = little
- Discontinue each item when **2** consecutive subitems are scored as 0 and go on to next item
- Discontinue test when subitems a and b of **2** consecutive items are scored as 0

**Item 7**

a. L 1 5 5 ⓪ ① ②

b. R 1 5 5 ⓪ ① ②

c. L 1 5 5 1 ⓪ ① ②

d. R 1 5 5 1 ⓪ ① ②

e. L 1 5 5 1 5 ⓪ ① ②

f. R 1 5 5 1 5 ⓪ ① ②

**Item 8**

a. R 1 2 1 ⓪ ① ②

b. L 1 2 1 ⓪ ① ②

c. R 1 2 3 1 ⓪ ① ②

d. L 1 2 3 1 ⓪ ① ②

e. R 1 2 3 1 3 ⓪ ① ②

f. L 1 2 3 1 3 ⓪ ① ②

**Item 9**

a. L 1 2 1 3 ⓪ ① ②

b. R 1 2 1 3 ⓪ ① ②

c. L 1 2 1 3 1 4 ⓪ ① ②

d. R 1 2 1 3 1 4 ⓪ ① ②

e. L 1 2 1 3 1 4 1 5 ⓪ ① ②

f. R 1 2 1 3 1 4 1 5 ⓪ ① ②

SENSORY INTEGRATION AND PRAXIS TESTS (SIPT)

# wps TEST REPORT

Western Psychological Services • 12031 Wilshire Blvd. • Los Angeles, CA 90025-1251

# 12. ORAL PRAXIS (OPr)

Protocol Sheet

A. Jean Ayres, Ph.D.

**REQUIRED TRANSMITTAL NUMBER**

| | | | | | | |
|---|---|---|---|---|---|---|
| 0 | 0 | 0 | 0 | 0 | 0 | 0 |
| 1 | 1 | 1 | 1 | 1 | 1 | 1 |
| 2 | 2 | 2 | 2 | 2 | 2 | 2 |
| 3 | 3 | 3 | 3 | 3 | 3 | 3 |
| 4 | 4 | 4 | 4 | 4 | 4 | 4 |
| 5 | 5 | 5 | 5 | 5 | 5 | 5 |
| 6 | 6 | 6 | 6 | 6 | 6 | 6 |
| 7 | 7 | 7 | 7 | 7 | 7 | 7 |
| 8 | 8 | 8 | 8 | 8 | 8 | 8 |
| 9 | 9 | 9 | 9 | 9 | 9 | 9 |

**CHILD'S NAME:**

**DIRECTIONS:** Sit opposite child. Item 1: **"Watch my mouth and do this."** Score response. If child's response is not well executed (does not receive score of 2): **"Try to do this just as I do it"** and demonstrate again. Item 2: **"Wait until I finish before you do these. Do it as many times as I do it."** (Refer to scoring criteria in Manual.)

Administer all items. Give no verbal description except on Item 12. To get child's attention before items: **"Watch me."** Before each item, make sure child is watching your face. If child watches any portion of item, item is not readministered. If child does not watch or begins execution before demonstration completed: **"Wait until I finish; then you do it"** and repeat demonstration.

Fill in the 0, 1, or 2 circle corresponding to the child's item score:

0 = Unable to perform action

1 = Executed with poor quality, or adequately with sufficient number of movements but with poor sequencing, or begun incorrectly but then corrected

2 = Executed well with good sequencing

Nonmirrored or reversed response is still considered correct if other criteria are met (except Items 13 and 14).

• Administer all items

| Item | Incorrect | Poor Quality | Well Executed |
|---|---|---|---|
| 1. Stick out tongue. (Commend child on performance.) | 0 | 1 | 2 |
| 2. Click teeth: 3x (1 sec) (Reinforce need to complete sufficient number of actions.) | 0 | 1 | 2 |
| 3. Pucker lips with obvious protrusion. | 0 | 1 | 2 |
| 4. With lips together, puff out cheeks. | 0 | 1 | 2 |
| 5. Smack lips. | 0 | 1 | 2 |
| 6. With mouth open so that lips do not touch, cover teeth with lips. | 0 | 1 | 2 |
| 7. Make "pkt" with audible but not voice sound: 2x (2 secs) | 0 | 1 | 2 |
| 8. Stick out tongue, retract, close mouth: 3x (3 secs) | 0 | 1 | 2 |
| 9. Put tongue in right cheek, then put tongue in left cheek: 2x (2 secs) | 0 | 1 | 2 |
| 10. Move jaw to right side, then move jaw to left side: 2x (2 secs) | 0 | 1 | 2 |
| 11. Put tongue in left cheek. | 0 | 1 | 2 |

| Item | Incorrect | Poor Quality | Well Executed |
|---|---|---|---|
| 12. Put tongue in your cheek. Push briefly. Say: **"Now hold it there."** Push on child's tongue. | 0 | 1 | 2 |
| 13. Start tongue midline upper lip, move tongue to right and lick lips all the way around. | 0 | 1 | 2 |
| 14. Start tongue midline upper lip, move tongue to left and lick lips all the way around. (Must be opposite Item 13.) | 0 | 1 | 2 |
| 15. Touch tongue to upper lip, then to lower lip: 2x (2 secs) | 0 | 1 | 2 |
| 16. Pucker (or purse) lips to right side of face, then to left side: 2x (2 secs) | 0 | 1 | 2 |
| 17. Bite lower lip, then bite upper lip: 2x (2 secs) | 0 | 1 | 2 |
| 18. Stick out lower lip, close lips: 2x (2 secs) | 0 | 1 | 2 |
| 19. Protrude jaw forward with open mouth, then retract jaw closing lips: 2x (2 secs) | 0 | 1 | 2 |

This Computerized Test Product
AUTHORIZED
By The Test Publisher

4 5 6 7 8 9
76429-321
Printed in U.S.A.
W-260B(12)

**For WPS Use Only**

| | | | | | | |
|---|---|---|---|---|---|---|
| 1 | 2 | 3 | 4 | 5 | 6 | 7 |
| 8 | 9 | 10 | 11 | 12 | 13 | 14 |
| 15 | 16 | 17 | 18 | 19 | 20 | 21 |

DO NOT WRITE

ON THIS SIDE

OF SHEET

**wps TEST REPORT**™

Western Psychological Services • 12031 Wilshire Blvd. • Los Angeles, CA 90025-1251

SENSORY INTEGRATION AND PRAXIS TESTS (SIPT)

# 13. MANUAL FORM PERCEPTION (MFP)

Protocol Sheet

A. Jean Ayres, Ph.D.

**REQUIRED TRANSMITTAL NUMBER**

| | | | | | | |
|---|---|---|---|---|---|---|
| 0 | 0 | 0 | 0 | 0 | 0 | 0 |
| 1 | 1 | 1 | 1 | 1 | 1 | 1 |
| 2 | 2 | 2 | 2 | 2 | 2 | 2 |
| 3 | 3 | 3 | 3 | 3 | 3 | 3 |
| 4 | 4 | 4 | 4 | 4 | 4 | 4 |
| 5 | 5 | 5 | 5 | 5 | 5 | 5 |
| 6 | 6 | 6 | 6 | 6 | 6 | 6 |
| 7 | 7 | 7 | 7 | 7 | 7 | 7 |
| 8 | 8 | 8 | 8 | 8 | 8 | 8 |
| 9 | 9 | 9 | 9 | 9 | 9 | 9 |

**CHILD'S NAME:**

## PART I

**DIRECTIONS FOR PART I:** Sit to right of child. **"I'm going to put one of these blocks in your hand."** Place shield and response card in position. **"You point to the picture here** (point to response card) **of the block in your hand. I'll put the first block in this hand** (touch child's right) **and you point to its picture here** (point to card) **with this hand** (touch child's left)." Item 1: **"Which one is this?."** If Item 1 and/or 2 is incorrect, bring child's hand with the form from under shield and reinstruct. Then readminister (but do not score second attempt). Following Items 1 and 2, turn response card over. **"Now the block will be one of these."**

Place forms in same orientation as shown below. Start timing when form is placed in child's hand and stop timing when child points to printed form (maximum = 30 seconds). Record time and accuracy (0 = incorrect, 1 = correct). After the test, fill in the circles corresponding to the number of seconds. Enter time as two digits (e.g., 6 seconds as 06).

Discontinue Part I after four consecutive items are scored as 0 for accuracy.

- Maximum time per item: **30** seconds
- Discontinue after **4** consecutive items are scored 0 for accuracy

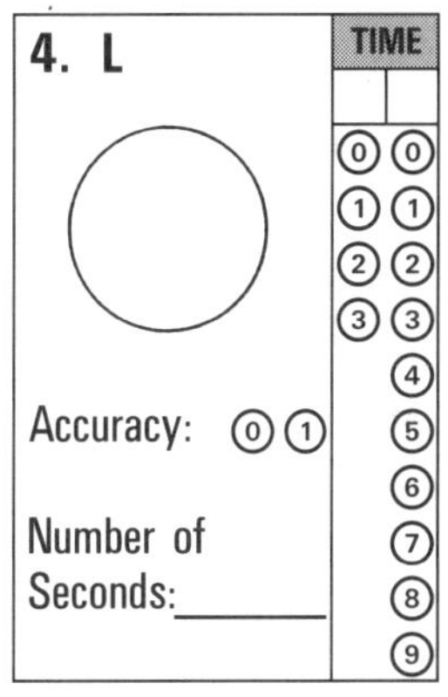

**Turn over response card before Item 3**

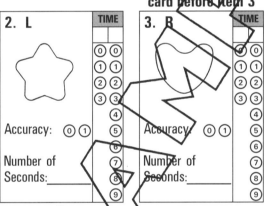

| Item | Accuracy | Number of Seconds | TIME (tens) | TIME (units) |
|---|---|---|---|---|
| 1. R | 0 1 | ______ | 0 1 2 3 | 0 1 2 3 4 5 6 7 8 9 |
| 2. L | 0 1 | ______ | 0 1 2 3 | 0 1 2 3 4 5 6 7 8 9 |
| 3. R | 0 1 | ______ | 0 1 2 3 | 0 1 2 3 4 5 6 7 8 9 |
| 4. L | 0 1 | ______ | 0 1 2 3 | 0 1 2 3 4 5 6 7 8 9 |
| 5. R | 0 1 | ______ | 0 1 2 3 | 0 1 2 3 4 5 6 7 8 9 |
| 6. L | 0 1 | ______ | 0 1 2 3 | 0 1 2 3 4 5 6 7 8 9 |
| 7. R | 0 1 | ______ | 0 1 2 3 | 0 1 2 3 4 5 6 7 8 9 |
| 8. L | 0 1 | ______ | 0 1 2 3 | 0 1 2 3 4 5 6 7 8 9 |
| 9. R | 0 1 | ______ | 0 1 2 3 | 0 1 2 3 4 5 6 7 8 9 |
| 10. L | 0 1 | ______ | 0 1 2 3 | 0 1 2 3 4 5 6 7 8 9 |

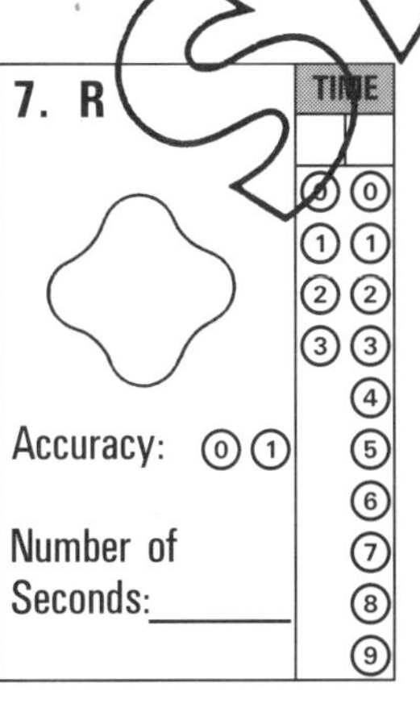

**Remove response card**

**CONTINUE ON BACK**

This Computerized Test Product
AUTHORIZED
By The Test Publisher

4 5 6 7 8 9 Printed in U.S.A.
76431-321
W-260B(13)

**For WPS Use Only**

1 2 3 4 5 6 7
8 9 10 11 12 13 14
15 16 17 18 19 20 21

## PART II

**DIRECTIONS FOR PART II:** Do not administer to 4-year-olds, or if five or more items on Part I were scored as 0 for accuracy. Items that follow a 4th consecutive zero for accuracy and items with a time score greater than 30 are considered to have an accuracy score of 0.

Place two-form plastic base with hole on right. Trial I: **"Now both hands will feel blocks at the same time. This hand will feel this block** (place child's right hand on stimulus form) **and the other hand will find one like it over here** (lead left hand across response forms). **Here's one and here's one.** (Place left hand on table.) **Keep this hand** (tap right) **on this block and find a block just like it with this hand** (tap left). **Tell me when you find the one that's the same. Be sure to feel each block. Begin here** (place left hand on first response form)....**This time we can look."** Remove shield and reinforce. Return shield. Turn plastic base. Trial II: **"Now your other hand will feel this block** (left hand on stimulus form) **and this hand will find one like it over here** (right hand rubs response forms). **Be sure to feel each block. Begin here** (right hand on first response form)." If response is incorrect, ask child to feel other block. Whether correct or not, remove shield and reexplain nature of task.

Place five-form plastic base with hole on right. Item 1: **"Now we'll do another set. Feel this block with this hand** (place hand) **and find one like it over here. Now there are more** (rub child's left hand across five response forms). **Be sure to feel all the blocks. Begin here** (place child's left hand of first response form)." Before Item 2: **"Now we will do another one. Feel this block with this hand** (place hand) **and find one like it with this hand** (tap left). **Be sure to feel all the blocks. Begin here."**

Before Item 6, reverse plastic base. **"Now your other hand will feel this block** (place hand) **and this hand will find one like it over here** (rub right hand over response blocks). **Be sure to feel all the blocks. Begin here."** Administer all items.

Record time (maximum = 45 seconds) and accuracy (0 = incorrect, 1 = correct).

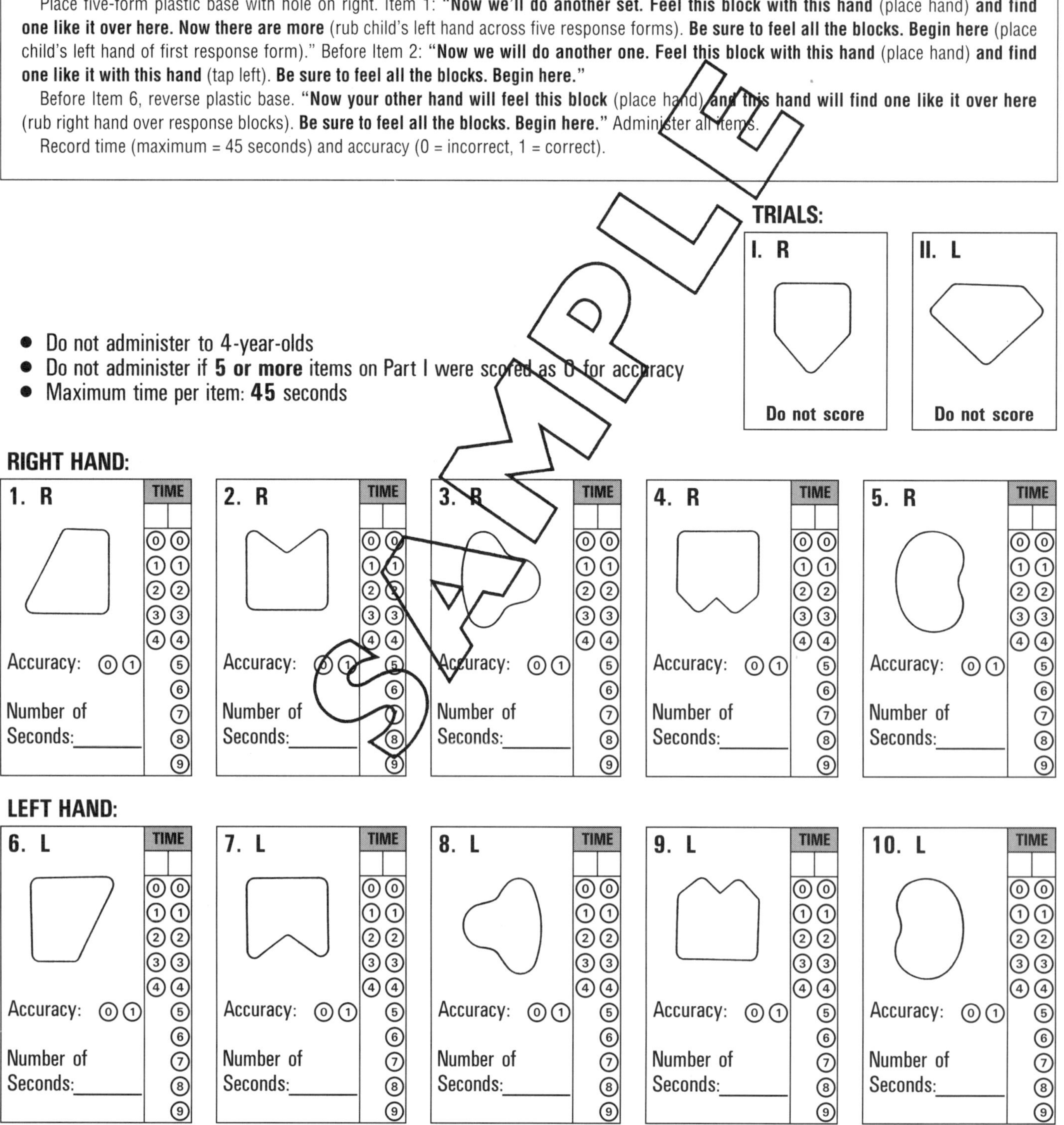

TRIALS:
I. R
Do not score
II. L
Do not score
• Do not administer to 4-year-olds
• Do not administer if 5 or more items on Part I were scored as 0 for accuracy
• Maximum time per item: 45 seconds
RIGHT HAND:
1. R
2. R
3. R
4. R
5. R
TIME
Accuracy: 0 1
Number of Seconds:
LEFT HAND:
6. L
7. L
8. L
9. L
10. L
SAMPLE


**DO NOT WRITE HERE**

**wps TEST REPORT**™

Western Psychological Services • 12031 Wilshire Blvd. • Los Angeles, CA 90025-1251

SENSORY INTEGRATION AND PRAXIS TESTS (SIPT)

# 14. KINESTHESIA (KIN)

Protocol Sheet

A. Jean Ayres, Ph.D.

**DIRECTIONS:** Sit opposite child. Tape test sheet to table. Trial A: **"We are going to play a game called 'going visiting.' I will take your finger to different 'pretend' houses. Point your finger like this."** Place child's right index finger at beginning of line: **"This is where you live. I'm going to take you to House A. Think how it feels to go there so you can come back to House A by yourself.** (Move finger to end of line.) **This is where House A is. Remember where House A is so you can come back to it. Leave your finger here awhile.** (Wait 3 seconds.) **I'll take you home.** (Move to beginning of line.) **This is where you live. Now put your finger on House A."** If off target: **"Place the tip of your finger exactly on the arrow."**

Trial B: **"Now let's see if your other hand can play the game without your eyes helping it. To do that, I will hold this shield here. It will be easier for you to feel where your finger is if you close your eyes."** Then repeat above directions, substituting "House B." After response: **"Leave your finger on the spot until I finish measuring.** (Mark location.) **On this one we can look to see how close you came to the house."**

**"That part of the game was practice for you to learn how. Now, for the rest of the game, your hands will play without your eyes helping them. We will go to different houses."** Give similar directions as for trial items. Refer to houses as "first house, second house," etc. Record distance to nearest tenth centimeter.

After Item 10, readminister the two most erroneous items. Enter all scores below as two digits (e.g., 0.9). Also, enter the item numbers of the readministered items and their item scores.

**REQUIRED TRANSMITTAL NUMBER**

| | | | | | | |
|---|---|---|---|---|---|---|
| 0 | 0 | 0 | 0 | 0 | 0 | 0 |
| 1 | 1 | 1 | 1 | 1 | 1 | 1 |
| 2 | 2 | 2 | 2 | 2 | 2 | 2 |
| 3 | 3 | 3 | 3 | 3 | 3 | 3 |
| 4 | 4 | 4 | 4 | 4 | 4 | 4 |
| 5 | 5 | 5 | 5 | 5 | 5 | 5 |
| 6 | 6 | 6 | 6 | 6 | 6 | 6 |
| 7 | 7 | 7 | 7 | 7 | 7 | 7 |
| 8 | 8 | 8 | 8 | 8 | 8 | 8 |
| 9 | 9 | 9 | 9 | 9 | 9 | 9 |

**CHILD'S NAME:**

- Administer all items
- Readminister **2** most erroneous items



## Item Scores (cm)

| Item 1 | | Item 2 | | Item 3 | | Item 4 | | Item 5 | |
|---|---|---|---|---|---|---|---|---|---|
| | . | | . | | . | | . | | . |
| 0 | 0 | 0 | 0 | 0 | 0 | 0 | 0 | 0 | 0 |
| 1 | 1 | 1 | 1 | 1 | 1 | 1 | 1 | 1 | 1 |
| 2 | 2 | 2 | 2 | 2 | 2 | 2 | 2 | 2 | 2 |
| 3 | 3 | 3 | 3 | 3 | 3 | 3 | 3 | 3 | 3 |
| 4 | 4 | 4 | 4 | 4 | 4 | 4 | 4 | 4 | 4 |
| 5 | 5 | 5 | 5 | 5 | 5 | 5 | 5 | 5 | 5 |
| 6 | 6 | 6 | 6 | 6 | 6 | 6 | 6 | 6 | 6 |
| 7 | 7 | 7 | 7 | 7 | 7 | 7 | 7 | 7 | 7 |
| 8 | 8 | 8 | 8 | 8 | 8 | 8 | 8 | 8 | 8 |
| 9 | 9 | 9 | 9 | 9 | 9 | 9 | 9 | 9 | 9 |

| Item 6 | | Item 7 | | Item 8 | | Item 9 | | Item 10 | |
|---|---|---|---|---|---|---|---|---|---|
| | . | | . | | . | | . | | . |
| 0 | 0 | 0 | 0 | 0 | 0 | 0 | 0 | 0 | 0 |
| 1 | 1 | 1 | 1 | 1 | 1 | 1 | 1 | 1 | 1 |
| 2 | 2 | 2 | 2 | 2 | 2 | 2 | 2 | 2 | 2 |
| 3 | 3 | 3 | 3 | 3 | 3 | 3 | 3 | 3 | 3 |
| 4 | 4 | 4 | 4 | 4 | 4 | 4 | 4 | 4 | 4 |
| 5 | 5 | 5 | 5 | 5 | 5 | 5 | 5 | 5 | 5 |
| 6 | 6 | 6 | 6 | 6 | 6 | 6 | 6 | 6 | 6 |
| 7 | 7 | 7 | 7 | 7 | 7 | 7 | 7 | 7 | 7 |
| 8 | 8 | 8 | 8 | 8 | 8 | 8 | 8 | 8 | 8 |
| 9 | 9 | 9 | 9 | 9 | 9 | 9 | 9 | 9 | 9 |

## Readministered Items

| Item No. | → | Score | |
|---|---|---|---|
| | | | . |
| 1 | | 0 | 0 |
| 2 | | 1 | 1 |
| 3 | | 2 | 2 |
| 4 | | 3 | 3 |
| 5 | | 4 | 4 |
| 6 | | 5 | 5 |
| 7 | | 6 | 6 |
| 8 | | 7 | 7 |
| 9 | | 8 | 8 |
| 10 | | 9 | 9 |

| Item No. | → | Score | |
|---|---|---|---|
| | | | . |
| 1 | | 0 | 0 |
| 2 | | 1 | 1 |
| 3 | | 2 | 2 |
| 4 | | 3 | 3 |
| 5 | | 4 | 4 |
| 6 | | 5 | 5 |
| 7 | | 6 | 6 |
| 8 | | 7 | 7 |
| 9 | | 8 | 8 |
| 10 | | 9 | 9 |

This Computerized Test Product AUTHORIZED By The Test Publisher

**For WPS Use Only**

| | | | | | | |
|---|---|---|---|---|---|---|
| 1 | 2 | 3 | 4 | 5 | 6 | 7 |
| 8 | 9 | 10 | 11 | 12 | 13 | 14 |
| 15 | 16 | 17 | 18 | 19 | 20 | 21 |


4 5 6 7 8 9
76432-321
Printed in U.S.A.
W-260B(14)

DO NOT WRITE

ON THIS SIDE

OF SHEET

**wps TEST REPORT**™

Western Psychological Services • 12031 Wilshire Blvd. • Los Angeles, CA 90025-1251

SENSORY INTEGRATION AND PRAXIS TESTS (SIPT)

# 15. FINGER IDENTIFICATION (FI)

Protocol Sheet

A. Jean Ayres, Ph.D.

**DIRECTIONS:** Sit opposite the child. Child's hands are palms down on table. Vision is *not* occluded on trial items. Trial I: **"Touch that finger."** If incorrect: **"Touch it this way** (touch child's right index finger to left middle finger)." Trial II: **"Touch that finger."** On Trial III, give no verbal directions unless child fails to respond: **"Touch those."** After response on Trial III: **"(That's right.) That time I touched two of your fingers at the same time."** On Trial IV, give no verbal directions unless no response: **"Touch that one."** After response on Trial IV: **"(That's right.) That time I touched one of your fingers in two different places."**

For test items, use shield to occlude vision while giving stimuli; remove it for child's response. Item 1: **"Now let's see if you can tell which finger I touch when you cannot see the finger that I touch."**

Fill in the circle corresponding to the score for each item (0 = incorrect, 1 = correct).

**REQUIRED TRANSMITTAL NUMBER**

| | | | | | | |
|---|---|---|---|---|---|---|
| 0 | 0 | 0 | 0 | 0 | 0 | 0 |
| 1 | 1 | 1 | 1 | 1 | 1 | 1 |
| 2 | 2 | 2 | 2 | 2 | 2 | 2 |
| 3 | 3 | 3 | 3 | 3 | 3 | 3 |
| 4 | 4 | 4 | 4 | 4 | 4 | 4 |
| 5 | 5 | 5 | 5 | 5 | 5 | 5 |
| 6 | 6 | 6 | 6 | 6 | 6 | 6 |
| 7 | 7 | 7 | 7 | 7 | 7 | 7 |
| 8 | 8 | 8 | 8 | 8 | 8 | 8 |
| 9 | 9 | 9 | 9 | 9 | 9 | 9 |

**CHILD'S NAME:**

- Administer all items
- On trial items, do not occlude vision
- On test items, occlude vision during stimulation

**TRIALS:** **I.** L middle **II.** R ring **III.** R index and R middle **IV.** 2 stimuli to L index

| RIGHT HAND | Incorrect | Correct | LEFT HAND | Incorrect | Correct |
|---|---|---|---|---|---|
| **1.** R middle | 0 | 1 | **2.** L middle and L ring | 0 | 1 |
| **3.** 2 stimuli to R little | 0 | 1 | **4.** L ring | 0 | 1 |
| **5.** R middle and R ring | 0 | 1 | **6.** L index and L ring | 0 | 1 |
| **7.** R ring | 0 | 1 | **8.** 2 stimuli to L little | 0 | 1 |
| **9.** 2 stimuli to R middle | 0 | 1 | **10.** L ring | 0 | 1 |
| **11.** R index and R little | 0 | 1 | **12.** L middle | 0 | 1 |
| **13.** R middle and R ring | 0 | 1 | **14.** L ring and L little | 0 | 1 |
| **15.** R ring | 0 | 1 | **16.** L middle and L ring | 0 | 1 |

This Computerized Test Product
AUTHORIZED
By The Test Publisher

4 5 6 7 8 9
76433-321
Printed in U.S.A.

For WPS Use Only

| | | | | | | |
|---|---|---|---|---|---|---|
| 1 | 2 | 3 | 4 | 5 | 6 | 7 |
| 8 | 9 | 10 | 11 | 12 | 13 | 14 |
| 15 | 16 | 17 | 18 | 19 | 20 | 21 |

W-260B(15)

DO NOT WRITE

ON THIS SIDE

OF SHEET

**wps TEST REPORT**

Western Psychological Services • 12031 Wilshire Blvd. • Los Angeles, CA 90025-1251

SENSORY INTEGRATION AND PRAXIS TESTS (SIPT)

# 16. GRAPHESTHESIA (GRA)

Protocol Sheet

A. Jean Ayres, Ph.D.

**REQUIRED TRANSMITTAL NUMBER**

| | | | | | | |
|---|---|---|---|---|---|---|
| 0 | 0 | 0 | 0 | 0 | 0 | 0 |
| 1 | 1 | 1 | 1 | 1 | 1 | 1 |
| 2 | 2 | 2 | 2 | 2 | 2 | 2 |
| 3 | 3 | 3 | 3 | 3 | 3 | 3 |
| 4 | 4 | 4 | 4 | 4 | 4 | 4 |
| 5 | 5 | 5 | 5 | 5 | 5 | 5 |
| 6 | 6 | 6 | 6 | 6 | 6 | 6 |
| 7 | 7 | 7 | 7 | 7 | 7 | 7 |
| 8 | 8 | 8 | 8 | 8 | 8 | 8 |
| 9 | 9 | 9 | 9 | 9 | 9 | 9 |

**CHILD'S NAME:**

**DIRECTIONS:** Sit across the table from child. Trial: **"I am going to draw some designs (pictures) on the back of your hand** (point to child's left hand) **with my finger. You draw the same thing in the same place** (point). **Draw with your finger** (point to child's right index finger). **I will show you how with this one.** (Draw trial design without shield.) **Draw what I drew. Draw it here** (point to left hand). **Draw it carefully."**

Item 1: **"Now I'll draw something different on your other hand. I'll draw without your watching me."** Position shield and draw Item 1 design. Then remove shield. If child does not automatically draw line: **"You draw the same thing. Draw it here** (point.) **Draw carefully."**

Replicate child's drawings in blank boxes below. Although the drawings may be scored later according to the criteria listed in the Manual, estimate whether each item would receive a score of 0 (not even partially correct). Discontinue the test after four consecutive items are scored as 0.

**TRIAL:** Left

- Discontinue test after **4** consecutive items are scored as 0

| RIGHT HAND | | | LEFT HAND | | |
|---|---|---|---|---|---|
| **1.** | | 0 1 2 | **2.** | | 0 1 2 |
| **3.** | | 0 1 2 | **4.** | | 0 1 2 |
| **5.** | | 0 1 2 | **6.** | | 0 1 2 |
| **7.** | | 0 1 2 | **8.** | | 0 1 2 |
| **9.** | | 0 1 2 | **10.** 1 2 | | 0 1 2 |
| **11.** 1 2 3 | | 0 1 2 | **12.** | | 0 1 2 |
| **13.** 1 2 | | 0 1 2 | **14.** 1 2 3 | | 0 1 2 |

This Computerized Test Product
AUTHORIZED
By The Test Publisher

4 5 6 7 8 9 Printed in U.S.A.
76435-321
W-260B(16)

**For WPS Use Only**

| | | | | | | |
|---|---|---|---|---|---|---|
| 1 | 2 | 3 | 4 | 5 | 6 | 7 |
| 8 | 9 | 10 | 11 | 12 | 13 | 14 |
| 15 | 16 | 17 | 18 | 19 | 20 | 21 |

DO NOT WRITE

ON THIS SIDE

OF SHEET

SENSORY INTEGRATION AND PRAXIS TESTS (SIPT)

# WPS TEST REPORT 17. LOCALIZATION OF TACTILE STIMULI (LTS)

Western Psychological Services • 12031 Wilshire Blvd. • Los Angeles, CA 90025-1251

Protocol Sheet

A. Jean Ayres, Ph.D.

**DIRECTIONS**: Sit across table from child. Trial: **"I am going to touch you lightly with this pen. Put your finger where I touch you. Put your finger there** (touch dorsum of child's left hand)." If necessary: **"Put your finger exactly on the spot I touched."**

Position shield. Item I: **"Now let's see how close you can put your finger when you can't see where I touch you. Put your finger here** (touch back of child's right hand and wait for response.) **If your finger doesn't land on the right spot, move it until it is on the right spot. . . . I'm going to measure you with this ruler. Leave your finger there until I finish."** Before each item: **"Put your finger here."**

Record distance to nearest tenth centimeter as two digits (e.g., 0.4). After Item 12, readminister the two most erroneous items and write the second scores in the spaces inside the parentheses. After the test, fill in the circles corresponding to the item scores, and record clinical observations.

REQUIRED TRANSMITTAL NUMBER

- Administer all items
- Readminister **2** most erroneous items

**TRIAL:** Dorsum left hand



**RIGHT** | **LEFT**

**PRONATED**

RIGHT: A — 5. __.__ (__.__) ; B — 3. __.__ (__.__) ; C — 1. __.__ (__.__)

LEFT: A — 2. __.__ (__.__) ; B — 6. __.__ (__.__) ; C — 4. __.__ (__.__)

**SUPINATED**

RIGHT: A — 11. __.__ (__.__) ; B — 9. __.__ (__.__) ; C — 7. __.__ (__.__)

LEFT: A — 8. __.__ (__.__) ; B — 12. __.__ (__.__) ; C — 10. __.__ (__.__)

**CHILD'S NAME:**

**OBSERVATIONS:**

**CLINICAL OBSERVATIONS**

- ○ No apparent tactile defensiveness
- ○ Possible tactile defensiveness
- ○ Definite tactile defensiveness

**Item Scores (cm)**

Item 1 | Item 2 | Item 3 | Item 4 | Item 5 | Item 6 | Item 7 | Item 8 | Item 9 | Item 10 | Item 11 | Item 12

**Readministered Items**

Item No. → Score | Item No. → Score

For WPS Use Only

AUTHORIZED — This Computerized Test Product Authorized By The Test Publisher



4 5 6 7 8 9
76436-321
Printed in U.S.A.
W-260B(17)

DO NOT WRITE
ON THIS SIDE
OF SHEET
SAMPLE

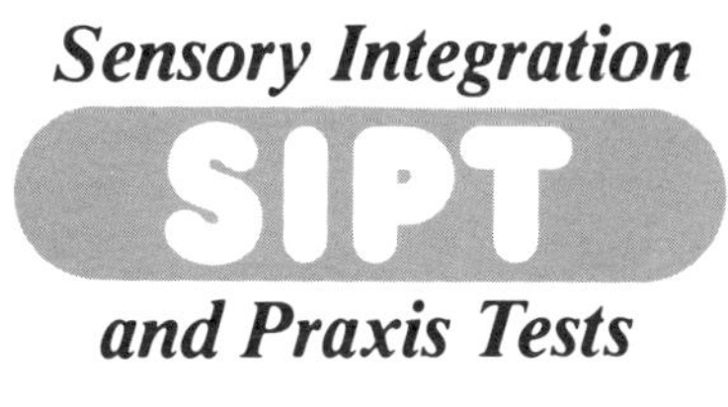


## Design Copying Test

Part I

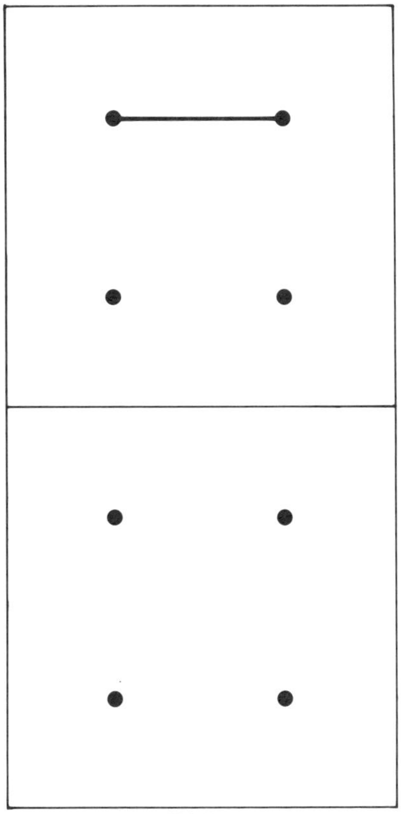

Trial

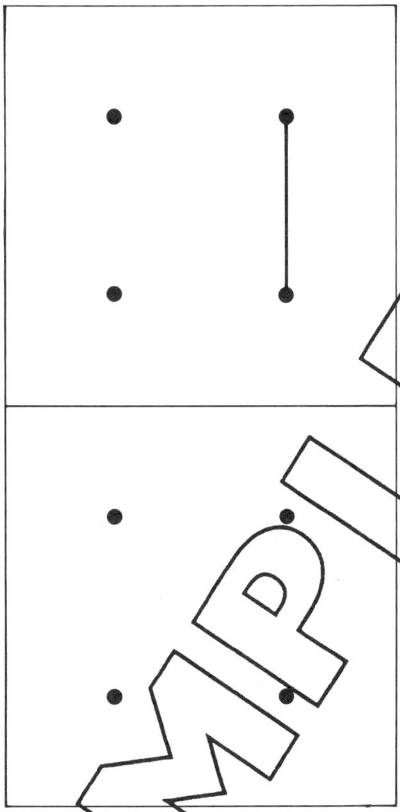

1

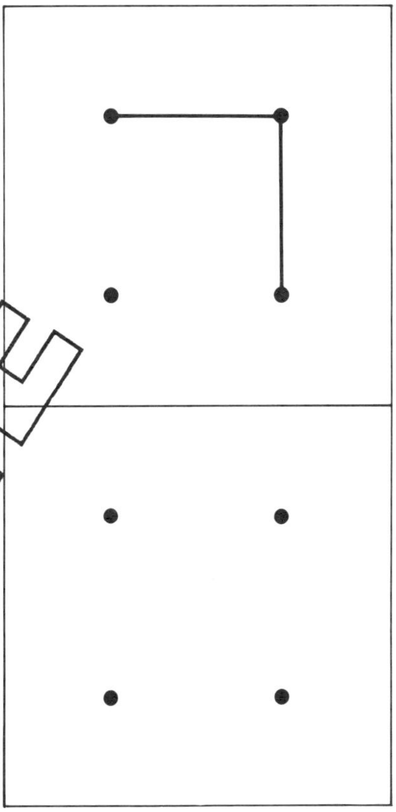

2

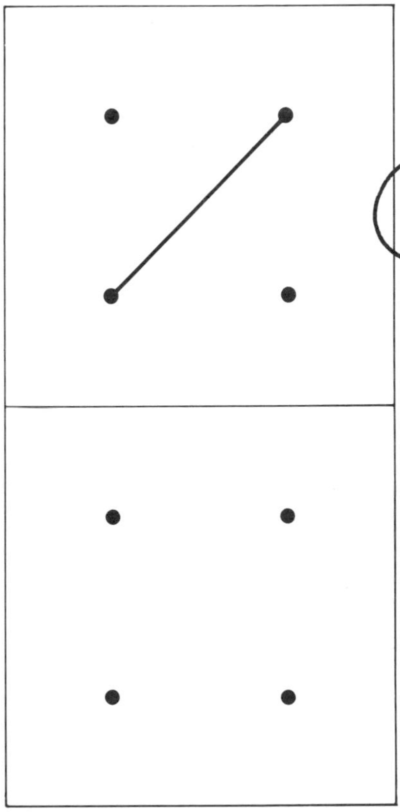

3

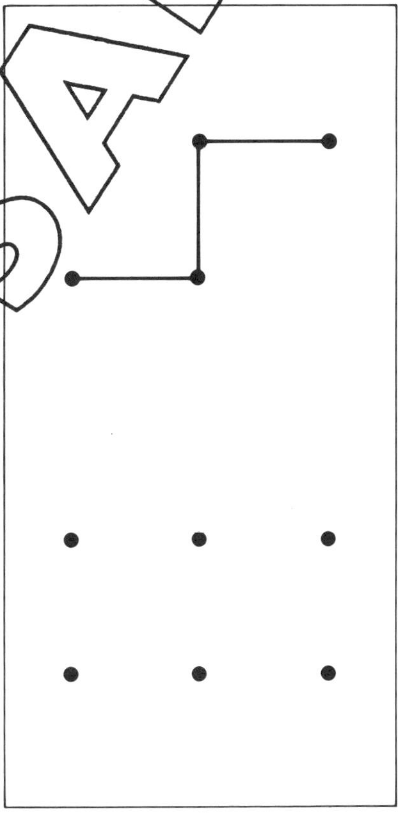

4

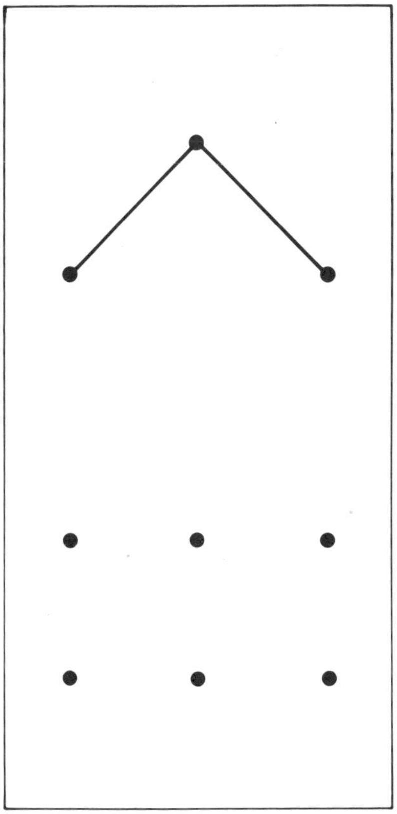

5

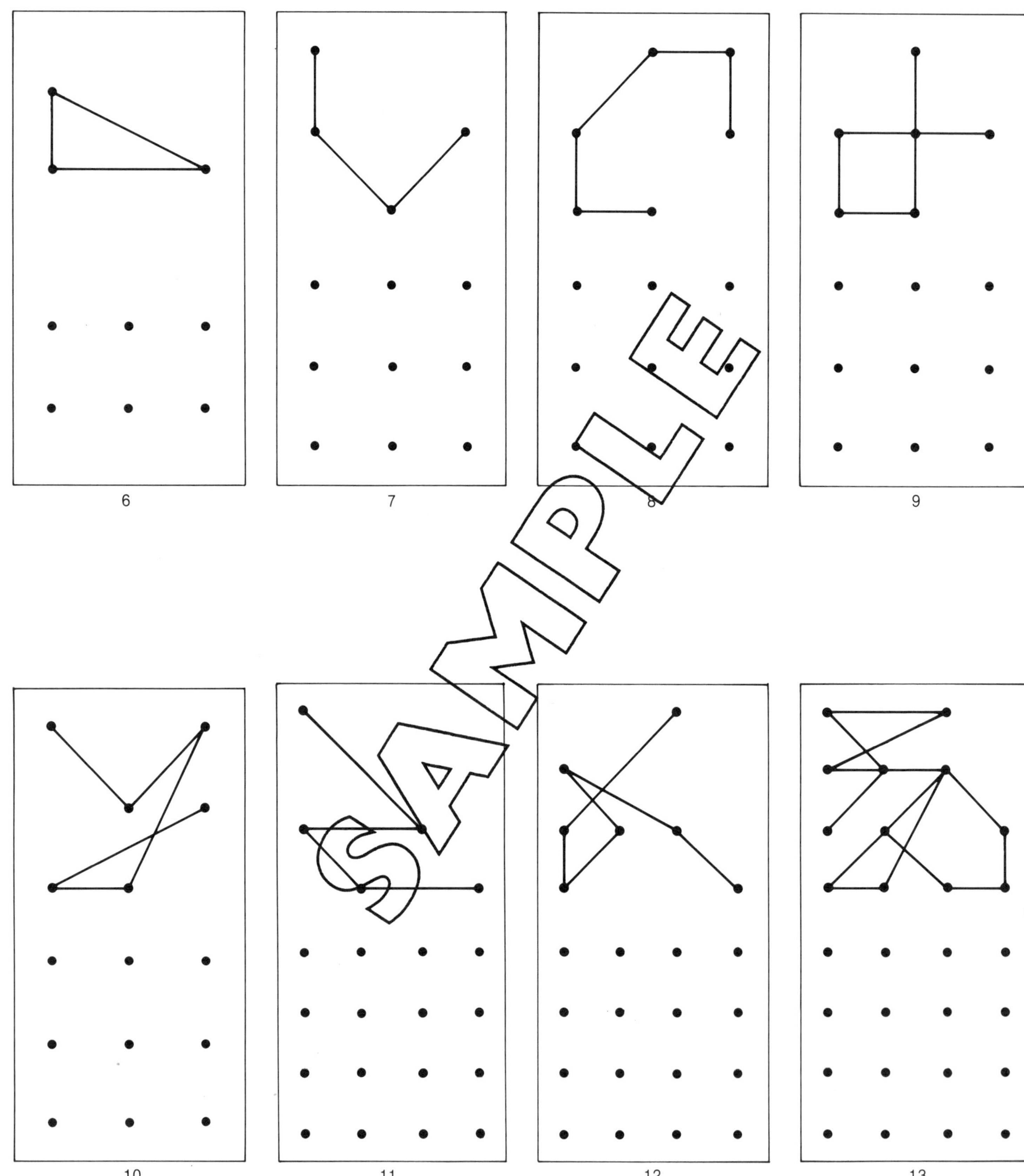
6
7
8
9
10
11
12
13

Part II

14

15

16

17

18

19

20

21

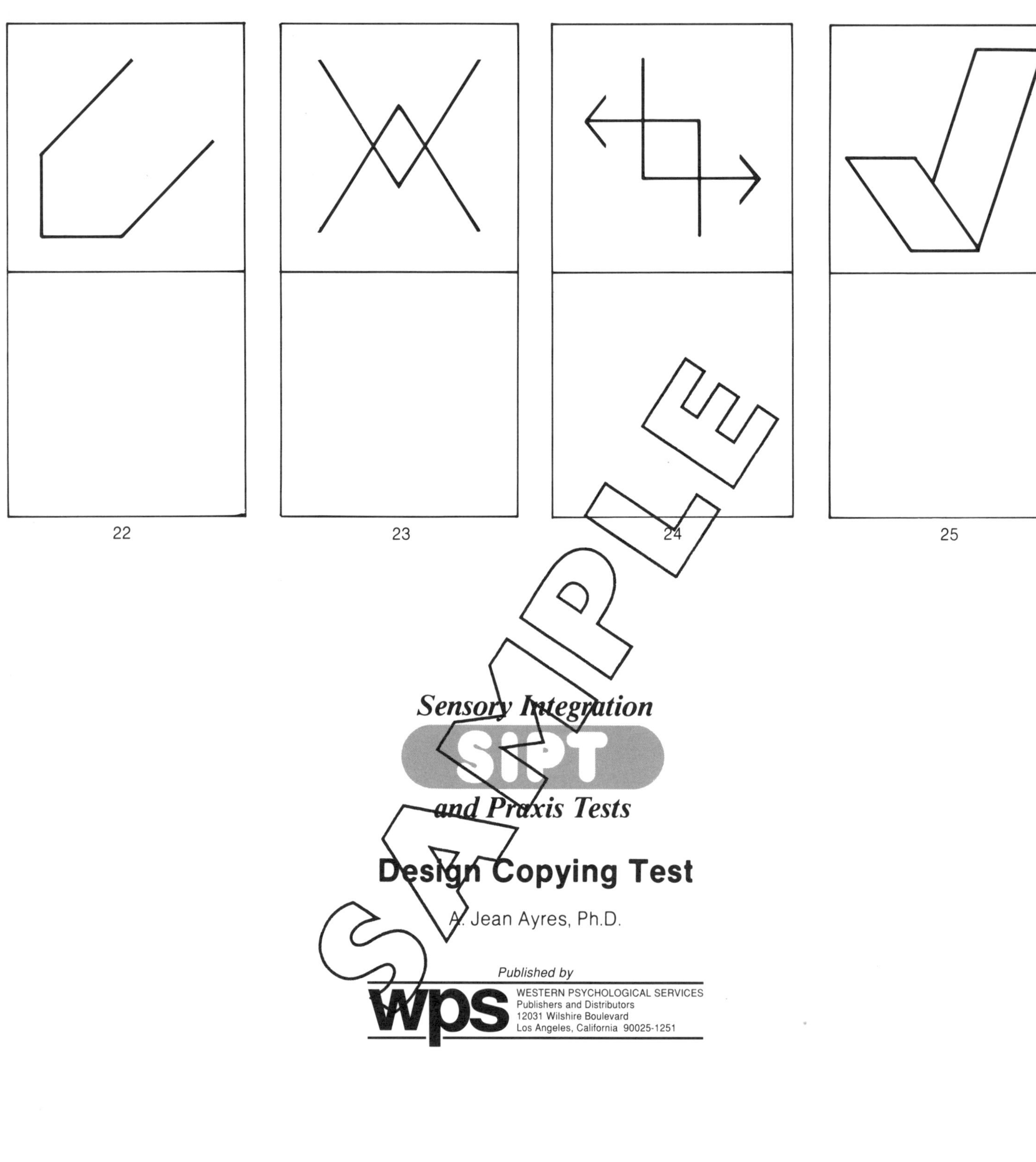




*Sensory Integration*

SIPT

*and Praxis Tests*

# Design Copying Test

A. Jean Ayres, Ph.D.

*Published by*

wps WESTERN PSYCHOLOGICAL SERVICES
Publishers and Distributors
12031 Wilshire Boulevard
Los Angeles, California 90025-1251

**Name:** ____________________

**Identification No.:** ____________________

 4 5 6 7 8 9 Printed in U.S.A.

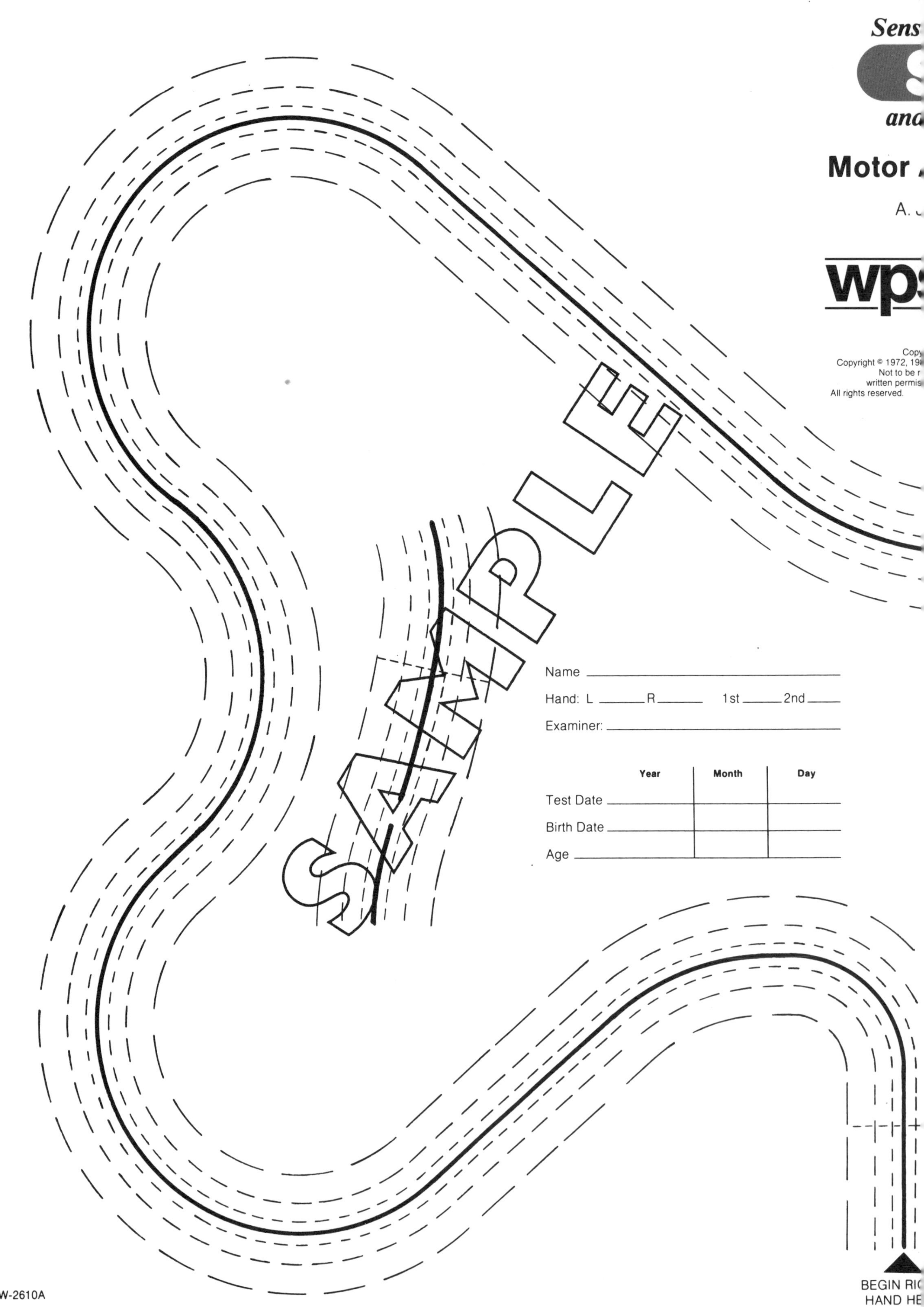

*Sens*

*and*

**Motor**

A. J

wps





Name ____________________

Hand: L _______ R _______ 1st _______ 2nd _______

Examiner: ____________________

| | Year | Month | Day |
|---|---|---|---|
| Test Date | | | |
| Birth Date | | | |
| Age | | | |

BEGIN RI
HAND HE

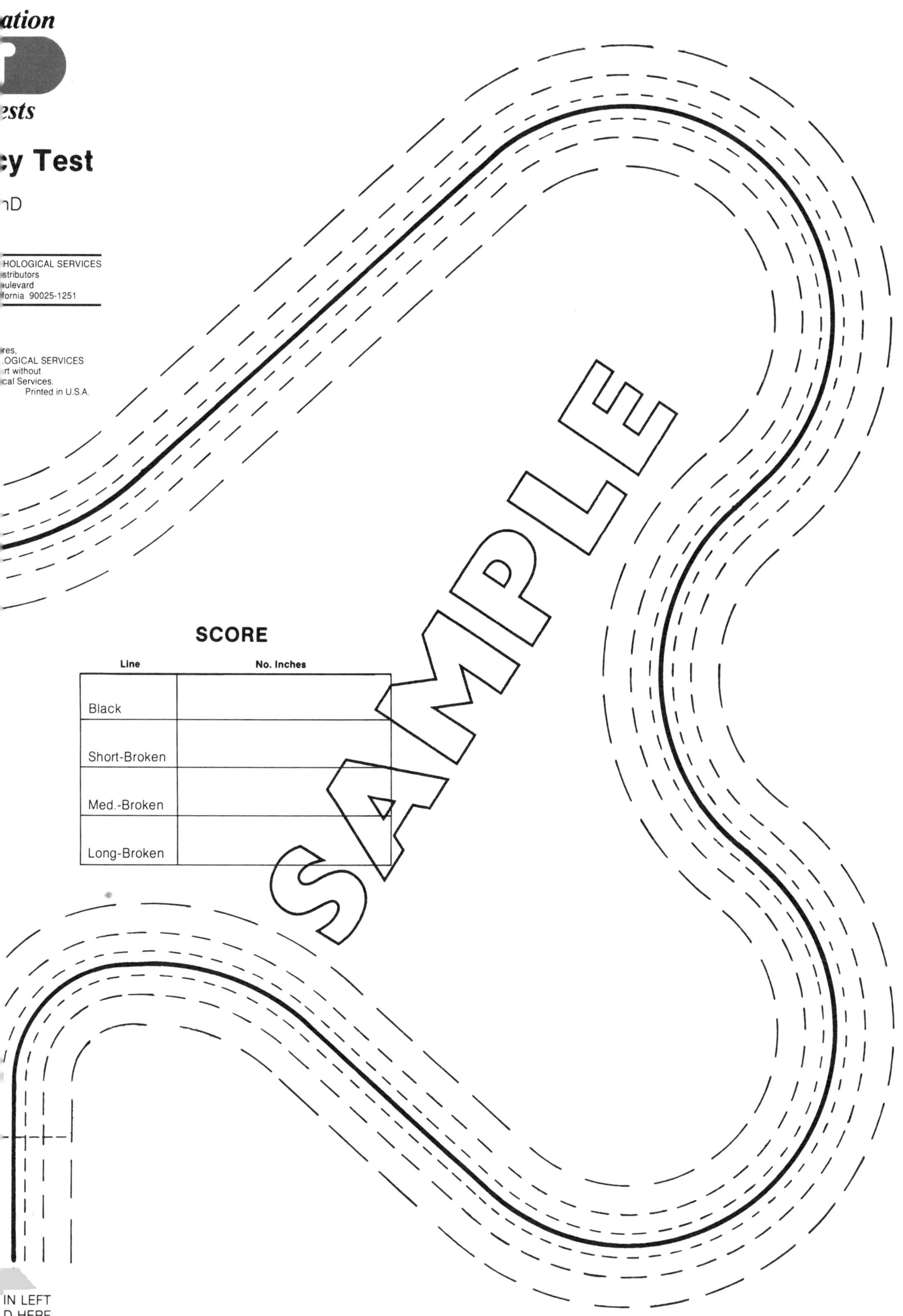


| Line | No. Inches |
|---|---|
| Black | |
| Short-Broken | |
| Med.-Broken | |
| Long-Broken | |

# Kinesthesia Test Sheet

A. Jean Ayres, Ph.D.

*Published by*

**wps** WESTERN PSYCHOLOGICAL SERVICES
Publishers and Distributors
12031 Wilshire Boulevard
Los Angeles, California 90025-1251

Name: ______________________________

ID Number: ______________________________

**DIRECTIONS:** Sit opposite child. Tape test sheet to table. Trial A: "We are going to play a game called 'going visiting.' I will take your finger to different 'pretend' houses. Point your finger like this." Place child's right index finger at beginning of line: "This is where you live. I'm going to take you to House A. Think how it feels to go there so you can come back to House A by yourself. (Move finger to end of line.) This is where House A is. Remember where House A is so you can come back to it. Leave your finger here awhile. (Wait 3 seconds.) I'll take you home. (Move to beginning of line.) This is where you live. Now put your finger on House A." If off target: "Place the tip of your finger exactly on the arrow."

Trial B: "Now let's see if your other hand can play the game without your eyes helping it. To do that, I will hold this shield here. It will be easier for you to feel where your finger is if you close your eyes." Then repeat above directions, substituting "House B." After response: "Leave your finger on the spot until I finish measuring. (Mark location.) On this one we can look to see how close you came to the house.

"That part of the game was practice for you to learn how. Now, for the rest of the game, your hands will play without your eyes helping them. We will go to different houses." Turn test sheet over to administer test items.



5 7 8 9

Printed in U.S.A.

W-2614A

.INE

:

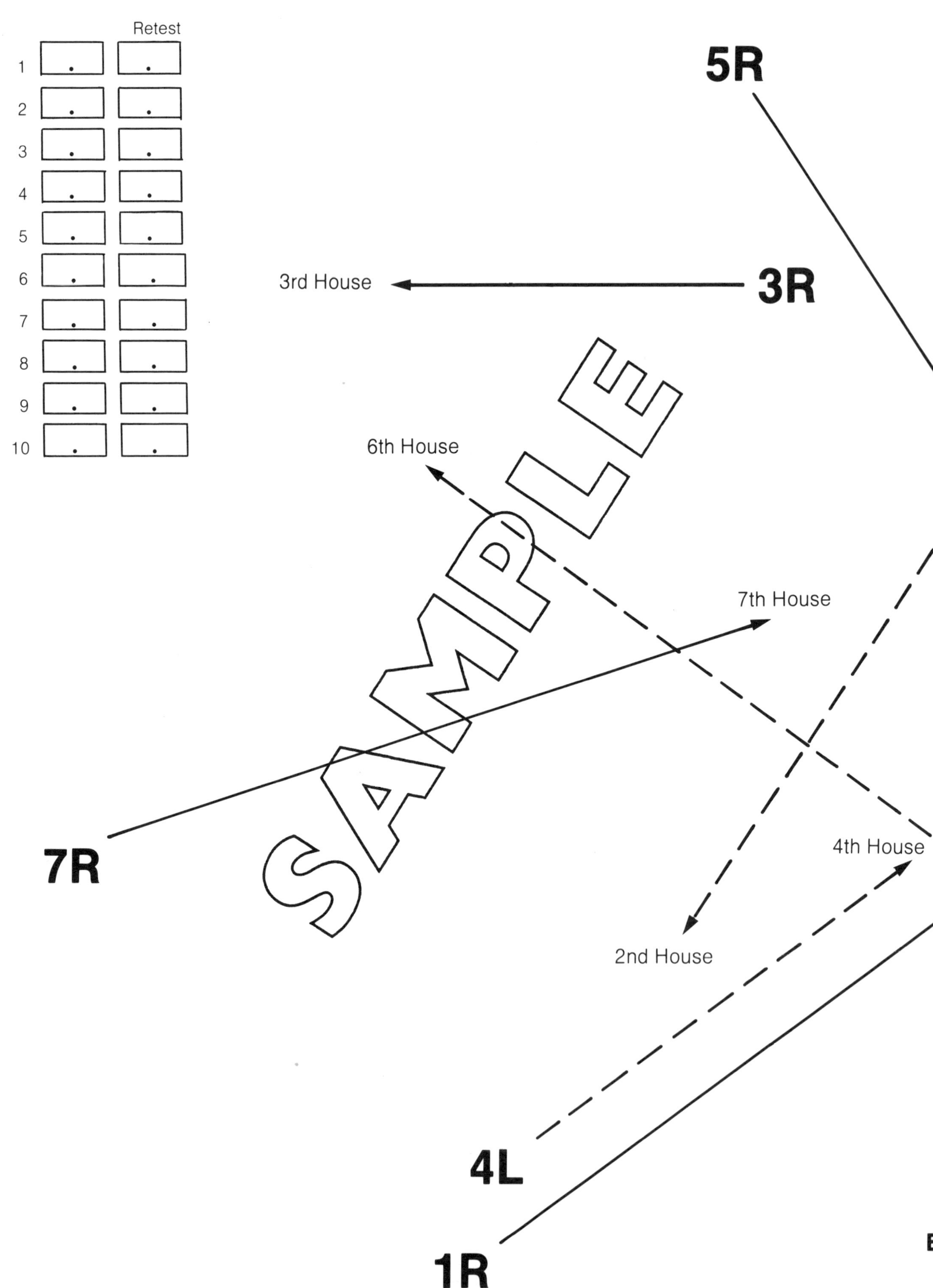
Scores in Centimeters
Retest
1
2
3
4
5
6
7
8
9
10
5R
3rd House
3R
6th House
SAMPLE
7th House
7R
4th House
2nd House
4L
1R
E

**DIRECTIONS:** Sit opposite child. Tape test sheet to table. Cover the test sheet with the shield. Give similar directions as for trial items: "This is where you live. I'm going to take you to the first house. Think how it feels to go there so you can come back to the first house by yourself. (Move finger to end of line.) This is where the first house is. Leave your finger here awhile. (Wait 3 seconds.) I'll take you home. (Move to beginning of line.) This is where you live. Now put your finger on the first house. Leave your finger on the spot until I finish measuring how close you are."

Refer to houses as "first house, second house," etc. Record distance to nearest tenth centimeter. After item 10, readminister the two most erroneous items. Enter all scores here and on the Protocol Sheet as two digits (e.g., 0.9). Also, enter the item numbers of the readministered items and their item scores.

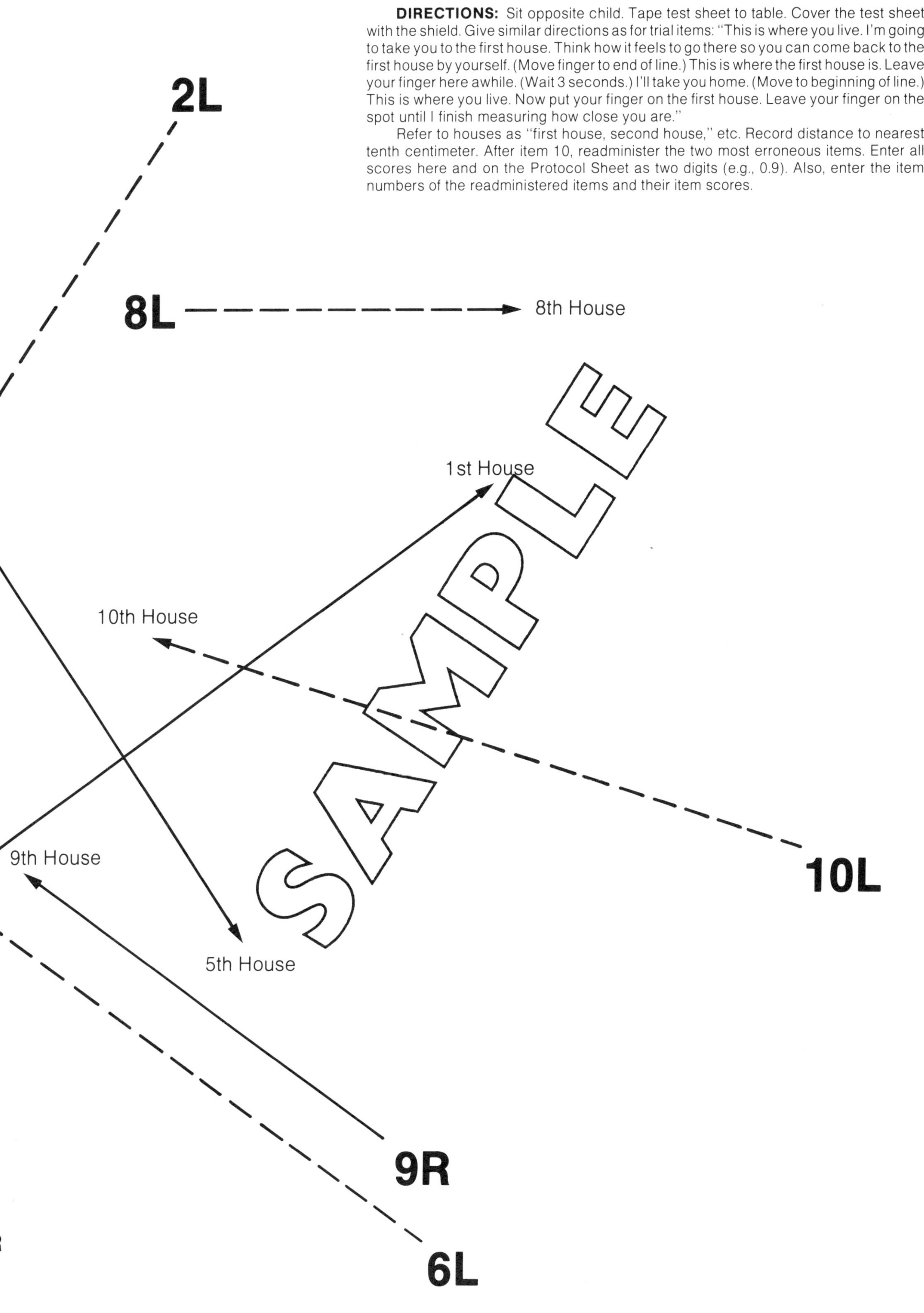